ELECTRONIC ENGINEERING SYSTEMS SERIES

Series Editor: **J. K. FIDLER**, *University of York*

Asssociate Series Editor: **PHIL MARS**, *University of Durham*

THE ART OF SIMULATION USING PSPICE - ANALOG AND DIGITAL
Bashir Al-Hashimi, Staffordshire University

FUNDAMENTALS OF NONLINEAR DIGITAL FILTERING
Jaakko Astola and Pauli Kuosmanen, Tampere University of Technology

WIDEBAND CIRCUIT DESIGN
Herbert J. Carlin, Cornell University and Pier Paolo Civalleri, Turin Polytechnic

PRINCIPLES AND TECHNIQUES OF ELECTROMAGNETIC COMPATIBILITY
Christos Christopoulos, University of Nottingham

OPTIMAL AND ADAPTIVE SIGNAL PROCESSING
Peter M. Clarkson, Illinois Institute of Technology

KNOWLEDGE-BASED SYSTEMS FOR ENGINEERS AND SCIENTISTS
Adrian A. Hopgood, The Open University

LEARNING ALGORITHMS: THEORY AND APPLICATIONS IN SIGNAL PROCESSING, CONTROL AND COMMUNICATIONS
Phil Mars, J. R. Chen, and Raghu Nambiar
University of Durham

DESIGN AUTOMATION OF INTEGRATED CIRCUITS
Ken G. Nichols, University of Southampton

INTRODUCTION TO INSTRUMENTATION AND MEASUREMENTS
Robert B. Northrop, University of Connecticut

CIRCUIT SIMULATION METHODS AND ALGORITHMS
Jan Ogrodzki, Warsaw University of Technology

WIDEBAND CIRCUIT DESIGN

Herbert J. Carlin • Pier Paolo Civalleri

CRC Press
Boca Raton Boston New York Washington, D.C. London

Library of Congress Cataloging-in-Publication Data

Carlin, Herbert J.
Wideband circuit design / Herbert J. Carlin and Pier Paolo Civalleri.
p. cm. -- (Electronic engineering systems series)
Includes bibliographical references and index.
ISBN 0-8493-7897-4 (alk. paper)
1. Linear integrated circuits--Design and construction. 2. Linear time invariant systems. I. Civalleri, Pier Paolo. II. Title.
III. Series.
TK7874.C36 1997
621.3815—dc21
for Library of Congress 97-26966
CIP

International Standard Book Number 0-8493-7897-4
Library of Congress Card Number 97-26966
Printed in the United States of America 1 2 3 4 5 6 7 8 9 0
Printed on acid-free paper

For their patience, fortitude, willing ears,
and endless encouragement.
To Mariann and Ketty
with love.

Preface

This book is about theoretical aspects of network theory as applied to wideband circuit design, yet in some sense it follows the scheme of a novel. At the outset we are presented with background material which subsequently leads to more complex developments as, so to speak, the plot unfolds. Modest beginnings move forward inevitably to advanced topics and we have tried to lay out the material as a self sufficient whole, so that at any stage of the presentation the reader will find underlying concepts described in an earlier portion of the text. We have also endeavored to be "user friendly" in our discussions and proofs. Our purpose has been to make the explanations as complete as possible, so that difficult points are not left hanging in mid-air.

Many of the traditional topics included in the text are approached from a novel (no pun intended) point of view. For example, the exponential, considered as an eigenfunction of a linear time invariant operator, provides the mechanism for steady state a.c. phasor analysis and, at the same time, leads inexorably to the role played by convolution, and Fourier and Laplace transforms in general time domain analysis. These mathematical ideas find their ultimate justification in translating the absolutely essential notions of causality, passivity and losslessness into performance criteria for physical systems.

Any foundational discussion of linear analog circuit theory must incorporate the Darlington representation of a lossless two-port and its many ramifications. We have used a somewhat different approach to this topic than ordinarily encountered. Our starting point is the well-known Belevitch representation of a lossless scattering matrix. Then by using Foster's Theorem in conjunction with the idea of transmission zeros, we arrive in a relatively simple and straightforward fashion to a complete derivation of the basic components of cascade transducer design including Type C, Brune, type D, and Richards (distributed case) sections. The widely used Butterworth, Chebyshev, and elliptic transducer gain functions are deduced in a unified fashion using complex mappings of analytic functions. This includes Amstutz' lucid and elegant account of elliptic filters. The final sections of the book deal with gain-bandwidth theory, and both the now classic analytic Fano-Youla procedure and the numerically based and broadly applicable Real Frequency Technique (RFT) are described. It may be noted that here, as well as throughout the text, illustrative examples are copiously interpolated.

It is perhaps not out of order in a Preface to mention that we feel especially comfortable with the topics undertaken in this book because of our long association as research colleagues in the areas covered herein. One other word about the general flavor of the text; although Professor Dante Youla of the Polytechnic University of New York was never explicitly consulted in connection with the manuscript, we have enjoyed a warm personal as well as professional relationship with him for many years, and we wholeheartedly acknowledge the underlying influence of Dan's ideas. We are also glad to express our thanks to Professor Chris Heegard and the many colleagues who helped us at the Cornell University School of Electrical Engineering and at the Department of Electronics of the Politecnico di Torino. Our security blanket was to know we could always count on their assistance in answering the variety of questions we continually came up with. To Professor L. Pandolfi, our appreciation for his solutions to some troubling mathematical difficulties. Finally, we are especially grateful to Professor Claudio Beccari for his many suggestions on technical matters and for his invaluable assistance in straightening out the numerous vexing programming problems in the preparation of the text.

HERBERT J. CARLIN
J. Preston Levis Professor of Engineering, Emeritus
Cornell University, Ithaca, NY

PIER PAOLO CIVALLERI
Professor of Electrotechnics
Politecnico di Torino, Turin, Italy

Contents

Preface

1 General Properties of Linear Circuits and Systems **1**
1.1 Operator Representation 1
1.2 Linear Time Invariant Systems and Operators 3
1.3 Causality 6
1.4 Power, Energy, and Passivity 8
1.5 Passivity, Linearity and Causality 17

2 LTI System Response to Exponential Eigenfunctions **23**
2.1 Solution of Operator Equations 23
2.2 LTI Operator Eigenfunctions 27
2.3 Homogeneous Solution of LTI Operator Equations 31
2.4 The Particular Solution under Exponential Excitation . . . 37
2.5 Conditions for Pure Eigenfunction Response 40
2.6 Phasors and A.C. Analysis 44
2.7 Network Geometry 46
2.8 Topology and Kirchhoff's Laws 53
2.9 Nodal Analysis 58
2.10 Mesh and Loop Analysis 66
2.11 Cut Set Analysis 74
2.12 Transfer Functions and n-Ports 78
2.13 Incidence Matrices and Network Equations 89
2.14 Tellegen's Theorem, Reciprocity, and Power 95

3 Impulses, Convolution, and Integral Transforms **105**
3.1 The Impulse Function 105
3.2 The Fourier Integral Theorem 110
3.3 Impulse Response and Convolution 121
3.4 Real–Imaginary Part Relations; The Hilbert Transform . . 126
3.5 Causal Fourier Transforms 132

3.6 Minimum Immittance Functions 140
3.7 Amplitude-Phase Relations 143
3.8 Numerical Evaluation of Hilbert Transforms 148
3.9 Operational Rules and Generalized Fourier Transforms . . . 152
3.10 Laplace Transforms and Eigenfunction Response 159

4 The Scattering Matrix and Realizability Theory 171
4.1 Physical Properties of n-Ports 171
4.2 General Representations of n-Ports 173
4.3 The Scattering Matrix Normalized to Positive Resistors . . 179
4.4 Scattering Relations for Energy and Power 186
4.5 Bounded Real Scattering Matrices 188
4.6 Positive Real Immittance Matrices 200
4.7 The Degree of a One-Port 210

5 One-Port Synthesis 213
5.1 Introduction 213
5.2 Lossless One-Port Synthesis 214
5.3 RC and RL One-Port Synthesis 227
5.4 The Scattering Matrix of a Lossless Two-Port 231
5.5 The Immittance Matrices of a Lossless Two-Port 238
5.6 Transmission Zeros 240
5.7 Darlington's Procedure of Synthesis 242
5.8 An Example 249
5.9 Cascade Synthesis: Type A and B Sections 252
5.10 Cascade Synthesis: Brune's Section 254
5.11 Cascade Synthesis: Darlington's C-Section 262
5.12 Cascade Synthesis: Darlington's D-Section 267
5.13 Ladder Synthesis; Fujisawa's Theorem 274
5.14 Transmission Zeros All Lying at Infinity and/or the Origin 279

6 Insertion Loss Filters 283
6.1 The Concept of a Filter and the Approximation Problem . 283
6.2 Synthesis of doubly terminated filters 286
6.3 Impedance Scaling, Frequency Transformations 289
6.4 Specifications for Amplitude Approximation 296
6.5 Butterworth Approximation 299
6.6 Chebyshev Approximation 304
6.7 Elliptic Approximation 312
6.8 Phase Equalization 321
6.9 Allpass C-Section Phase Equalizers 323
6.10 Allpass D-Section Phase Equalizers 326
6.11 Bessel Approximation 329
6.12 Synthesis of Single-Terminated Filters 332

7 Transmission Lines **337**
7.1 The TEM Line 337
7.2 The Unit Element (UE); Richards' Transformation 341
7.3 Richards' Theorem: UE Reactance Functions 350
7.4 Doubly Terminated UE Cascade 352
7.5 Stepped Line Gain Approximations 356
7.6 Transfer Functions for Stepped Lines and Stubs 365
7.7 Coupled UE Structures 373

8 Broadband Matching I: Analytic Theory **383**
8.1 The Broadbanding Problem 383
8.2 The Chain Matrix of a Lossless Two-Port 385
8.3 Complex Normalization 386
8.4 The Gain-Bandwidth Restrictions 391
8.5 The Gain-Bandwidth Restrictions in Integral Form 401
8.6 Example: Double Zero of Transmission 406
8.7 Double Matching 408

9 Broadband Matching II: Real Frequency Technique **415**
9.1 Introduction 415
9.2 Single Matching 419
9.3 Transmission Line Equalizers 425
9.4 Double Matching 429
9.5 Double Matching of Active Devices 433

Appendices **439**

A Analytic Functions **439**
A.1 General Concepts 439
A.2 Integration of Analytic Functions 442
A.3 The Cauchy Integral Formula 444
A.4 Laurent and Taylor Expansions 445
A.5 The Theorem of Residues 448
A.6 Zeros, Poles and Essential Singularities 449
A.7 Some Theorems on Analytic Functions 451
A.8 Classification of Analytic Functions 452
A.9 Multivalued Functions 453
A.10 The Logarithmic Derivative 455
A.11 Functions with a Finite Number of Singularities 456
A.12 Analytic Continuation 457
A.13 Calculus of Definite Integrals by the Residue Method 459

B Linear Algebra **465**
B.1 General Concepts 465
B.2 Geometrical Interpretation 467
B.3 Linear Simultaneous Equations 471
B.4 Eigenvalues and Eigenvectors 477

Index **483**

1

General Properties of Linear Circuits and Systems

1.1 Operator Representation

In this introductory chapter we examine the basic properties of linear time invariant (LTI) systems based on the physical postulates such systems satisfy in the time domain. The treatment is from an elementary point of view, nevertheless, in order to lend some generality to the discussion an operator formalism is used. Consider the simple circuit of a capacitor C in parallel with a resistor $R = 1/G$. An input current source (a function of time t) $\jmath(t) \equiv x(t)$ excites the circuit, and the response or output is the voltage time function, $v(t) \equiv y(t)$, across the RC combination. The response of this single- input-single-output system is determined by the *inhomogeneous* differential equation

$$\jmath(t) = C\frac{dv(t)}{dt} + Gv(t) \tag{1.1.1}$$

If this is written using operator notation, we have

$$x(t) = S\{y(t)\} \tag{1.1.2}$$

where S is the operator $G + C\,d/dt$. In general, eq. (1.1.2) simply means that operator S transforms a function $y(t)$ into another function $x(t)$.

The operator S in eq. (1.1.2) has special properties reflecting the LTI nature of the system and may, of course, contain other operations such as integration besides the derivative and constant multiplier operations of eq. (1.1.1). In this book we will be particularly concerned with LTI operators and in the next section will examine some of their properties. In the inhomogeneous equation (1.1.2) $x(t)$ is a given forcing function or input, and any output $y(t)$, which satisfies the equation, is a solution which must be determined. The form of eq. (1.1.2) (input equals an operation on

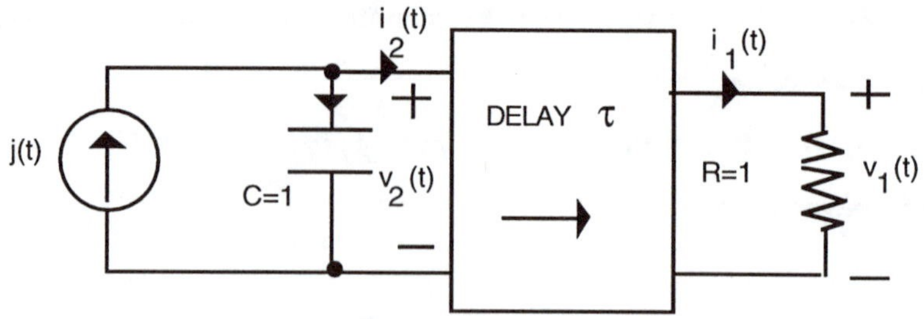

FIGURE 1.1.1
Circuit with time shift.

the output) is particularly useful in the analysis of circuit problems similar to that corresponding to eq. (1.1.1), since the operators are conveniently defined. There are also many problems in which the operator equation is written in a form inverse to that of eq. (1.1.2), i.e., where the operator transforms a given input $x(t)$ into the output $y(t)$, $y(t) = S^{-1}\{x(t)\}$.

As a preparation for the next example, as well as for general use in the following, we define here the *shift operator* T_τ

$$T_\tau\{f(t)\} = f(t-\tau) \tag{1.1.3}$$

whose effect, when applied to a function $f(t)$, is that of *delaying* it by an amount τ.

Example 1.1.1

For the circuit of Fig. 1.1.1 write the operator equation $x(t) = S\{y(t)\}$, where $x(t) = j(t)$, $y(t) = v_1(t)$. The effect of the delay line is to translate or shift the time variable so that the output signal is a delayed version of the input according to eq. (1.1.3).

Solution Referring to the figure, v_1 is delayed τ with respect to v_2, or v_2 is delayed $-\tau$,

$$v_2(t) = v_1(t+\tau) = T_{-\tau}\{v_1(t)\}$$

Here $T_{-\tau}$ is the time shift operator, which introduces an *advance* of τ. Similarly

$$i_2(t) = T_{-\tau}\{i_1(t)\}$$

Writing the differentiation operation as $D = d/dt$ and working our way back from the output resistor, we obtain the required operator equation

$$x(t) = CDT_{-\tau}\{y(t)\} + 1/RT_{-\tau}\{y(t)\} = (CD + 1/R)\,T_{-\tau}\{y(t)\} = S\{y(t)\}$$

or with $C = R = 1$

$$S = \left(\frac{d}{dt} + 1\right) T_{-\tau}$$

We have assumed the distributive law under addition which is easily verified for the operators involved in this example. □

1.2 Linear Time Invariant Systems and Operators

Qualitatively speaking, a linear system is one in which the law of superposition holds. That is if two individual signals are superposed, then the respective responses to these signals are similarly superposed to give the total response. Otherwise stated, the response is proportional to the stimulus. This simple property has very deep consequences, and even more so does its negation (nonlinearity).

In a time invariant system, the physical properties remain invariant with time, hence if a signal is time delayed, i.e., t replaced by $t - \tau$, then the response is similarly time delayed. Thus if a time invariant system starts in a completely deenergized state when excited, its response will always be the same measured from the time of initiation of the input signal. These properties can be precisely phrased in terms of the operators which reflect the physical properties of the system. At the outset we assume that the functions of time under discussion are members of a function space which is appropriately defined. For example, the space might consist of all real absolutely integrable functions. In general the members of the space can be combined according to the usual rules of arithmetic. In the discussions that follow when we say all "admissible functions" or simply "all functions" we mean functions in the defined space. The operator maps functions in the defined space (*the domain of the operator*) into the space of the transformed functions (*the range of the operator*). The domain and range need not coincide. Furthermore, although our interest is essentially confined to LTI operators, it should be noted that the concepts of domain and range apply to all operators, linear or not.

DEFINITION 1.2.1 *A* real *operator is one which maps every real function of the real variable t into a real function of t.*

DEFINITION 1.2.2 *A linear operator is one which satisfies*

$$\begin{aligned} S\{ax(t)\} &= aS\{x(t)\} && \text{(homogeneity)} \\ S\{x_1(t) + x_2(t)\} &= S\{x_1(t)\} + S\{x_2(t)\} && \text{(additivity)} \end{aligned} \tag{1.2.1}$$

or equivalently

$$S\{a_1x_1(t) + a_2x_2(t)\} = a_1S\{x_1(t)\} + a_2S\{x_2(t)\}$$

for all functions $x_k(t)$ *and arbitrary constants (real or complex)* a_k, $(k = 1, 2)$.

DEFINITION 1.2.3 *A time invariant operator* S *is one for which* $x(t) = S\{y(t)\}$ *implies* $x(t-\tau) = S\{y(t-\tau)\}$ *for any* τ, *i.e., the response to an input delayed by some amount is the response to the original input delayed by the same amount.*

Definition 1.2.3 is equivalent to the statement that a time invariant operator S is one which commutes with the time shift operator T_τ for all τ.

$$T_\tau S = S T_\tau$$

In fact, given that $x(t) = S\{y(t)\}$ implies $x(t-\tau) = S\{y(t-\tau)\}$, we have

$$x(t-\tau) = T_\tau\{x(t)\} = T_\tau S\{y(t)\}$$

and

$$x(t-\tau) = S\{y(t-\tau)\} = S T_\tau\{y(t)\}$$

Comparing the right sides of the equations above, the result follows.

DEFINITION 1.2.4 *A real LTI system is one whose operator satisfies the above Definitions 1.2.1, 1.2.2, and 1.2.3.*

It is important to note that to test whether a system satisfies the basic definitions, it will be assumed quiescent or in the zero state (i.e. all voltages and currents zero) when excitation is initiated.

Example 1.2.1
Determine whether the following real operators are linear and/or time invariant.

(a) $S\{f(t)\} = D\{f(t)\} = \dfrac{df(t)}{dt}$

Linearity is satisfied since

$$D\{af_1(t) + bf_2(t)\} = aD\{f_1(t)\} + bD\{f_2(t)\}$$

The derivative operator is time invariant since

$$T_\tau D\{f(t)\} = T_\tau\{f'(t)\} = f'(t-\tau) = D\{f(t-\tau)\} = DT_\tau\{f(t)\}$$

so that T_τ and D commute.

(b) $S\{f(t)\} = \int_{-\infty}^{t} f(u)\,du$

By direct application of the definitions as in (a), the operator is linear provided the integral exists. Time invariance is checked as follows.

$$T_\tau S\{f(t)\} = \int_{-\infty}^{t-\tau} f(u)\,du = \int_{-\infty}^{t} f(v-\tau)\,dv =$$

$$= \int_{-\infty}^{t} f_\tau(v)dv = S\{f_\tau(t)\} = ST_\tau\{f(t)\}$$

Note the substitution $u = v - \tau$ taking place across the third equality sign. Comparing the first and the last terms in the equation chain above, we obtain $T_\tau S = ST_\tau$, and time invariance is established. In general, the time invariant property is not satisfied if the lower limit of the integral defining the operator is real and finite, although reality and linearity would not be affected.

(c) $S\{f(t)\} = f(t) + C$, C a real constant

Time invariance is easily verified. Apply the first of eqs. (1.2.1) as a check of linearity.

$$S\{af(t)\} = af(t) + C, \text{ but } aS\{f(t)\} = a[f(t) + C] \neq af(t) + C$$

so that surprisingly the operator is nonlinear.

(d) $S\{f(t)\} = f(-t)$ time reversal

Clearly S is linear. For time invariance

$$T_\tau S\{f(t)\} = T_\tau\{f(-t)\} = f(-t+\tau)$$

$$ST_\tau\{f(t)\} = S\{f(t-\tau)\} = f(-t-\tau)$$

Since $f(-t+\tau) \neq f(-t-\tau)$, we have $T_\tau S \neq ST_\tau$ and the time reversal operator is not time invariant.

(e) In the space of complex signals $f(t) = u(t) + jv(t)$, consider the operator which selects the real part of $f(t)$

$$S\{f(t)\} = \Re f(t)$$

Time invariance is easily established. For linearity we can check homogeneity by the first of eqs. (1.2.1) using an arbitrary complex constant $\alpha = a + jb$,

$$S\{\alpha f(t)\} = \Re\{(a+jb)[u(t)+jv(t)]\} = au(t) - bv(t) \neq \alpha \Re f(t)$$

so that the operator is nonlinear.

(f) $S\{f(t)\} = T_\tau\{f(t)\} = f(t-\tau)$, τ real, time translation

The operator is evidently linear and time invariant.

(g) $C\{f(t)\} = \int_{-\infty}^{\infty} h(t-\tau)f(\tau)\,d\tau$, τ real

In the above equation $h(t)$ is a prescribed real function. C is the *convolution* operator and the equation is usually written $(Cf)(t) = h(t) * f(t)$. Direct application of the definitions shows that C is LTI (see Corollary 3.3.2). □

The material so far discussed has concerned single input-single-output systems and scalar operators and functions have been used. When we consider multi-input multi-output systems, the scalar functions are replaced by vectors whose elements are scalar functions. The operators are replaced by matrices whose elements are scalar operators. Thus we would write

$$\boldsymbol{x}(t) = \boldsymbol{S}\{\boldsymbol{y}(t)\} \tag{1.2.2}$$

If there are m inputs and and n outputs, then $\boldsymbol{x}$ is a column vector of m components, and the column vector $\boldsymbol{y}$ has n entries. The matrix operator $\boldsymbol{S}$ is of dimensions $m \times n$, i.e., m rows and n columns. Boldface lower case type is used to denote vectors; matrices are identified by boldface upper case fonts. Usually we will be concerned with n-port networks which have a voltage and current defined at each of n ports of access. In this case the number of inputs and outputs will be the same ($m = n$); voltage or current can play the role of either input or output and the matrix operators would be square and of dimension $n \times n$. With this understood, the definitions for reality, linearity, and time invariance have precisely the same form as earlier given, except vectors replace scalar functions and matrix operators replace scalar operators. Finally, unless explicitly excluded, the property of reality will be implied whenever we refer to an LTI system or operator.

1.3 Causality

The systems and circuits we are generally concerned with possess the property of *causality*, or are often referred to as deterministic or nonpredictive. Qualitatively we can think of the *input(s)* at the access ports of the system as the action of the external world as measured by the independent signal variables at these ports, and the *output(s)* as the system reaction (response variables) to the input. In a causal system the reaction at the output cannot anticipate events at the input. Since this concept plays a major role in the analysis of LTI systems we discuss it further in this section.

DEFINITION 1.3.1 *A* causal *operator S with $x(t) = S\{y(t)\}$ is one in which given any two signal vectors $x_1(t)$, $x_2(t)$ such that*

$$x_1(t) = x_2(t), \quad t \leq t_0$$

it follows that

$$y_1(t) = y_2(t), \quad t \leq t_0$$

The causality property takes on a particularly simple form in the case of a linear system as expressed by the following theorem.

THEOREM 1.3.1
(Restricted causality) *Given a* linear *system described by $x(t) = S\{y(t)\}$. The system is causal if and only if for any real t_0 and any signal which satisfies*

$$x(t) = 0, \quad t \leq t_0$$

the response satisfies

$$y(t) = 0, \quad t \leq t_0$$

PROOF For sufficiency, consider two signals $x_1(t)$ and $x_2(t)$, and their responses $y_1(t)$ and $y_2(t)$ satisfying the hypothesis, such that

$$x_1(t) = x_2(t), \quad t \leq t_0$$

Construct a new signal $x(t)$

$$x(t) = x_1(t) - x_2(t) = 0, \quad t \leq t_0$$

Then by hypothesis the response corresponding to $x(t)$

$$y(t) = 0, \quad t \leq t_0$$

but by linearity

$$y(t) = y_1(t) - y_2(t) = 0$$

or

$$y_1(t) = y_2(t), \quad t \leq t_0$$

Thus x and y satisfy Definition 1.3.1 and S is causal. Next assume the causality definition is satisfied and the converse (necessity) follows in a similar fashion, so the theorem is proved. The result is clearly valid if n-port vectors, $\boldsymbol{x}(t)$ and $\boldsymbol{y}(t)$ are used instead of scalar signals. ■

For future reference we shall often refer to a signal which vanishes for $t \leq t_0$ as a *causal signal (vector)*. By time invariance an LTI system satisfies the following theorem.

THEOREM 1.3.2
An LTI system is causal if and only if a causal response results when the excitation vectors are causal for some particular t_0 *(often we choose* $t_0 = 0$*).*

Example 1.3.1
Consider the nonlinear system with $x(t) = S\{y(t)\}$ defined as follows

$$\text{If } x(t) \leq 1 \quad \text{for all } t, \qquad \text{then } y(t) = 0.5x(t)$$

$$\text{If } x(t) > 1 \quad \text{for some } t, \quad \text{then } y(t) = 2x(t)$$

Determine whether the system satisfies either general causality, Definition 1.3.1, or restricted causality, Theorem 1.3.1.

Solution Clearly for any $x(t) = 0$, $t \leq t_0$ then $y(t) = 0$, $t \leq t_0$, so the system satisfies the causality property of Theorem 1.3.1. On the other hand suppose $x_1(t) = \epsilon^t$, $t \leq 0$; $x_1(t) = 1$, $t > 0$; and $x_2(t) = \epsilon^t$, $t \leq 0$; $x_2(t) = 3$, $t > 0$, i.e. $x_1(t) = x_2(t)$, $t \leq 0$. Then $y_1(t) = .5\epsilon^t$, $t \leq 0$ and $y_2(t) = 2\epsilon^t \neq y_1(t)$ for $t \leq 0$. The system violates Definition 1.3.1 and is thus noncausal, verifying the fact that the restricted form of causality is not sufficient for a nonlinear system. □

1.4 Power, Energy, and Passivity

Consider an n-port system with n component *real* voltage and current column vectors $\boldsymbol{v}(t)$, $\boldsymbol{i}(t)$ measured at the ports. At the k-th port the scalar voltage and current are $v_k(t)$, $i_k(t)$. The power $p(t)$ delivered to the n-port is the sum of the individual port input powers

$$p(t) = \sum_{k=1}^{n} v_k(t)\, i_k(t) = \boldsymbol{i}'(t)\, \boldsymbol{v}(t) \tag{1.4.1}$$

In eq. (1.4.1) the operation $'$ indicates vector or matrix transpose, i.e., rows and columns interchanged, so that a column vector $\boldsymbol{i}$ becomes a row vector $\boldsymbol{i}'$ and vice versa. Thus eq. (1.4.1) designates the scalar product of $\boldsymbol{v}$ and $\boldsymbol{i}$. Furthermore we are assuming that these vectors are related by an operator equation of the form of eq. (1.2.2). If we integrate the instantaneous power from $t = -\infty$, when the system is assumed quiescent, we get the total energy $w(t)$ delivered to the n-port up to time t.

$$w(t) = \int_{-\infty}^{t} \boldsymbol{i}'(\tau)\, \boldsymbol{v}(\tau)\, d\tau$$

If the signals are complex, eq. (1.4.1) is replaced by the following

$$p(t) = \Re \sum_{k=1}^{n} v_k(t)\, i_k^*(t) = \Re\, \boldsymbol{i}^\dagger(t)\, \boldsymbol{v}(t) \tag{1.4.2}$$

In eq. (1.4.2) we have used the notation $[(\ )^*]' = (\ )^\dagger$, where $(\ )^*$ means complex conjugate. It is clear that in the case of real signals, eq. (1.4.2) reduces to eq. (1.4.1).

Eq. (1.4.2) is essentially a definition that will be justified on physical grounds later on. The total energy delivered to the n-port up to time t is still obtained by integrating the instantaneous power in eq. (1.4.2) from $-\infty$ to t

$$w(t) = \Re \int_{-\infty}^{t} \boldsymbol{i}^\dagger(\tau)\, \boldsymbol{v}(\tau)\, d\tau \tag{1.4.3}$$

A passive system is one that always acts as a sink of energy. That is, the total energy into the system measured at any time t beyond the moment of initial excitation (when the system is quiescent) cannot be negative, i.e., no net energy is returned to the source for any t. This implies that all energy entering the system is either stored inside (in the form of magnetic or dielectric energy) or is exchanged irreversibly (usually as heat) with the surroundings (through interfaces different from the n electrical ports); in fact, a reversible exchange could be used to invert the energy flow and to supply the source with net energy. It thus follows that for a passive system

$$w(t) \geq 0 \quad \forall t$$

DEFINITION 1.4.1 *A system is* passive *if the energy $w(t)$ delivered to the system, defined as in eq. (1.4.3), is always nonnegative, i.e.,*

$$w(t) = \Re \int_{-\infty}^{t} \boldsymbol{i}^\dagger(\tau)\, \boldsymbol{v}(\tau)\, d\tau \geq 0 \quad \forall t \tag{1.4.4}$$

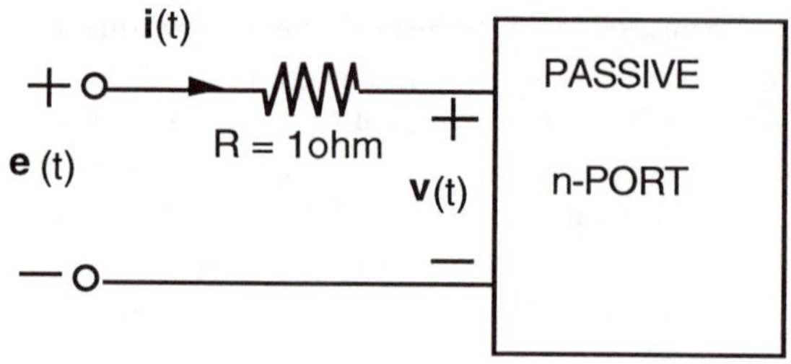

FIGURE 1.4.1
Augmented n-port.

If signals are real, eq. (1.4.4) reduces to

$$w(t) = \int_{-\infty}^{t} \boldsymbol{i}'(\tau)\,\boldsymbol{v}(\tau)\,d\tau \geq 0 \quad \forall t \tag{1.4.5}$$

The following theorem shows that for linear n-ports passivity with respect to real signals and passivity with respect to complex signals are equivalent.

THEOREM 1.4.1
In a linear *system, passivity with respect to real signals, eq. (1.4.5), implies passivity with respect to complex signals, eq. (1.4.4), and conversely.*

PROOF Let $\boldsymbol{v}(t) = \boldsymbol{S}\{\boldsymbol{i}(t)\}$. Then with $\boldsymbol{v}(t) = \boldsymbol{v}_r(t) + j\boldsymbol{v}_i(t)$, $\boldsymbol{i}(t) = \boldsymbol{i}_r(t) + j\boldsymbol{i}_i(t)$ it follows that since the operator $\boldsymbol{S}$ is linear, and real, we may equate real and imaginary parts in the above operator equation and obtain $\boldsymbol{v}_r(t) = \boldsymbol{S}\{\boldsymbol{i}_r(t)\}$ and $\boldsymbol{v}_i(t) = \boldsymbol{S}\{\boldsymbol{i}_i(t)\}$. By hypothesis, eq. (1.4.5) holds for the real signal pair $\boldsymbol{v}_r(t)$, $\boldsymbol{i}_r(t)$ and for the real pair $\boldsymbol{v}_i(t)$, $\boldsymbol{i}_i(t)$. Now

$$\Re\,\boldsymbol{i}^{\dagger}(t)\,\boldsymbol{v}(t) = \boldsymbol{i}_r'(t)\,\boldsymbol{v}_r(t) + \boldsymbol{i}_i'(t)\,\boldsymbol{v}_i(t) \tag{1.4.6}$$

According to eq. (1.4.5), when both terms on the right side of eq. (1.4.6) are integrated, the results are nonnegative, so that eq. (1.4.4) follows and the theorem is proved. (The converse is evident.) ■

If a given n-port is augmented by adding a positive series resistor to each port, the new structure is known as the *(series) augmented n-port.* Fig. 1.4.1 shows a schematic diagram for an augmented n-port. For convenience the single port of the diagram represents all n ports.

Let the resistances R_k be arranged in a diagonal resistance matrix $\boldsymbol{R}_0$. The vectors $\underline{\boldsymbol{v}}$ and $\underline{\boldsymbol{i}}$ defined by the following equations

$$\underline{\boldsymbol{v}} = \boldsymbol{R}_0^{-\frac{1}{2}}\boldsymbol{v} \tag{1.4.7}$$

$$\underline{\boldsymbol{i}} = \boldsymbol{R}_0^{\frac{1}{2}}\boldsymbol{i} \tag{1.4.8}$$

are termed the "voltage" and "current" vectors *normalized* to the port resistances.[1]

The equation of the terminating one-ports

$$\boldsymbol{e} = \boldsymbol{v} + \boldsymbol{R}_0 \boldsymbol{i}$$

after normalizing $\boldsymbol{e}$ as

$$\underline{\boldsymbol{e}} = \boldsymbol{R}_0^{-\frac{1}{2}} \boldsymbol{e} \tag{1.4.9}$$

becomes

$$\underline{\boldsymbol{e}} = \underline{\boldsymbol{v}} + \underline{\boldsymbol{i}} \tag{1.4.10}$$

Corresponding to the *normalized* n-port, the diagonal matrix of series resistors should form the unit matrix $\boldsymbol{I}_n$.

The following Theorem 1.4.2, under the conditions illustrated in Definition 1.4.2, provides a sufficient condition for the existence of the energy integral for all t.

DEFINITION 1.4.2 *An n-component signal vector $\boldsymbol{w}(t)$ is square integrable, i.e., in L_n^2, if*

$$\int_{-\infty}^{+\infty} \boldsymbol{w}^{\dagger}(t)\, \boldsymbol{w}(t)\, dt = \int_{-\infty}^{+\infty} |\boldsymbol{w}(t)|^2\, dt < +\infty \tag{1.4.11}$$

THEOREM 1.4.2

For a passive n-port, for which the admissible voltage and current vectors $\boldsymbol{v}(t)$, $\boldsymbol{i}(t)$ are both square integrable, the energy integral built on such a pair exists for all t.

PROOF Assume, according to eqs. (1.4.7) and (1.4.8), that the state at the ports be described by the normalized voltage and current vectors, $\underline{\boldsymbol{v}}$ and $\underline{\boldsymbol{i}}$. Note that since $|\underline{\boldsymbol{v}} - \underline{\boldsymbol{i}}|^2 \geq 0$ the inequality

$$|\underline{\boldsymbol{v}}(t)|^2 + |\underline{\boldsymbol{i}}(t)|^2 \geq 2\Re\, \underline{\boldsymbol{i}}(t)^{\dagger} \underline{\boldsymbol{v}}(t) \tag{1.4.12}$$

is valid. Since $\Re\, \underline{\boldsymbol{i}}^{\dagger}(t)\, \underline{\boldsymbol{v}}(t) = \Re\, \boldsymbol{i}^{\dagger}(t)\, \boldsymbol{v}(t)$ (as can be seen immediately from eqs. (1.4.7) and (1.4.8)), the left hand side of eq. (1.4.12) is always nonnegative, so that when it is integrated from $-\infty$ to t the integral is bounded $\forall t$ because $\underline{\boldsymbol{v}}(t)$ and $\underline{\boldsymbol{i}}(t)$ satisfy eq. (1.4.11). The integral of the right hand side is just twice the energy integral, hence by virtue of the inequality the energy integral exists $\forall t$. ∎

[1] This terminology is somewhat misleading, since such vectors both have the physical dimensions of the square root of a power; however, no better terms have been introduced in the literature so far.

The following Theorem 1.4.3 shows that for an augmented passive n-port, the mapping excitation $\to$ response is from L_n^2 into itself.

THEOREM 1.4.3

If a passive n-port is augmented as shown in Fig. 1.4.1 and the n-component excitation voltage vector $\mathbf{e}(t)$ is square integrable, then the current and voltage vectors, $\boldsymbol{i}(t)$, $\boldsymbol{v}(t)$ are also square integrable.

PROOF From eq. (1.4.10), we have $\underline{e}(t) = \underline{v}(t) + \underline{i}(t)$. Thus

$$|\underline{e}(t)|^2 = |\underline{v}(t)|^2 + |\underline{i}(t)|^2 + 2\Re\, \underline{i}^\dagger(t)\, \underline{v}(t)$$

Integrate both sides of the preceding equation

$$\int_{-\infty}^{\infty} |\underline{e}(t)|^2\, dt = \int_{-\infty}^{\infty} (|\underline{v}(t)|^2 + |\underline{i}(t)|^2)\, dt + \int_{-\infty}^{\infty} 2\Re\, \underline{i}^\dagger(t)\, \underline{v}(t)\, dt \quad (1.4.13)$$

By hypothesis, the left hand side of eq. (1.4.13) is bounded and, by passivity, the last term on the right hand side of eq. (1.4.13) is nonnegative. Therefore, since the first term on the right is nonnegative, it must be bounded, so the theorem is proved and both $\underline{v}(t)$ and $\underline{i}(t)$, thus $\boldsymbol{v}(t)$ and $\boldsymbol{i}(t)$ are square integrable. ■

Example 1.4.1

The linear time invariant resistor, inductor, and capacitor are passive.

Solution For an LTI resistor defined by

$$v = Ri \qquad R > 0$$

we have

$$w(t) = R \int_{-\infty}^{t} i^2(\tau)\, d\tau \geq 0$$

because the integrand is nonnegative. The integrand exists for all t, since we take $i \in L^2$.

For an LTI inductor defined by

$$v = L\frac{di}{dt} \qquad L > 0$$

we have

$$w(t) = L \int_{-\infty}^{t} \frac{di(\tau)}{d\tau}\, i(\tau)\, d\tau = \frac{1}{2} L i^2(t) \geq 0$$

For an LTI capacitor defined by

$$i = C\frac{dv}{dt} \qquad C > 0$$

we have

$$w(t) = C\int_{-\infty}^{t}\frac{dv(\tau)}{d\tau}\,v(\tau)\,d\tau = \frac{1}{2}Cv^2(t) \geq 0$$

Note that the energy into an inductor (a capacitor) at time t only depends on the value of the current (the voltage) at that time. □

In steady state alternating current (a.c.) problems, the port voltages and currents are trigonometric functions (not square integrable). The steady state average power delivered to the system is independent of the number of elapsed full periods. Suppose that for a one-port system the voltage and associated current are periodic and defined by

$$v(t) = \sqrt{2}|V|\cos(\omega t + \psi) = \Re\sqrt{2}\,V\,\epsilon^{j\omega t}, \ V = |V|\,\epsilon^{j\psi}$$

$$i(t) = \sqrt{2}|I|\cos(\omega t + \theta) = \Re\sqrt{2}\,I\,\epsilon^{j\omega t}, \ I = |I|\,\epsilon^{j\theta}$$

We have used Euler's formula, $\epsilon^{jx} = \cos x + j\sin x$, to express the signals in complex form. The quantities V, I are complex phasors with r.m.s. amplitudes. The period of $v(t)$ and $i(t)$ is $T = 2\pi/\omega$. The instantaneous power is given by

$$p(t) = v(t)i(t) = 2|V||I|\cos(\omega t + \psi)\cdot\cos(\omega t + \theta)$$

Using the identity $\cos x\,\cos y = (1/2)[\cos(x+y)+\cos(x-y)]$, we immediately obtain

$$p(t) = |V||I|[\cos(2\omega t + \psi + \theta) + \cos(\psi - \theta)]$$

The period of $p(t)$ is half the period of voltage and current and, since the average value of $\cos(2\omega t + \psi + \theta)$ over any number of fundamental periods $kT = k2\pi/\omega$ is zero, the average power P delivered to the one-port is given by

$$P = |V||I|\cos\varphi$$

where $\varphi = \psi - \theta$ and $\cos\varphi$ is the *power factor.*

The quantity

$$A = VI^* = |V||I|\,\epsilon^{j\varphi} = |V||I|\cos\varphi + j|V||I|\sin\varphi = P + jQ$$

is termed the *complex power*; its real part is the average or *active* power P, its imaginary part is the *reactive* power Q.

The above concepts are readily extended to n-ports by defining the complex power entering the n-port as

$$A = \boldsymbol{I}^\dagger\boldsymbol{V} = \Re\,\boldsymbol{I}^\dagger\boldsymbol{V} + j\,\Im\,\boldsymbol{I}^\dagger\boldsymbol{V} = P + jQ \tag{1.4.14}$$

In eq. (1.4.14), $\boldsymbol{V}$ and $\boldsymbol{I}$ are column vectors whose components are the n phasors associated with the corresponding ports. It will be shown in Section 2.13 that for a passive n-port $Q = 2\omega(\overline{W}_m - \overline{W}_e)$ where $\overline{W}_m$ and $\overline{W}_e$ are the average values of magnetic and electric energy stored inside the n-port. Note that, while the definition of the active power P is invariant whether the complex power is defined as $\boldsymbol{I}^\dagger\boldsymbol{V}$ (as we did in eq. (1.4.2)) or as $\boldsymbol{V}^\dagger\boldsymbol{I}$, the sign of the reactive power Q changes; thus a definite convention must be adopted.

The definition of complex power used for phasors can be considered as a particular case of a more general definition holding for general complex time functions.

$$a(t) = \boldsymbol{i}^\dagger(t)\boldsymbol{v}(t) = \Re\,\boldsymbol{i}^\dagger(t)\boldsymbol{v}(t) + j\,\Im\,\boldsymbol{i}^\dagger(t)\boldsymbol{v}(t) = p(t) + jq(t)$$

which includes eq. (1.4.14) as a particular case, if we take as complex voltage and current vectors the following

$$\boldsymbol{v}(t) = \boldsymbol{V}\epsilon^{j\omega t} \tag{1.4.15}$$

$$\boldsymbol{i}(t) = \boldsymbol{I}\epsilon^{j\omega t} \tag{1.4.16}$$

Note that in eqs. (1.4.15) and (1.4.16) the components of vectors $\boldsymbol{V}$ and $\boldsymbol{I}$ are r.m.s., not peak values.

When an LTI system is initially excited (at $t = -\infty$) by trigonometric signals, it will eventually operate under a.c. steady state conditions and the voltage and current vectors will be as specified by the two previous equations (1.4.15) and (1.4.16) for all times $t > T$, provided $t - T$ is sufficiently large so that all transients are negligible. If the system is passive, then the energy $w(T)$ delivered to the system over the epoch τ, $-\infty < \tau \leq T$ is nonnegative and finite. The total energy $w(t)$ for any $t > T$ is given by eq. (1.4.3), and by passivity this must be nonnegative. Thus

$$w(t) = w(T) + \Re\,\boldsymbol{I}^\dagger\boldsymbol{V}\int_T^t d\tau \geq 0$$

and therefore

$$w(t) = w(T) + (t - T)\,\Re\,\boldsymbol{I}^\dagger\boldsymbol{V} \geq 0$$

Since $w(T) \geq 0$ and finite, and $(t - T) \geq 0$, for sufficiently large t, $w(t)$ will have the sign of $\Re\,\boldsymbol{I}^\dagger\boldsymbol{V}$.

The conclusion is stated in the following theorem.

THEOREM 1.4.4

A necessary *condition for passivity is*

$$P = \Re\,\boldsymbol{I}^\dagger(j\omega)\boldsymbol{V}(j\omega) \geq 0 \quad \forall\,\omega$$

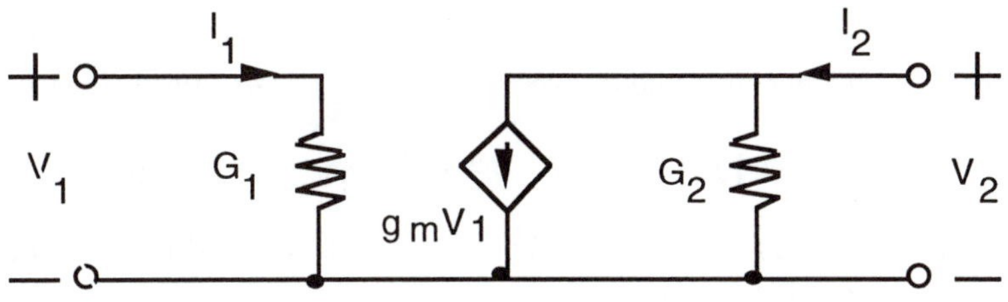

FIGURE 1.4.2
Simple model of an amplifier, $g_m > 0$.

that is, under steady state a.c. operation, the active power entering the n-port must be nonnegative at all frequencies.

We finally note that complex signals can receive different physical interpretations.

For example the equations

$$\begin{aligned} \boldsymbol{v}(t) &= \boldsymbol{V}(t)\epsilon^{j\omega t} \\ \boldsymbol{i}(t) &= \boldsymbol{I}(t)\epsilon^{j\omega t} \end{aligned}$$

where $\boldsymbol{V}(t)$ and $\boldsymbol{I}(t)$ are complex time functions, describe a sinusoidal carrier amplitude modulated by a (slowly) varying signal. Power calculated according to eq. (1.4.2) is the average modulation envelope power.

We can also view the complex signal as two-phase symmetrical components, consisting of a pair (real and imaginary parts) of orthogonal time functions.

Example 1.4.2

For the amplifier model shown, find the bounds on transconductance g_m so that the system is nonpassive, i.e. active and capable of power gain.

Solution The V_k, I_k in Fig. 1.4.2 are phasors. It is easy to see that

$$I_1 = G_1 V_1, \quad I_2 = g_m V_1 + G_2 V_2$$

Write the average power inequality that determines passivity

$$\Re\, \boldsymbol{I}^{\dagger}\boldsymbol{V} = \Re\left(I_1 \; I_2 \right)^{*} \begin{pmatrix} V_1 \\ V_2 \end{pmatrix} = \Re\left(V_1 I_1^{*} + V_2 I_2^{*}\right)$$

or, taking into account the above expressions for currents in terms of voltages,

$$\Re\, \boldsymbol{I}^{\dagger}\boldsymbol{V} = G_1|V_1|^2 + G_2|V_2|^2 + \frac{g_m}{2}V_1V_2^{*} + \frac{g_m}{2}V_1^{*}V_2 \geq 0$$

since $\Re\, V_1V_2^{*} = (1/2)(V_1V_2^{*} + V_1^{*}V_2)$. It is easily verified that the preceding passivity inequality is identical to

$$\Re\, \boldsymbol{I}^{\dagger}\boldsymbol{V} = G_1\left(V_1 + V_2\frac{g_m}{2G_1}\right)\left(V_1^{*} + V_2^{*}\frac{g_m}{2G_1}\right) + |V_2|^2\left(G_2 - \frac{g_m^2}{4G_1}\right)$$

Thus passivity requires

$$G_1\left|V_1 + V_2\frac{g_m}{2G_1}\right|^2 + |V_2|^2\left(G_2 - \frac{g_m^2}{4G_1}\right) \geq 0 \qquad (1.4.17)$$

Evidently the second term of eq. (1.4.17) must be nonnegative (the first is automatically so) for the inequality to hold for all V_1, V_2. Therefore, passivity requires $g_m \leq 2\sqrt{G_1G_2}$. Thus the system will be active if

$$g_m > 2\sqrt{G_1G_2}$$

□

We now turn to the important concept of a *lossless* system.

DEFINITION 1.4.3 *A system is said to be* lossless *if it is passive and moreover*

$$w(+\infty) = \Re\int_{-\infty}^{+\infty} \boldsymbol{i}^{\dagger}(\tau)\, \boldsymbol{v}(\tau)\, d\tau = 0$$

The physical idea behind the concept of a lossless system is that of a passive system in which no irreversible exchanges with the surroundings can take place (and hence in thermal equilibrium no dissipation is possible); since reversible exchanges are prevented by passivity, all energy into the system must be stored in magnetic and/or electric form. Since port voltage and current vectors belong to L_n^2, they tend to zero for $t \to +\infty$. Since the stored energy only depends on the excitation, it must tend to the same value for $t \to +\infty$ as for $t \to -\infty$, i.e., zero. Thus the total energy exchanged through the ports must eventually sum to zero, as stated in Definition 1.4.3.

Example 1.4.3
The linear time invariant inductor and capacitor are lossless.

Solution In Example 1.4.1 we have shown that both one-ports are passive. Since for each element, v and i belong to L^2 we have $lim_{t\to+\infty}i(t) = 0$ and $lim_{t\to+\infty}v(t) = 0$. Using w of the referenced Example, $w(+\infty) = 0$. □

1.5 Passivity, Linearity and Causality

We will show in this section that the passivity and linearity properties of a system operator are strongly tied to the causality property. Consider a linear passive n-port. For convenience take $\boldsymbol{v}(t)$ as excitation, $\boldsymbol{i}(t)$ as response. First note that in some special cases the response is not unique. For example a one-port short circuit only admits $v(t) = 0$, but the current may be arbitrary. Or more generally suppose that some of the ports, considered as branches, form a closed loop G. In this case the port voltages sum to zero regardless of the port currents. The port voltages are linearly dependent and the port currents are not unique for they may contain an arbitrary nonzero component circulating in G and independent of the port voltages. The causal properties of such a system cannot be exhibited when the voltages are taken as impressed signal variables. Therefore in our consideration of the relation between causality, linearity, and passivity we will exclude as input port variables cases where these variables (voltage or current as the case may be) are forced by the system to be linearly dependent.

DEFINITION 1.5.1 *An n-port is voltage* nondegenerate *if it admits n linearly independent port voltage vectors. That is for such an n-port we can always find an $n \times n$ matrix of voltage column vectors*

$$\boldsymbol{M} = (\boldsymbol{v}_1(t), \boldsymbol{v}_2(t)...., \boldsymbol{v}_n(t))$$

of rank n. Each $\boldsymbol{v}_k(t)$, $j = 1, 2, ..., n$ is an n-element column vector whose j-th component $v_{kj}(t)$ represents a voltage across port j.

THEOREM 1.5.1
Given a linear, passive n-port. If the n-port is voltage nondegenerate it must be causal (in the sense of Theorem 1.3.1) under voltage excitation (v,i causal).

PROOF Let $\boldsymbol{v}(t)$ be a signal such that $\boldsymbol{v}(t) = 0$, $t < t_0$, $\boldsymbol{i}(t) = \boldsymbol{S}\{\boldsymbol{v}(t)\}$.[2]

[2]Note that here the operator S is an input-output mapping, rather than an output-input mapping as in the previous sections.

We wish to show that in a nondegenerate n-port, if $\boldsymbol{S}$ is linear and passive, then the response $\boldsymbol{i}(t)$ vanishes for $t < t_0$.

Let $\boldsymbol{v}_0(t)$ be any other admissible voltage vector with $\boldsymbol{i}_0(t) = \boldsymbol{S}\{\boldsymbol{v}_0(t)\}$. As defined below choose another voltage $\boldsymbol{e}_0(t)$, with response current $\boldsymbol{j}_0(t) = \boldsymbol{S}\{\boldsymbol{e}_0(t)\}$

$$\boldsymbol{e}_0(t) = \boldsymbol{v}_0(t) + \alpha\boldsymbol{v}(t) \quad (\boldsymbol{e}_0(t) = \boldsymbol{v}_0(t) \quad t < t_0)$$

In the above relation α is an arbitrary constant. By linearity $\boldsymbol{j}_0(t) = \boldsymbol{i}_0(t) + \alpha\boldsymbol{i}(t)$. Now consider the passivity inequality

$$\Re \int_{-\infty}^{t} \boldsymbol{j}_0^{\dagger}(\tau)\, \boldsymbol{e}_0(\tau)\, d\tau \geq 0 \quad \forall t$$

Substitute for $\boldsymbol{e}_0(t)$, with $t < t_0$ and α real.

$$\Re \int_{-\infty}^{t} \boldsymbol{i}_0^{\dagger}(\tau)\, \boldsymbol{v}_0(\tau)\, d\tau + \alpha\Re \int_{-\infty}^{t} \boldsymbol{i}^{\dagger}(\tau)\, \boldsymbol{v}_0(\tau)\, d\tau \geq 0 \quad t < t_0$$

By passivity the first integral is nonnegative $\forall t$. Hence for any assigned $t < t_0$ we can always choose an α to make the sum of the two integrals negative and thus contradict the inequality unless

$$\Re \int_{-\infty}^{t} \boldsymbol{i}^{\dagger}(\tau)\, \boldsymbol{v}_0(\tau)\, d\tau = 0 \quad \forall t < t_0$$

It therefore follows that $\Re\, \boldsymbol{i}^{\dagger}(t)\boldsymbol{v}_0(t) = 0$, $\forall t < t_0$. We can repeat all the steps using as arbitrary constant the pure imaginary quantity $j\alpha$ instead of α. The result would then be $\Im\, \boldsymbol{i}^{\dagger}(t)\, \boldsymbol{v}_0(t) = 0$. We conclude therefore

$$\boldsymbol{i}^{\dagger}(t)\, \boldsymbol{v}_0(t) = 0 \quad t < t_0$$

Since the n-port is nondegenerate, by proceeding exactly as described above we can choose n linearly independent voltage vectors $\boldsymbol{v}_1(t), \ldots, \boldsymbol{v}_n(t)$ and obtain n relations of the form $\boldsymbol{i}^{\dagger}(t)\, \boldsymbol{v}_k(t) = 0$, $k = 1, 2, \ldots n$, $t < t_0$. It follows that using the $n \times n$ matrix $\boldsymbol{M}(t)$ whose columns are the $\boldsymbol{v}_k(t)$ we have

$$\boldsymbol{M}^{\dagger}(t)\, \boldsymbol{i}(t) = 0, \quad t < t_0$$

Since $\boldsymbol{M}$ is of full rank, it is nonsingular, whence it follows that $\boldsymbol{i}(t) = 0$, $t < t_0$ or the system is causal. ∎

The result has been proved using voltage as excitation, current as response i.e. (v, i) causality. The dual theorem is clearly valid; A linear passive n-port is (i, v) causal (current excitation, voltage response) if the system admits n linearly independent current vectors. It is evident that all linear passive one-ports are (v, i) causal except the short circuit; and of course all linear passive one-ports are (i, v) causal except the open circuit.

THEOREM 1.5.2
Referring to Fig. (1.4.1), all linear passive augmented n-ports are (v, i) causal.

PROOF The normalized voltage vector $\underline{e}(t)$ at the ports of the augmented structure is related to the signals at the ports of the given n-port by

$$\underline{e}(t) = \underline{v}(t) + \underline{i}(t)$$

Choose $\underline{e}(t)$ as the signal, $\underline{e}(t) = 0$, $t < t_0$. First we show that the responses $\underline{v}(t)$, and $\underline{i}(t)$ vanish $t < t_0$. The above equation yields $\underline{v}(t) = -\underline{i}(t)$, $t < t_0$. Hence invoking the energy integral

$$\Re \int_{-\infty}^{t} \underline{i}^{\dagger}(\tau)\, \underline{v}(\tau)\, d\tau = -\int_{-\infty}^{t} \underline{v}^{\dagger}(\tau)\underline{v}(\tau)\, d\tau \leq 0 \qquad t < t_0$$

This clearly violates passivity unless $\underline{v}(t) = 0$, $t < t_0$. Thus $\underline{v}(t)$ is a causal signal vector and so too is $\underline{i}(t)$, so restricted causality is proved. Up to this point we have not assumed linearity, but now introduce the linearity hypothesis and by the restricted causality Theorem 1.3.1 it follows that the augmented n-port must be causal. ∎

Theorem 1.5.1 shows that if an n-port operator is linear and passive then (excluding the degenerate case) the system must be causal. This means that if the system is initially in the quiescent state a causal forcing function or vector must yield a unique solution of the operator equation which is also causal. Passivity is the key restriction, so that not all LTI systems need be causal. We demonstrate this with the following example.

Example 1.5.1
Find the solution of the following operator equation subject to the conditions that (a) the system is initially quiescent, and (b) the response as $t \to +\infty$ is bounded.

$$v(t) = \mathrm{u}(t) = \frac{di(t)}{dt} - i(t) = S\{i(t)\}$$

Solution Clearly the operator S is LTI, and the causal input signal is the unit step defined by

$$\mathrm{u}(t) = 0,\ t < 0 \quad \mathrm{u}(t) = 1,\ t \geq 0$$

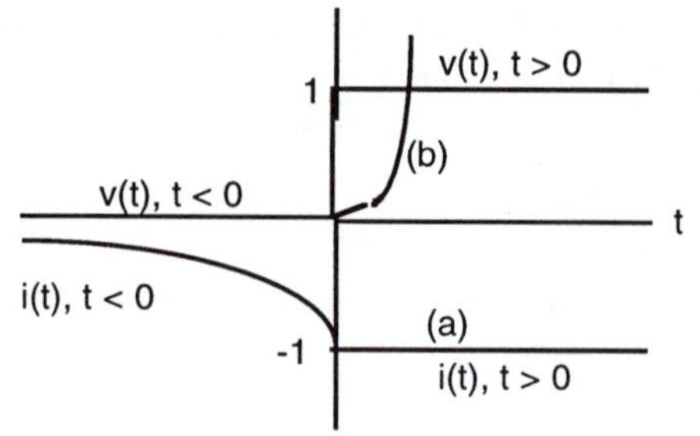

FIGURE 1.5.1
Responses to $v(t) = \mathrm{u}(t)$: (a) Noncausal, (b) Unstable.

so that

$$\mathrm{u}(-t) = 1,\ t \leq 0 \quad \mathrm{u}(t) = 0,\ t > 0$$

The value of the step at $t = 0$ has been chosen as 1, but this is rather arbitrary. The value at the origin is *a priori* not defined. One could designate the value as 1/2, which is consistent with the value $1/2[\mathrm{u}(0+) + \mathrm{u}(0-)]$ which would arise in Fourier analysis.

For $t < 0$ the general solution of the operator equation is $i(t) = A\epsilon^t$, A an arbitrary constant. Note that $i(t) \to 0$ as $t \to -\infty$ which is consistent with initial quiescence. For $t > 0$, the solution is $i(t) = B\epsilon^t - 1$, B another constant. The requirement of a bounded solution as $t \to +\infty$ yields $B = 0$. Finally the presence of the derivative in the operator demands that the solution $i(t)$ be continuous $\forall t$ and this determines the value of $A = -1$, so that there be no discontinuity at t = 0. The complete solution is therefore

$$i(t) = -[\epsilon^t\,\mathrm{u}(-t) + \mathrm{u}(t)]$$

The excitation $v(t) = \mathrm{u}(t)$, and response $i(t)$ curve (a), are plotted in Fig. 1.5.1.

It is clear by inspecting the figure that although the system is LTI, it is not causal since the response exists prior to the excitation. In order to check passivity let us find the energy $w(t)$ by integrating the instantaneous power $p(t) = v(t)i(t)$ directly from the figure. Thus

$$w(t) = -t < 0, \quad t > 0$$

and the system is active, so that a noncausal response is theoretically permissible. It may be noted that we can change the statement of the problem by imposing the boundary condition of causality on the response i.e. $i(t) = 0$ for $t < 0$. In this case $A = 0$, and continuity at the origin determines $B = 1$. The causal response (plotted in Fig. 1.5.1 as curve (b)) is therefore

$$i(t) = (\epsilon^t - 1)\,\mathrm{u}(t)$$

Of the two solutions the stable (bounded) solution is noncausal and the causal solution is unstable. Implicit in this problem are many of the basic properties of LTI systems and these will be investigated in Chapter 2. □

2

LTI System Response to Exponential Eigenfunctions

2.1 Solution of Operator Equations

The response of an LTI circuit or system to an exponential driving signal is of particular importance since the general operation of such systems in both the frequency and time domains is completely determined by the response to an exponential time signal. First we consider some general aspects of the time domain response of LTI systems to arbitrary forcing functions. The system is initially quiescent or in the zero initial state and the LTI inhomogeneous operator equation (scalar form) is given by eq. (2.1.1)

$$x(t) = S\{y(t)\} \tag{2.1.1}$$

If $x(t) = 0$ in eq. (2.1.1), we have the homogeneous form of the operator equation. Suppose the most general solution of the homogeneous equation is $y = y_h(t)$. That is

$$0 = S\{y_h(t)\}$$

Given $y_h(t)$, we seek the general solution $y(t)$ of eq. (2.1.1) over an interval I when $x(t)$ is prescribed and continuous on I. Let $y = y_p(t)$ be any particular solution of the inhomogeneous equation valid on the interval I. Thus

$$x(t) = S\{y(t)\} = S\{y_p(t)\}$$

and invoking linearity $S\{y(t)\} - S\{y_p(t)\} = S\{y(t) - y_p(t)\} = 0$. The operand $y(t) - y_p(t)$ must therefore be equal to the most general homogeneous solution $y_h(t)$. Thus the most general solution of the inhomogeneous equation eq. (2.1.1) on the interval I is

$$y(t) = y_h(t) + y_p(t) \tag{2.1.2}$$

When the boundary conditions are substituted into eq. (2.1.2), the arbitrary constants may be evaluated and the complete solution valid on the

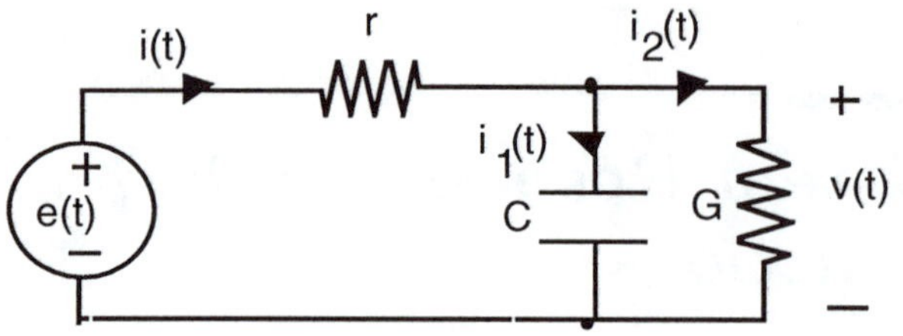

FIGURE 2.1.1
RC circuit with two resistors.

interval I is obtained. If eq. (2.1.1) is an n-th order constant coefficient differential equation, in general, there are n such constants to be determined. As an illustration of the above procedure and to provide motivation for further discussion, consider the following example of an RC circuit.

Example 2.1.1

For the passive circuit of Fig. 2.1.1 let the complex signal excitation $e(t) = E\epsilon^{st}$, $s = \sigma + j\omega$ be applied at time $t = T$ with the circuit initially quiescent, i.e., $e(t) = E\epsilon^{st}\ \mathrm{u}(t - T)$, and $v(-\infty) = 0$. Find the output voltage $v(t)$ and the input current $i(t)$ for $t > -\infty$.

Solution Using the Kirchhoff Current Law (KCL)

$$i(t) = i_1(t) + i_2(t) = C\frac{dv(t)}{dt} + Gv(t)$$

The Kirchhoff Voltage Law (KVL) around the input loop gives

$$e(t) = ri(t) + v(t) \tag{2.1.3}$$

so that substituting for $i(t)$

$$e(t) = r[i_1(t) + i_2(t)] + v(t) = rC\frac{dv(t)}{dt} + (rG + 1)v(t)$$

and the operator equation has the form

$$e(t) = S\{v(t)\} = a\frac{dv(t)}{dt} + bv(t) \quad a = rC,\ b = rG + 1 \tag{2.1.4}$$

First we find the general solution of the homogeneous equation

$$\frac{dv}{dt} + cv = 0, \quad c = \frac{b}{a} \tag{2.1.5}$$

We seek the solution of eq. (2.1.5) in the form

$$v(t) = A\epsilon^{pt}$$

where A is an arbitrary constant and p a complex constant to be determined. By substituting the expression above into eq. (2.1.5) we obtain

$$pA\epsilon^{pt} + cA\epsilon^{pt} = 0$$

or

$$(p + c)A\epsilon^{pt} = 0$$

Since $\epsilon^{pt} \neq 0, -\infty < t < +\infty$, and for all (complex) values of p, the equation above reduces to either $A = 0$ (the so called *trivial* solution) or

$$p + c = 0 \tag{2.1.6}$$

the *characteristic equation* that gives for the value of p the real negative number $p = -c$. Thus the general solution of the homogeneous equation (2.1.5) is

$$v(t) = A\epsilon^{-ct} \tag{2.1.7}$$

and includes the trivial solution as a particular case. The function in eq. (2.1.7) is a *natural mode*, the solution of eq. (2.1.6) $p = -c$ is a *natural frequency.*

The complete solution of eq. (2.1.4) may be described over the two intervals $t < T$, and $t \geq T$. The solution $v(t) = 0$ for $t < T$ (the interval over which $e(t) = 0$) follows immediately since the circuit of Fig. 2.1.1 is linear and passive and therefore must yield a causal response to a causal signal (Theorem 1.5.1). In the interval $t \geq T$ it is easy to show that a particular solution has the form $B\epsilon^{st}$, B another constant. By substituting this expression in eq. (2.1.4), we find

$$e(t) = E\epsilon^{st} = aBs\epsilon^{st} + bB\epsilon^{st}, \qquad t \geq T$$

so that, by deleting the nonzero common factor ϵ^{st}, the value of B is determined as $B = E/(as + b)$ and, therefore, $[E/(as + b)]\,\epsilon^{st}$ is a particular solution. Referring to eq. (2.1.2) all solutions of eq. (2.1.4) must therefore have the form

$$v(t) = A\epsilon^{-ct} + \frac{E}{as + b}\epsilon^{st}, \qquad t \geq T$$

To evaluate the arbitrary constant A we use the boundary condition stemming from $v(t) = 0$, $t < T$. Denote the time just beyond $t = T$ as $t = T^+$. Then, since the voltage across the capacitor must be continuous else the current $i_1(t) = Cdv/dt$ would have an infinite spike, we have $v(T) = v(T^+)_{T^+ \to T} = 0$, or

$$v(T) = A\epsilon^{-cT} + E/(as + b)\epsilon^{sT} = 0, \quad A = -\frac{E}{as + b}\epsilon^{(s+c)T}$$

We can therefore write the complete solution for $v(t)$ using the values of a and b from eq. (2.1.4)

$$v(t) = \begin{cases} 0 & t < T \\ E\,\dfrac{\epsilon^{st} - \epsilon^{(s+c)T}\,\epsilon^{-ct}}{rC(s+c)} & t \geq T \end{cases} \qquad s \neq -c \tag{2.1.8}$$

The current $i(t)$ is found directly by substituting $v(t)$ into eq. (2.1.3)

$$i(t) = \begin{cases} 0 & t < T \\ E\,\dfrac{[rC(s+c)-1]\epsilon^{st} + \epsilon^{(s+c)T}\,\epsilon^{-ct}}{r^2C(s+c)} & t \geq T \end{cases} \qquad s \neq -c \tag{2.1.9}$$

Referring to eqs. (2.1.8) and (2.1.9), when $s = -c$ the response takes on the indeterminate form $\frac{0}{0}$. To evaluate $v(t)$ in this case, factor ϵ^{-ct} from eq. (2.1.8) and then allow $s \to -c$, using $\epsilon^{x} = 1 + x$, when $x \to 0$. Thus

$$v(t) = \lim_{s \to -c} E\epsilon^{-ct}\,\frac{\epsilon^{(s+c)t} - \epsilon^{(s+c)T}}{rC(s+c)} = \lim_{s \to -c} \frac{E(s+c)(t-T)}{rC(s+c)}\epsilon^{-ct}$$

or

$$v(t) = E\epsilon^{-ct}\frac{(t-T)}{rC} \tag{2.1.10}$$

It is important to note that in this case there is a term proportional to t. Substitute eq. (2.1.10) into eq. (2.1.3)

$$i(t) = \frac{E}{r}\epsilon^{-ct}(1 - \frac{t-T}{rC}) \tag{2.1.11}$$

We can check the boundary condition for $i(T)$. The voltage across the capacitor at $t = T$ is zero. Thus, referring to Fig. 2.1.1, $i(T) = e(T)/r = E\epsilon^{sT}/r$. This is verified by substituting $t = T$ into the solutions eqs. (2.1.9) and (2.1.11). □

Example 2.1.2

For the system of Fig. 2.1.1 suppose $e(t) = E\epsilon^{st}$, $t > -\infty$, sometimes termed an *eternal exponential.* Find the responses $v(t)$, $i(t)$ and determine the conditions under which these outputs are directly proportional to ϵ^{st}, i.e., such that the natural mode stemming from the homogeneous solution does not appear in the complete solution.

Solution We proceed by allowing $T \to -\infty$ in the preceding solutions. Using eq. (2.1.8) the output voltage $v(t)$ for $t \geq T$ is

$$v(t) = H\epsilon^{st} - H\epsilon^{(s+c)T}\epsilon^{-ct}, \quad H = \frac{E}{rC(s+c)}, \quad t \geq T \tag{2.1.12}$$

We now allow $T \to -\infty$ and seek for a condition such that the second term of eq. (2.1.12) is negligible compared to the first over the infinite domain $-\infty < t < +\infty$. We therefore let $T \to -\infty$ and examine the limit of the ratio (F) of the two terms.

$$F = \lim_{T\to-\infty} \frac{H\epsilon^{(s+c)T}\epsilon^{-ct}}{H\epsilon^{st}} = \lim_{T\to-\infty} \epsilon^{(s+c)(T-t)}, \quad t > T \tag{2.1.13}$$

The magnitude of the exponential depends only on $\Re(s+c)$ since $|\epsilon^{jx}| = 1$. Furthermore, we note that as $T \to -\infty$, $(T-t) \to -\infty$ as well for any $t > T$. Thus $F \to 0$ only if $\Re(s+c) > 0$, i.e., if $\Re s > -c$, for then the exponent of eq. (2.1.13) becomes negatively infinite. In this case, therefore, the homogeneous solution is negligible compared to the particular solution and, in the limit, the final complete response to the eternal exponential is simply $H\epsilon^{st}$, or

$$E\epsilon^{st} = S\{H(s)\epsilon^{st}\}, \quad -\infty < t < +\infty; \quad \Re s > -c = -\frac{rG+1}{rC}$$

A similar result holds for $i(t)$.

We note that the coefficient of the natural mode is $H\epsilon^{(s+c)T}$ and this quantity also goes to zero as $T \to -\infty$, again provided $\Re s > -c$. If $\Re s < -c$, the limiting value of the coefficient does not exist and there is no solution for the eternal exponential excitation. For the case $s = -c$, referring to eqs. (2.1.10), (2.1.11), there is *never* a solution when $T \to -\infty$ i.e. to the eternal exponential. □

The considerations above can be easily generalized to the case in which p is no longer a real negative number, but a complex constant. In such a case, inequality $\Re s > -c$ is replaced by inequality $\Re s > \Re p$.

Thus we can state a simple and important theorem.

THEOREM 2.1.1

Let the exponential signal $x(t) = X(s)\,\epsilon^{st}$, $t > -\infty$, excite an LTI causal circuit possessing only a single natural frequency p. Then the complete response is of the form $y(t) = Y(s)\,\epsilon^{st}$, $t > -\infty$, if and only if $\Re s > \Re p$. (If the circuit satisfies "reality", p is purely real, though of course s need not be.)

2.2 LTI Operator Eigenfunctions

We have seen in the previous section that, for the case of a simple first order constant coefficient linear differential equation, a particular solution to an exponential input function is itself a similar exponential function. We now show that this is a property of *all* LTI operator equations, whether of scalar or matrix character. The latter case corresponds to a set of simultaneous scalar operator equations. An example of a matrix operator equation is shown below,

$$\begin{pmatrix} 1+\dfrac{d}{dt} & \dfrac{d^2}{dt^2} \\ \\ \dfrac{d^2}{dt^2} & 2+3\dfrac{d}{dt} \end{pmatrix} \begin{pmatrix} v_1 \\ \\ v_2 \end{pmatrix} = \begin{pmatrix} f_1 \\ \\ f_2 \end{pmatrix}$$

Here the 2×2 array is a matrix operator, operating on the column vector $\boldsymbol{v}(t)$.

We can represent the general LTI matrix operator equation by

$$\boldsymbol{S}\{\boldsymbol{v}(t)\} = \boldsymbol{f}(t) \tag{2.2.1}$$

with $\boldsymbol{S}$ an $n \times n$ LTI matrix operator whose components S_{jk} perform LTI operations on each element of the column vector $\boldsymbol{v}$ thus yielding the components of $\boldsymbol{f}$. Now suppose that the vector $\boldsymbol{v}(t) = \epsilon^{st}\boldsymbol{y}(s)$ where $\boldsymbol{y}(s)$ is a column vector whose components depend on s, but not on t. In effect, $\boldsymbol{y}(s)$ is a constant vector as far as the operator equation is concerned. We wish to determine $\boldsymbol{f}(t) = S\{\epsilon^{st}\boldsymbol{y}(s)\}$. Applying the linearity and time invariant properties of S

$$\epsilon^{s\tau}\boldsymbol{f}(t) = S\{\epsilon^{s\,(t+\tau)}\boldsymbol{y}(s)\} = \boldsymbol{f}(t+\tau)$$

where τ is a real constant. This equation is valid for all t and τ. In particular we may choose $t = 0$

$$\epsilon^{s\tau}\,\boldsymbol{f}(0) = S\{\epsilon^{s\tau}\boldsymbol{y}(s)\} = \boldsymbol{f}(\tau)$$

Replacing τ by t changes nothing, so that

$$\boldsymbol{f}(t) = \epsilon^{st}\boldsymbol{x}(s) = S\{\epsilon^{st}\boldsymbol{y}(s)\}, \qquad \boldsymbol{x}(s) = \boldsymbol{f}(0) \tag{2.2.2}$$

The vectors $\boldsymbol{x}(s)$ and $\boldsymbol{y}(s)$ are both time independent.

Comparing eq. (2.2.2) with eq. (2.2.1) we make the following observation; for *any* LTI system, a vector exponential excitation (implying a multiple input system) $\boldsymbol{f}(t) = \epsilon^{st}\boldsymbol{x}(s)$ has as particular solution the vector $\epsilon^{st}\boldsymbol{y}(s)$, i.e., to within multiplicative constants, all the responses have exactly the same exponential functional form as the excitation.

Let us put this result into sharper focus.

DEFINITION 2.2.1 Eigenfunction and eigenvalue matrix of an operator S: *If, with excitation vector $g(t)\boldsymbol{x}(s)$, the operator equation satisfies*

$$g(t)\boldsymbol{x}(s) = \boldsymbol{S}\{g(t)\boldsymbol{y}(s)\} = \boldsymbol{\Lambda}(s)g(t)\boldsymbol{y}(s) \tag{2.2.3}$$

where $\boldsymbol{x}(s)$ *and* $\boldsymbol{y}(s)$ *are time independent column vectors and* $\boldsymbol{\Lambda}(s)$ *is a time independent matrix, then* $g(t)$ *is an* eigenfunction *of* $\boldsymbol{S}$*, and* $\boldsymbol{\Lambda}(s)$ *is an* eigenvalue matrix.

For a scalar operator eq. (2.2.3) becomes

$$g(t)x(s) = S\{g(t)y(s)\} = \lambda(s)g(t)y(s)$$

where $\lambda(s)$ is a time independent scalar and is termed an *eigenvalue.* By virtue of eq. (2.2.2) we can readily show that all LTI systems have ϵ^{st} as an eigenfunction, and the eigenvalue matrix is readily found. If in eq. (2.2.2) we take advantage of the fact that $\boldsymbol{y}(s)$ is time independent and $\boldsymbol{S}$ is a linear operator, then

$$\epsilon^{st}\boldsymbol{x}(s) = \boldsymbol{S}\{\epsilon^{st}\boldsymbol{y}(s)\} = \boldsymbol{S}\{\epsilon^{st}\}\boldsymbol{y}(s) \tag{2.2.4}$$

and evidently to satisfy this equation it must be

$$\boldsymbol{S}\{\epsilon^{st}\}\boldsymbol{y}(s) = \epsilon^{st}\boldsymbol{\Lambda}(s)\boldsymbol{y}(s) \tag{2.2.5}$$

where

$$\boldsymbol{x}(s) = \boldsymbol{\Lambda}(s)\boldsymbol{y}(s), \qquad \boldsymbol{\Lambda}(s) = \epsilon^{-st}\boldsymbol{S}\{\epsilon^{st}\} \tag{2.2.6}$$

and $\boldsymbol{\Lambda}(s)$ is a time independent $n \times n$ matrix. Notationally eq. (2.2.6) may be understood in the following way. Since ϵ^{st} is common to all elements of the $n \times n$ matrix $\boldsymbol{S}\{\epsilon^{st}\}$ after the operations are explicitly carried out, the (jk)-th element of $\boldsymbol{\Lambda}(s)$ is $\lambda_{jk}(s) = \epsilon^{-st}S_{jk}\{\epsilon^{st}\}$, which is tantamount to deleting the exponential ϵ^{st} after the operation is performed.

THEOREM 2.2.1

Every LTI operator (scalar or matrix) has ϵ^{st} *as an eigenfunction, and the corresponding eigenvalue matrix is the time independent array* $\boldsymbol{\Lambda}(s) = \epsilon^{-st}\boldsymbol{S}\{\epsilon^{st}\}$.

Two commonly used LTI scalar operators often contained in $\boldsymbol{S}$ are the derivative and integral operators

$$D^n = \frac{d^n}{dt^n}, \quad D^{-1} = \int_{-\infty}^{t} (\,)\,d\tau, \quad DD^{-1} = 1 \tag{2.2.7}$$

The corresponding eigenvalues, using eq. (2.2.6) are

$$\lambda(s) = \epsilon^{-st}\frac{d^n}{dt^n}\epsilon^{st} = s^n, \quad \lambda(s) = \epsilon^{-st}\int_{-\infty,\ \Re s>0}^{t} (\,)\epsilon^{s\tau}d\tau \; = \frac{1}{s} \tag{2.2.8}$$

We consider the following example to illustrate how all this works out.

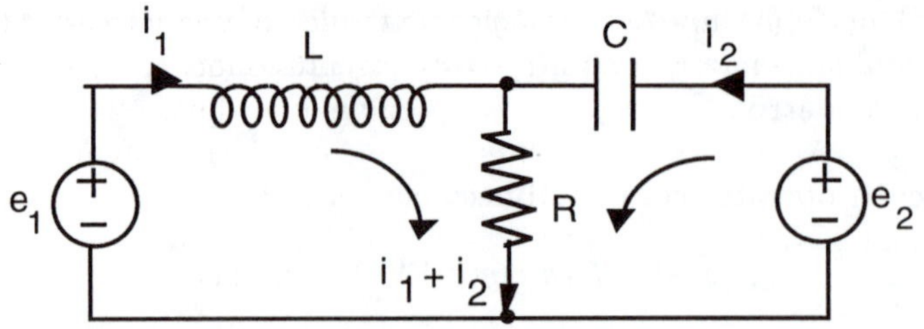

FIGURE 2.2.1
Circuit for eigenvalue matrix calculation.

Example 2.2.1
For the circuit of Fig. 2.2.1, let

$$\begin{pmatrix} e_1 \\ e_2 \end{pmatrix} = \boldsymbol{S}\left\{\begin{pmatrix} i_1 \\ i_2 \end{pmatrix}\right\}$$

Find the eigenvalue matrix for $\boldsymbol{S}$.

Solution Applying KVL to the two meshes shown (see Section 2.10)

$$e_1 = LDi_1 + Ri_1 + Ri_2$$

$$e_2 = Ri_1 + \tfrac{1}{C}D^{-1}i_2 + Ri_2$$

or

$$e_1 = (LD + R)i_1 + Ri_2$$

$$e_2 = Ri_1 + \left(\tfrac{1}{C}D^{-1} + R\right) i_2$$

Inspection of the two preceding equations gives the operator matrix $\boldsymbol{S}$. Thus

$$\boldsymbol{S} = \begin{pmatrix} LD + R & R \\ R & \dfrac{1}{C}D^{-1} + R \end{pmatrix}$$

Applying eqs. (2.2.6), (2.2.7), (2.2.8) to $\boldsymbol{S}$, the eigenvalue matrix is determined as

$$\boldsymbol{\Lambda}(s) = \epsilon^{-st}\boldsymbol{S}\{\epsilon^{st}\} = \begin{pmatrix} Ls + R & R \\ R & \dfrac{1}{sC} + R \end{pmatrix} \qquad (2.2.9)$$

□

In the above example the eigenvalue matrix is identical to the mesh impedance matrix of the circuit of Fig. 2.2.1. Indeed, it should be clear that for general problems of mesh analysis the eigenvalue matrix is simply determined by replacing each branch inductance of the circuit, L_k by an impedance sL_k, each capacitance C_k by an impedance $1/sC_k$, and each resistor by impedance R_k, and then proceeding according to the usual rules of mesh analysis for resistor circuits where, in a formal manner, branch impedances are topologically combined as though they were resistors.

Thus, taking $\boldsymbol{E}$, $\boldsymbol{I}$ as column vectors whose components are, respectively, the E_k, and I_k

$$\boldsymbol{E}(s) = \boldsymbol{Z}(s)\boldsymbol{I}(s), \quad \boldsymbol{\Lambda}(s) \equiv \boldsymbol{Z}(s) \tag{2.2.10}$$

with $\boldsymbol{Z}$(s) an impedance matrix.

Similarly we can use branch admittances $1/sLk$, sCk, $1/R_k$ for a representation of the form

$$\boldsymbol{I}(s) = \boldsymbol{Y}(s)\boldsymbol{E}(s), \quad \boldsymbol{\Lambda}(s) \equiv \boldsymbol{Y}(s) \tag{2.2.11}$$

and $\boldsymbol{Y}(s)$ is an admittance matrix.

The eigenfunction response is a particular solution of an LTI system under exponential excitation. But the major significance of the eigenfunction solution is that under appropriate, and not unduly restrictive, conditions it becomes the *total* response to an exponential forcing function. This is discussed in the next three sections.

2.3 Homogeneous Solution of LTI Operator Equations

We have shown that the exponential ϵ^{st} is an eigenfunction of every LTI operator, that, is to say, under exponential excitation it is a particular solution of the operator equation. The important question is to determine when the particular solution is identical with the complete solution. This means that we seek conditions under which the homogeneous solution is negligible when the "eternal exponential" is the forcing function. An example of this was presented in Section 2.1 for a system with a single natural frequency. We seek a generalization of Theorem 2.1.1 given in that section, and towards this end we first examine the homogeneous solution of the general LTI operator equation. Consider the scalar case of eq. (2.2.5), so that $\boldsymbol{\Lambda}(s) \equiv \lambda(s)$. When $\lambda(s) = 0$ we obtain a homogeneous operator equation, and the *characteristic equation* is

$$\lambda(s) = 0 \tag{2.3.1}$$

The roots $s = p_k$ which satisfy eq. (2.3.1) are the *natural frequencies* and, again referring to eq. (2.2.5), the function $\epsilon^{p_k t}$ satisfies the homogeneous equation, and is a *natural mode.* If $\lambda(s)$ is a polynomial of degree n and there are n simple roots, i.e., all the root factors have the form $(s - p_k)$, then we have precisely n natural modes. If eq. (2.3.1) has a repeated root corresponding to a factor $(s - p_k)^r$, then, counting the multiplicity of the root, we will still have n roots, but the number of distinct zeros is $n - r$ and we must seek other than purely exponential natural modes to account for a total of n modes.

Write the equation (2.2.6) for the scalar case and a natural frequency of multiplicity $r > 1$. Then since $\lambda(s) = (s - p_k)^r q(s)$, with $q(p_k) \neq 0$, we obtain

$$S\{\epsilon^{st}\} = \lambda(s)\,\epsilon^{st} = (s - p_k)^r\, q(p_k)\epsilon^{st} \tag{2.3.2}$$

Define

$$\lambda^{[j]}(s) \equiv \frac{d^j \lambda(s)}{ds^j}$$

then clearly

$$\lambda^{[j]}(p_k) = \lambda^{[j]}(s)\Big|_{s=p_k} = 0 \qquad j = 0, 1, \ldots, r-1$$

Hence, if the product rule for a derivative is employed and the right side of eq. (2.3.2) is differentiated $r - 1$ times at $s = p_k$, it follows that these derivatives are all zero. Now differentiate the left side of eq. (2.3.2) with respect to s. Since S operates on ϵ^{st} as a function of the variable t (not s), we can differentiate under the operator and obtain as the j-th derivative at $s = p_k$

$$\frac{d^{(j)} S\{\epsilon^{st}\})}{ds^j} = S\{t^j \epsilon^{st}\}\big|_{s=p_k}$$

Finally, therefore,

$$S\{t^j \epsilon^{p_k t}\} = 0, \qquad j = 0, 1, \ldots, r-1$$

In other words, the $t^j \epsilon^{p_k t}$, $j = 0, 1, \ldots, r-1$ are all natural modes since they satisfy the homogeneous operator equation. It follows that if the characteristic equation has m distinct roots, counted according to multiplicity, there are a total of n natural modes (the *mode spectrum*), and these are of the form $\epsilon^{p_k t}$ or $t^j \epsilon^{p_k t}$. Each of these modes satisfies the homogeneous operator equation. They are all *linearly independent*, that is, no one of them can be expressed as a linear combination of the others. Furthermore, by linearity, any linear combination of the natural modes is also a solution. Thus designating the natural modes $t^j \epsilon^{p_k t}$ by $\mu_{kj}(t)$, we have

$$v(t) = \sum_{k=1}^{m} \sum_{j=0}^{r_k - 1} a_{kj} \mu_{kj}(t) \tag{2.3.3}$$

where the a_{kj} are arbitrary constants. The linear combination $v(t)$ shown in eq. (2.3.3) is the most general solution of the homogeneous operator equation. The following theorem (which can be rigorously deduced from the theory of linear ordinary constant coefficient differential equations, if finding the roots of $\lambda(s)$ reduces to the factorization of an n-th degree polynomial[1]) summarizes the above discussion.

THEOREM 2.3.1

The most general solution of a scalar LTI homogeneous operator equation valid on an interval is the linear combination of the n natural modes as given by eq. (2.3.3).

The solution of the homogeneous matrix operator equation, eq. (2.2.4), is treated in a manner similar to that used for the scalar equation. If we refer to eq. (2.2.5) we obtain a homogeneous solution by considering

$$\boldsymbol{S}\{\epsilon^{st}\}\boldsymbol{y}(s) = \boldsymbol{\Lambda}(s)\,\epsilon^{st}\boldsymbol{y}(s) = \mathbf{0} \tag{2.3.4}$$

where $\boldsymbol{y}(s)$ is a column vector of n components, corresponding to an $\boldsymbol{S}$ matrix of dimensions $n \times n$.

A nontrivial solution $\boldsymbol{y}(s) \not\equiv \mathbf{0}$ is obtained only if the matrix $\boldsymbol{\Lambda}(s)$ is singular, or equivalently if $\det \boldsymbol{\Lambda}(s) \equiv \Delta(s) = 0$. Thus the natural frequencies, p_k, are the roots of the characteristic equation

$$\Delta(p_k) = 0 \tag{2.3.5}$$

and eq. (2.3.5) indicates that corresponding to a root p_k of multiplicity one, $\epsilon^{p_k t}\boldsymbol{y}(p_k)$ is a vector natural mode. In this case of a discrete natural frequency p_k, the constant vector $\boldsymbol{y}(p_k) = (y_1, y_2, ...y_n)^T$ is a solution of the set of n homogeneous algebraic equations $\boldsymbol{\Lambda}(p_k)\boldsymbol{y}(p_k) = \mathbf{0}$.

Just as in the scalar case, we can also demonstrate natural mode functions with time dependence of the form $t^j\epsilon^{p_k t}$, when a root of the characteristic equation has a multiplicity greater than 1.

To show this we will first assume two nearly equal single multiplicity natural frequencies. Thus $\Delta(s) = (s - p_k)[s - (p_k + \delta)]q(s)$, $q(p_k) \neq 0$. As we allow $\delta \to 0$, $\Delta(s)$ acquires a double multiplicity natural frequency at $s = p_k$. Corresponding to the roots p_k, $p_k + \delta$, the system exhibits two vector natural modes, $\epsilon^{p_k t}\boldsymbol{y}(p_k)$ and $\epsilon^{(p_k+\delta)t}\boldsymbol{y}(p_k+\delta)$, each of which satisfy the n homogeneous operator equations. Then by linearity the following solution of the homogeneous matrix equation is valid

$$\lim_{\delta \to 0} \boldsymbol{S}\left\{\frac{1}{\delta}\left[\epsilon^{(p_k+\delta)t}\boldsymbol{y}(p_k+\delta) - \epsilon^{p_k t}\boldsymbol{y}(p_k)\right]\right\} = \mathbf{0}$$

[1]E. A. Coddington, "An Introduction to Ordinary Differential Equations", Englewood Cliffs, NJ, Prentice Hall, 1961, Ch. 2.

or

$$\boldsymbol{S}\left\{\left(\frac{d}{ds}[\epsilon^{st}\boldsymbol{y}(s)]\right)_{s=p_k}\right\} = \mathbf{0}$$

Therefore, if the characteristic equation has a double root at $s = p_k$, a corresponding homogeneous solution is

$$\boldsymbol{y}_h(t) = \left[\epsilon^{st}\frac{d\boldsymbol{y}(s)}{ds} + t\epsilon^{st}\boldsymbol{y}(s)\right]_{s=p_k}$$

or

$$\boldsymbol{y}_h(t) = \epsilon^{p_k t}\boldsymbol{y}_1 + t\epsilon^{p_k t}\boldsymbol{y}_0 \tag{2.3.6}$$

where $\boldsymbol{y}_0 = \boldsymbol{y}(p_k)$, $\boldsymbol{y}_1 = \boldsymbol{y}^{[1]}(p_k)$ are constant vectors. It should be noted that, unlike the scalar case, the t dependent mode by itself cannot in general provide a homogeneous solution.

If the characteristic equation has a root of multiplicity $r > 1$, a linear combination of vectors whose time dependence is $t^j\epsilon^{p_k t}$, $j = 0, \ldots, r-1$, will be a homogeneous solution. To see this we can start with r distinct roots in the characteristic equation and let them coalesce into a root of multiplicity r. In the limit we obtain

$$\boldsymbol{S}\left\{\left(\frac{d^{r-1}}{ds^{r-1}}[\epsilon^{st}\boldsymbol{y}(s)]\right)_{s=p_k}\right\} = \mathbf{0} \tag{2.3.7}$$

so that, similar to the case of a double root, eq. (2.3.6), the homogeneous solution corresponding to p_k of multiplicity r becomes

$$\boldsymbol{y}_h(t) = \epsilon^{p_k t}\boldsymbol{y}_{r-1} + t\epsilon^{p_k t}\boldsymbol{y}_{r-2} + \cdots + t^{r-2}\epsilon^{p_k t}\boldsymbol{y}_1 + t^{r-1}\epsilon^{p_k t}\boldsymbol{y}_0 \tag{2.3.8}$$

where the $\boldsymbol{y}_j = \boldsymbol{y}^{[j]}(p_k)$ are constant vectors. Note that for this calculation it is not necessary to know the explicit expression for the vector function $\boldsymbol{y}(s)$.

In order to obtain a solution for the $\boldsymbol{y}_j$, substitute the homogeneous solution eq. (2.3.8) into eq. (2.3.7). Thus if the multiplicity is $r = 2$

$$\boldsymbol{S}\{\epsilon^{p_k t}\boldsymbol{y}_1 + t\epsilon^{p_k t}\boldsymbol{y}_0\} = \mathbf{0}$$

or

$$\boldsymbol{S}\{\epsilon^{p_k t}\boldsymbol{y}_1\} + \left[\frac{d}{ds}\boldsymbol{S}\{\epsilon^{p_k t}\boldsymbol{y}_0\}\right]_{s=p_k} = \mathbf{0}$$

Referring to eq. (2.3.4), we have

$$\begin{aligned}\boldsymbol{\Lambda}(p_k)\epsilon^{p_k t}\boldsymbol{y}_1 + \frac{d}{ds}[\boldsymbol{\Lambda}(s)\epsilon^{st}\boldsymbol{y}_0]_{s=p_k} &= \\ \epsilon^{p_k t}\left[\boldsymbol{\Lambda}(p_k)\boldsymbol{y}_1 + \boldsymbol{\Lambda}^{[1]}(p_k)\boldsymbol{y}_0\right] + t\epsilon^{p_k t}\boldsymbol{\Lambda}(p_k)\boldsymbol{y}_0 &= \mathbf{0}\end{aligned} \tag{2.3.9}$$

It follows that

$$\boldsymbol{\Lambda}(p_k)\boldsymbol{y}_0 = \boldsymbol{0} \tag{2.3.10}$$

$$\boldsymbol{\Lambda}(p_k)\boldsymbol{y}_1 + \boldsymbol{\Lambda}^{[1]}(p_k)\boldsymbol{y}_0 = \boldsymbol{0} \tag{2.3.11}$$

Since, by eq. (2.3.4), $\boldsymbol{\Lambda}(p_k)$ is singular, we solve the homogeneous algebraic equations (2.3.10) for $\boldsymbol{y}_0$, and substitute the result into eq. (2.3.11) to find $\boldsymbol{y}_1$. The resulting homogeneous solution will contain $r = 2$ arbitrary constants for evaluation when the complete solution under initial conditions is calculated. Evidently the successive solution process just described for second order roots can be extended to natural frequencies of arbitrary multiplicity.[2]

As an illustration, consider the following example.

Example 2.3.1

For the circuit of Fig. 2.2.1 with eigenvalue matrix given by eq. (2.2.9) find the general homogeneous solution when $L = C = 1$, $R = 1/2$.

Solution Referring to eq. (2.2.9) and using the given element values, the eigenvalue matrix and the characteristic equation are

$$\boldsymbol{\Lambda}(s) = \begin{pmatrix} s + \dfrac{1}{2} & \dfrac{1}{2} \\ \dfrac{1}{2} & \dfrac{1}{s} + \dfrac{1}{2} \end{pmatrix}$$

$$\Delta(s) = \frac{1}{2s} + \frac{s}{2} + 1 = 0$$

so the characteristic equation is

$$s^2 + 2s + 1 = (s+1)^2 = 0$$

Thus $n = 2$, and there is a single natural frequency $s = p_1 = -1$ of multiplicity $r = 2$. Referring to eq. (2.3.6), the solution has the form

$$\boldsymbol{i}_h(t) = \epsilon^{p_1 t}\boldsymbol{y}_1 + t\epsilon^{p_1 t}\boldsymbol{y}_0$$

[2] An alternate and powerful procedure is available. This requires that the defining operator equations be transformed to a system of first order differential equations, $dx(t)/dt = Ax(t)$, the so-called state equations. The method is not described here since for our purposes it is important to remain directly associated with the immittance description of the circuit.

As discussed above, we now determine the equations (2.3.10) (2.3.11) obtained by expanding $\boldsymbol{S}\{\boldsymbol{i}_h\} = \mathbf{0}$. From eq. (2.3.10), using $p_1 = -1$

$$\frac{1}{2}\begin{pmatrix} -1 & 1 \\ 1 & -1 \end{pmatrix}\begin{pmatrix} y_{01} \\ y_{02} \end{pmatrix} = \begin{pmatrix} 0 \\ 0 \end{pmatrix} \tag{2.3.12}$$

The y_{0j} are the components of $\boldsymbol{y}_0$. Since the coefficient matrix $\boldsymbol{\Lambda}(p_1)$ is singular (the two simultaneous equations defined by (2.3.12) are linearly dependent) we can solve eq. (2.3.12). Choose $y_{02} = a$, an arbitrary constant, then, since $-y_{01} + y_{02} = 0$, it follows that $y_{01} = a$, or

$$\boldsymbol{y}_0 = \begin{pmatrix} a \\ a \end{pmatrix}$$

Now apply eq. (2.3.11)

$$\boldsymbol{\Lambda}(-1)\boldsymbol{y}_1 + \boldsymbol{\Lambda}^{[1]}(-1)\boldsymbol{y}_0 = 0$$

or

$$\frac{1}{2}\begin{pmatrix} -1 & 1 \\ 1 & -1 \end{pmatrix}\begin{pmatrix} y_{11} \\ y_{12} \end{pmatrix} + \begin{pmatrix} 1 & 0 \\ 0 & -1 \end{pmatrix}\begin{pmatrix} a \\ a \end{pmatrix} = \begin{pmatrix} 0 \\ 0 \end{pmatrix}$$

The result is two linearly dependent equations

$$\begin{aligned} -y_{11} + y_{12} &= -2a \\ y_{11} - y_{12} &= 2a \end{aligned}$$

Therefore we may choose $y_{11} = b$, an arbitrary constant, with the result that $y_{12} = b - 2a$. Thus

$$\boldsymbol{y}_1 = \begin{pmatrix} b \\ b - 2a \end{pmatrix}$$

The general homogeneous solution is therefore

$$i_{h1}(t) = b\epsilon^{-t} + at\epsilon^{-t}$$

$$i_{h2}(t) = (b - 2a)\epsilon^{-t} + at\epsilon^{-t}$$

The two arbitrary constants a and b can be evaluated once the form of the complete solution is determined and the initial conditions introduced. If $a = 0$ the only mode present is proportional to ϵ^{-t} but there is no possible choice of constants which allows a solution consisting only of the t dependent mode. □

2.4 The Particular Solution under Exponential Excitation

The exponential ϵ^{st} is an eigenfunction, as well as generally a particular solution of any LTI operator equation when the scalar (or vector) forcing function is proportional to the eternal exponential ϵ^{st}, $t > -\infty$. We have also noted in Theorem 2.1.1 that under appropriate conditions the *complete* solution of an exponentially excited system possessing a single natural frequency can correspond to an eigenfunction. In order to determine the constraints under which this result can be extended to the general LTI system, we examine the particular solution of the general LTI matrix operator equation with exponential excitation. Using the general form of the operator equation (2.2.4) with excitation vector $\epsilon^{st}\boldsymbol{x}(s)$

$$\boldsymbol{S}\{\boldsymbol{v}(t)\} = \epsilon^{st}\boldsymbol{x}(s) \tag{2.4.1}$$

A particular solution of eq. (2.4.1) is $\epsilon^{st}\boldsymbol{y}(s)$, and for this case eq. (2.4.1) becomes (as in eq. (2.2.5))

$$\boldsymbol{S}\{\epsilon^{st}\}\boldsymbol{y}(s) = \epsilon^{st}\boldsymbol{\Lambda}(s)\boldsymbol{y}(s) = \epsilon^{st}\boldsymbol{x}(s)$$

Suppose first that the excitation frequency s is distinct from all the natural frequencies. Then $\det \boldsymbol{\Lambda}(s) \neq 0$, so that $\boldsymbol{\Lambda}(s)$ is nonsingular. Solve for $\boldsymbol{y}$

$$\boldsymbol{y}(s) = \boldsymbol{\Lambda}^{-1}(s)\boldsymbol{x}(s), \quad \boldsymbol{v}(t) = \epsilon^{st}\boldsymbol{\Lambda}^{-1}(s)\boldsymbol{x}(s) \tag{2.4.2}$$

and a particular solution, $\boldsymbol{v}(t)$, is completely determined as a vector with eigenfunction time dependence.

Next consider the case where p_k is a root of the characteristic equation of order r_k, coincident with the excitation frequency, i.e., $s = p_k$. Since the eigenvalue matrix $\boldsymbol{\Lambda}(p_k)$ is now singular, the solution given in eq. (2.4.2) is invalid. However, there is now a particular solution of the form

$$\boldsymbol{y}_p(t) = \epsilon^{st}\left(\boldsymbol{y}_{r_k} + t\boldsymbol{y}_{r_k-1} + \cdots + t^{r_k}\boldsymbol{y}_0\right), \quad s = p_k \tag{2.4.3}$$

This is similar to the case of a homogeneous solution, eq. (2.3.8), when there is a root of multiplicity r_k+1. The constant vectors $\boldsymbol{y}_j$ are once again evaluated by the successive solution procedure used in the homogeneous case (Section 2.3), but the right side, instead of being zero as in eq. (2.3.9), is $\epsilon^{p_k t}\boldsymbol{x}$.

If, say, the multiplicity of the natural frequency p_k is $r_k = 2$, and the excitation vector is $\epsilon^{p_k t}\boldsymbol{x}$, then using eq. (2.4.3)

$$\boldsymbol{S}\{\boldsymbol{y}_p(t)\} = \left(\boldsymbol{S}\{\epsilon^{st}\}\boldsymbol{y}_2 + \frac{d}{ds}\boldsymbol{S}\{\epsilon^{st}\}\boldsymbol{y}_1 + \frac{d^2}{ds^2}\boldsymbol{S}\{\epsilon^{st}\}\boldsymbol{y}_0\right)_{s=p_k} = \epsilon^{p_k t}\boldsymbol{x}$$

Applying the eigenvalue equation (2.2.4)

$$\epsilon^{p_k t}\boldsymbol{x} = \left[\epsilon^{st}\boldsymbol{\Lambda}(s)\boldsymbol{y}_2 + \frac{d}{ds}\left(\epsilon^{st}\boldsymbol{\Lambda}(s)\boldsymbol{y}_1\right) + \frac{d^2}{ds^2}\left(\epsilon^{st}\boldsymbol{\Lambda}(s)\boldsymbol{y}_0\right)\right]_{s=p_k} \tag{2.4.4}$$

Collecting terms in the $t^j\epsilon^{p_k t}$ on the right hand side of eq. (2.4.4) and setting the coefficients equal to the corresponding terms on the left

$$\boldsymbol{\Lambda}(p_k)\boldsymbol{y}_0 = \boldsymbol{0} \quad (\text{coef. of } t^2\epsilon^{p_k t}) \tag{2.4.5}$$

$$\boldsymbol{\Lambda}(p_k)\boldsymbol{y}_1 + 2\boldsymbol{\Lambda}^{[1]}(p_k)\boldsymbol{y}_0 = \boldsymbol{0} \quad (\text{coef. of } t\epsilon^{p_k t}) \tag{2.4.6}$$

$$\boldsymbol{\Lambda}(p_k)\boldsymbol{y}_2 + \boldsymbol{\Lambda}^{[1]}(p_k)\boldsymbol{y}_1 + \boldsymbol{\Lambda}^{[2]}(p_k)\boldsymbol{y}_0 = \boldsymbol{x} \quad (\text{coef. of } \epsilon^{p_k t}) \tag{2.4.7}$$

Successively solving eqs. (2.4.5),(2.4.6),(2.4.7), gives the vectors $\boldsymbol{y}_0$, $\boldsymbol{y}_1$, $\boldsymbol{y}_2$.

To illustrate the procedure, we have the following example.

Example 2.4.1

Consider the same LTI system as in Example 2.3.1. The characteristic equation has a double root (natural frequency) at $p_1 = -1$. Let the given forcing function be $\epsilon^{p_1 t}\boldsymbol{e}$, with a coincident excitation frequency. Find a particular solution, $\boldsymbol{i}_p(t)$.

Solution The particular solution has the form of eq. (2.4.3)

$$\boldsymbol{i}_p(t) = \epsilon^{p_k t}(\boldsymbol{y}_2 + t\boldsymbol{y}_1 + t^2\boldsymbol{y}_0), \qquad s = p_k = -1$$

Now apply eqs. (2.4.5), (2.4.6), (2.4.7) to find $\boldsymbol{y}_0$, $\boldsymbol{y}_1$, $\boldsymbol{y}_2$ using $\boldsymbol{\Lambda}(s)$ of Example 2.3.1 to obtain

$$\boldsymbol{\Lambda}(-1) = \frac{1}{2}\begin{pmatrix} -1 & 1 \\ 1 & -1 \end{pmatrix}, \ \boldsymbol{\Lambda}^{[1]}(-1) = \begin{pmatrix} 1 & 0 \\ 0 & -1 \end{pmatrix}, \ \boldsymbol{\Lambda}^{[2]}(-1) = \begin{pmatrix} 0 & 0 \\ 0 & -2 \end{pmatrix}$$

Apply eq. (2.4.5), which is the same as eq. (2.3.12), so that with arbitrary constant α

$$\boldsymbol{y}_0 = \begin{pmatrix} \alpha \\ \alpha \end{pmatrix}$$

Substitute into eq. (2.4.6) and we have a relation for the components y_{1j}

$$\frac{1}{2}\begin{pmatrix} -y_{11} + y_{12} \\ y_{11} - y_{12} \end{pmatrix} + \begin{pmatrix} 2\alpha \\ -2\alpha \end{pmatrix} = \begin{pmatrix} 0 \\ 0 \end{pmatrix}$$

This defines two linearly dependent equations, so let $y_{11} = \beta$, an arbitrary constant, and the result is $y_{12} = \beta - 4\alpha$. Thus

$$\boldsymbol{y}_1 = \begin{pmatrix} \beta \\ \beta - 4\alpha \end{pmatrix}$$

Finally, the previous results are substituted into eq. (2.4.7), ($\boldsymbol{x} \equiv \boldsymbol{e}$ with components e_1, e_2) so as to determine y_{21}, y_{22}.

$$-\frac{1}{2}(y_{21} - y_{22}) = e_1 - \beta$$
$$\frac{1}{2}(y_{21} - y_{22}) = e_2 - 4\alpha + \beta$$

These equations can only be consistent and linearly dependent if

$$e_1 - \beta = -(e_2 - 4\alpha + \beta), \quad \alpha = \frac{e_1 + e_2}{4}$$

so that choosing $y_{21} = \gamma$, an arbitrary constant, then $y_{22} = 2e_1 - 2\beta + \gamma$ and

$$\boldsymbol{y}_2 = \begin{pmatrix} \gamma \\ 2e_1 - 2\beta + \gamma \end{pmatrix}$$

According to eq. (2.1.2) when we seek the complete solution, the general homogeneous solution is added to *any single* particular solution. For the problem at hand, with arbitrary values of β and γ (α has been eliminated), the resultant $\boldsymbol{y}_j$ vectors generate valid particular solutions. For example, we may make the simple choice $\beta = \gamma = 0$ and obtain a particular solution

$$\boldsymbol{i}_p(t) = \epsilon^{-t} \begin{pmatrix} 0 \\ 2e_1 \end{pmatrix} + t\epsilon^{-t} \begin{pmatrix} 0 \\ -(e_1 + e_2) \end{pmatrix} + \frac{1}{4}t^2\epsilon^{-t} \begin{pmatrix} e_1 + e_2 \\ e_1 + e_2 \end{pmatrix}$$

The complete solution is obtained by adding $\boldsymbol{i}_p(t)$, as above, to the general homogeneous solution $\boldsymbol{i}_h(t)$ of Example 2.3.1.

$$i_1(t) = b\epsilon^{-t} + at\epsilon^{-t} + \tfrac{1}{4}(e_1 + e_2)t^2\epsilon^{-t}$$

$$i_2(t) = (b - 2a + 2e_1)\epsilon^{-t} + (a - e_1 - e_2)t\epsilon^{-t} + \tfrac{1}{4}(e_1 + e_2)t^2\epsilon^{-t}$$

The two arbitrary constants, a and b, can be evaluated by introducing the initial conditions. □

Example 2.4.2

Suppose, in the previous problem (Example 2.4.1, Fig. 2.2.1), that the excitation, $\epsilon^{p_1 t}\boldsymbol{e}$ is applied at $t = 0$, with $i_1(0) = 0$, and the capacitor uncharged so that its initial voltage is zero. $L = C = 1$, $R = 1/2$. Then $i_2(0) = e_2/R = 2e_2$. Find the constants a and b, and the complete solution.

Solution Introducing $t = 0$ into the two previous equations for the complete solution gives $b = 0$, $a = e_1 - e_2$. The final solution is therefore

$$i_1(t) = (e_1 - e_2)t\epsilon^{-t} + \tfrac{1}{4}(e_1 + e_2)t^2\epsilon^{-t}, \quad t \geq 0$$

$$i_2(t) = 2e_2\epsilon^{-t} - 2e_2 t\epsilon^{-t} + \tfrac{1}{4}(e_1 + e_2)t^2\epsilon^{-t}, \quad t \geq 0$$

As a check we can compute the inductor voltage $v_L(0)$

$$v_L(t) = L\frac{di_1}{dt} = (e_1 - e_2)\epsilon^{-t} - \frac{1}{2}(e_1 - 3e_2)t\epsilon^{-t} - \frac{e_1 + e_2}{4}t^2\epsilon^{-t}$$

so that at $t = 0$

$$v_L(0) = e_1 - e_2$$

which is easily verified by inspection of Fig. 2.2.1. □

2.5 Conditions for Pure Eigenfunction Response

We now consider the possibility that the complete response to an eternal exponential forcing function proportional to ϵ^{st}, $t > -\infty$, be a pure eigenfunction with all natural modes suppressed. The LTI system equation is

$$\epsilon^{st}\boldsymbol{x}(s) = \boldsymbol{S}\{\boldsymbol{y}(t)\}, \qquad t > -\infty$$

and summing the homogeneous and particular solutions for the complete response, we obtain

$$\boldsymbol{y}(t) = \boldsymbol{y}_h(t) + \boldsymbol{y}_p(t)$$

We seek constraints so that $\boldsymbol{y}_h(t) = \boldsymbol{0}$, and $\boldsymbol{y}_p(t) = \epsilon^{st}\boldsymbol{b}(s)$. If we first assume the excitation applied at time T, then referring to eq. (2.4.3) in Section 2.4, an immediate restriction is evident; the complex excitation frequency s must differ from any of the p_k natural frequencies otherwise the particular solution will not be a pure exponential. With this condition satisfied the complete solution for any one of the scalar components of $\boldsymbol{y}(t)$ has the form

$$y_j(t) = b_j\epsilon^{st} + \sum_{k=1}^{m}\sum_{i=0}^{r_k-1} a_{ki}\mu_{ki}(t)$$

where the $n = \sum_{k=1}^{m} r_k$ natural modes are the $\mu_{ki}(t)$. According to Section 2.3 each of these may be written

$$\mu_{kq}(t) = a_{kq}t^q\epsilon^{p_k t} \tag{2.5.1}$$

The system is assumed causal and is quiescent when the excitation is applied at time T. Later we allow $T \to -\infty$. One other important practical assumption is also introduced at this point, that of stability.

DEFINITION 2.5.1 *An LTI system is stable if the homogeneous solution remains bounded as $t \to +\infty$. Otherwise, it is unstable.*

For the moment unstable systems are excluded from our discussion. As we have seen in eq. (2.5.1), the homogeneous solution has terms proportional to $t^q \epsilon^{p_k t}$. Using a variant of l'Hospital's rule,[3] we easily verify

$$\lim_{t \to +\infty} t^q \epsilon^{p_k t} = 0, \quad q = 0, 1, \dots r_k, \quad \text{if } \Re p_k < 0 \qquad (2.5.2)$$

If $\Re p_k = 0$ so that $p_k = j\omega$ is purely imaginary, then the associated natural mode $t^q \epsilon^{j\omega t}$ is clearly unbounded as $t \to +\infty$ unless $q = 0$. That is, a stable system may not have purely imaginary natural frequencies whose multiplicity exceeds unity. Thus,

THEOREM 2.5.1
An LTI system with natural frequencies p_k is stable if and only if $\Re p_k \leq 0$, with p_k of unit multiplicity when $\Re p_k = 0$.

We can visualize the conditions for suppressing the homogeneous solution in a stable system excited by an eternal exponential by using the idea of dominance.

DEFINITION 2.5.2 *Let $\alpha(t) = at^m \epsilon^{st}$ and $\beta(t) = bt^n \epsilon^{p_k t}$, with $\Re p_k \leq 0$. Then $\beta(t)$ dominates $\alpha(t)$ as $t \to -\infty$ if $\lim_{t \to -\infty} (|\beta(t)| - |\alpha(t)|) = +\infty$.*

As an immediate consequence of the definition it follows that $\beta(t)$ dominates $\alpha(t)$ if and only if $\beta(t)$ increases more rapidly than $\alpha(t)$ as $t \to -\infty$.

$$\lim_{t \to -\infty} \left| \frac{\beta(t)}{\alpha(t)} \right| = \left| \frac{b}{a} t^{(n-m)} \epsilon^{(p_k - s)t} \right| = +\infty \qquad (2.5.3)$$

Referring to eq. (2.5.2), and noting that the exponent $\Re p_k t > 0$ for $t < 0$, we require this exponent to be larger than that of $\alpha(t)$ for $t \to -\infty$. We conclude, therefore, that $\beta(t)$ dominates $\alpha(t)$ $\forall\,(m, n)$ if and only if $\Re s > \Re p_k$, $k = 1, 2, \dots, m$. By the same token, it is evident that, as t increases the amplitude of the dominant function $|\beta(t)|$, decreases more rapidly than $|\alpha(t)|$.

We can spot the natural frequencies p_k in the complex plane as indicated schematically in Fig. 2.5.1. Suppose the excitation (and, of course, the particular solution as well) is proportional to ϵ^{st}, $s \neq p_k$. Then if s is located on a vertical line, such as RS, to the right of all natural modes, all components of the homogeneous solution dominate the particular solution. On the other hand, if s is located on line LM, then the particular solution dominates some of the natural modes.

[3] Let $f(a) = \infty$, $g(a) = \infty$, then $\lim_{t \to a} f(t)/g(t) = \lim_{t \to a} f'(t)/g'(t)$, and use repeated differentiation with $a = \infty$, and $f(t) = t^q$, $g(t) = \epsilon^{-p_k t}$.

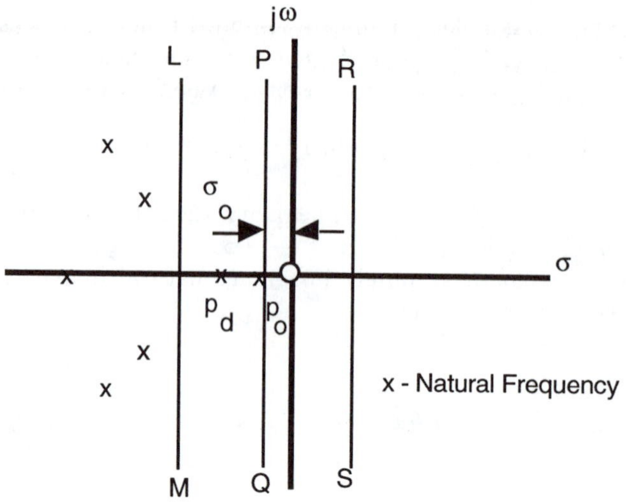

FIGURE 2.5.1
A distribution of natural frequencies in the complex plane. σ_0 is the abscissa of convergence.

Suppose the excitation frequency s is located on the line RS in Fig. 2.5.1 so that the particular solution is dominated by the natural modes. If we impose the initial condition $y_j(T) = 0$, then as the initial excitation time T becomes very large in magnitude and negative, the particular solution tends to become small compared to each of the dominating natural mode terms. The zero initial condition can, therefore, only be satisfied if the undetermined coefficients of the homogeneous response are made arbitrarily small compared to the coefficients of the particular solution. Furthermore, referring to the discussion of eq. (2.5.3), for any given $T > -\infty$, the homogeneous solution can only decrease with respect to the particular solution as $t > T$ increases, so if it is arbitrarily small to begin with, it is only more so at later times. On the other hand, if the particular solution dominates some of the natural modes, then the arbitrary coefficients of these natural modes must be large to satisfy the zero initial conditions, and the homogeneous solution cannot be neglected. (Example 2.1.2 in Section 2.1 is an illustration of what occurs.) Therefore, if the excitation frequency is to the right of all the natural frequencies (e.g., on line RS in Fig. 2.5.1), the homogeneous solution will be negligible with respect to the particular solution for $t > T > -\infty$ when the boundary conditions of an initially quiescent system, excited by an eigenfunction at time $T \to -\infty$, are imposed.

Now suppose that s, as shown in Fig. 2.5.1, is located on line PQ instead of to the right of it. The natural frequency p_0 is on the same line. In this case $\Re s = \Re p_0$, $s \neq p_0$. Except for $b_0 \epsilon^{p_0 t}$, all the natural modes will dominate. With this one exception, as discussed above, the quiescent initial conditions will force the natural mode coefficients to be arbitrarily small in order to satisfy initial conditions as $T \to -\infty$. The final initial condition equation, therefore, need only include one mode and has the form

$$\lim_{T\to-\infty} \left(a_0\epsilon^{st} + b_0\epsilon^{p_0t}\right)_{t=T} = 0, \qquad \Re s = \Re p_0$$

Since $s - p_0 = j\Omega$, a purely imaginary quantity, the coefficient b_0 of the homogeneous response is given by

$$b_0 = \lim_{T\to-\infty} \left(-\frac{1}{a_0}\epsilon^{j\Omega T}\right)$$

The term at the right has no limit as $T \to -\infty$, although it is bounded. Thus the coefficient b_0 cannot be arbitrarily small. In other words, the homogeneous solution will not be suppressed if the real parts of the excitation and natural frequencies are equal.

As a consequence of the previous discussion, we can state the following result.

THEOREM 2.5.2

Let an LTI system, quiescent initially at time $t = T$, be excited by $\epsilon^{st}\boldsymbol{x}(s)$ applied at $t = T$, with $T > T_0$, $T_0 \to -\infty$. Then given any t_0 we can always choose $T < t_0$, so that the homogeneous response, $t > t_0$, becomes negligible compared to the forced eigenfunction response, provided the exponential eigenfunction excitation is dominated by all the natural modes, i.e., provided $\Re s > \Re p_0$. The quantity p_0 is the furthest right natural frequency, or equivalently, the natural frequency with largest real part, $\Re p_0 = \sigma_0$, the abscissa of convergence.

In effect, provided $\Re s > \sigma_0$, we can always choose the signal initiation time, T, early enough so that the complete response to the eternal exponential eigenfunction, $\epsilon^{st}\boldsymbol{x}$, $t > -\infty$, is also an eigenfunction. This is true, in the sense of Theorem 2.5.2, even if $\Re s < 0$.

In certain problems it would be desirable to find some way of using the eigenfunction approach even though $\Re s = \Re p_0$. One important example is that of a.c. analysis in which the exciting function is proportional to $\epsilon^{j\omega t}$, i.e., $s = j\omega$, and we wish to assure an eigenfunction response even with systems that have natural frequencies on the $j\omega$ boundary. The following device can be applied to justify eigenfunction analysis in cases of this sort without violating the constraints of Theorem 2.5.2.

Let $p_0 = j\omega_0$, and $s = \delta + j\omega$, with $\omega \neq \omega_0$ and $\delta > 0$. The constraints on s are satisfied so that the response is proportional to $\epsilon^{st} = \epsilon^{\delta t}\epsilon^{j\omega t}$. Consequently, for any prescribed finite range of t, we can always choose δ small enough so that $\epsilon^{\delta t}$ approximates unity within an arbitrarily small tolerance and both the excitation and response may, therefore, be taken as proportional to $\epsilon^{j\omega t}$ over the prescribed range of t. For practical purposes we can, therefore, take the excitation and response to be proportional to $\epsilon^{j\omega t}$ in a.c. problems even when there are natural frequencies on the $j\omega$ boundary. The following theorem relates the range of t to the value of δ.

THEOREM 2.5.3

(**A.c. analysis**) *Let the excitation of a stable LTI system with a natural frequency at* $j\omega_0$ *be* ϵ^{st}, $s = \delta + j\omega$, $\omega \neq \omega_0$. *Then with* $\tau > 0$, *there is a* δ, $0 < \delta << 1/\tau$, *so that the excitation and response will be proportional to* $\epsilon^{j\omega t}$ *to within an arbitrarily small tolerance over the range* $-\tau < t < \tau$.

That is, we can have eigenfunction operation over any desired large range of t simply by assuming that δ is very small. In effect the complex excitation frequency s is allowed to approach arbitrarily close to the $j\omega$ axis from the right half plane, except at the point $j\omega_0$.

2.6 Phasors and A.C. Analysis

In problems of a.c. analysis, the coefficients (real or complex) of the voltage and current response ϵ^{st} eigenfunctions are called phasors and were briefly employed in connection with the average power calculation of Section 1.4. Given the eigenfunction, it is clearly the computation of these phasors that is required for determining circuit operation under eigenfunction conditions (Theorems 2.5.2, 2.5.3). The phasors are found by using the eigenvalue matrix as in eqs. (2.2.10), (2.2.11). In many a.c. problems we are directly interested in the response to the real functions $\cos(\omega t + \varphi)$ and $\sin(\omega t + \varphi)$, and complex phasor analysis with $s = j\omega$ can also be employed for such real excitations by using the following Theorem.

THEOREM 2.6.1

Let a linear system which satisfies "reality" be excited by the complex signal $x(t) = \Re x(t) + j\Im x(t)$; $\Re x(t)$, $\Im x(t)$ *are real. Then, if the system equation is* $S\{y(t)\} = x(t)$, *it follows that* $y(t)$ *is generally complex and*

$$\Re x(t) = S\{\Re y(t)\}, \quad \Im x(t) = S\{\Im y(t)\}$$

In other words, if we find the response to a complex signal, then the real and imaginary parts of the response are, respectively, the responses to the

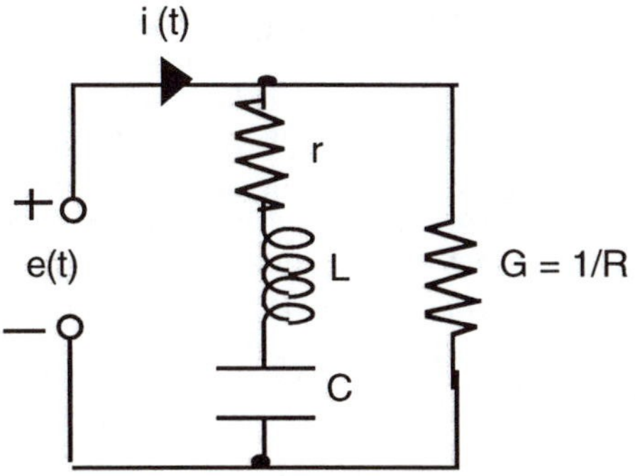

FIGURE 2.6.1

Circuit for a.c. analysis. $e(t) = \sqrt{2}\,V\cos(\omega t + \varphi)$, $t > -\infty$.

real and imaginary parts of the signal. This is easily shown. By linearity,

$$\begin{aligned} x(t) = \Re x(t) + j\Im x(t) \quad &= S\{y(t)\} = \\ = S\{\Re y(t) + j\Im y(t)\} &= S\{\Re y(t)\} + jS\{\Im y(t)\} \end{aligned}$$

By reality, $S\Re y(t)$, $S\Im y(t)$ are real, so equating real and imaginary parts, the Theorem follows.

Example 2.6.1
The RLC circuit of Fig. 2.5.1 is excited by the real voltage signal $e(t) = \sqrt{2}\,V\cos(\omega t+\varphi)$, $t > -\infty$. Find the response current $i(t)$ and,in particular, its value as $r \to 0$.

Solution Since by Euler's formula $\epsilon^{j(\omega t+\varphi)} = \cos(\omega t + \varphi) + j\sin(\omega t + \varphi)$, we have

$$e(t) = \sqrt{2}\,\Re\,E\epsilon^{j\omega t},\ E = |E|\epsilon^{j\varphi},\ |E| = V$$

where the coefficient E is a (complex r.m.s.) phasor.

Now compute the response $\sqrt{2}\,I\epsilon^{j\omega t}$ to the exponential eigenfunction $\sqrt{2}\,E\epsilon^{j\omega t}$. Here $s = j\omega$

$$\sqrt{2}\,E\epsilon^{st} = \sqrt{2}\,S\{I\epsilon^{st}\} = \sqrt{2}\,\lambda(s)I\epsilon^{st}$$

Referring to eq. (2.2.10), the eigenvalue $\lambda(s)$ is simply the input impedance $Z(s)$.

$$\lambda(s) \equiv Z(s) = \frac{1}{Y(s)} = \left[G + \frac{s}{L(s^2 + \tau_0 s + \omega_0^2)}\right]^{-1}$$

The time constant $\tau_0 = r/L$ and $\omega_0^2 = 1/LC$. The characteristic equation $\lambda(s) = 0$ is $Z(p_k) = 0$, or

$$s^2 + \tau_0 s + \omega_0^2 = 0$$

and the two natural frequencies are given by

$$p_{1,2} = \frac{1}{2}\left(-\tau_0 \pm \sqrt{\tau_0^2 - 4\omega_0^2}\right)$$

Since $\tau_0 > 0$, both roots (natural frequencies) have negative real parts, i.e., are to the left of the $j\omega$ axis. Thus with $s = j\omega$, Theorem 2.5.2 is satisfied and the response to $\sqrt{2}\,E\epsilon^{j\omega t}$ is $\sqrt{2}\,I\epsilon^{j\omega t}$ where,

$$I = YE = \left[G + \frac{j\omega}{L(\omega_0^2 - \omega^2 + j\tau_0\omega)}\right]E \tag{2.6.1}$$

When $\tau_0 \to 0$, the natural frequencies move to the $j\omega$ axis and the complex phasor I in the limit is

$$I = \left[G + \frac{j\omega}{L(\omega_0^2 - \omega^2)}\right]E$$

provided $\omega \neq \omega_0$. The natural frequencies are at $\pm j\omega_0$. This would be precisely the result if Theorem 2.5.3 were applied directly to the circuit with $r =\ 0$, i.e.,if $\Re\, s = \Re\, p_k = 0$. However, if the excitation and natural frequencies coincide, $s = p_k$ (resonance), that is, $\tau_0 = 0$, $\omega = \omega_0$, the response will contain a component proportional to $t\epsilon^{st}$ and will not be an eigenfunction. In this case, the admittance $Y(j\omega)_{\omega=\omega_0} = \infty$ and phasor analysis is not applicable.

Returning to the example, the real response to $e(t) = \sqrt{2}\,|E|\cos(\omega t + \varphi)$ is readily calculated. First express eq. (2.6.1) in polar form

$$Y = |Y|\,\epsilon^{-j\theta} = g + jb$$

$$g = G + \frac{\tau_0\omega^2}{L[(\omega_0^2 - \omega^2)^2 + (\tau_0\omega)^2]}, \qquad b = \frac{\omega(\omega_0^2 - w2)}{L[(\omega_0^2 - \omega^2)^2 + (\tau_0\omega)^2]}$$

$$\theta = -\tan^{-1}\frac{b}{g}, \qquad |Y| = \sqrt{g^2 + b^2}$$

Then, since $e(t) = \Re\sqrt{2}\,E\epsilon^{j\omega t}$, $E = |E|\,\epsilon^{j\varphi}$, and $I = YE$, applying Theorem 2.6.1

$$i(t) = \Re\sqrt{2}\,I\epsilon^{j\omega t} = \Re\left(|Y|\epsilon^{-j\theta}\sqrt{2}\,|E|\epsilon^{j\varphi}\epsilon^{j\omega t}\right) = |Y|\sqrt{2}\,|E|\cos(\omega t - \theta + \varphi)$$

which is valid $tau_0 \geq 0$, provided $\omega \neq \omega_0$. □

2.7 Network Geometry

There are a number of different methods available to systematically carry out the eigenfunction analysis of an electric circuit. Essentially these are all concerned with the determination of the eigenvalue matrix $\mathbf{\Lambda}(s)$. If the excitation and response signals have the same dimensions, then $\mathbf{\Lambda}(s)$ is dimensionless. If the operator relation has the form $\boldsymbol{e}(t) = \boldsymbol{S}\{\boldsymbol{i}(t)\}$, then $\mathbf{\Lambda}(s)$ is an impedance matrix, and if $\boldsymbol{e}$ and $\boldsymbol{i}$ are interchanged, $\mathbf{\Lambda}(s)$ is an admittance matrix. See eqs. (2.2.10), (2.2.11).

Among the most useful techniques for network analysis are *Nodal Analysis* and *Mesh Analysis*, which, for planar networks, are mutually related by duality. *Cut Set Analysis* and *Loop Analysis* generalize nodal and mesh analysis and retain dual properties, although in a more abstract sense.

These concepts will be introduced and illustrated in the following Sections of this chapter. It will be seen that at the core of the various methods is the eigenvalue matrix $\mathbf{\Lambda}(s)$, represented in turn by the *nodal admittance matrix*, the *mesh impedance matrix*, the *cut set admittance matrix*, and the *loop impedance matrix*. These matrices can be calculated from *branch admittance* and *impedance matrices*, which contain all the information on the physical nature of the elements comprising the network and, in effect, list the *constitutive* data for the network structure. These matrices are combined with *incidence matrices*, which describe the connection properties of the network independently of the elements that are interconnected.

Therefore, we initially describe the geometric properties of network structures in a manner which is independent of the specific nature of their constitutive elements. Such properties are invariant under continuous deformations (representing the various ways, often apparently very different from each other, of drawing a circuit diagram on a piece of paper), and hence fall within the province of *Topology*. At the outset we need some definitions.

DEFINITION 2.7.1 *A* branch *is a two terminal circuit described by a box accessed by two terminals, or when only topological properties are of interest, by a line segment or arc connecting the terminals. We usually assume all branches are* directed, *that is are marked with direction arrows to indicate current polarity. The branch voltage (drop) polarity (unless specifically indicated otherwise) is also determined by the arrow; plus (+) at the tail of the arrow, minus (−) at the head of the arrow.*

Note that the box may contain only a single element, e.g., a resistor R, or an independent source element, or we may choose the contents of the branch to be an interconnection of circuit components, e.g., R, L, and C elements connected in parallel, again with two accessible terminals. Such detailed structures appear in the branch immittance[4] matrix, but are irrelevant when only branch connection properties are of concern.

[4]We use the word "immittance" to mean "impedance and/or admittance".

DEFINITION 2.7.2 *A* node *is a point to which the terminals of one or more branches are connected. A node may also be an isolated point. A branch is said to be* incident *on a terminating node. Other branches may also be incident on the same node, making electrical contact at that node. A node is also referred to as a* junction *or* vertex.

Isolated nodes usually represent environments having no predetermined electric relation with the given network. The case of nodes incident on a single branch occurs in some instances, for example, the ground node. When only two branches are incident on a single node, the branches are series-connected but are kept distinct rather than being replaced with one resultant branch. In all other cases a node is connected to at least three branches.

DEFINITION 2.7.3 *A (network)* graph *is a topologic structure consisting of branches incident on a set of nodes. The contents of the branches, which give rise to the* constitutive relations *between branch voltage and current are not relevant to the graph and are usually omitted in topologic analysis. When branch polarity arrows are included, the result is a* directed graph.

DEFINITION 2.7.4 *A* simply connected graph *is one in which there is a continuous branch path between every pair of nodes. A* multiply connected graph *consists of two or more simply connected graphs. Each such part has no nodes in common with any other part and no continuous branch path can be drawn between nodes belonging to different parts.*

DEFINITION 2.7.5 *A* planar graph *is one that can be drawn on a plane without requiring branch crossings.*

DEFINITION 2.7.6 *A* cut set *is a set of branches which when deleted separates a simply connected graph into two parts, such that when any deleted branch is restored, the graph becomes simply connected.*

A cut set can be always visualized as the set of branches cut by a closed surface that divides the graph into two parts; this set, therefore, consists of the branches that connect nodes of one part with nodes of the other.

DEFINITION 2.7.7 *A* loop *is any closed path of branches on a given graph. It may be directed in either of two senses. A planar graph may always be drawn so that its loops are laid out in a contiguous fashion like* *window panes. Such loops are called* meshes; *each mesh can be directed clockwise or counterclockwise.*

Note that meshes can be put into one-to-one correspondence with the circle, thus justifying the existence of a clockwise and a counterclockwise orientation. On the other hand, a general loop, having, say, a figure 8 shape can still be directed with either of two opposite orientations, but these are not necessarily purely clockwise or counterclockwise.

DEFINITION 2.7.8 *A* separable *or* hinged *simply connected graph is one that consists of two subgraphs which share only a single common node. If the graph does not possess this property, it is* nonseparable.

DEFINITION 2.7.9 *A* nondegenerate graph *is one in which every branch is contained in at least one loop.*

A degenerate graph either has one or more "dangling branches" and includes nodes with only one incident branch, or can be divided into two subgraphs connected by a single branch. In the latter case, shrinking that branch would produce a separable graph.

For network applications a separable and/or degenerate graph may always be replaced by a set of separated subgraphs, each of which is nonseparable and nondegenerate. Thus in the following, a simply connected graph will always be considered as nonseparable and nondegenerate.

DEFINITION 2.7.10 *On a simply connected graph, a* tree *is a simply connected subgraph consisting of a set of branches connecting every node of the graph, but not forming any loops.*

DEFINITION 2.7.11 *The* cotree *associated with a given tree on a simply connected graph is the subgraph consisting of the set of branches left out of the tree. Each branch of a cotree is termed a* link.

DEFINITION 2.7.12 *The* rank ρ *of a graph is the total number of branches in the set of trees lying in each of its separate parts.*

It will be shown below that for a graph consisting of n nodes and m separate parts, the rank is an invariant independent of the specific choice of branches in the trees, $\rho = n - m$.

DEFINITION 2.7.13 *The* nullity ν *of a network graph is the total* *number of links in the set of cotrees lying in each of its separate parts.*

It will be shown below that for a graph consisting of n nodes and m separate parts, the nullity is $\nu = b - n + m$.

THEOREM 2.7.1
On a simply connected graph containing n nodes, every tree contains $\rho = n - 1$ branches, where ρ is the rank of the graph.

PROOF One can always construct a tree on a simply connected graph of n nodes and b branches by the following process. The first branch (chosen arbitrarily) is incident on two nodes. Thereafter choose each succeeding branch to be incident on one new node until all nodes are exhausted. The procedure results in no closed loops, and the resultant simply connected subgraph (tree) consists of the interconnection of all n nodes by $n - 1$ branches. This is the number of branches in every tree of the simply connected graph. ∎

COROLLARY 2.7.1
The rank of a graph consisting of n nodes and m separate parts is $\rho = n - m$.

PROOF On any of the m separate parts, say part i, the tree has $n_i - 1$ branches (Theorem 2.7.1). The sum of the tree branches of all parts is thus $\sum_{i=1}^{m}(n_i - 1) = n - m$. ∎

COROLLARY 2.7.2
The number of links on a graph containing n nodes, b branches, and m separate parts is $\nu = b - n + m$, where ν is the nullity of the graph.

PROOF On any of the m separate parts, say part i, the cotree has $b_i - n_i + 1$ branches (Definition 2.7.11 and Theorem 2.7.1). The sum of the tree branches of all parts is thus $\sum_{i=1}^{m}(b_i - n_i + 1) = b - n + m$. ∎

Fig. 2.7.1 shows a planar simply connected graph with two different trees indicated. The total number of branches is $b = 9$, the number of nodes is $n = 6$. The number of tree branches is $n - 1 = 5$ and the number of links in each cotree is $b - n + 1 = 4$.

A tree does not contain loops; the following theorem states the dual property for a cotree.

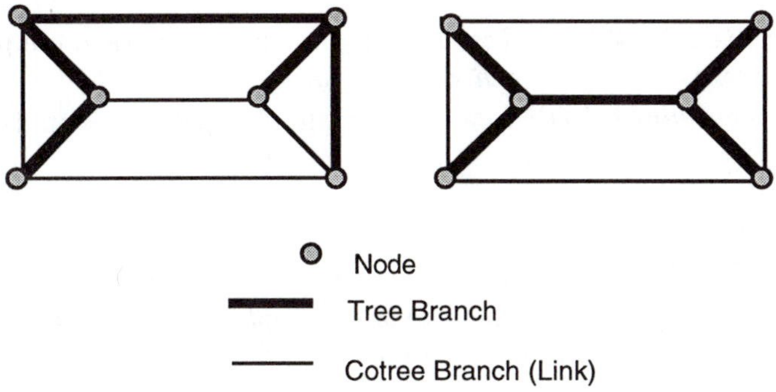

FIGURE 2.7.1
Two trees and associated cotrees on a given graph.

THEOREM 2.7.2

Given any tree $\mathcal{T}$ on a simply connected graph, then its associated cotree, $\mathcal{C}$, contains no cut sets.

PROOF Delete the links of $\mathcal{C}$. The remaining subgraph is the tree $\mathcal{T}$, which by definition has a branch path between each pair of nodes, hence is still simply connected. The graph therefore remains in one part even when the links are deleted, so the cotree contains no cutsets. ∎

THEOREM 2.7.3

In a simply connected graph, choose any tree $\mathcal{T}$ with its associated cotree $\mathcal{C}$. Then given any branch B of $\mathcal{T}$, a unique cut set can always be constructed consisting of B plus appropriate branches of $\mathcal{C}$.

PROOF If $\mathcal{C}$ is deleted, the remaining subgraph $\mathcal{T}$ is simply connected. Deleting any one branch, B, of $\mathcal{T}$ opens the only connecting path between the terminating nodes of B, and hence the resulting subgraph is in two parts. Thus B and selected links of $\mathcal{C}$ form a cut set. Suppose this cut set is not unique. Then each of two separate cut set surfaces, both containing B but different links of $\mathcal{C}$, split the given graph in two. Direct the two cut set surface normals in opposite directions and form the union, $\mathcal{U}$, of the two cut set surfaces. The closed surface $\mathcal{U}$ no longer has B as an incident branch, yet still cuts the graph into two parts, i.e., $\mathcal{U}$ is a cut set surface, but now the cut set so formed consists only of links in $\mathcal{C}$. This contradicts Theorem 2.7.2. ∎

We close this Section by introducing the important concepts of fundamental cut sets and fundamental loops.

DEFINITION 2.7.14 *For a simply connected graph, a* fundamental cut set *consists of one branch of a tree and a unique set of links chosen from the associated cotree.*

THEOREM 2.7.4

For any chosen tree on a simply connected graph of n nodes, there are $n-1$ fundamental cut-sets, one for each branch of the tree.

PROOF Each cut set is constructed by drawing a closed surface dividing the graph in two parts and cutting one and only one tree branch. The operation is repeated for all tree branches, thus constructing $n-1$ closed surfaces (which can always be chosen to be mutually nonintersecting). Each surface defines a cut set as the tree branch and the links traversing the surface. Hence there are precisely $n-1$ cut sets, each containing a single tree branch. ■

DEFINITION 2.7.15 *On a simply connected graph on which a tree has been chosen, a* fundamental loop *consists of one link of the associated cotree plus the branches from the chosen tree, which connect the end points of the link and therefore close the loop.*

THEOREM 2.7.5

For a simply connected graph of b branches, n nodes, and rank $\rho = n-1$, there are precisely $\nu = b - \rho = b - n + 1$ fundamental loops based on a given tree.

PROOF Choose a tree $\mathcal{T}$. For any link $\mathcal{L}$, there is only one unique set of branches from the tree which can complete a loop, i.e., a unique fundamental loop. For if there were a second closed path consisting of $\mathcal{L}$ and a different set of branches from $\mathcal{T}$ the tree itself would contain a closed loop. This is because $\mathcal{L}$ and its two terminating nodes would be common to both loops, hence with $\mathcal{L}$ removed a closed path remains. Thus the assertion that $\mathcal{T}$ is a tree would be contradicted. Therefore, there are precisely $\nu = b - \rho$ unique fundamental loops for a given tree, one for each link. ■

2.8 Topology and Kirchhoff's Laws

The geometric concepts discussed in the previous Section can be combined with the Kirchhoff current and voltage laws (KCL and KVL) and the constitutive equations relating voltage and current in the branches to yield a complete analysis of the physical properties and operation of the most general LTI electric circuit.

We state at the outset the mode for applying KCL and KVL to the cut sets and loops of a network.

Given a network with directed branches; to any cut set of the network graph we associate a KCL equation according the following rule.

RULE 2.8.1

Direct the normal to the cut set surface. Then the KCL equation states that the algebraic sum of the currents in the directed branches incident on the cut set surface is zero. A branch current term of the sum takes a (+) sign when its branch direction is the same as that of the cut set surface normal; (−) when the branch and cut set normal directions are opposite.

Moreover to any loop in the directed graph we associate a KVL equation according the following rule.

RULE 2.8.2

Choose an orientation for the loop. Then the KVL equation states that the algebraic sum of the voltages across the directed branches traversed by the loop is zero. A branch current term of the sum takes a (+) sign when its branch direction is the same as the loop orientation ; (−) if the branch and loop directions are opposite.

The following theorem establishes a procedure for writing a maximal set of KCL equations for a given network, and relates the number of equations to the rank ρ of the network graph.

THEOREM 2.8.1

In a simply connected network of n nodes choose any tree. Then the maximum number R of linearly independent KCL equations for a given network is obtained by applying the KCL (Rule 2.8.1) to the fundamental cut sets associated with the tree. The number of independent equations equals the rank ρ of the network graph and is, therefore, $R = n - 1$, 1 less than the number of nodes. If the graph has m parts and a total of n nodes, the result is $R = \rho = n - m$.

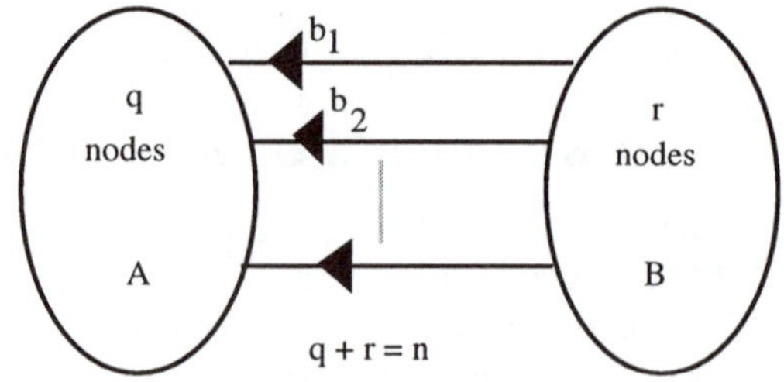

FIGURE 2.8.1
Simply connected n node graph. $1 \leq q \leq n-1$, $r \geq 1$.

PROOF We consider first the case where KCL is applied to the n nodes of a simply connected graph, that is each cut set surface surrounds a single node. The result is then extended to the fundamental cut sets.

Consider node k of a simply connected graph possessing n nodes and b branches. The KCL equation for this node can be written as $J_k = 0$, where J_k represents an appropriate linear combination of all branch currents in the graph. The current coefficients in the sum are $1, -1, or 0$. Take the normal direction of the surface surrounding each node as away from the node. Then if a branch is directed away from node k, a plus (+) sign is assigned to that branch current; if towards k, the sign is minus (−); if a branch is not incident on node k, its current gets a zero multiplier. There are n equations, one for each node. To examine the linear independence of these KCL equations, we assume arbitrary values of the branch currents and seek to determine whether or not one or more of the J_k is expressible as a linear combination of the others. To do this, form the following equation consisting of the general linear combination of all the J_k set equal to zero.

$$a_1 J_1 + a_2 J_2 + \ldots + a_n J_n = 0 \tag{2.8.1}$$

To test for linear dependence of the J_k we must determine whether eq. (2.8.1) has a solution with at least one of the a_k nonzero. On the other hand, if the only admissible solution is for all $a_k = 0$, the n equations $J_k = 0$ are linearly independent. The test is simple to carry out. J_k is the outward flux of current through the nodal cut set surface for node k. For any physically permissible distribution of currents in the branches, $\sum_{k=1}^{n_i} J_k$ is the total outward flux of current through a surface enclosing all the nodes and branches, i.e., surrounding the entire simply connected graph. This flux is, of course, zero. In other words, eq. (2.8.1) is satisfied when all the $a_k = 1$. Therefore the set is linearly dependent. The number of linearly independent equations, or equivalently the rank ρ, of this simply connected graph must, therefore, be at least one less than the total number of nodes. We now show that ρ is precisely $n-1$.

Fig. 2.8.1 schematically represents a simply connected graph (Definition 2.7.4). Suppose there are less than $n-1$ linearly independent KCL equations. This is equivalent to assuming that a solution of eq. (2.8.1) exists with nonzero coefficients $a_1, a_2, \ldots a_q$, $1 \leq q \leq n-1$ and the remaining $r \geq 1$ coefficients set to zero. Let section A of the graph contain the nodes 1 through q, with the remaining r nodes in section B. Since the graph is simply connected (and therefore, according to our conventions, nonseparable and nondegenerate), there are two or more branches $b_1, b_2, \ldots$ connecting sections A and B. Each of these branches contributes its branch current to only one of the J_k in section A. Thus, in

$$a_1 J_1 + a_2 J_2 + \ldots + a_q J_q = 0$$

one of the currents in a branch connecting A and B occurs in one of the nodal sums J_k, and in no other, so there is no possibility of cancellation with all nonzero coefficients. Therefore a solution requires that for the node in question, $a_k = 0$. This contradicts the assumption of nonzero values for all the q coefficients associated with the nodes in section A. Since this reasoning applies to each node in section A, we conclude that $a_k = 0$, $k = 1, 2, \ldots, q$. It follows that any q nodal equations, $q = 1, 2, \ldots, n-1$ are linearly independent, i.e., there must be *at least* $n - 1$ independent nodal equations. But it was shown earlier that the n nodal equations are linearly dependent. We have, therefore, proved that the number of linearly independent node equations for a simply connected graph is precisely $n-1$. If the graph is not simply connected but has m separate parts, with n the total number of nodes for all parts, then clearly the rank is $\rho = R = n - m$, which is equal to the number of tree branches for the entire graph.

The case in which KCL is applied to a set of $n - m$ fundamental cut sets (the number of tree branches) can now be settled. Each cut set equation sums the current flux across a cut set surface, hence it may be obtained by summing the nodal equations at the nodes contained within the cut set surface; thus the cut set equations are linearly dependent on node equations. It follows that the number of linearly independent fundamental cut set equations cannot exceed $n - m$. But the total set is linearly independent since by construction each equation includes a single tree branch current excluded from the others. Hence the number of linearly independent cut set equations is exactly $\rho = n - m$. ■

The following theorem establishes a procedure for writing a maximal set of KVL equations on a given network and relates their number to the nullity ν of its graph.

THEOREM 2.8.2

In a simply connected network, a maximal set of N linearly independent KVL equations is obtained by applying KVL (Rule 2.8.2) to the fundamental loops associated with a cotree. The number of equations equals the nullity *of the network graph* $N = \nu = b - n + 1$*. If the graph has* m *parts,* $N = \nu = b - n + m$*.*

PROOF Each fundamental loop equation of a simply connected graph has only a single link voltage, and this does not appear in any other loop (Corollary 2.7.5). Hence a general linear combination of fundamental loop KVL sums, Σ, cannot produce internal cancellations of the link voltages (one per loop equation) when nonzero multipliers are employed. (See eq. (2.8.1).) Thus analogous to the case for KCL cut sets, the solution of the equations used to test linear independence $\Sigma = 0$ requires that each of the coefficients multiplying the fundamental loop KVL sums be zero. We therefore conclude that the KVL equations taken around the $\nu = b - n + 1$ fundamental loops are linearly independent. To show that this is the maximum number, consider now the equation obtained by applying KVL to any set of loops (including nonfundamental ones). Each of these must contain one or more links, but each link voltage is linearly dependent on tree voltages (the dependence is given by a fundamental loop equation), thus *any* loop equation is linearly dependent on fundamental loop equations. Using a basic principle of linear algebra, the maximum number of independent loop equations must be equal to the number of linearly independent *fundamental* loop equations. i.e., $N = \nu = b - n + 1$ and, for a graph of m parts $N = \nu = b - n + m$. ∎

COROLLARY 2.8.1

The ν *fundamental loops include all the branches of a simply connected nondegenerate graph.*

PROOF We demonstrate the result by showing that the ν KVL equations for a complete set of fundamental loops involve every branch voltage of the associated tree. Clearly these tree branches plus the links in the fundamental loops comprise all the branches of the graph. Suppose one or more tree branch voltages do not appear anywhere in the set of fundamental KVL loop equations. Since each link of the cotree formed on the given tree closes a unique loop through the tree, and since the graph is nondegenerate, any new loop equation containing one of the left over tree branch voltages must include at least *two* links. Furthermore, since a fundamental loop contains only one link, any link voltage can be expressed as a linear combination of branches of the tree, exclusive of the leftover branches. Now look at a loop containing leftover branches and two of the link voltages. Replace one of these by its representation in terms of tree branch voltages and you have a loop equation which contains only one link voltage plus tree branch voltages (including leftover ones); in other words a

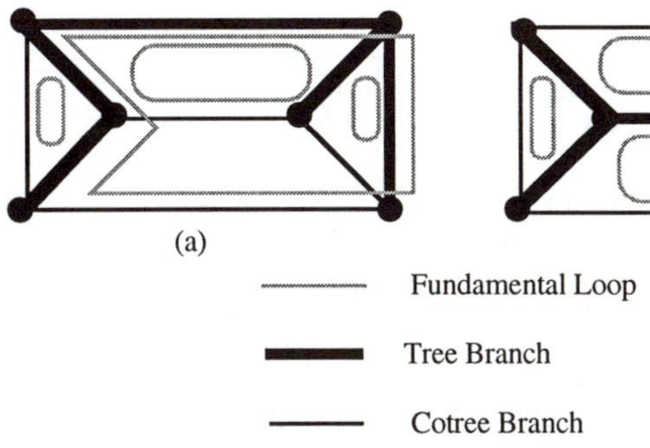

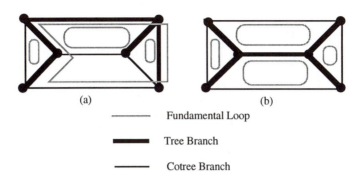

FIGURE 2.8.2
Fundamental loops for a prescribed tree, $\nu = 4$.

fundamental loop equation. The so-called leftover tree branches are already in a fundamental loop, and so simply do not qualify as "leftover". ■

It is possible to choose a set of loop equations which are linearly independent and do not constitute a set of fundamental loops. However, as discussed in the proof of Theorem 2.8.2 the maximum number of independent loop equations is equal to the number of fundamental loops. A dual argument applies to cut sets. To highlight this result we have the following theorem.

THEOREM 2.8.3

(a) The maximum number of linearly independent cut set equations (whether fundamental or not) in a network is equal to $\rho = n - m$, the rank of the graph. That the cut sets be fundamental is sufficient for ν independent KCL equations, but not necessary.

(b) The maximum number of linearly independent loop equations (whether fundamental or not) in a network is equal to $\nu = b - n + m$, the nullity of the graph. That the loops be fundamental is sufficient for ν independent KVL equations, but not necessary.

Example 2.8.1

For the graph of Fig. 2.7.1, redrawn on Fig. 2.8.2, with nullity $\nu = 4$, show the four fundamental loops corresponding to the two trees.

Solution The two sets of fundamental loops are sketched in Fig. 2.8.2. Note there is only one link per loop, and the remaining branches belong to a tree. In the case of Fig. 2.8.2, the loops are meshes (Definition 2.7.7) as well. □

2.9 Nodal Analysis

Nodal analysis is a frequently used general method for combining the topologic properties of a network with the constitutive relations defining voltage and current interdependence in the branches, so as to obtain an eigenvalue matrix for the complete system under eigenmode operation. The exponential, ϵ^{st}, of course does not appear explicitly in the analysis, which is carried out in terms of phasors. The initial step of nodal analysis is to choose any convenient set of $n-1$ nodes. The remaining node, n, is designated the ground or *datum* node. The excitation phasors appear as nodal current sources at the $n-1$ nodes, with the exponential eigenfunction ϵ^{st} omitted. The current return is at the datum node. At the k-th node, $k \neq n$, the net nodal current source term $J_k(s)$ is a linear combination of branch source currents incident on node k. The output or response quantities are the potential drops between each of the selected $n-1$ nodes and the ground node. These are the nodal voltages $V_1, V_2, \dots, V_{n-1}$. These nodal voltages are across branches which form a tree (some of the tree branches may be open circuits and have zero admittance), and connect the $n-1$ vertices to the datum node. If a solution for these tree branch potentials is obtained, the KVL equations for the fundamental loops associated with the chosen node to ground tree immediately yield all the link voltages. A knowledge of the nodal potentials, therefore, suffices to determine by linear combination *all* the branch voltages of the network.

Nodal analysis yields a set of $n-1$ relations involving the J_k and the V_k phasors. Thus

$$\boldsymbol{J}(s) = \boldsymbol{\Lambda}(s)\boldsymbol{V}(s) \equiv \boldsymbol{Y}(s)\boldsymbol{V}(s) \tag{2.9.1}$$

In eq. (2.9.1) $\boldsymbol{J}(s)$ and $\boldsymbol{V}(s)$ are the $(n-1)$-component vectors of nodal excitation currents and nodal response potentials, respectively. The eigenvalue matrix is designated as $\boldsymbol{Y}(s)$ and termed the *nodal admittance matrix.* Its dimensions are $(n-1)\times(n-1)$.

The scenario for nodal analysis is straightforward. At the $n-1$ chosen nodes we write the linearly independent KCL equations. They are linearly independent. The branch currents are then replaced by the branch voltages by using the constitutive relations for the branches in the form $I_{Bk}(s) = Y_k(s)V_{Bk}(s)$, where I_{Bk}, V_{Bk}, Y_k are branch k current, branch k voltage, and branch k admittance, respectively. This assumes that there is no branch to branch coupling, e.g., due to magnetic fields or controlled sources, a restriction which will later be removed.

Once the branch currents are replaced by branch voltages, a further step permits all branch voltages to be expressed in terms of the node to datum voltages, which are across branches of the node to datum tree. Those branch voltages already in this category need no further attention. The remaining

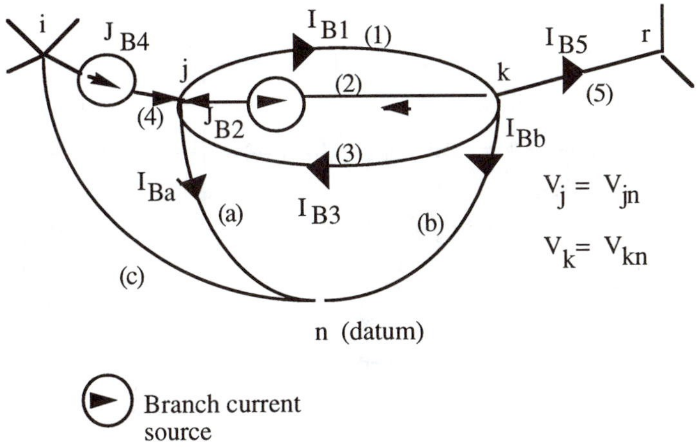

FIGURE 2.9.1
Portion of a graph to illustrate nodal analysis.

branch voltages must be links and hence, as discussed in Section 2.8, may be written as linear combinations of the nodal (tree) voltages. When the nodal current source terms are transposed to one side and the response terms (node to datum voltages) to the other, we have the $n-1$ KCL equations of nodal analysis eq. (2.9.1), but now with the nodal voltages as unknowns.

We can illustrate the procedure using Fig. 2.9.1 which shows a portion of a graph.

Write the KCL equation for node j.

$$I_{B1} + J_{B2} - I_{B3} - J_{B4} + I_{Ba} = 0$$

Put the branch source current terms (J_{Bk}) on one side, and replace the remaining branch currents by branch voltages, using constitutive relations, with Y_k as branch k admittance and V_{jk} as the branch voltage drop from node j to k.

$$-J_{B2} + J_{B4} \equiv J_j = Y_1 V_{jk} - Y_3 V_{kj} + Y_a V_{jn} \qquad (2.9.2)$$

Express all branch voltages in terms of the node voltages, which as noted above, are across branches that form a tree. Thus for any branch connecting nodes j and k (the datum potential is taken as $V_n = 0$)

$$V_{jk} = -V_{kj} = V_j - V_k$$

Collecting terms in the nodal potentials, the KCL equation (2.9.2) for node j becomes

$$J_j = (Y_1 + Y_3 + Y_a)\, V_j - (Y_1 + Y_3)\, V_k \qquad (2.9.3)$$

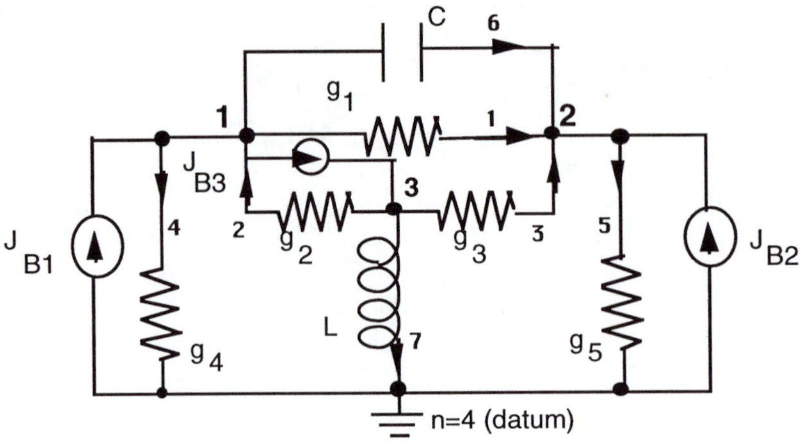

FIGURE 2.9.2
Four node circuit for nodal analysis.

We may determine the KCL equation for node k in a similar fashion and obtain

$$J_{B2} \equiv J_k = (Y_1 + Y_3 + Y_5 + Y_b)V_k - (Y_1 + Y_3)V_j - Y_5 V_r \tag{2.9.4}$$

The $n-1$ nodal analysis equations, each written as in eqs. (2.9.3) and (2.9.4), result in the matrix formulation eq. (2.9.1).

We can readily deduce an algorithm for the elements y_{ij} of the nodal admittance matrix $\boldsymbol{Y}(s)$ by extrapolating from eqs. (2.9.3), (2.9.4). Thus

$$\begin{aligned} y_{ii} &= +\Sigma \text{ all branch admittances incident on node } i \\ y_{ij} &= -\Sigma \text{ all branch admittances connecting nodes } i \text{ and } j \end{aligned} \tag{2.9.5}$$

Example 2.9.1
Fig. 2.9.2 shows a four node RLC circuit. Find the equations of nodal analysis.

Solution Since the rank $\rho = n - 1 = 3$, the nodal admittance matrix $\boldsymbol{Y}$(s) has dimensions 3×3, and is found using eq. (2.9.5). For example, $y_{11} = sC + g_1 + g_2 + g_4$, the sum of all branch admittances terminating on node 1, and $y_{12} = -(sC + g_1)$, the negative sum of admittances incident on both nodes 1 and 2. Continuing in this fashion

$$\boldsymbol{Y}(s) = \begin{pmatrix} sC + g_1 + g_2 + g_4 & -(g_1 + sC) & -g_2 \\ -(g_1 + sC) & sC + g_1 + g_3 + g_4 & -g_3 \\ -g_2 & -g_3 & g_2 + g_3 + \dfrac{1}{sL} \end{pmatrix}$$

The matrix $\boldsymbol{Y}(s)$ is symmetric about the main diagonal ($y_{ij} = y_{ji}$). This is always the case for passive *reciprocal*[5] networks such as the RLC circuit of Fig. 2.9.2, and also includes the case of networks with nongyromagnetic mutual coupling between branches. Such structures satisfy reciprocity, but others, some of whose elements involve such phenomena as the Hall effect or the Faraday effect with magnetized ferrites at microwave frequencies, can still be passive and yet nonreciprocal. Active networks with controlled sources in which branch-to-branch coupling is asymmetric are generally nonreciprocal circuits and $y_{ij} \neq y_{ji}$ may occur, though not necessarily. Returning to the example, the nodal admittance equations are

$$\boldsymbol{J} = \begin{pmatrix} J_{B1} - J_{B3} \\ J_{B2} \\ J_{B3} \end{pmatrix} = \boldsymbol{Y}(s) \begin{pmatrix} V_1 \\ V_2 \\ V_3 \end{pmatrix} = \boldsymbol{Y}(s)\boldsymbol{V}$$

Note that on the source side, a branch current source into the node gets a plus (+) sign, directed away, a minus (−) sign. □

The process of writing the nodal admittance matrix $\boldsymbol{Y}$ can be mechanized by starting with the branch admittance matrix, $\boldsymbol{Y}_b$. This matrix is really an inventory of the numbered admittances y_j, in the branches; these are placed on the main diagonal, the other elements being zero. If there are h branches, excluding the independent source branches, then

$$\boldsymbol{Y}_b = \text{diag}(y_1, y_2, \ldots, y_h) \tag{2.9.6}$$

The nodal admittance matrix, $\boldsymbol{Y}$, is obtained by combining $\boldsymbol{Y}_b$ with a *Branch-node Incidence Matrix*, $\boldsymbol{A}$, whose elements a_{ij} are defined as follows

$$a_{ij} = \begin{cases} -1 & \text{branch } i \text{ incident towards node } j \\ 1 & \text{branch } i \text{ incident away from node } j \\ 0 & \text{branch } i \text{ not incident on node } j \end{cases} \tag{2.9.7}$$

We omit the details here and merely give the final expression for the nodal admittance matrix $\boldsymbol{Y}$

$$\boldsymbol{Y} = \boldsymbol{A}'\boldsymbol{Y}_b\boldsymbol{A} \tag{2.9.8}$$

The notation $'$ means matrix transpose, i.e., rows and columns of the matrix interchanged.

Refer to Fig. 2.9.2 to illustrate the procedure. Thus

$$\boldsymbol{Y}_b = \text{diag}\left(g_1, g_2, g_3, g_4, g_5, \frac{1}{sL}, sC\right)$$

[5]The concept of reciprocity will be fully discussed in Section 2.13.

and

$$
\boldsymbol{A}' = \begin{pmatrix} 1 & -1 & 0 & 1 & 0 & 1 & 0 \\ -1 & 0 & -1 & 0 & 1 & -1 & 0 \\ 0 & 1 & 1 & 0 & 0 & 0 & 1 \end{pmatrix}
$$

The final result for $\boldsymbol{Y}(s)$ given in Example 2.9.1 is readily verified by applying eq. (2.9.8). The direction signs for the branches cancel out in the process of expanding eq. (2.9.8) so that the nodal admittance matrix, as indicated by eq. (2.9.5), is independent of the sense of the branch arrows.

In the previous discussion, nodal analysis was carried out assuming that the V_j, I_j constitutive relations for a given branch only involved variables in that branch. We can readily remove this restriction to permit coupled branch constitutive relations, for example, due to magnetic coupling between branches, or active element controlled source coupling between branches, or indeed any arbitrary branch coupling component. One device that can be used is to initially treat all *coupled branch* currents as though they were independent (but fictitious) sources. After the nodal analysis equations are set up under this assumption (at this stage involving only the response in the uncoupled branches), the current-voltage equations for the coupled branches are transformed by substituting the relations between the coupled branch voltages and the node to datum potentials. The fictitious current source terms (now expressed in terms of the nodal response voltages) are then transposed to the response side of the equations and the result is the nodal analysis relations $\boldsymbol{J}(s) = \boldsymbol{Y}(s)\boldsymbol{V}(s)$, where $\boldsymbol{J}(s)$ is the true independent nodal current source vector.

If the branch to branch coupling is purely due to conventional magnetically coupled coils (but not for anisotropic gyromagnetic coupling), reciprocity is preserved and the nodal admittance matrix remains symmetric about its main diagonal, i.e., $\boldsymbol{Y}' = \boldsymbol{Y}$ or $y_{ij} = y_{ji}$. On the other hand, if branches are coupled asymmetrically, as is generally the case when controlled sources are present, then the system is *nonreciprocal* and $\boldsymbol{Y}' \neq \boldsymbol{Y}$. The following example will demonstrate the method just described.

Example 2.9.2

Fig. 2.9.3 is a circuit with branch coupling due to both a *Voltage Controlled Current Source (VCCS)* and a pair of magnetically coupled coils. Find the nodal admittance relations assuming eigenmode operation and using phasors.

Solution Referring to Fig. 2.9.3, and using $L_{12} = L_{21}$, since the magnetic coupling satisfies reciprocity, the constitutive equations for branch magnetic coupling are

$$
\begin{pmatrix} V_a \\ V_b \end{pmatrix} = s \begin{pmatrix} L_{11} & L_{12} \\ L_{12} & L_{22} \end{pmatrix} \begin{pmatrix} I_a \\ I_b \end{pmatrix}
$$

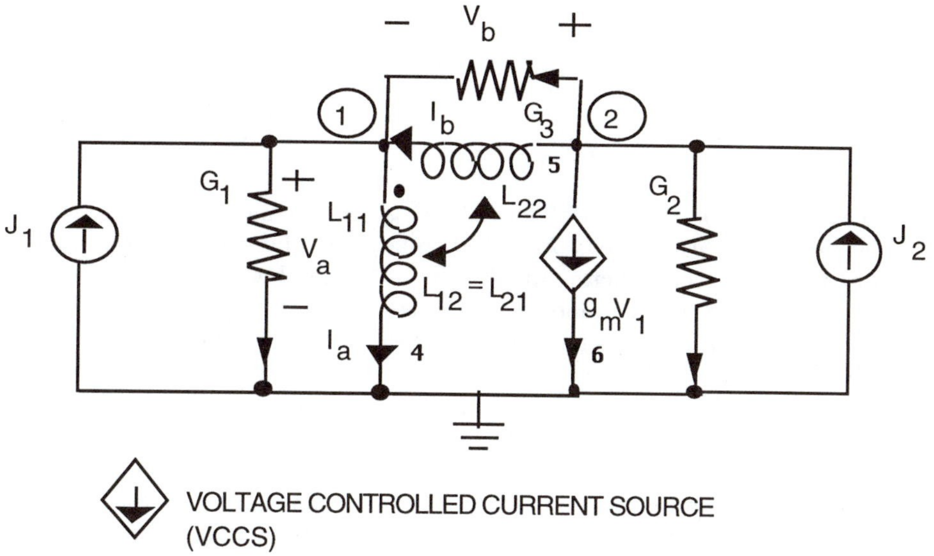

FIGURE 2.9.3
Circuit with controlled source and magnetic coupled branches.

or

$$\begin{pmatrix} I_a \\ I_b \end{pmatrix} = \frac{1}{s} \begin{pmatrix} L_{11} & L_{12} \\ L_{12} & L_{22} \end{pmatrix}^{-1} \begin{pmatrix} V_a \\ V_b \end{pmatrix} = \frac{1}{s} \begin{pmatrix} \Gamma_{11} & \Gamma_{12} \\ \Gamma_{21} & \Gamma_{22} \end{pmatrix} \begin{pmatrix} V_a \\ V_b \end{pmatrix} \tag{2.9.9}$$

where assuming $\boldsymbol{L}$, the inductance matrix, is nonsingular, $\boldsymbol{\Gamma} = \boldsymbol{L}^{-1}$, the inverse inductance matrix (also symmetric), is defined by

$$\Gamma_{11} = \frac{L_{22}}{\Delta} \qquad \Gamma_{22} = \frac{L_{11}}{\Delta} \qquad \Gamma_{12} = \Gamma_{21} = \frac{-L_{12}}{\Delta}$$

$$\Delta = L_{11}L_{22} - L_{12}^2 \neq 0$$

We now introduce the explicit nodal potentials, V_1, V_2 of Fig. 2.9.3, into the branch constitutive relations (2.9.9). Thus $V_a = V_1$ and $V_b = -V_1 + V_2$. Rewrite eq. (2.9.9) under this substitution,

$$\begin{pmatrix} I_a \\ I_b \end{pmatrix} = \boldsymbol{\Gamma} \begin{pmatrix} 1 & 0 \\ -1 & 1 \end{pmatrix} \begin{pmatrix} V_1 \\ V_2 \end{pmatrix} = \frac{1}{s} \begin{pmatrix} \Gamma_{11} - \Gamma_{12} & \Gamma_{12} \\ \Gamma_{12} - \Gamma_{22} & \Gamma_{22} \end{pmatrix} \begin{pmatrix} V_1 \\ V_2 \end{pmatrix} \tag{2.9.10}$$

Designate $\boldsymbol{J}_f$ as the nodal source vector including the coupled branch currents as fictitious sources. Referring to Fig. 2.9.3

$$\boldsymbol{J}_f = \begin{pmatrix} J_1 - I_a + I_b \\ J_2 - g_m V_1 - I_b \end{pmatrix}$$

Substitute for I_a and I_b using eq. (2.9.10)

$$\boldsymbol{J}_f = \begin{pmatrix} J_1 \\ J_2 \end{pmatrix} - \begin{pmatrix} \dfrac{\Gamma_{11} - 2\Gamma_{12} + \Gamma_{22}}{s} & \dfrac{\Gamma_{12} - \Gamma_{22}}{s} \\ g_m + \dfrac{\Gamma_{12} - \Gamma_{22}}{s} & \dfrac{\Gamma_{22}}{s} \end{pmatrix} \begin{pmatrix} V_1 \\ V_2 \end{pmatrix} \tag{2.9.11}$$

The first term on the right of eq. (2.9.11) is the true independent source nodal excitation, $\boldsymbol{J}$, whereas $\boldsymbol{J}_f$, which consists of the fictitious sources, is a nodal source vector acting on the circuit formed by omitting all coupled branches (both active and magnetic, and temporarily assumed to be sources) from Fig. 2.9.3. Therefore, using eq. (2.9.5), which applies to a graph with uncoupled branches

$$\boldsymbol{J}_f = \begin{pmatrix} J_{f1} \\ J_{f2} \end{pmatrix} = \begin{pmatrix} G_1 + G_3 & -G_3 \\ -G_3 & G_2 + G_3 \end{pmatrix} \begin{pmatrix} V_1 \\ V_2 \end{pmatrix}$$

Substitute the above result into eq. (2.9.11) and transpose the true independent sources $\boldsymbol{J}$ to the left, and the response terms (operating on $\boldsymbol{V}$) to the right side

$$\boldsymbol{J} \equiv \begin{pmatrix} J_1 \\ J_2 \end{pmatrix} = \begin{pmatrix} G_1 + G_3 + \dfrac{\Gamma_{11} - 2\Gamma_{12} + \Gamma_{22}}{s} & -G_3 + \dfrac{\Gamma_{12} - \Gamma_{22}}{s} \\ g_m - G_3 + \dfrac{\Gamma_{12} - \Gamma_{22}}{s} & G_2 + G_3 + \dfrac{\Gamma_{22}}{s} \end{pmatrix} \begin{pmatrix} V_1 \\ V_2 \end{pmatrix}$$
$$= \boldsymbol{Y}\boldsymbol{V} \tag{2.9.12}$$

It is clear that the final nodal admittance matrix is not symmetric, i.e., $\boldsymbol{Y}(s) \neq \boldsymbol{Y}'(s)$ so that the system is nonreciprocal, but inspection of eq. (2.9.12) indicates that the asymmetry is due to the presence of the VCCS and not due to the magnetic coupling between branches.

Incidence matrices can also be employed to determine the nodal admittance matrix when there is coupling between branches exactly as discussed in eqs. (2.9.7), (2.9.8). Indeed, especially for complicated structures, the incidence technique may be preferable since it readily lends itself to computer programming. The only modification in the procedure is in the formulation of the branch admittance matrix, $\boldsymbol{Y}_b$. This matrix is no longer purely diagonal, but has off-diagonal terms to take account of branch-to-branch coupling. All response branches, coupled and uncoupled, are directed and numbered including those with controlled sources, and those containing magnetically coupled coils. Other more general forms of coupling can also be included. Thus referring to the branch numbers and polarities in Fig. 2.9.3, the branch coupling admittance equations for the magnetically coupled branches are given by eq. (2.9.9), with subscripts a and b replaced by $b4$ and $b5$. The branch coupling equation for the controlled source is $I_{b5} = g_m V_{b1}$. Again referring to Fig. 2.9.3 for branch numbering and orientation, the branch constitutive relations are shown below.

The branch admittance matrix $\boldsymbol{Y}_b$ has dimensions 6×6 (the independent source branches are excluded)

$$\boldsymbol{I}_b = \boldsymbol{Y}_b \boldsymbol{V}_b = \begin{pmatrix} G_1 & 0 & 0 & 0 & 0 & 0 \\ 0 & G_2 & 0 & 0 & 0 & 0 \\ 0 & 0 & G_3 & 0 & 0 & 0 \\ 0 & 0 & 0 & s\Gamma_{11} & s\Gamma_{12} & 0 \\ 0 & 0 & 0 & s\Gamma_{12} & s\Gamma_{22} & 0 \\ g_m & 0 & 0 & 0 & 0 & 0 \end{pmatrix} \begin{pmatrix} V_{b1} \\ V_{b2} \\ V_{b3} \\ V_{b4} \\ V_{b5} \\ V_{b6} \end{pmatrix} \tag{2.9.13}$$

Finally for the circuit of Fig. 2.9.3, the branch-node incidence matrix transposed (i.e., node-branch) of dimensions 2×6, is

$$\boldsymbol{A}' = \begin{pmatrix} 1 & 0 & -1 & 1 & -1 & 0 \\ 0 & 1 & 1 & 0 & 1 & 1 \end{pmatrix} \tag{2.9.14}$$

If eqs. (2.9.13) and (2.9.14) are substituted into eq. (2.9.8), the result given by eq. (2.9.12) follows. It should be clear that the incidence procedure will handle arbitrary branch coupling (on an admittance basis), even when more than two branches are coupled. One simply formulates the appropriate $\boldsymbol{Y}_b$ matrix and employs eq. (2.9.8). □

The solution of the nodal analysis equations, eqs. (2.9.8), (2.9.12), gives the components of the node-to-datum voltage vector, $\boldsymbol{V}$

$$\boldsymbol{V} = \boldsymbol{Y}^{-1}\boldsymbol{J}$$

The branch voltages and currents follow as

$$\boldsymbol{V}_b = \boldsymbol{A}\boldsymbol{V} \quad \boldsymbol{I}_b = \boldsymbol{Y}_b \boldsymbol{V}_b \tag{2.9.15}$$

In some situations voltage sources may be present (either independent or controlled) in a circuit which is to be handled by nodal analysis. In such cases a simple device that usually takes care of dealing only with current sources is to use Thévenin's theorem to change each voltage source to a current source. Thus a voltage source E in series with an impedance Z becomes a current source V/Z shunted by an admittance $1/Z$ (similarly by Norton's theorem a current source J in parallel with admittance Y becomes voltage source J/Y in series with $1/Y$). If the source is controlled by current (instead of voltage), the control current must be expressed in terms of node to datum voltages. Using this idea, except for some special cases, all sources become currents and all controlled sources are treated like VCCS sources.

Nodal analysis then proceeds as described above. Source equivalence, of course, cannot be used when the voltage source is not in series with an impedance. Special devices may be employed to handle this situation to obtain current sources or voltage sources with series impedances. If the voltage source is in parallel with an impedance or a current source, the shunting element is replaced by an open circuit. If a current source is in series with an impedance or a voltage source, the series element is replaced by a short circuit. Finally, if none of the above, "source transportation" can be used. This involves applying Kirchhoff's Laws so that the voltage source without series impedance becomes two voltage sources each moved to a new position, but each now in series with an impedance (or a current source becomes two current sources each in parallel with an admittance). In all these special cases, application of KVL and/or KCL shows that the transformation of the sources does not affect response elsewhere in the circuit. Details are discussed in the cited reference.[6] One note of warning; when circuit models are devised which employ ideal independent or controlled sources, special attention must be paid to avoid contriving a system which is inconsistent with the Kirchhoff laws. An example is forming a closed loop of independent voltage sources which do not add to zero; of course, more subtle possibilities are available to trap the unwary designer.

In some situations, voltage sources may be present (either independent or controlled) in a circuit which is to be handled by nodal analysis. In such cases, a simple device that often takes care of the requirement for all current sources is to use Thévenin's theorem to change each voltage source to a current source. Thus a voltage source E in series with an impedance Z becomes a current source E/Z shunted by an admittance $1/Z$ (similarly by Norton's theorem a current source J in parallel with admittance Y becomes a voltage source J/Y in series with an impedance $1/Y$). If the source is controlled by current (instead of voltage), the control current must be expressed in terms of node to datum voltages. When this procedure is applicable, all sources become independent current sources and all controlled sources are transformed into VCCS elements. Nodal analysis then proceeds as described above. A more general approach which does not require only current sources and handles all the possibilities of branch-to-branch coupling is described in Section 2.13.

2.10 Mesh and Loop Analysis

Nodal analysis is based on a set of linearly independent KCL equations written at $n-1$ of the n nodes of a network. In the case of a *planar* network

[6] L. O. Chua, C. A. Desoer, and E. S. Kuh, *Linear and Nonlinear Circuits*, New York, McGraw Hill, 1987, p. 696 et seq.

with ν meshes (see Definitions 2.7.5, 2.7.7), nodal analysis can be transformed by *duality* into mesh analysis where ν KVL mesh equations replace the nodal KCL equations. The entire mesh analysis procedure is, in fact, precisely deducible from nodal analysis, provided the given graph is planar and drawn with the meshes laid out. A further assumption, analogous to the choice of current sources in nodal analysis, is that all the sources are either independent voltage generators or current controlled voltage sources, CCVS. (See discussion at the end of Section 2.9.)

In the following analysis we will restrict the planar graphs considered to those that are simply connected (Definition 2.7.4), nondegenerate (Definition 2.7.9), and nonseparable (Definition 2.7.8). The assumption of planarity is crucial, the other assumptions are simply for convenience, and simplicity, and not strictly necessary.

DEFINITION 2.10.1 *In a planar graph laid out to exhibit the meshes, the* outer loop *consists of the closed path of* p *branches forming the outside boundary of the graph.*

First we show that in a planar graph $\mathcal{G}$ the ν mesh KVL equations are linearly independent. Perform the following construction. After drawing the graph in mesh form, place a single node (a q-type node) inside each mesh of the given planar graph. Add one more node, designated $m = \nu + 1$, outside $\mathcal{G}$ for a total of $\nu + 1$ dual nodes. Construct a *dual* graph $\mathcal{G}'$ by connecting each pair of the ν q-type internal nodes with a single branch, which in turn only crosses a single branch of $\mathcal{G}$. If the number of outer loop branches is p, then from node m draw p branches each of which crosses a single outer loop branch, crosses no other branches, and is incident on a q-type node (in an outer *mesh*). Fig. 2.10.1 shows this construction for a graph of $\nu = 3$ meshes.

Consider any mesh KVL equation of the planar graph $\mathcal{G}$ made up of a sum of the branch voltages taken counterclockwise around the mesh. Each of these branches is crossed by a single branch of $\mathcal{G}'$ one end of which terminates at the internal q-type node of the mesh. For example, in Fig. 2.10.1 branch 1 of $\mathcal{G}$, mesh A, is crossed by branch a of $\mathcal{G}'$. Assuming normal crossing angles, direct the branches of $\mathcal{G}'$ so that each direction arrow is rotated 90° clockwise from the arrow of the crossed branch of $\mathcal{G}$. This is illustrated on Fig. 2.10.1. Now any KVL mesh equation in $\mathcal{G}$ can be transformed into a KCL equation at the q-type node inside the mesh by replacing each branch voltage of $\mathcal{G}$ by the branch current of the crossing branch of $\mathcal{G}'$. For example, KVL for mesh A is $V_1 - V_4 - V_2 = 0$, and KCL at the associated q-type node is $I_a - I_d - I_b$. Therefore, since the ν nodal equations of $\mathcal{G}'$ are a complete set of linearly independent KCL equations (i.e., lead to the determination of all branch voltages and currents of $\mathcal{G}'$),

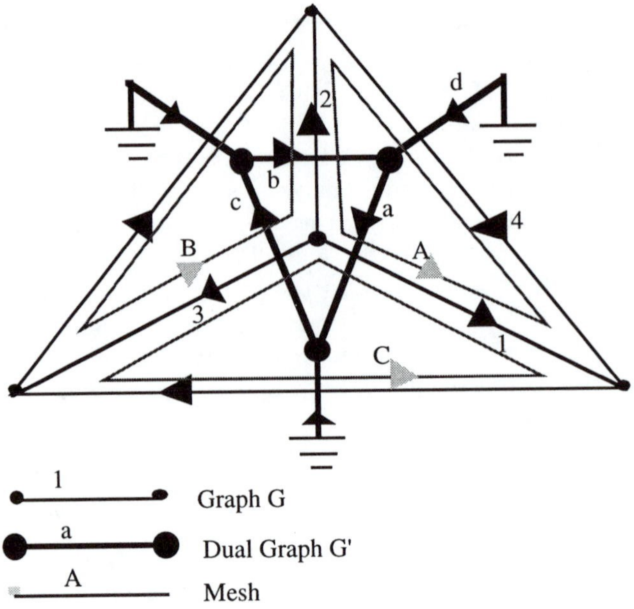

FIGURE 2.10.1
Example of the dual of a planar graph.

so too are the precisely equivalent KVL mesh equations of $\mathcal{G}$, which have the same number of branches as $\mathcal{G}'$.

Every quantity associated with the graph $\mathcal{G}$ has a dual with respect to $\mathcal{G}'$, e.g., meshes of G and nodes of G', branch voltages in $\mathcal{G}$ and branch currents in $\mathcal{G}'$, the nullity of $\mathcal{G}$ and the rank of $\mathcal{G}'$, etc. To obtain the dual in $\mathcal{G}$ of the node to datum voltages of $\mathcal{G}'$ we define the *mesh currents.* With each mesh of $\mathcal{G}$ associate a mesh current symbolized by a directed closed loop, e.g., the mesh current I_A associated with mesh A shown in Fig. 2.10.1. The mesh current is defined so that the signed algebraic sum of mesh currents traversing a branch gives the branch current. Thus in Fig. 2.10.1, $I_1 = I_A - I_C$, $I_2 = I_B - I_A$, etc. These equations have precisely the same form as the relations between node potentials and branch voltages (Section 2.9, just below eq. (2.9.2)) and the mesh currents are the duals of the nodal potentials. In mesh analysis therefore the response quantities are the mesh currents. Once these are determined from the mesh equations, all branch currents can be calculated. This is the exact dual of the determination of branch voltages in terms of nodal potentials as a consequence of nodal analysis.

Guided by duality, we can set up the mesh analysis equations as follows: (a) write the ν KVL equations for the meshes; (b) use the branch constitutive impedance relations to replace branch voltages by branch currents; (c) replace the branch currents by the appropriate linear combinations of mesh currents. The following theorem summarizes the preceding discussion.

THEOREM 2.10.1
The equations of mesh analysis have the form

$$\boldsymbol{E}(s) = \boldsymbol{Z}(s)\boldsymbol{I}(s)$$

where $\boldsymbol{E}(s)$ *is the net independent voltage driving source vector for the meshes (dual to* $\boldsymbol{J}(s)$*),* $\boldsymbol{Z}(s)$ *is the* mesh impedance matrix, *and* $\boldsymbol{I}(s)$ *is the vector of mesh currents, i.e., the response to* $\boldsymbol{E}(s)$.

In a planar graph there are ν *linearly independent KVL mesh equations, and the dimensions of* $\boldsymbol{Z}(s)$ *are* $\nu \times \nu$*, where* ν *is the nullity of* $\mathcal{G}$.

If $\mathcal{G}$ *has* b *branches and* n *nodes, then* $\nu = b - n + 1$.

In other words, you have only to count the meshes in a planar graph to ascertain the nullity. Theorem 2.10.1 is valid, even if the simply connected planar graph is degenerate and separable.

The mesh impedance matrix $\boldsymbol{Z}(s)$ is the dual of the nodal admittance matrix $\boldsymbol{Y}(s)$. Thus referring to eq. (2.9.5) we have the following algorithm for determining the elements of $\boldsymbol{Z}$.

THEOREM 2.10.2
In a planar graph with directed meshes, i.e., clockwise or counterclockwise, and all of whose (uncoupled) constitutive branch relations are of the form $V_k = Z_k I_k$*, where* V_k, I_k, Z_k *are the branch voltages, branch currents, and branch impedances, respectively, the elements of the mesh impedance matrix are given by*

$$\begin{aligned} z_{ii} &= \phantom{\mu_{ij}}\ \Sigma \text{ all branch impedances traversed by mesh } i \\ z_{ij} &= \mu_{ij}\ \Sigma \text{ all branch impedances common to meshes } i \text{ and } j \end{aligned} \tag{2.10.1}$$

$\mu_{ij} =\ \ 1$ *if meshes* i *and* j *traverse their common branches in the same direction;*

$\mu_{ij} = -1$ *if meshes* i *and* j *traverse their common branches in opposite directions;*

$\mu_{ij} =\ \ 0$ *if meshes* i *and* j *have no common branches.*

The sign only depends on the relative direction of the meshes and does not depend on branch orientation. In the case where meshes are all oriented in the same direction, meshes always traverse their common branches in opposite directions. Thus $\mu_{ij} = -1$ for all meshes with common branches, and the result is the exact dual of nodal analysis, i.e., all nonzero off-diagonal elements are negative.

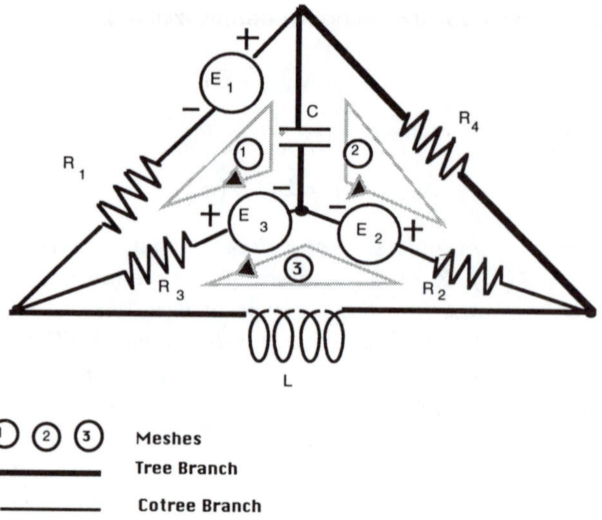

FIGURE 2.10.2
Circuit for mesh and fundamental loop analysis.

As in nodal analysis, mesh analysis can be extended to the case of circuits with branch-to-branch coupling. This case will be illustrated at the end of this Section by using incidence matrices. Mesh analysis is also readily generalized to loop analysis, in which case the graph need not be planar, and the procedure is no longer the dual of nodal analysis. It suffices to state that the number of independent KVL loop equations in a nonplanar graph is still $\nu = b - n + 1$, and the ν fundamental loops always constitute an independent set (Theorem 2.8.2). An example of fundamental loop analysis will be given later in this Section.

Example 2.10.1
Fig. 2.10.2 shows a 3 mesh network. Find the mesh impedance equations.

Solution For the indicated choice of meshes in Fig. 2.10.2, the mesh impedance matrix can be written by inspection using Theorem 2.10.2.

$$\boldsymbol{Z}(s) = \begin{pmatrix} \dfrac{1}{sC} + R_1 + R_3 & \dfrac{1}{sC} & R_3 \\ \dfrac{1}{sC} & \dfrac{1}{sC} + R_2 + R_4 & -R_2 \\ R_3 & -R_2 & sL + R_2 + R_3 \end{pmatrix}$$

The z_{12} and z_{13} terms are positive because meshes 1 and 2 are in the same direction along branch C. The z_{23} term is negative because meshes 2 and 3 traverse branch R_2 in opposite directions, etc.

The mesh excitation vector $\boldsymbol{E}$, on the source side of the mesh equations, is found as the algebraic sum of the independent branch voltage sources around the mesh. A term of the sum is positive if it is directed minus $(-)$ to plus $(+)$ in the direction of the mesh, and vice versa. Thus the final mesh equations for Fig. 2.10.2 can be written in terms of the mesh currents I_k

$$\boldsymbol{E}(s) = \begin{pmatrix} E_1 + E_3 \\ E_2 \\ -E_2 + E_3 \end{pmatrix} = \boldsymbol{Z}(s) \begin{pmatrix} I_1 \\ I_2 \\ I_3 \end{pmatrix} = \boldsymbol{Z}(s)\boldsymbol{I}(s)$$

□

In principle, loop analysis is carried out in the same fashion as mesh analysis; however, it may be employed for a nonplanar graph. One simply chooses a set of ν loops to define a set of ν linearly independent KVL equations, and then applies Theorem 2.10.2 replacing the word "mesh" by "loop". The meshes (in a planar network) provide a simple method of choosing a set of ν linearly independent equations. So too do the fundamental loops. In the following example the circuit analysis equations are written for a set of fundamental loops. Once the fundamental loop currents (analogous to the mesh currents) are determined as the solution of the loop equations, it is a straightforward matter to compute all branch voltages and currents.

Example 2.10.2

Choose a set of fundamental loops in Fig. 2.10.2 and write the equations of loop analysis. Express the branch currents in the C branch and the L branch in terms of the loop currents.

Solution Referring to Fig. 2.10.2, set up the fundamental loops by first selecting a tree. This is shown as branches C, R_4, L. Since the nullity $\nu = 3$, the cotree has 3 links, R_1, R_2, R_3. The fundamental loops (one link per loop), with directions in the order of the specified branches, are 1. (R_1, R_4, L), 2. (R_3, C, R_4, L), 3. (R_2, R_4, C). Theorem 2.10.2 can now be used for the fundamental loops. For example, the sum of branch impedances in loop 1 is $z_{11} = R_1 + R_4 + sL$. Loops 1 and 2 have common branches R_4, L traversed in the same direction so that $\mu_{12} = 1$, and $z_{12} = R_4 + sL$, etc. The final loop equations are

$$\begin{pmatrix} E_1 \\ -E_3 \\ E_2 \end{pmatrix} = \begin{pmatrix} R_1 + R_4 + sL & R_4 + sL & -R_4 \\ R_4 + sL & R_3 + R_4 + sL + \dfrac{1}{sC} & -(R_4 + \dfrac{1}{sC}) \\ -R_4 & -(R_4 + \dfrac{1}{sC}) & R2 + R_4 + \dfrac{1}{sC} \end{pmatrix} \begin{pmatrix} I_{L1} \\ I_{L2} \\ I_{L3} \end{pmatrix}$$

or

$$\boldsymbol{E}_L = \boldsymbol{Z}_L \boldsymbol{I}_L \tag{2.10.2}$$

where the subscript L has been used to distinguish the loop quantities from the earlier employed mesh quantities. $\boldsymbol{E}_L$ in eq. (2.10.2) is the net loop driving voltage vector. Each component is the algebraic sum of the independent voltage sources taken around and in the direction of the loop. If an independent source in the loop is $(-)$ to $(+)$ in the direction of the loop, it gets a $(+)$ sign in the loop sum. If directed $(+)$ to $(-)$ in the loop direction, it gets a $(-)$ sign.

The branch currents in the C and L branches may be expressed as the algebraic sum of loop currents traversing these branches. Thus, assuming I_C is directed upward, and I_L from right to left,

$$I_C = I_{L2} - I_{L3}$$

$$I_L = I_{L1} + I_{L2}$$

□

In the discussion of nodal analysis a method to mechanize the process was introduced. This involved the use of a branch-node incidence matrix, $\boldsymbol{A}$, operating on the branch admittance matrix, as indicated in eqs. (2.9.6), (2.9.7), (2.9.8). The same sort of technique is applicable to mesh and loop analysis but the appropriate tool is the *Branch-Loop (Mesh) Incidence Matrix.* (Hereafter we shall generally use the generic term "loop" and consider a mesh to be a specific form of loop, restricted to planar graphs.)

Number and direct the branches (excluding *independent* sources) and the chosen loops of the graph. Then define the elements b_{ij} of the branch-loop incidence matrix $\boldsymbol{B}$ as follows (compare with eq. (2.9.7))

$$b_{ij} = \begin{cases} -1 & \text{branch } i \text{ traversed by loop } j \text{ in opposite directions} \\ 1 & \text{branch } i \text{ traversed by loop } j \text{ in same direction} \\ 0 & \text{branch } i \text{ not traversed by loop } j \end{cases} \tag{2.10.3}$$

If there are b numbered branches and ν loops, the dimensions of $\boldsymbol{B}$ are $b \times \nu$. The branch impedance matrix $\boldsymbol{Z}_b = \boldsymbol{Y}_b^{-1}$ lists, in numerical order, the uncoupled branch impedances on its main diagonal. It also includes submatrices representing branch-to-branch coupling, and controlled sources enter the description at this point. The dimensions of $\boldsymbol{Z}_b$ are $b \times b$.

Again with derivation omitted, the final expression for the loop impedance matrix $\boldsymbol{Z}_L$ is

$$\boldsymbol{Z}_L = \boldsymbol{B}' \boldsymbol{Z}_b \boldsymbol{B} \tag{2.10.4}$$

The loop impedance matrix is substituted into eq. (2.10.2) to obtain the final equations formulating loop analysis. The solution of these simultaneous equations gives the loop currents

$$\boldsymbol{I}_L = \boldsymbol{Z}_L^{-1} \boldsymbol{E}_L \tag{2.10.5}$$

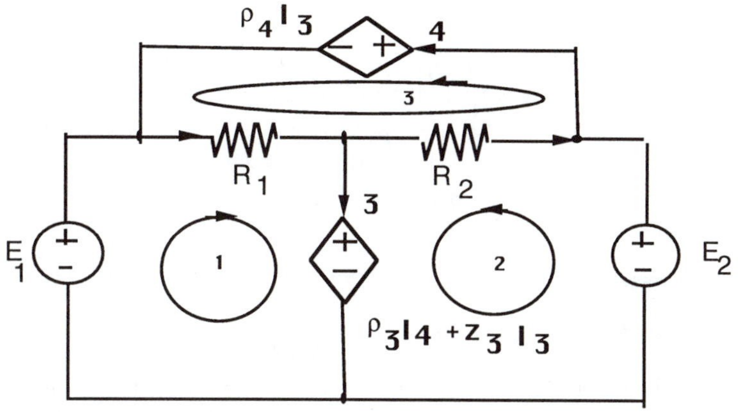

FIGURE 2.10.3
3 mesh circuit with 2 coupled branches.

Once the loop currents are found it may readily be shown that the branch currents are determined as (see also eq. (2.9.15))

$$\boldsymbol{I}_b = \boldsymbol{B}\boldsymbol{I}_L \tag{2.10.6}$$

The following example illustrates the use of incidence matrices for a circuit with coupled branches.

Example 2.10.3
For the three mesh circuit of Fig. 2.10.3, use the incidence matrix $\boldsymbol{B}$ for obtaining the mesh impedance matrix.

Solution The mesh-branch incidence matrix $\boldsymbol{B}'$ is given by

$$\boldsymbol{B}' = \begin{pmatrix} 1 & 0 & 1 & 0 \\ 0 & -1 & 1 & 0 \\ 1 & 1 & 0 & 1 \end{pmatrix}$$

For example, mesh 2 is incident on branch 2, but mesh and branch are oppositely directed and element [22] of $\boldsymbol{B}'$ is -1; mesh 2 is incident on branch 3 in the same direction, so the [23] element is 1; etc.

The branch impedance matrix $\boldsymbol{Z}_b$ is 4×4 since there are 4 branches. Branches 3 and 4 are coupled. By inspection of Fig. 2.10.3 the branch equations are

$$\boldsymbol{Z}_b\boldsymbol{I}_b = \begin{pmatrix} R_1 & 0 & 0 & 0 \\ 0 & R_2 & 0 & 0 \\ 0 & 0 & z_3 & \rho_3 \\ 0 & 0 & \rho_4 & 0 \end{pmatrix} \begin{pmatrix} I_{b1} \\ I_{b2} \\ I_{b3} \\ I_{b4} \end{pmatrix}$$

Then by eq. (2.10.4)

$$\boldsymbol{Z} = \boldsymbol{B}'\boldsymbol{Z}_b\boldsymbol{B} = \begin{pmatrix} R_1 + z_3 & z_3 & R_1 + \rho_3 \\ z_3 & R_2 + z_3 & -R_2 + \rho_3 \\ R_1 + \rho_4 & -R_2 + \rho_4 & R_1 + R_2 \end{pmatrix}$$

□

2.11 Cut Set Analysis

Mesh analysis focuses on a particular set of KVL equations, those taken around the meshes of a planar network. Loop analysis generalizes this formulation to include any set of linearly independent loops, and is not limited to planar networks. Similarly nodal analysis, though not limited to planar graphs, considers only one of the available means for writing the KCL equations; those written for the nodes of the graph. Writing the nodal KCL equations is particularly convenient, especially since any set of $n-1$ nodal equations (n, the total number of nodes) is linearly independent. However, just as in loop analysis, it is often useful to generalize the formulation of KCL equations at other than the nodes of the graph. This generalization is carried out at *cut sets* and allows the choice of arbitrary node pair potentials as the variables of response, rather than being restricted exclusively to the node to datum voltage of nodal analysis.

DEFINITION 2.11.1 *The* cut set potential *is a voltage (with respect to a reference point) assigned to a cut set surface and its associated cut set of branches. If a directed branch is incident on several cut set surfaces, the polarized branch voltage can be expressed as a linear combination of the cut set voltages. A cut set voltage gets a (+) sign in the sum if its normal points in the same direction as the branch, a (−) sign if oppositely directed.*

The physical visualization of a loop current is as a fictitious circulating current. A branch current is the superposition of such loop currents. Analogously the cut set surface is visualized as a fictitious equipotential surface. All nodes inside a cut set surface are at the cut set potential. The potential of a node enclosed by more than one surface has a potential which is the algebraic superposition of the cut set surface potentials. Since the voltage drop across a branch is the difference in potential between its two terminal nodes, it follows that this voltage is evaluated as given in Definition 2.11.1.

The node to datum potentials are special cases of cut set potentials. Imagine that each of $n-1$ cut set surfaces surrounds a single node and is directed away from it; the cut set potential is then identical to the node to datum potential, and the nodal KCL equations are the same as the cut set KCL equations for the $n-1$ cut sets.

The cut set potentials are the unknown response variables of cut set analysis and play a dual role to the loop currents. In fundamental cut set analysis, the node pair voltage drop across a tree branch is identical to the potential of the one cut set surface upon which it is incident. Thus an appropriate group of desired node pair potentials can be chosen as the response variables.

Cut set analysis proceeds analogously to loop analysis. The KCL equations are written for $n-1$ fundamental cut sets in terms of branch currents. The net cut set independent current source vector $\boldsymbol{J}_C$ is placed on one side of the equation. The branch voltages are introduced using the constitutive V, I relations for the branches to eliminate the branch currents. Then applying Definition 2.11.1, the cut set (node-pair) potentials V_{Ck} replace the branch voltages. The result, after collecting terms in the V_{Ck}, is

$$\boldsymbol{J}_C = \boldsymbol{Y}_C \boldsymbol{V}_C \tag{2.11.1}$$

which should be compared with the dual expression for loop analysis, eq. (2.10.2); $\boldsymbol{Y}_C$ is the cut set or node pair admittance matrix. Note that a current in $\boldsymbol{J}_C$ on the left (source) side of eq. (2.11.1) is (+) if *opposed* in direction to the cut set surface normal.

All the equations of loop analysis, eqs. (2.10.3) to (2.10.6), are transformed by duality into the equations of cut set analysis. This includes, for example, $\boldsymbol{A}_C$, the branch-cut set incidence matrix which is analogous to $\boldsymbol{B}_L$, the branch-loop incidence matrix. Essentially the word "loop" in eq. (2.10.3) is replaced by "cut set surface". Then $\boldsymbol{Y}_C$ can be directly related to the branch admittance matrix, $\boldsymbol{Y}_b$

$$\boldsymbol{Y}_C = \boldsymbol{A}'_C \boldsymbol{Y}_b \boldsymbol{A}_C \tag{2.11.2}$$

The cut set admittance matrix can readily be written by inspection for networks without branch-to-branch coupling by referring to eq. (2.10.1). Accordingly the elements Y_{ij} of the cut set admittance matrix $\boldsymbol{Y}_C$ are given by

$$\begin{aligned} y_{ii} &= \quad \Sigma \text{ all branch admittances incident on cut set surface } i; \\ y_{ij} &= \mu_{ij}\, \Sigma \text{ all branch admittances incident on cut set surfaces } i \text{ and } j. \end{aligned} \tag{2.11.3}$$

$\mu_{ij} = 1$ if the normals to cut sets i and j are pointed along their common incident branches in the same direction;

$\mu_{ij} = -1$ if the normals to cut sets i and j are pointed along their common incident branches in opposite directions;

$\mu_{ij} = 0$ if cut sets i and j have no common branches.

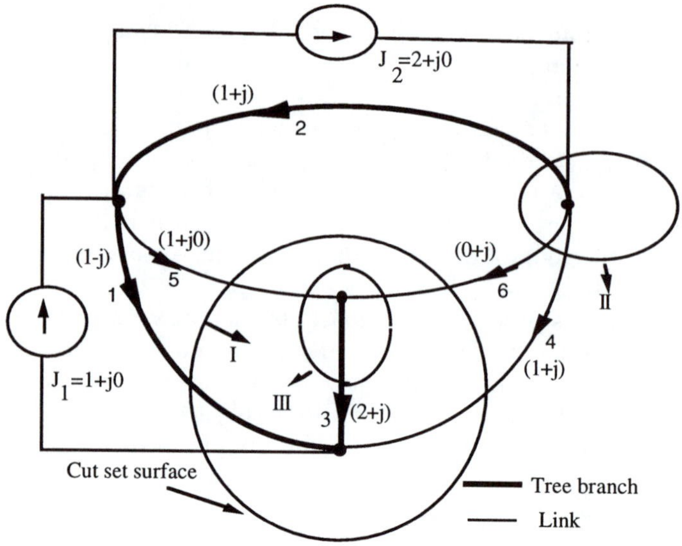

FIGURE 2.11.1
Circuit graph for cut set analysis.

The sign μ_{ij} depends only on the relative direction of the cut set surface normals and does not depend on branch direction.

The following example illustrates cut set analysis.

Example 2.11.1

Refer to the 6 branch, 4 node (rank $\rho = 3$) circuit graph of Fig. 2.11.1.

(a) Determine the cut set admittance matrix for the fundamental cut sets associated with the tree shown on the graph.

(b) For the branch admittances and independent sources indicated, find (numerically) the cut set potentials and the voltage drops across the 6 directed branches.

Solution

(a) For the three cut set surfaces (I,II,III), as shown on Fig. 2.11.1, one and only one tree branch (i.e., branches 1,2,3) is incident per cut set surface. The remaining branches incident on each cut set surface are in the cotree, so the cut sets formed are fundamental. Use eq. (2.11.3) to write the cut set admittance matrix by inspection. For example, cut set surfaces 1 and 3 have their normals oriented in opposite directions, and branches 5 and 6 are incident on both; accordingly, the (1,3) element is $-(y_5 + y_6)$. Cut set surface III has branches 3, 5, 6 incident (these branches constitute cut set III) so the (3,3) element is $(y_3 + y_5 + y_6)$, etc.

$$\boldsymbol{Y}_C = \begin{pmatrix} y_1 + y_4 + y_5 + y_6 & y_4 + y_6 & -(y_5 + y_6) \\ y_4 + y_6 & y_2 + y_4 + y_6 & -y_6 \\ -(y_5 + y_6) & -y_6 & y_3 + y_5 + y_6 \end{pmatrix}$$

The reader may wish to verify the above result using eq. (2.11.2) and the incidence matrix $\mathbf{A}_C$. For the graph of Fig. 2.11.1 with three fundamental cut sets and six branches

$$\boldsymbol{A}'_C = \begin{pmatrix} 1 & 0 & 0 & 1 & 1 & 1 \\ 0 & 1 & 0 & 1 & 0 & 1 \\ 0 & 0 & 1 & 0 & -1 & -1 \end{pmatrix}$$

For instance, branch 5 is incident on cut set surface III, but opposite in direction to the normal; thus the (3,5) element of $\boldsymbol{A}'_C$ is -1, etc.

(b) The matrix cut set equation (2.11.1) with numerical values, as given on Fig. 2.11.1, substituted for the admittances of $\boldsymbol{Y}_C$, is

$$\boldsymbol{J}_C = \begin{pmatrix} 1 \\ 2 \\ 0 \end{pmatrix} = \begin{pmatrix} 3+j & 1+j2 & -1-j \\ 1+j2 & 2+j3 & -j \\ -1-j & -j & 3+j2 \end{pmatrix} = \begin{pmatrix} V_{C1} \\ V_{C2} \\ V_{C3} \end{pmatrix} = \boldsymbol{Y}_C \boldsymbol{V}_C$$

Here, as mentioned earlier, the sources, J_{C1}, J_{C2} take a (+) sign since they are opposed to the cut set surface normals. The V_{Ck} are the cut set potentials and are readily computed by inverting $\boldsymbol{Y}_C$. Thus

$$\boldsymbol{V}_C = \boldsymbol{Y}_C^{-1} \boldsymbol{J}_C = \begin{pmatrix} -0.0252 - j0.0420 \\ 0.3529 - j0.4118 \\ 0.1429 \end{pmatrix}$$

The six branch voltages are given by (see eq. (2.10.6)) for the dual relation)

$$\boldsymbol{V}_b = \boldsymbol{A}_C \boldsymbol{V}_C = \begin{pmatrix} -0.0252 - j0.0420 \\ 0.3529 - j0.4118 \\ 0.1429 \\ 0.3277 - j0.4538 \\ -0.1681 - j0.0420 \\ 0.1849 - j0.4538 \end{pmatrix}$$

The first three branch voltages, those of the tree, coincide with the cut set potentials.

□

2.12 Transfer Functions and n-Ports

The transfer function is a basic constituent of circuit operation. It generally depends on frequency and focuses on the input-output response(s) needed for specific design purposes. The system is presumed to be operating in an eigenmode. Initially assume a single input, then the transfer function is the ratio of voltage or current phasor at some selected output point in the system to the independent input signal source phasor. Thus

$$H(s) \equiv \text{transfer function} = \frac{\text{output response phasor}}{\text{input signal source phasor}} \tag{2.12.1}$$

The computation of $H(s)$ may be carried out by solving the network analysis equations, e.g., nodal or mesh equations, though in specific cases a direct calculation may suffice. $H(s)$ is either dimensionless (voltage or current ratio) or has the dimensions of impedance or admittance.

Under eigenmode operation, consider the relation between sources and responses in terms of the eigenvalue matrix assumed square and nonsingular almost everywhere (a.e.), i.e., except for a denumerable number of discrete values of s. Then if $\boldsymbol{x}, \boldsymbol{y}$ are, respectively, the vectors of input and output phasors

$$\boldsymbol{x}(s) = \boldsymbol{\Lambda}(s)\boldsymbol{y}(s)$$

If we designate $\boldsymbol{H} = \boldsymbol{\Lambda}^{-1}$, then

$$\boldsymbol{y}(s) = \boldsymbol{\Lambda}^{-1}(s)\boldsymbol{x}(s) = \boldsymbol{H}(s)\boldsymbol{x}(s) \tag{2.12.2}$$

Designate the elements of $\boldsymbol{H}$ as h_{ij}, i.e., $\boldsymbol{H} = (h_{ij})$. Then, using the rule for matrix inverse

$$h_{ij}(s) = \frac{M_{ji}(s)}{\Delta(s)}$$

where $\Delta(s) = \det \boldsymbol{\Lambda}(s)$, $M_{ji}(s)$ is the (ji)-th *cofactor* or signed minor of $\boldsymbol{\Lambda}$

$$M_{ji} = (-1)^{i+j} A_{ji}$$

and A_{ji} is the resulting determinant (*minor*) formed by deleting row j and column i from $\boldsymbol{\Lambda}$.

The solution for the elements y_k of $\boldsymbol{y}$ obtained from eq. (2.12.2) when $\boldsymbol{\Lambda}$ is an $n \times n$ matrix is

$$y_k = \sum_{j=1}^{n} h_{kj} x_j = \sum_{j=1}^{n} \frac{M_{jk}}{\Delta} x_j, \qquad k = 1, 2, \ldots, n \tag{2.12.3}$$

Suppose all the input forcing terms x_j except the i-th are set to zero. Then

$$y_k = h_{ki} x_i = \frac{M_{ik}}{\Delta} x_i, \qquad x_j = 0,\ j \neq i$$

The h_{ki} are the transfer functions, $H(s)$, defined by eq. (2.12.1), and eq. (2.12.3) expresses a superposition of the effects of a multiplicity of inputs on the output response; clearly a consequence of linearity. The individual transfer functions taken from the above equation are

$$h_{ij}(s) = \frac{M_{ji}(s)}{\Delta(s)}$$

In an important class of LTI systems an additional physical property is satisfied, namely that the each of the interconnected individual impedance elements of the system are constant with frequency or are proportional to either s or $1/s$. Such a system is *lumped* corresponding to a physical configuration in which the components are very large compared to the wavelength of the excitation so that the resultant electromagnetic fields can be presumed concentrated in discrete "lumps" surrounding each circuit element. (There are also *distributed* systems in which this assumption is not satisfied.) Since only a finite number of arithmetic operations are involved in computing the transfer functions, i.e., sums, products, and quotients of constants, s, and $1/s$, it is clear that for lumped systems these transfer functions will be rational.

Each transfer function $h_{ij}(s)$ is therefore a ratio of polynomials $M_{ji}(s)$ and $\Delta(s)$. The zeros of the numerator correspond to complex (angular) frequencies at which an input at port j yields no output at port i; they are *transmission zeros.* The zeros of the denominator (or poles of the function) correspond to complex (angular) frequencies at which an output at access terminal pair or port i may appear with no input at port j; they are *natural frequencies.* For an n-port (i.e., n-access terminal pairs; a formal definition of "n-port" is given later) satisfying the reality postulate, both numerator and denominator polynomials have real coefficients and, therefore, in a "real" system the transmission zeros as well as the natural frequencies either are real or occur in conjugate complex pairs.

When a transfer function is used to compute the eigenfunction response, Theorem 2.5.2 (and the special case of Theorem 2.5.3) must be kept in mind, namely that the eigenfunction response is valid only for exponential forcing functions ϵ^{st}, when $\Re s > \sigma_o$. (σ_o is the abscissa of convergence.) That is, the domain of $H(s)$ must be to the right of the farthest right denominator root of $h_{ij}(s)$ as plotted in the complex s plane.

We summarize the properties of a general (rational) transfer function $H(s)$ for lumped systems in the following theorem.

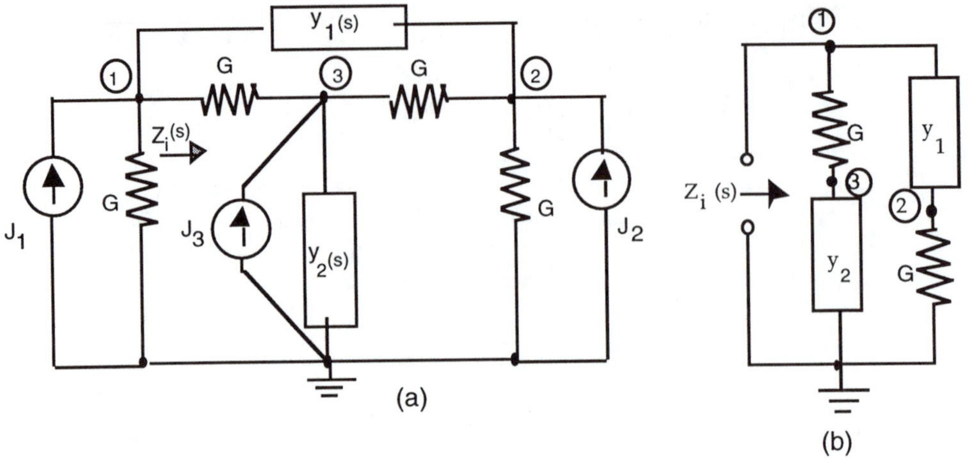

FIGURE 2.12.1
(a) Bridged Tee: $y_1 = 1/y_2$, $J_2 = J_3 = 0$. (b) Branch 2-3 removed.

THEOREM 2.12.1
If an LTI system is lumped and satisfies reality, then $H(s)$ is a rational function with real coefficients, whose zeros (transmission zeros) *and poles* (natural frequencies) *are either real or occur in conjugate complex pairs. For eigenfunction operation the complex excitation frequency must satisfy $\Re s > \sigma_o$, the abscissa of convergence, defined by the farthest right natural frequency as plotted in the complex s-plane.*

Example 2.12.1
Consider the *Bridged Tee* of Fig. 2.12.1 (a), under eigenmode operation, with $y_2(s) = 1/y_1(s)$, $J_2 = J_3 = 0$. Take $G = 1$.

(a) Find the input impedance $z_i(s)$ and the transfer function $H(s) = V_2(s)/V_1(s)$.

(b) Find $H(s)$ when $y_1(s)$ is the input admittance of the circuit consisting of $L = 1$, $C = 1$, in parallel, and determine the abscissa of convergence for eigenmode operation.

(c) How do $H(s)$ and z_i change if all admittances of Fig. 2.12.1 are multiplied by G?

Solution

(a) The nodal admittance matrix $\boldsymbol{Y}(s)$ for the bridged-T can be written by inspection using eq. (2.9.5).

$$\boldsymbol{Y}(s) = \begin{pmatrix} 2+y_1(s) & -y_1(s) & -1 \\ -y_1(s) & 2+y_1(s) & -1 \\ -1 & -1 & 2+y_2(s) \end{pmatrix}$$

so that

$$\boldsymbol{J}(s) = \begin{pmatrix} J_1(s) \\ J_2(s) \\ J_3(s) \end{pmatrix} = \boldsymbol{Y}(s) \begin{pmatrix} V_1(s) \\ V_2(s) \\ V_3(s) \end{pmatrix}$$

Since we are interested in the transfer function with input at 1 and output at 2, set $J_2 = J_3 = 0$ and assume $J_1 \neq 0$. Solve for V_2, V_3, V_1 using Cramer's rule. (To solve for V_j, replace column j of $\boldsymbol{Y}(s)$ by $\boldsymbol{J}$. Then V_j is the determinant of the resulting matrix divided by Δ.)

$$V_2 = \frac{1}{\Delta} \begin{vmatrix} 2+y_1(s) & J_1 & -1 \\ -y_1(s) & 0 & -1 \\ -1 & 0 & 2+y_2(s) \end{vmatrix} = \frac{-J_1}{\Delta}[-y_1(2+y_2)-1]$$

where $\Delta = \det \boldsymbol{Y}(s)$ and the vertical bars | | designate a determinant. Now using $y_1 y_2 = 1$, we obtain for V_2

$$V_2 = \frac{2J_1}{\Delta}(y_1+1)$$

We can solve for V_3 and V_1 in a similar fashion and obtain

$$V_1 = \frac{J_1}{\Delta}[(2+y_1)(2+y_2)-1] = \frac{2J_1}{\Delta}(2+y_1+y_2)$$

and

$$V_3 = \frac{2J_1}{\Delta}(y_1+1) = V_2$$

Since $V_2 = V_3$, zero current flows in the resistor connecting nodes 2 and 3. The resistor can therefore be open-circuited (or short-circuited) without affecting system operation. The circuit is actually a balanced bridge and the branch from 2 to 3 can be visualized as the galvanometer arm. We can redraw the circuit with this branch removed as in Fig. 2.12.1 (b). The input admittance (when $G = 1$) is then given by $y_1 + 1$ in parallel with $y_2 + 1$. Using $y_1 y_2 = 1$

$$Y_i(s) = \frac{1}{Z_i(s)} = \frac{y_2}{y_2+1} + \frac{y_1}{y_1+1} = \frac{1}{y_1+1} + \frac{y_1}{y_1+1} = 1$$

(The same result is obtained if the resistor is short circuited.) In other words, assuming the inverse relation between y_1 and y_2, with $G = 1$, the impedance z_i is simply a constant resistance of one ohm.

The impedance facing the J_1 source is half unit (two unit resistances in parallel), so the input voltage is $V_1 = J_1/2$. Using our earlier expression for V_1

$$V_1 = \frac{2J_1(2 + y_1 + y_2)}{\Delta} = \frac{J_1}{2}$$

which gives an easy way to find Δ

$$\Delta = 4(2 + y_1 + y_2)$$

We can now evaluate the transfer function $H(s)$

$$H(s) = \frac{V_2(s)}{V_1(s)} = \frac{2J_1(s)}{\Delta}(1 + y_1)\,\frac{2}{J_1} = \frac{1 + y_1}{2 + y_1 + y_2}$$

and substituting $y_1 = 1/y_2$

$$H(s) = \frac{1 + y_2}{y_2^2 + 2y_2 + 1} = \frac{1}{1 + y_2}$$

(b) The admittance of unit capacitance and unit inductance in parallel is $y_1(s) = s + 1/s$, so that $y_2 = 1/y_1 = s/s^2 + 1$. From item (a), the transfer function is therefore

$$H(s) = \frac{1}{1 + y_2} = \frac{s^2 + 1}{s^2 + s + 1}$$

The roots of denominator are the natural frequencies

$$p_{1,2} = -\frac{1}{2} \pm \frac{j\sqrt{3}}{2}$$

and therefore the abscissa of convergence is

$$\sigma_o = -\frac{1}{2}$$

Notice that the transfer function varies with frequency ($s = j\omega$) and has a zero at $s = \pm j$, i.e., at $\omega = \pm 1$. Nevertheless, the input impedance remains a constant resistance at all frequencies.

(c) Since all admittances are multiplied by the same constant G, all voltage ratios are unchanged and, therefore, $H(s)$ is unaffected. On the other hand, any branch admittance, hence that measured across a node pair, must be multiplied by G. Therefore, the impedance z_i is changed from unity to $z_i(s) = 1/G$.

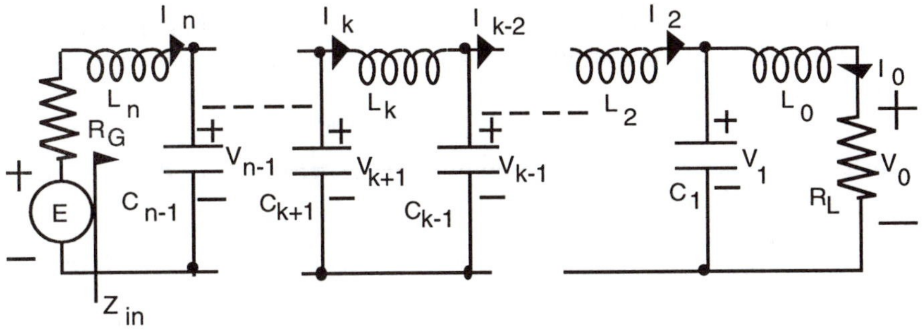

FIGURE 2.12.2
Low-Pass LC ladder; $m = n + 1$ reactive elements.

□

Example 2.12.2
Fig. 2.12.2 shows a *low-pass LC ladder circuit* with $m = n + 1$ reactive elements.

(a) Show that the transfer function $H(s) = I_0(s)/E(s)$ has the form $H(s) = 1/P_m(s)$ where $P_m(s)$ is a polynomial of degree m in s.

(b) Show that

$$\Re\, Z_{in}(j\omega) \equiv R(\omega) = \frac{1}{Q(\omega^2)} \tag{2.12.4}$$

where $Q(\omega^2)$ is a real, even, nonnegative polynomial in ω^2.

Solution

(a) A direct approach to this problem is simpler than mesh or nodal analysis. Since the circuit is LTI, $H(s)$ is unaffected no matter how we choose the output current phasor. For simplicity let it be a constant I_0, and then determine the associated input voltage $E(s)$. Their ratio gives the transfer function. We first state the following.

If, referring to Fig. 2.12.2, $I_{k-2}/I_0 = P_{k-2}$ and $V_{k-1}/I_0 = P_{k-1}$ where P_{k-2} and P_{k-1} are polynomials of the indicated (subscript) degree, then the voltage and current ratios will have the form $I_k/I_0 = P_k$ and $V_{k+1}/I_0 = P_{k+1}$, i.e., polynomials raised in degree by 2 corresponding to stepping past an inductor-capacitor pair.

For the proof of this induction property, we begin with I_{k-2} and V_{k-1} in Fig. 2.12.2. By inspection of the figure

$$I_k = I_{k-2} + sC_{k-1}V_{k-1}$$

By hypothesis I_{k-2}/I_0 and V_{k-1}/I_0 are polynomials of degrees indicated by their subscripts. Then

$$I_k = P_{k-2}I_0 + sC_{k-1}P_{k-1}I_0 = P_kI_0 \tag{2.12.5}$$

a polynomial in s of degree k, since it is the sum of polynomials of degree $k-2$ and k. Similarly

$$V_{k+1} = sL_kI_k + V_{k-1} = sL_kP_kI_0 + P_{k-1}I_0$$

a polynomial of degree $k+1$ in s and the induction hypothesis is proved. Referring to Fig. 2.12.2 at the load side of the ladder

$$I_0 = I_0 \qquad V_1 = sL_0I_0$$

polynomials of degree 0, 1, and 2 in s, respectively. For the ladder shown in Fig. 2.12.2, apply induction as we move down the ladder from inductor to inductor. At each step the current polynomial degree increases by 2, so that at the generator I_n/I_0 is a polynomial of degree n. Similarly starting with V_1, a polynomial of degree 1, the voltage V_{n-1}/I_0 is a polynomial of degree $n-1$. Therefore

$$\begin{aligned} E(s) &= (R_G + sL_n)I_n + V_{n-1} = (R_GP_n + sL_nP_n + P_{n-1})I_0 \\ &= P_{n+1}(s)I_0 \end{aligned}$$

Thus, we have

$$H(s) = \frac{I_0}{E(s)} = \frac{1}{P_m(s)}$$

corresponding to $m = n+1$ ladder elements, which is the required result. Evidently the result is the same whether the ladder starts and/or finishes with a series L or shunt C element. Also, the polynomial property holds for ladder transfer functions V_0/E, I_0/I_n which are dimensionless, or for V_0/I_n, I_0/E whose dimensions are (transfer) impedance and admittance.

(b) The average power delivered to the load R_L, when $s = j\omega$, is

$$P_L(\omega) = R_L|I_0|^2$$

But, since the ladder is purely reactive (no dissipation), the power to the load is equal to the input power

$$P_{in}(\omega) = P_L(\omega) = \Re\, Z_{in}(j\omega)\, |I_n(j\omega)|^2$$

Thus, using eq. (2.12.5)

$$\Re Z_{in}(j\omega) = R_L \left| \frac{I_0}{I_n(j\omega)} \right|^2 = \frac{R_L}{P_n(j\omega)P_n(-j\omega)}$$

Since P_n is a polynomial in $s = j\omega$ with real coefficients, it can be written as the sum of its even powers, $e(\omega)$, and odd powers $jo(\omega)$, e and o both real. Then,

$$P_n(j\omega)P_n(-j\omega) = [e(\omega) + jo(\omega)][e(\omega) - jo(j\omega)] = e^2(\omega) + o^2(\omega)$$

Since e is even and o is odd, their squares are both even and real and the sum is an even, real, positive polynomial. We may therefore write

$$R(\omega) \equiv R(\omega^2) = \frac{R_L}{P_n(j\omega)P_n(-j\omega)} = \frac{1}{Q(\omega^2)} \geq 0$$

which is eq. (2.12.4). □

Of particular interest, once the transfer function is determined, is the *frequency response.* This is the amplitude and phase of $H(s)$ on the $j\omega$ axis. That is, the frequency response is specified as

$$H(j\omega) = |H(j\omega|\ \epsilon^{j\phi(\omega)}$$

(The phase response is often defined as $-\phi(\omega)$, but, for the moment, it will be less confusing if we stick to the plus sign.)

The frequency response is a direct expression of the signal processing properties of the LTI circuit. Clearly for $H(j\omega)$ to exist, the system must operate in an eigenmode when $s = j\omega$. A causal unstable system with natural frequencies in the right half of the s plane or with multiplicity greater than 1 on the $j\omega$ axis does not satisfy this criterion (Theorems 2.5.2 and 2.5.3), so it is necessary that a system be stable to possess a frequency response.

Example 2.12.3
Find the frequency response of the Bridged-T, Example 2.12.1 (b), Fig. 2.12.1. The arm y_1 is a parallel resonant circuit $L = 1$, $C = 1$, $G = 1$.

Solution The transfer function as computed in Example 2.12.1 (b) is

$$H(s) = \frac{1}{1 + y_2} = \frac{s^2 + 1}{s^2 + s + 1}$$

The system is stable and the frequency response is

$$H(j\omega) = \frac{-\omega^2 + 1}{-\omega^2 + j\omega + 1} = |H(j\omega)|\epsilon^{j\phi(\omega)}$$

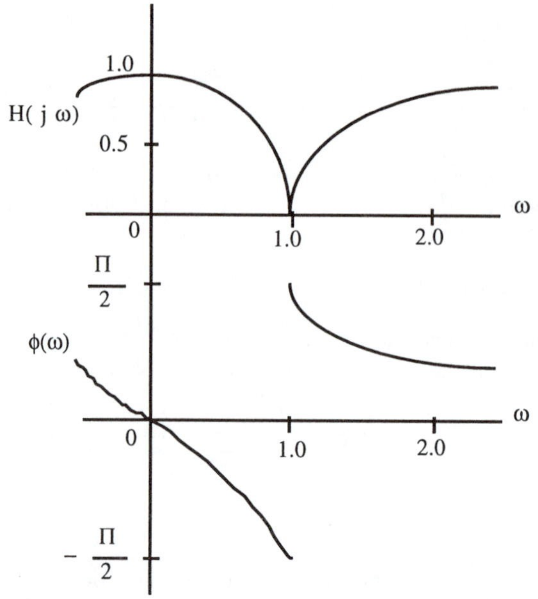

FIGURE 2.12.3
Frequency response of Bridged-T filter.

where

$$|H(j\omega)| = \frac{|1-\omega^2|}{\sqrt{(1-\omega^2)^2+\omega^2}}, \quad \phi(\omega) = -\arctan\frac{\omega}{1-\omega^2} + \alpha(\omega)\pi$$

and

$$\alpha(\omega) = 0, \ -1 < \omega < 1; \quad \alpha(\omega) = 1, \ 1 \leq \omega; \quad \alpha(\omega) = -1, \ -1 \geq \omega$$

Fig. 2.12.3 shows plots of the amplitude and phase response, which are, respectively, even and odd in ω. The signal frequency $\omega = 1$ is totally eliminated at the output and neighboring frequencies attenuated. Nevertheless, the input remains matched to unit resistance (Example 2.12.1 (a)). The frequency response is that of a simple *constant resistance filter.* □

Generally, circuit design involves the response of a network at one or more terminal pairs which function as input or output access ports to a fixed structure. This is illustrated in Examples 2.12.1, 2.12.2, and 2.12.3, where the focal points for system operation are across two distinct node pairs, (1-n, 2-n) in Fig. 2.10.3 (a), the input to inductor L_n, and the output of L_0 in Fig. 2.11.1.

DEFINITION 2.12.1 *An n-*port network *is a circuit structure in which only specified terminal pairs are accessible for applying input signal sources* *or extracting output signals, or for connecting other components. These accessible terminal pairs are* ports *to a fixed structure and constitute the node pairs available for observing voltage and/or current.*

An n-port may have a large number of nodes and branches making up its internal structure but relatively few accessible ports and its operation is defined by properties measured at the ports. A schematic diagram for an n-port is shown in Fig. 2.12.4. Two useful (but not exclusive) means for characterizing an n-port are its *Open Circuit Impedance Matrix* $\boldsymbol{Z}(s)$, and its *Short Circuit Admittance Matrix*, $\boldsymbol{Y}(s)$. If the admittance formulation is used, then, referring to Fig. 2.12.4

$$\boldsymbol{I} = \boldsymbol{Y}\boldsymbol{V} \tag{2.12.6}$$

or

$$\begin{pmatrix} I_1 \\ I_2 \\ \vdots \\ I_n \end{pmatrix} = \begin{pmatrix} y_{11} & y_{12} & \cdots & y_{1n} \\ y_{21} & y_{22} & \cdots & y_{2n} \\ \vdots & \vdots & \cdots & \vdots \\ y_{n1} & y_{n2} & \cdots & y_{nn} \end{pmatrix} \begin{pmatrix} V_1 \\ V_2 \\ \vdots \\ V_n \end{pmatrix} \tag{2.12.7}$$

The matrix elements can be determined by short circuiting appropriate ports and exciting others. To evaluate y_{12}, for example, refer to Fig. 2.12.4 and place short circuits across all ports except port 2, which is to be excited with V_2. Then, since all the $V_j = 0,\ j \neq 2$, eq. (2.12.7) tells us

$$y_{12} = \left.\frac{I_1}{V_2}\right|_{V_j=0,\, j\neq 2}$$

Each element of the $\boldsymbol{Y}$ matrix can be evaluated in this fashion. In general,

$$y_{jk} = \left.\frac{I_j}{V_k}\right|_{V_j=0,\, j\neq k} \tag{2.12.8}$$

In other words, with the polarity designations of Fig. 2.12.4, place a source at port k, short all other ports, and measure the short circuited port currents. The ratios of eq. (2.12.8) are the y_{jk}.

Similarly in the impedance formalism

$$\boldsymbol{V} = \boldsymbol{Z}\boldsymbol{I} \tag{2.12.9}$$

Comparing eqs. (2.12.6) and (2.12.9)

$$\boldsymbol{Z} = \boldsymbol{Y}^{-1}$$

Instead of short circuit measurements, the elements $\boldsymbol{Z} = (z_{jk})$ are measured with appropriate ports open circuited.

$$z_{jk} = \left.\frac{V_j}{I_k}\right|_{I_j=0,\, j\neq k} \tag{2.12.10}$$

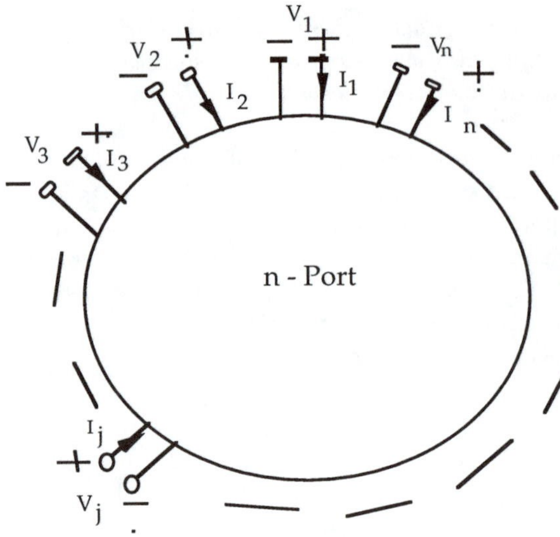

FIGURE 2.12.4
Schematic diagram of an n-port.

In other words, with the polarity designations of Fig. 2.12.4, place a source at port k, open all other ports, and measure the open circuited port voltages. The ratios of eq. (2.12.10) are the z_{jk}.

Example 2.12.4
Refer to Fig. 2.12.1 (a) ($G = 1$), under the conditions of Example 2.12.1, and define ports 1 and 2 across terminal pairs node 1 to datum, and node 2 to datum. J_1 and J_2 are external independent current sources and $J_3 = 0$. Node 3 is not accessible. Find the two-port open circuit impedance matrix.

Solution The results of Example 2.12.1 (a) can be used. Thus, with $J_2 = 0$, (port 2 open), $V_1/J_1 = z_{11} = 1/2$. Also

$$z_{21} = \left.\frac{V_2}{J_1}\right|_{J_2=0} = \left.\frac{V_2 z_{11}}{V_1}\right|_{J_2=0} = \frac{H(s)}{2} = \frac{1}{2(1+y_2)}$$

Since the structure of Fig. 2.12.1 (a) is symmetric with respect to its two ports, $z_{11} = z_{22}$. Furthermore, the circuit is reciprocal so that $z_{12} = z_{21}$. The two-port impedance matrix is therefore

$$\boldsymbol{Z}(s) = \begin{pmatrix} 1/2 & \dfrac{1}{1+y_2(s)} \\ \dfrac{1}{1+y_2(s)} & 1/2 \end{pmatrix}$$

Once the $\boldsymbol{Z}$ and/or $\boldsymbol{Y}$ n-port matrices are known, it is a simple matter to compute the various transfer functions associated with the ports viewed as inputs or outputs to the system. □

2.13 Incidence Matrices and Network Equations

Incidence matrices can provide special insights into the properties of networks. Furthermore, they are useful tools for writing general computer programs which perform automated analysis of electric circuits. The following discussion describes how the incidence matrices introduced in Sections 2.9, 2.10, and 2.11 can be used to express KCL and KVL analysis.[7]

On a network, $\mathcal{N}$ branches may have any complexity, the only restriction being that each branch be two-terminal. However, coupling between branches is allowed. Usually (but not always) each branch is identified with a single circuit element, e.g., R, L, C, or the primary or secondary of a pair of coupled coils, or a controlled source, to transfer all the network complexity into the connections. We denote the branch voltage and current vectors as $\boldsymbol{V}_b$ and $\boldsymbol{I}_b$ whatever the specific nature of the branches.

On the graph $\mathcal{G}$ associated with $\mathcal{N}$, we choose any set of ν independent loops (e.g., the fundamental loops). Then the KVL equation for the j-th loop has the form $\sum_i b_{ji} v_{bj} = 0$, where b_{ji} is $+1, -1, 0$ depending on whether branch i and loop j are oriented in the same or opposite directions, or the branch is not incident on the loop. Referring to eq. (2.10.3), it is clear that the b_{ji} are loop-branch incidence coefficients and a complete set of KVL equations for $\mathcal{G}$ is

$$\boldsymbol{B}'\boldsymbol{V}_b = \boldsymbol{0} \tag{2.13.1}$$

and $\boldsymbol{B}$ is the branch-loop incidence matrix.

We can formalize the loop current concept discussed in Section 2.11 by noting that the j-th branch current i_{bj} may be expressed in terms of the loop currents i_{Li} as $i_{bj} = \sum_i c_{ij} i_{Li}$, where c_{ij} is ± 1 or 0 depending on whether loop i is incident in the same sense, opposite sense, or not at all on branch j. For graph $\mathcal{G}$, with all branch and loop numbering and orientation retained, it is evident that the $c_{ij} \equiv b_{ij}$ and

$$\boldsymbol{I}_b = \boldsymbol{B}\boldsymbol{I}_L \tag{2.13.2}$$

Dual results for cut set analysis are

$$\boldsymbol{A}'_C \boldsymbol{I}_b = \boldsymbol{0} \tag{2.13.3}$$

[7]Most of this section applies to general time dependent voltage and current vectors, but in the discussion of average power and energy, the voltages and currents are phasors.

$$\boldsymbol{V}_b = \boldsymbol{A}_C \boldsymbol{V}_C \tag{2.13.4}$$

where $\boldsymbol{A}_C$ is the branch-cutset incidence matrix and $\boldsymbol{V}_C$ is the cutset potential vector. In the above, the subscripts "b","L", and "C" refer to "branch", "loop", and "cut set", respectively.

We now prove the following theorem, relating matrices $\boldsymbol{A}_C$ and $\boldsymbol{B}$ in the case of fundamental cut sets and loops.

THEOREM 2.13.1

Given an oriented and numbered graph $\mathcal{G}$, choose a tree $\mathcal{T}$ whose cotree is $\mathcal{L}$. The associated incidence matrices are $\boldsymbol{B}$ for the fundamental loops which each contain a unique branch of $\mathcal{L}$, and $\boldsymbol{A}_C$ for the fundamental cutsets which each contain a unique branch of $\mathcal{T}$. Then,

$$\boldsymbol{A}'_C \boldsymbol{B} = \boldsymbol{0} \tag{2.13.5}$$

PROOF Since each fundamental loop consists of a single link and some branches of the tree, by KVL the link voltages (subscript "l") are linear combinations of the tree voltages (subscript "t"), $\boldsymbol{V}_l = \boldsymbol{F}\boldsymbol{V}_t$, where the matrix $\boldsymbol{F}$ defining the linear combination has dimensions "nullity $\times$ rank" ($\nu \times \rho$). Thus, $\boldsymbol{V}_l - \boldsymbol{F}\boldsymbol{V}_t = \boldsymbol{0}$, (the branch voltage vector is $\boldsymbol{V}_b = (\boldsymbol{V}_t \ \boldsymbol{V}_l)'$), and with $\boldsymbol{I}_\rho$ the ρ-columned identity matrix

$$(-\boldsymbol{F} \ \boldsymbol{I}_\rho) \begin{pmatrix} \boldsymbol{V}_t \\ \boldsymbol{V}_l \end{pmatrix} = \boldsymbol{0}$$

Referring to eq. (2.13.1), it is clear that

$$\boldsymbol{B} = \begin{pmatrix} -\boldsymbol{F}' \\ \boldsymbol{I}_\rho \end{pmatrix}$$

If we now apply eq. (2.13.2), we have $-\boldsymbol{F}'\boldsymbol{I}_l = \boldsymbol{I}_t$; hence, $(\boldsymbol{I}_\rho \ \boldsymbol{F}')(\boldsymbol{I}_t \ \boldsymbol{I}_l)' = \boldsymbol{0}$. Comparing this with eq. (2.13.3), $\boldsymbol{A}'_C = (\boldsymbol{I}_\rho \ \boldsymbol{F}')$, the final result is

$$\boldsymbol{A}'_C \boldsymbol{B} = (\boldsymbol{I}_\rho \ \boldsymbol{F}') \begin{pmatrix} -\boldsymbol{F}' \\ \boldsymbol{I}_\rho \end{pmatrix} = -\boldsymbol{F}' + \boldsymbol{F}' = \boldsymbol{0} \qquad \blacksquare$$

We now present a general formalism for network equations. The equations of *unconnected* branches can be written in the general form

$$\begin{aligned} \boldsymbol{I}_{b1} + \boldsymbol{J}_{b1} &= \boldsymbol{Y}_{b1}\boldsymbol{V}_{b1} + \boldsymbol{H}_b \boldsymbol{I}_{b2} \\ \boldsymbol{V}_{b2} + \boldsymbol{E}_{b2} &= \boldsymbol{K}_b \boldsymbol{V}_{b1} + \boldsymbol{Z}_{b2}\boldsymbol{I}_{b2} \end{aligned} \tag{2.13.6}$$

where branches 1 are called *shunt* branches and branches 2 *series* branches because in the first case the independent source currents $\boldsymbol{J}_{b1}$ are in shunt with the admittance elements, $\boldsymbol{Y}_{b1}$, with response currents $\boldsymbol{I}_{b1}$, while in the second case the voltage sources $\boldsymbol{E}_{b2}$ are in series with the impedance elements, with response voltages $\boldsymbol{V}_{b2}$. Matrix $\boldsymbol{Z}_{b2}$ includes resistor, inductor, and capacitor impedances on its main diagonal; nondiagonal terms represent (symmetric) magnetic couplings and (generally unsymmetric) current controlled voltage sources (CCVS). Matrix $\boldsymbol{Y}_{b1}$ includes resistor, inductor, and capacitor admittances on its main diagonal; nondiagonal terms still represent (symmetric) magnetic couplings and (generally unsymmetric) voltage controlled current sources (VCCS). Matrix $\boldsymbol{K}_b$ takes into account voltage controlled voltage sources (VCVS) and possibly magnetic couplings; matrix $\boldsymbol{H}_b$ describes current controlled current sources (CCCS) and possibly magnetic couplings. Thus, the general branch includes an immittance (including coupled coil elements), a controlled source, and an independent source, the latter conventionally polarized opposite to the designated branch direction. Specific branches can omit one or more of these components. The branch current and voltage vectors $(\boldsymbol{I}_{b1}, \boldsymbol{V}_{b2})$ are measured with respect to the *entire* branch. An example of polarity conventions for a type 2 (series) branch is shown on Fig. 2.13.1. A type 1 branch may be treated by duality.

The case of an ideal n-winding ideal transformer admits the representation (2.13.6) with $\boldsymbol{Z}_{b2} = 0$, $\boldsymbol{Y}_{b1} = 0$, and $\boldsymbol{K}_b = -\boldsymbol{H}'_b$. When all the windings are threaded by a single common magnetic flux (the most common case), the sum of the ampere turns over all windings is zero (this fact defines $\boldsymbol{H}_b$), and the voltage per turn is the same for each winding (defines $\boldsymbol{K}_b$). In this case any one of the windings is placed in shunt branch set 1, the remaining windings in series branch set 2, and the dimensions of $\boldsymbol{H}_b$ are $1 \times (n-1)$. When the magnetic core includes parallel as well as series magnetic paths for the flux, the dimensions of $\boldsymbol{H}_b$ are $r \times s$ and the shunt set 1 contains r branches, the series set 2, s branches, $n = r + s$. Ideal voltage sources and short circuits, ideal current sources and open circuits are included as particular cases in branch sets 1 and 2, respectively. Matrices $\boldsymbol{A}'$ and $\boldsymbol{B}'$ can be partitioned according to the two sets of branches 1 and 2

$$\boldsymbol{A}' = (\boldsymbol{A}'_1 \quad \boldsymbol{A}'_2) \tag{2.13.7}$$

$$\boldsymbol{B}' = (\boldsymbol{B}'_1 \quad \boldsymbol{B}'_2) \tag{2.13.8}$$

and can be used to find from eqs. (2.13.6) a set of network equations in the cutset potentials and loop currents.

Two special cases often occur. If the branch set 1 is empty, the equations are those of impedance analysis as in eq. (2.10.4); if the branch set 2 is empty, we reduce to admittance analysis as in eq. (2.11.2).

The most important case uses as variables the node potentials for the branch set 1 and the branch currents for the branch set 2. We then derive the KCL equations by premultiplying the equations of branch set 1 times

$\boldsymbol{A}_1'$ and obtain KVL by expressing all branch voltages in terms of the nodal potentials

$$\begin{aligned} \boldsymbol{A}_1'\boldsymbol{I}_{b1} + \boldsymbol{A}_1'\boldsymbol{J}_{b1} &= \boldsymbol{A}_1'\boldsymbol{Y}_{b1}\boldsymbol{A}_1\boldsymbol{V} + \boldsymbol{A}_1'\boldsymbol{H}_b\boldsymbol{I}_{b2} \\ \boldsymbol{A}_2\boldsymbol{V} + \boldsymbol{E}_{b2} &= \boldsymbol{K}_b\boldsymbol{A}_1\boldsymbol{V} + \boldsymbol{Z}_{b2}\boldsymbol{I}_{b2} \end{aligned} \tag{2.13.9}$$

We can now eliminate currents $\boldsymbol{I}_{b1}$ by writing eq. (2.13.3) in expanded form using eq. (2.13.7

$$\boldsymbol{A}_1'\boldsymbol{I}_{b1} + \boldsymbol{A}_2'\boldsymbol{I}_{b2} = \boldsymbol{0}$$

and replacing $\boldsymbol{A}_1'\boldsymbol{I}_{b1}$ with $-\boldsymbol{A}_2'\boldsymbol{I}_{b2}$ in the first of eqs. (2.13.9). The result is

$$\begin{aligned} \boldsymbol{A}_1'\boldsymbol{J}_{b1} &= \boldsymbol{A}_1'\boldsymbol{Y}_{b1}\boldsymbol{A}_1\boldsymbol{V} + (\boldsymbol{A}_1'\boldsymbol{H}_b + \boldsymbol{A}_2')\boldsymbol{I}_{b2} \\ \boldsymbol{E}_{b2} &= (\boldsymbol{K}_b\boldsymbol{A}_1 - \boldsymbol{A}_2)\boldsymbol{V} + \boldsymbol{Z}_{b2}\boldsymbol{I}_{b2} \end{aligned}$$

or, with self-evident meaning of the symbols,

$$\begin{aligned} \boldsymbol{J}_1 &= \boldsymbol{Y}_1\boldsymbol{V} + \boldsymbol{H}\boldsymbol{I}_{b2} \\ \boldsymbol{E}_{b2} &= \boldsymbol{K}\boldsymbol{V} + \boldsymbol{Z}_{b2}\boldsymbol{I}_{b2} \end{aligned} \tag{2.13.10}$$

Eqs. (2.13.10) describe one form of the *Modified Node Analysis*, at present the most used technique for d.c. and a.c. automatic analysis. Note that the unknowns are the node potentials, $\boldsymbol{V}$, and a subset of the branch currents, while the forcing terms are the node-to-datum impressed currents and the branch voltage sources. One may wonder why branch currents have not been expressed through loop currents, and correspondingly branch voltage sources have not been replaced by loop voltage sources. The reason is that, in all practical cases, the number of branches in set 2 is smaller (and in many instances, much smaller) than the number of loops. Note also that when only independent current sources are present, and the coupling is limited to magnetic coils and VCCS elements, then eq. (2.13.10) takes the nodal analysis form $\boldsymbol{J}_1 = \boldsymbol{Y}_1\boldsymbol{V}$ discussed in Section 2.9.

Example 2.13.1

Write the Modified Node Equations for the network of Fig. 2.13.1.

Solution The branch matrices are the following

$$\boldsymbol{Y}_{b1} = \begin{pmatrix} G_1 & 0 & 0 & 0 & 0 \\ 0 & G_2 & 0 & 0 & 0 \\ 0 & 0 & G_3 & 0 & 0 \\ 0 & 0 & 0 & 0 & 0 \\ 0 & 0 & 0 & 0 & 0 \end{pmatrix}$$

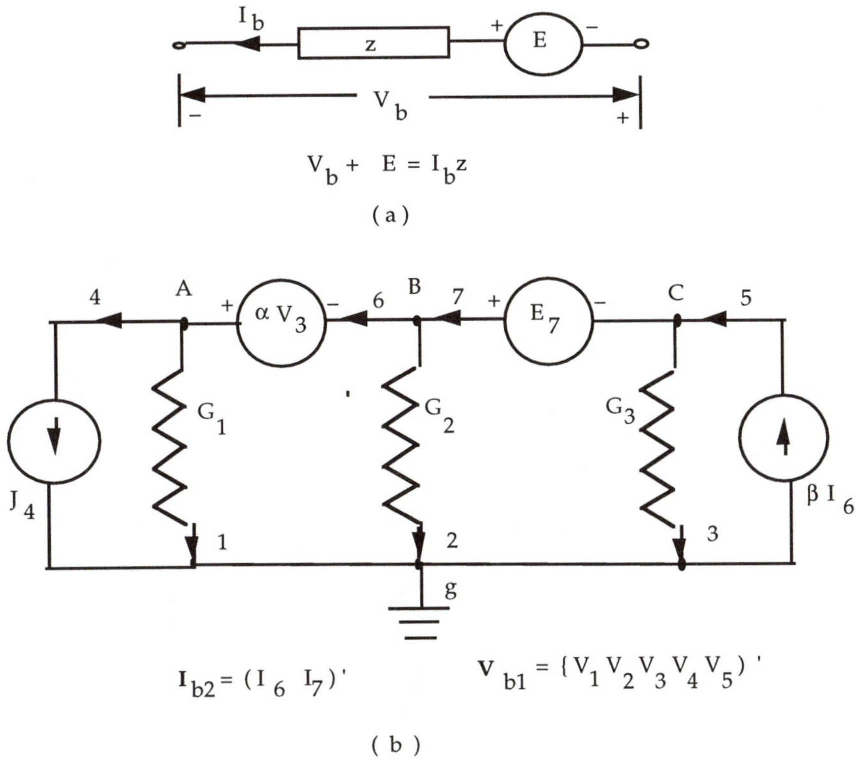

FIGURE 2.13.1
(a) Polarity conventions for series branch. (b) A network to be analyzed by Modified Node Equations.

$$\boldsymbol{H}_b = \begin{pmatrix} 0 & 0 \\ 0 & 0 \\ 0 & 0 \\ 0 & 0 \\ \beta & 0 \end{pmatrix}$$

$$\boldsymbol{K}_b = \begin{pmatrix} 0 & 0 & -\alpha & 0 & 0 \\ 0 & 0 & 0 & 0 & 0 \end{pmatrix}$$

$$\boldsymbol{Z}_{b2} = \begin{pmatrix} 0 & 0 \\ 0 & 0 \end{pmatrix}$$

The forcing vectors are[8]

$$\boldsymbol{J}_b = \begin{pmatrix} 0 \\ 0 \\ 0 \\ -J_4 \\ 0 \end{pmatrix}$$

and

$$\boldsymbol{E}_b = \begin{pmatrix} 0 \\ E_7 \end{pmatrix}$$

The (transposed) branch node incidence matrix is

$$\boldsymbol{A}' = \left(\begin{array}{ccccc|cc} 1 & 0 & 0 & 1 & 0 & -1 & 0 \\ 0 & 1 & 0 & 0 & 0 & 1 & -1 \\ 0 & 0 & 1 & 0 & -1 & 0 & 1 \end{array}\right)$$

By using eqs. (2.13.10), we obtain

$$\begin{pmatrix} -J_4 \\ 0 \\ 0 \\ 0 \\ E_7 \end{pmatrix} = \begin{pmatrix} G_1 & 0 & 0 & -1 & 0 \\ 0 & G_2 & 0 & 1 & -1 \\ 0 & 0 & G_3 & -\beta & 1 \\ 1 & -1 & -\alpha & 0 & 0 \\ 0 & 1 & -1 & 0 & 0 \end{pmatrix} \begin{pmatrix} V_A \\ V_B \\ V_C \\ I_6 \\ I_7 \end{pmatrix}$$

Once the nodal potentials have been computed, the analysis can be completed by computing all branch voltages using the equation $\boldsymbol{V}_b = \boldsymbol{A}\boldsymbol{V}$ and all branch currents by eqs. (2.13.6). Thus the analysis problem is completely solved. □

Network elements are defined in terms of derivatives, but the usual methods for treating circuit problems stem from nodal and mesh analysis and immittance formalisms, which generally lead to integro-differential equations in the time domain and rational fractions in the frequency domain. It is often desirable to choose procedures which yield purely differential equations in the time domain and polynomials in the frequency domain. We now briefly consider some of these techniques.

One way to deal directly with differential equations is to adopt a revised format for Modified Node Analysis. Thus, in branch eqs. (2.13.6) we now include capacitors in the branch set 1 and inductors (including magnetic couplings) in the branch set 2. Using this approach all coefficients of the

[8]The Kirchhoff equations have zero on the left side and branch voltage or current sums on the right. Conventionally the independent source terms are transposed to the left side; then a source term polarized *opposite* to branch direction gets a plus sign, otherwise a minus sign.

variables are either constant or linear functions of s. This property is clearly preserved through the transformations that bring eqs. (2.13.6) into the final modified nodal eqs. (2.13.10), and the resulting matrix is, therefore, *polynomial* in s.

Another way to obtain the same result is to write the branch equations in the form

$$\boldsymbol{M}\boldsymbol{V}_b - \boldsymbol{N}\boldsymbol{I}_b = \boldsymbol{f}$$

which clearly include the equations of all circuit elements so far introduced, with the proviso that inductors and capacitors are represented by the equations $v = Lsi$ and $i = Csv$, respectively, and vector $\boldsymbol{f}$ includes all independent generators. The complete network equations are

$$\begin{aligned} \boldsymbol{A}'\boldsymbol{I}_b &= 0 \\ \boldsymbol{B}'\boldsymbol{V}_b &= 0 \\ \boldsymbol{M}\boldsymbol{V}_b - \boldsymbol{N}\boldsymbol{I}_b &= \boldsymbol{f} \end{aligned}$$

These are termed the *Tableau equations*, have coefficients that either are constant or are linear functions of s, and the coefficient matrix is *polynomial.*

2.14 Tellegen's Theorem, Reciprocity, and Power

There are certain important general properties of an electric network which depend only on the applicability of KVL and KCL regardless of the contents of the branches. These properties are subsumed in *Tellegen's Theorem.* We start by introducing the concept of *compatible* systems of currents and voltages in a network.

DEFINITION 2.14.1 *Branch voltage or current vectors* $\boldsymbol{V}_b$ *or* $\boldsymbol{I}_b$ *are* compatible *if individually they satisfy eq. (2.13.1) (KVL:* $\boldsymbol{B}'\boldsymbol{V}_b = \boldsymbol{0}$*) or eq. (2.13.3) (KCL:* $\boldsymbol{A}'_C\boldsymbol{I}_b = \boldsymbol{0}$*), respectively.*

A compatible branch voltage vector satisfies eq. (2.13.4) and the current vector eq. (2.13.2).

Consider now a network graph $\mathcal{G}$ and associate with it a compatible branch voltage vector $\boldsymbol{V}_{b\alpha}$ and a compatible branch current vector $\boldsymbol{I}_{b\beta}$. Subscripts α and β denote the fact that such vectors are not necessarily related because no branch constitutive equations have been introduced. In a physical situation, $\boldsymbol{V}_{b\alpha}$ could be the branch voltage vector of a network $\mathcal{N}_1$ and $\boldsymbol{I}_{b\beta}$ the current vector of network $\mathcal{N}_2$, different networks but both having the same graph $\mathcal{G}$. Compatible branch voltages and currents on a network graph satisfy the following Theorem.

THEOREM 2.14.1
(Tellegen's Theorem) *Let $\boldsymbol{V}_{b\alpha}$ and $\boldsymbol{I}_{b\beta}$ be compatible branch voltage and current vectors on a network graph $\mathcal{G}$. Then the equation holds*

$$\boldsymbol{V}'_{b\alpha}\boldsymbol{I}_{b\beta} = 0 \tag{2.14.1}$$

PROOF We obtain the result immediately by introducing loop currents and cut set potentials as in eqs. (2.13.2), (2.13.4). Employing eq. (2.13.5) for a set of fundamental loops and a set of fundamental cut sets

$$\boldsymbol{V}'_{b\alpha}\boldsymbol{I}_{b\beta} = \boldsymbol{V}'_{C\alpha}\boldsymbol{A}'_C\boldsymbol{B}\boldsymbol{I}_{L\beta} = \boldsymbol{0}$$

since, according to eq. (2.13.5), $\boldsymbol{A}'_C\boldsymbol{B} = \boldsymbol{0}$. Thus the equation is valid for all compatible branch voltages $\boldsymbol{V}_{b\alpha}$ and currents $\boldsymbol{I}_{b\beta}$. ■

It should be noted that Tellegen's Theorem does not involve the branch constitutive properties, since the key relation eq. (2.13.5) is based only on topology, hence the network structure may be linear, nonlinear, time invariant, or time variable. The voltage and current vectors may be time dependent functions, or time independent phasors, or Fourier or Laplace transforms.

A first remarkable consequence of Tellegen's Theorem is the following corollary.

COROLLARY 2.14.1
In an electric network the sum of the instantaneous powers flowing into the branches is zero, $\sum_h v_h(t)i_h(t) = 0$, summed over all network branches.

PROOF Since the branch voltages and currents, constrained by the constitutive relations, obviously satisfy compatibility, the result follows immediately from eq. (2.14.1). ■

A second, no less remarkable, consequence of Tellegen's Theorem is the following corollary (often referred to in the European literature as Boucherot's Theorem).

COROLLARY 2.14.2
In an LTI electric network operating in the a.c. steady state, both the sum of the active powers and the sum of the reactive powers flowing into the branches are zero.

PROOF With reference to eq. (2.14.1), choose $\boldsymbol{V}_{b\alpha} = \boldsymbol{V}_b$ and $\boldsymbol{I}_{b\beta} = \boldsymbol{I}_b^\dagger$. These vectors satisfy compatibility and their components are assumed to be r.m.s. phasor quantities. Thus, eq. (2.14.1) gives

$$\boldsymbol{I}_b^\dagger \boldsymbol{V}_b = 0$$

Therefore, the sum of the average complex powers flowing into the network branches is zero, and so, separately, are the sums of the active (real part) and of the reactive (imaginary part) powers flowing into the same branches. ∎

The first part of Boucherot's Theorem can be rephrased by stating that given an n-port, the sum of the active powers injected into the network by the sources at the access ports equals the sum of the active powers consumed by the internal branches; this statement is by no means obvious, since the constitutive relations are not used in the proof, hence the Principle of Energy Conservation has not been directly invoked. The second part of the theorem explains why the concept of reactive power is important; reactive power is a quantity that is conserved.

We now apply Boucherot's Theorem to a passive n-port and obtain the following,

THEOREM 2.14.2

Let P and Q be the active and the reactive powers entering a passive n-port. Then the following relations are valid

$$\begin{aligned} P &\geq 0 \\ Q &= 2\omega(\overline{W}_m - \overline{W}_e) \end{aligned} \tag{2.14.2}$$

where $\overline{W}_m$ and $\overline{W}_e$ are the average magnetic energy and the average dielectric energy stored in the inductors, and in the capacitors respectively.

PROOF The voltages and currents are taken as r.m.s. phasors and, for the sake of simplicity, we assume that the internal branches are not coupled magnetically (it can be shown that the theorem applies to this case as well). Thus the structure only contains resistors, inductors, and capacitors. Boucherot's Theorem, in rephrased form, tells us that the sum of the port active powers P must equal the sum of the resistor branch dissipated or active powers; $\sum_h R_h |I_h|^2$ (h taken over the resistive branches). Since the last quantity is nonnegative, so too is P. Again the sum of the port reactive powers Q must equal the sum of the reactive powers into the inductors $\sum_h V_h I_h^* = \sum_h \omega L_h |I_h|^2$ (h taken over the inductive branches) plus the sum of the reactive powers into the capacitors $\sum_h V_h I_h^* = -\sum_h \omega C_h |V_h|^2$

(summed over the capacitive branches). But

$$\sum_h \omega L_h |I_h|^2 = 2\omega \sum_h \frac{1}{2} L_h |I_h|^2 = 2\omega \sum_h \overline{W}_{mh}$$

Analogously

$$-\sum_h \omega C_h |V_h|^2 = -2\omega \sum_h \frac{1}{2} C_h |V_h|^2 = -2\omega \sum_h \overline{W}_{eh}$$

The proof is completed by observing that the average magnetic and dielectric energies stored in the inductors and capacitors, respectively, sum to the total average magnetic energy and the average dielectric energy stored in the n-port. ∎

The real frequencies at which $\overline{W}_m = \overline{W}_e$, i.e., $Q = 0$ are called *resonant frequencies*; they separate the frequency bands in which the n-port is inductive from those in which it is capacitive.

Another important concept connected with compatible voltages and currents is that of *Reciprocity.*

DEFINITION 2.14.2 *A network N is said to be* reciprocal *if for any two systems of branch emf's and impressed currents,* $\boldsymbol{E}_{b\alpha}$, $\boldsymbol{J}_{b\alpha}$ *and* $\boldsymbol{E}_{b\beta}$, $\boldsymbol{J}_{b\beta}$, *the response vectors for the branches* $\boldsymbol{V}_{b\alpha}$, $\boldsymbol{I}_{b\beta}$ *and* $\boldsymbol{V}_{b\beta}$, $\boldsymbol{I}_{b\alpha}$ *satisfy*

$$\boldsymbol{I}'_{b\beta}\boldsymbol{E}_{b\alpha} + \boldsymbol{J}'_{b\beta}\boldsymbol{V}_{b\alpha} = \boldsymbol{I}'_{b\alpha}\boldsymbol{E}_{b\beta} + \boldsymbol{J}'_{b\alpha}\boldsymbol{V}_{b\beta} \tag{2.14.3}$$

taken over all the branches of N.

The definition finds its significance in the following,

THEOREM 2.14.3

(Reciprocity Theorem) *Given an electric network $\mathcal{N}$ which satisfies the following equations (see eqs. (2.13.6)),*

$$\begin{aligned} \boldsymbol{Y}_{b1} &= \boldsymbol{Y}'_{b1} \\ \boldsymbol{Z}_{b2} &= \boldsymbol{Z}'_{b2} \\ \boldsymbol{K}_b &= -\boldsymbol{H}'_b \end{aligned}$$

Then $\mathcal{N}$ is reciprocal, and the converse is also true.

PROOF

We construct from eqs. (2.13.6) the expressions $\boldsymbol{I}'_{b1\beta}\boldsymbol{V}_{b1\alpha}$, $\boldsymbol{I}'_{b2\beta}\boldsymbol{V}_{b2\alpha}$, and

$\boldsymbol{I}'_{b1\alpha}\boldsymbol{V}_{b1\beta}$, $\boldsymbol{I}'_{b2\alpha}\boldsymbol{V}_{b2\beta}$. Thus we have

$$\boldsymbol{I}'_{b1\beta}\boldsymbol{V}_{b1\alpha} + \boldsymbol{J}'_{b1\beta}\boldsymbol{V}_{b1\alpha} = \boldsymbol{V}'_{b1\beta}\boldsymbol{Y}'_{b1}\boldsymbol{V}_{b1\alpha} + \boldsymbol{I}'_{b2\beta}\boldsymbol{H}'_{b}\boldsymbol{V}_{b1\alpha}$$
$$\boldsymbol{I}'_{b2\beta}\boldsymbol{V}_{b2\alpha} + \boldsymbol{I}'_{b2\beta}\boldsymbol{E}_{b2\alpha} = \boldsymbol{I}'_{b2\beta}\boldsymbol{K}_{b}\boldsymbol{V}_{b1\alpha} + \boldsymbol{I}'_{b2\beta}\boldsymbol{Z}_{b2}\boldsymbol{I}_{b2\alpha}$$

and

$$\boldsymbol{I}'_{b1\alpha}\boldsymbol{V}_{b1\beta} + \boldsymbol{J}'_{b1\alpha}\boldsymbol{V}_{b1\beta} = \boldsymbol{V}'_{b1\alpha}\boldsymbol{Y}'_{b1}\boldsymbol{V}_{b1\beta} + \boldsymbol{I}'_{b2\alpha}\boldsymbol{H}'_{b}\boldsymbol{V}_{b1\beta}$$
$$\boldsymbol{I}'_{b2\alpha}\boldsymbol{V}_{b2\beta} + \boldsymbol{I}'_{b2\alpha}\boldsymbol{E}_{b2\beta} = \boldsymbol{I}'_{b2\alpha}\boldsymbol{K}_{b}\boldsymbol{V}_{b1\beta} + \boldsymbol{I}'_{b2\alpha}\boldsymbol{Z}_{b2}\boldsymbol{I}_{b2\beta}$$

The sum of the first and the second equation, on the one hand, and of the third and the fourth, on the other, gives

$$\boldsymbol{J}'_{b1\beta}\boldsymbol{V}_{b1\alpha} + \boldsymbol{I}'_{b2\beta}\boldsymbol{E}_{b2\alpha} = \boldsymbol{V}'_{b1\beta}\boldsymbol{Y}'_{b1}\boldsymbol{V}_{b1\alpha} + \boldsymbol{I}'_{b2\beta}\boldsymbol{H}'_{b}\boldsymbol{V}_{b1\alpha} + \\ + \boldsymbol{I}'_{b2\beta}\boldsymbol{K}_{b}\boldsymbol{V}_{b1\alpha} + \boldsymbol{I}'_{b2\beta}\boldsymbol{Z}_{b2}\boldsymbol{I}_{b2\alpha}$$

and

$$\boldsymbol{J}'_{b1\alpha}\boldsymbol{V}_{b1\beta} + \boldsymbol{I}'_{b2\alpha}\boldsymbol{E}_{b2\beta} = \boldsymbol{V}'_{b1\alpha}\boldsymbol{Y}'_{b1}\boldsymbol{V}_{b1\beta} + \boldsymbol{I}'_{b2\alpha}\boldsymbol{H}'_{b}\boldsymbol{V}_{b1\beta} + \\ + \boldsymbol{I}'_{b2\alpha}\boldsymbol{K}_{b}\boldsymbol{V}_{b1\beta} + \boldsymbol{I}'_{b2\alpha}\boldsymbol{Z}_{b2}\boldsymbol{I}_{b2\beta}$$

The terms $\boldsymbol{I}'_{b1\beta}\boldsymbol{V}_{b1\alpha}$, $\boldsymbol{I}'_{b2\beta}\boldsymbol{V}_{b2\alpha}$, and $\boldsymbol{I}'_{b1\alpha}\boldsymbol{V}_{b1\beta}$, $\boldsymbol{I}'_{b2\alpha}\boldsymbol{V}_{b2\beta}$ cancel in the sum due to Tellegen's Theorem. Taking the difference of the previous two equations after matrix transposition of the terms on the right hand (they are scalar products), we obtain

$$\boldsymbol{J}'_{b1\beta}\boldsymbol{V}_{b1\alpha} + \boldsymbol{I}'_{b2\beta}\boldsymbol{E}_{b2\alpha} - \boldsymbol{J}'_{b1\alpha}\boldsymbol{V}_{b1\beta} - \boldsymbol{I}'_{b2\alpha}\boldsymbol{E}_{b2\beta} =$$
$$= \boldsymbol{V}'_{b1\beta}(\boldsymbol{Y}'_{b1} - \boldsymbol{Y}_{b1})\boldsymbol{V}_{b1\alpha} + \boldsymbol{I}'_{b2\beta}(\boldsymbol{H}'_{b} + \boldsymbol{K}_{b})\boldsymbol{V}_{b1\alpha} - \boldsymbol{I}'_{b2\alpha}(\boldsymbol{H}'_{b} + \boldsymbol{K}_{b})\boldsymbol{V}_{b1\beta} +$$
$$+ \boldsymbol{I}'_{b2\beta}(\boldsymbol{Z}_{b2} - \boldsymbol{Z}'_{b2})\boldsymbol{I}_{b2\alpha}$$

If the conditions of the Theorem 2.14.3 are satisfied, the right hand side of the preceding equation is zero and, therefore, the network is reciprocal according to Definition 2.14.2.

The converse follows since, with reciprocity given, the last equation must be zero for arbitrary values of $\boldsymbol{I}_{b2\alpha}, \boldsymbol{I}_{b2\beta}, \boldsymbol{V}_{b1\alpha}, \boldsymbol{V}_{b1\beta}$. ■

Some important particular cases must be considered.

(a) A network all of whose branches are one-ports is reciprocal; in fact matrices $\boldsymbol{Y}_{b1}$ and/or $\boldsymbol{Z}_{b2}$ are diagonal while $\boldsymbol{H}_b = 0$ and $\boldsymbol{K}_b = 0$.

(b) A network whose branches are magnetically coupled is reciprocal; in fact, matrices $\boldsymbol{Y}_{b1}$ and/or $\boldsymbol{Z}_{b2}$ are symmetric.

(c) Referring to the discussion of ideal transformers after eq. (2.13.6), a multiport circuit element with no sources, $\boldsymbol{Y}_{b1} = 0$, $\boldsymbol{Z}_{b2} = 0$, and $\boldsymbol{K}_b = -\boldsymbol{H}'_b$, where $\boldsymbol{H}_b$ is an $r \times s$ real constant matrix, is reciprocal; it is called an $(r+s)$-port (or winding) ideal transformer with turn-ratio matrix $\boldsymbol{K}_b$; the ports 1 are called "shunt ports", the ports 2 "series ports".

The Reciprocity Theorem equally relates port voltage vectors $\boldsymbol{V}$ and port current vectors $\boldsymbol{I}$ belonging to different states α and β for n-ports whose internal branches are symmetrically coupled.

COROLLARY 2.14.3

(Reciprocity Theorem for n-ports) *For a LTI n-port, whose internal branches are symmetrically coupled and/or are (multiport) ideal transformers, the following equality holds*

$$\boldsymbol{I}'_\beta \boldsymbol{V}_\alpha = \boldsymbol{I}'_\alpha \boldsymbol{V}_\beta \tag{2.14.4}$$

where $\boldsymbol{V}_\alpha$ and $\boldsymbol{V}_\beta$ are the port voltage vectors, $\boldsymbol{I}_\alpha$ and $\boldsymbol{I}_\beta$ the port current vectors corresponding to the two states.

PROOF An LTI n-port can be viewed as a network in which the independent ideal voltage and current sources, (say a total of n, some of which may have emf's or impressed currents which are zero), are designated as accessible ports.[9] The remaining immittance branches are internal to the resulting n-port. Since Tellegen's Theorem is valid for the complete system (ports and internal branches), eq. (2.14.3) still holds. Rewrite the reciprocity relation, this time separating the port branches designated $\boldsymbol{V}, I$ from the internal branches $\boldsymbol{V}_b, \boldsymbol{I}_b$

$$\boldsymbol{I}'_\beta \boldsymbol{V}_\alpha + \boldsymbol{I}'_{b\beta} \boldsymbol{V}_{b\alpha} = \boldsymbol{I}'_\alpha \boldsymbol{V}_\beta + \boldsymbol{I}'_{b\alpha} \boldsymbol{V}_{b\beta}$$

Since, by hypothesis, the internal branches satisfy reciprocity, the second terms on each side of the equation are equal, and eq. (2.14.4) follows. But if reciprocity holds for the ports, the converse is not necessarily valid; in fact, the cancellation of the second terms on each side may occur only globally and not necessarily at the level of groups of branches defining a multiterminal element. ■

[9] The internal portion of the structure is presumed fixed, whereas the ports are available to provide access for voltage or current excitation or may be terminated in passive immittances.

The Corollary 2.14.3 states that for n-ports, internal reciprocity implies port reciprocity while the converse is not generally true.[10]

Example 2.14.1
The gyrator is a basic nonreciprocal circuit element with impedance (resistance) matrix

$$\boldsymbol{R} = \begin{pmatrix} 0 & r \\ -r & 0 \end{pmatrix} \tag{2.14.5}$$

Show that the two-port obtained by cascade connecting two gyrators of resistances r_1 and r_2 is an ideal transformer of ratio r_1/r_2.

Solution From eq. (2.14.5), by denoting as 1 and 3 the ports of the first gyrator and by 3 and 2 those of the second (note that the input port of the second is connected to the output port of the first), we obtain

$$\begin{pmatrix} v_1 \\ i_1 \end{pmatrix} = \begin{pmatrix} 0 & -r_1 \\ -\dfrac{1}{r_1} & 0 \end{pmatrix} \begin{pmatrix} v_3 \\ -i_3 \end{pmatrix}; \quad \begin{pmatrix} v_3 \\ i_3 \end{pmatrix} = \begin{pmatrix} 0 & -r_2 \\ -\dfrac{1}{r_2} & 0 \end{pmatrix} \begin{pmatrix} v_2 \\ -i_2 \end{pmatrix}$$

Thus, by eliminating the *internal* port 3 we have

$$\begin{pmatrix} v_1 \\ i_1 \end{pmatrix} = \begin{pmatrix} 0 & -r_1 \\ -\dfrac{1}{r_1} & 0 \end{pmatrix} \begin{pmatrix} 0 & -r_2 \\ -\dfrac{1}{r_2} & 0 \end{pmatrix} \begin{pmatrix} v_2 \\ -i_2 \end{pmatrix} = \begin{pmatrix} \dfrac{r_1}{r_2} & 0 \\ 0 & \dfrac{r_2}{r_1} \end{pmatrix} \begin{pmatrix} v_2 \\ -i_2 \end{pmatrix}$$

The resulting two-port is port reciprocal since it is an ideal transformer, but is not internally reciprocal since gyrators are nonreciprocal elements.

□

As a result of the previous discussion, the following Corollary is established for n-port impedance and admittance matrices, as well as for loop impedance matrices and cut set admittance matrices. (The last statement follows from the fact that the loop impedance matrix and the cut set admittance matrix can be viewed as the impedance and the admittance matrices

[10] Although satisfaction of the reciprocity relations at the ports does not necessarily imply internal branch reciprocity, nor for that matter does port passivity imply internal branch passivity (for example, a resistor of resistance $R > 0$ in series with a nonpassive resistor of resistance $-r < 0$ with $R > r$ yields an internally nonpassive, but port passive resistor), nevertheless, it is possible to show by a rather involved synthesis technique that an n-port satisfying port reciprocity and passivity may always be realized by a structure all of whose internal branches satisfy reciprocity and passivity. For example, see Y. Oono, "Application of Scattering Matrices to the Synthesis of n-Ports", *IRE Trans. on Circ. Th.*, vol. CT-3, no. 2, pp. 111-120, June 1956.

of n-ports, whose ports are in series with links and in shunt with tree branches, respectively.)

COROLLARY 2.14.4

If a network satisfies reciprocity (Definition 2.14.3, Theorem 2.14.3), the global immittance matrices associated with accessible ports (i.e., nodal, cut set, mesh, loop, n-port), are symmetric. Thus, when there are no internal independent sources, the global impedance and admittance matrices are symmetric whenever $\boldsymbol{K}_b = -\boldsymbol{H}_b'$ *and* $\boldsymbol{Z}_b = \boldsymbol{Z}_b'$ *or* $\boldsymbol{Y}_b = \boldsymbol{Y}_b'$.

We stress the fact that different choices of independent variables can yield different relations expressing reciprocity for their associated matrices when constitutive properties are included. The fundamental description of the reciprocity property, however, remains eq. (2.14.3), often termed *Lorentz reciprocity*, after its famous discoverer, who derived the result using electromagnetic fields.

Since a scalar is a 1×1 symmetric matrix, an obvious consequence of Corollary 2.14.3 is the following,

COROLLARY 2.14.5

Any LTI one-port possessing impedance z *or admittance* y *is reciprocal.*

Example 2.14.2

Discuss the instantaneous and complex power properties of the gyrator.

Solution With reference to eq. (2.14.5), the instantaneous power $p(t)$ entering the two ports is

$$p(t) = v_1(t)i_1(t) + v_2(t)i_2(t) = ri_2(t)i_1(t) - ri_1(t)i_2(t) = 0$$

The active power P, as the average of instantaneous power $p(t)$, is zero, too. However, the reactive power entering the two-port is not necessarily zero. In fact, suppose port 2 is terminated on a capacitor C; let its r.m.s. phasor voltage be v_2 and the current *into* C be $i_2' = -i_2$. Divide the second 2-port equation by the first

$$\frac{-ri_1}{v_1} = \frac{v_2}{-ri_2'} \quad \text{or} \quad y_1 = \frac{1}{r^2 y_2}$$

The gyrator inverts the capacitor load admittance $y_2 = j\omega C$, yielding an inductive input admittance $y_1 = 1/jr^2\omega C$. Then the reactive powers to the input and the load, Q_1, Q_2, are

$$Q_1 = |v_1|^2 y_1 = \frac{|v_1|^2}{r^2 y_2}, \quad Q_2 = |v_2|^2 y_2 = r^2|i_1|^2 y_2 = r^2|v_1|^2|y_1|^2 y_2 = \frac{|v_1|^2}{r^2 y_2^*}$$

and by Boucherot's Theorem the reactive power delivered to the gyrator is

$$Q = Q_1 - Q_2 = \frac{|v_1|^2}{r^2}\left[\frac{1}{y_2} - \frac{1}{y_2^*}\right] = \frac{-2j|v_1|^2}{r^2\omega C}$$

□

Thus the reactive power Q is not directly related to the instantaneous power $p(t)$. The importance of Q lies mainly in its conservation property as expressed by Boucherot's Theorem.

3

Impulses, Convolution, and Integral Transforms

3.1 The Impulse Function

The *impulse* or *Dirac delta* function and the eternal exponential are two mathematical objects which are about as dissimilar a pair as one can conceive, yet they are siblings and play a complementary role in the analysis of systems. The LTI operator eigenfunction ϵ^{st} is continuous and nonzero $-\infty < t < +\infty$. The impulse function $\delta(t)$ is zero everywhere except at the origin, where it has an infinite discontinuity.

The impulse function is not an ordinary function, because actual infinities are not allowed in classical analysis. However, it can be rigorously defined as a linear continuous mapping of the space of real continuous functions $\mathcal{C}^0$ into the set of real numbers $\mathcal{R}$, i.e., as a *functional*[1] or *generalized function.*[2]

We use the following definition of $\delta(t)$.

DEFINITION 3.1.1 *The impulse function $\delta(t)$ is defined as a linear continuous functional on $\mathcal{C}^0$ mapping each function ϕ into its value at the origin $\phi(0)$.*

$$\int_{-\infty}^{+\infty} \delta(t)\ \phi(t)\, dt = \phi(0) \qquad (3.1.1)$$

for all functions $\phi(t) \in \mathcal{C}^0$.

[1]A *functional* is a mapping of a vector space X into the set of complex numbers $\mathcal{C}$. Here we limit ourselves to the case where both the domain and the range of the functional are real.

[2]An excellent account of generalized functions, with applications to electric networks, is found in: A. H. Zemanian, *Distribution Theory and Transform Analysis*, New York, Dover, 1965.

It is stipulated that all operations on the δ function, including its approximation through sequences of ordinary functions, must be understood in the sense of eq. (3.1.1).[3] Thus, for example, an equation like

$$\lim_{n\to\infty} \delta_n(t) = \delta(t)$$

where the $\delta_n(t)$ are ordinary functions, must be understood as a shorthand writing of the equation

$$\lim_{n\to\infty} \int_{-\infty}^{+\infty} \delta_n(t)\,\phi(t)\,dt = \int_{-\infty}^{+\infty} \delta(t)\,\phi(t)\,dt = \phi(0)$$

for all $\phi \in C^0$.

Since $\phi(0) = 0$ for all functions ϕ whose support[4] does not contain the origin (in effect the definition integral only maps the single point $\phi(0)$, regardless of the value of ϕ elsewhere), we say that $\delta(t)$ is zero in the open set $t \neq 0$.

The preceding discussion outlines the mathematical foundations for the impulse function, but pragmatically we can visualize $\delta(t)$ by starting with a rectangular pulse initiated $t = -\Delta/2$ and terminated at $t = +\Delta/2$ (hence with a duration Δ) and with height $1/\Delta$, so the area of the pulse is unity. Then go to the limit as $\Delta \to 0$, and the result can be thought of as a pulse of unit area at the origin, but of infinite amplitude and zero duration. Thus let

$$\delta_\Delta(t) = \begin{cases} 1/\Delta, & \text{if } -\Delta/2 \le t \le \Delta/2 \\ 0, & \text{otherwise} \end{cases} \tag{3.1.2}$$

Then

$$\lim_{\Delta\to 0} \delta_\Delta(t) = \delta(t) \tag{3.1.3}$$

It is clear that the impulse, as the limit of a sequence of even functions, is even, i.e., $\delta(t) = \delta(-t)$, so that eq. (3.1.1) with $\phi = 1$ yields

$$\int_{-\infty}^{+\infty} \delta(t)\,dt = \int_{-\infty}^{+\infty} \delta(-t)\,dt = 1$$

Note that eq. (3.1.1) works as well if, instead of $\phi \in C^0$, we use any $f(t)$ continuous *at the origin.* Thus

$$\int_{-\infty}^{+\infty} f(t)\,\delta(t)\,dt = f(0) = f(0) \int_{-\infty}^{+\infty} \delta(t)\,dt$$

[3]This is called the *weak* sense; a great advantage is that in the weak sense such operations as integration and differentiation of sequences can be freely interchanged without the need of a specific justification.

[4]The *support of a function* is the closure of the set of points at which the function is nonzero.

Since the equation above holds for all $f(t)$ continuous at the origin, we have

$$f(t)\,\delta(t) = f(0)\,\delta(t) \tag{3.1.4}$$

By shifting the origin, it follows that

$$f(t)\,\delta(t - t_0) = f(t_0)\,\delta(t - t_0) \tag{3.1.5}$$

when $f(t)$ is continuous $t = t_0$. The operation in eq. (3.1.5) is often referred to as *impulse sampling.* From eq. (3.1.5), by replacing the constant t_0 with the variable τ and integrating, we obtain

$$\int_{-\infty}^{+\infty} f(\tau)\ \delta(t - \tau)\ d\tau = f(t) \tag{3.1.6}$$

Eq. (3.1.6) expresses the so-called *sifting property.* As we shall shortly see, "sifting", by selecting values of a given function at arbitrary times t, provides a useful representation of the function.

The unit step function defined in Section 1.5 is closely related to the impulse function, since

$$\int_{-\infty}^{t} \delta(\tau)\ d\tau = \begin{Bmatrix} 0, & t < 0 \\ 1, & t > 0 \end{Bmatrix} = \mathrm{u}(t) \tag{3.1.7}$$

The above equation is reasonable in view of the fact that, as pointed out earlier, we can take $\delta(t) = 0, t \neq 0$. Consequently the integral is zero, $t < 0$, and remains at unit value $\forall t > 0$.

If we differentiate both sides of eq. (3.1.7) we have a representation of $\delta(t)$ in terms of $\mathrm{u}(t)$.

$$\delta(t) = \frac{d\,\mathrm{u}(t)}{dt} \tag{3.1.8}$$

This result, as well as the assignment of the value of 1/2 to $\mathrm{u}(0)$ as a "natural" (though not necessary) choice, is consistent with eq. (3.1.7) in view of the following visualization of $\mathrm{u}(t)$.

Fig. 3.1.1 shows a function $u_\Delta(t)$ which is zero, $t < -\Delta/2$, then has a straight line ramp of slope $1/\Delta$ from $t = -\Delta/2$ until $t = \Delta/2$ where the function is 1, and thereafter remains unity. We can write

$$\mathrm{u}(t) = \lim_{\Delta \to 0} u_\Delta(t)$$

Now take the derivative of $\mathrm{u}(t)$ as

$$\frac{d\,\mathrm{u}(t)}{dt} = \lim_{\Delta \to 0} \frac{du_\Delta(t)}{dt}$$

But referring to eq. (3.1.2), and the function $u_\Delta(t)$ as shown in Fig. 3.1.1, it is clear that

$$\delta_\Delta(t) = \frac{du_\Delta(t)}{dt}$$

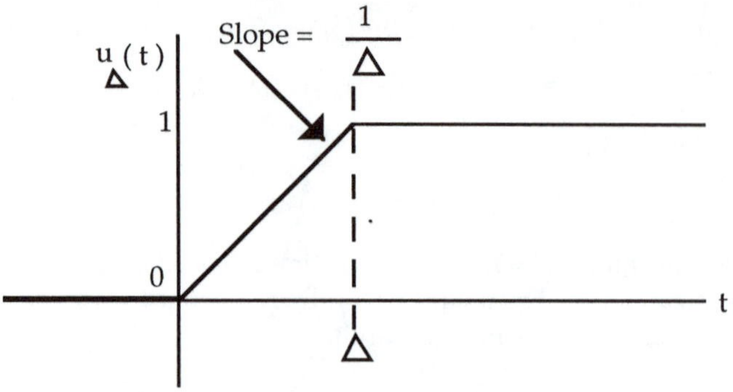

FIGURE 3.1.1
Approximation of u(t), the unit step function.

so that

$$\delta(t) = \lim_{\Delta\to 0} \delta_\Delta(t) = \lim_{\Delta\to 0} \frac{du_\Delta(t)}{dt} = \frac{d\,\mathrm{u}(t)}{dt}$$

Example 3.1.1
Find a representation, using the impulse function, for the derivative of a function $f(t)$ continuous in the interval I: $t_1 \le t \le t_2$, except at the point t_0 in I where there is a discontinuity $f(t_0^+) - f(t_0^-) = b - a$.

Solution Fig. 3.1.2 shows the function. Referring to the figure and noting that, except at $t = t_0$, $f'(t) = f_1'(t)$, the derivative $f'(t)$ in the interval I is

$$f'(t) = f'(t)|_{t\neq t_0} + (b-a)\frac{d\,\mathrm{u}(t-t_0)}{dt}$$

or using eq. (3.1.8)

$$f'(t) = f'(t)|_{t\neq t_0} + (b-a)\delta(t-t_0)$$

□

Example 3.1.2
Verify the following expressions:

(a)

$$\delta(at) = \frac{1}{|a|}\,\delta(t), \quad a \neq 0 \tag{3.1.9}$$

(b)

$$\delta(t^2 - a^2) = \frac{1}{2a}\big(\delta(t-a) + \delta(t+a)\big), \quad a > 0 \tag{3.1.10}$$

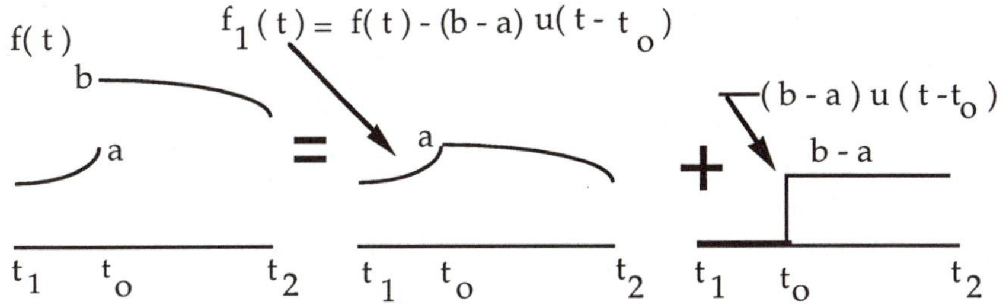

FIGURE 3.1.2
Example of discontinuous function.

(c)

$$\int_{-\infty}^{\infty} \delta(\tau - a)\ \delta(t - \tau - b)\ d\tau = \delta(t - a - b)$$

Solution

(a) The equation given can be demonstrated by testing both sides of eq. (3.1.9) to see whether each yields the same result in the sifting integral (3.1.6). Thus for the right hand side of eq. (3.1.9) we have $\int_{-\infty}^{\infty} 1/|a|\, f(\tau)\delta(\tau)\ d\tau = 1/|a|\ f(0)$. Now substitute the left hand side of eq. (3.1.9) into the sifting integral using the change of variable $u = a\tau$

$$\int_{-\infty}^{\infty} f(\tau)\ \delta(a\tau)\ d\tau = \frac{1}{|a|}\int_{-\infty}^{\infty} f\left(\frac{u}{|a|}\right)\ \delta(u)\ du = \frac{1}{|a|} f(0)$$

(since $\delta(t)$ is even), the same result as with the right hand side, and eq. (3.1.9) is verified.

(b) If we check the right hand side of eq. (3.1.10) in the sifting integral (3.1.6), the result is $1/2a\ [f(-a) + f(a)]$. Now evaluate the sifting integral using the left hand side of eq. (3.1.10).

$$\int_{-\infty}^{+\infty} f(\tau)\,\delta(\tau^2 - a^2)\,d\tau = \int_{-\infty}^{0} f(\tau)\,\delta[-2a(\tau + a)]\,d\tau +$$
$$+ \int_{0}^{+\infty} f(\tau)\,\delta[2a(\tau - a)]\,d\tau$$

Since $\delta(\tau^2 - a^2)$ is zero except at $\tau = \pm a$, we have used the value of $\tau^2 - a^2 = (\tau - a)(\tau + a)$ in the neighborhood of $\tau = \pm a$, i.e., $2\tau(\tau - a), -2\tau(\tau + a)$. The two integrals on the right hand side of the above equation are then each evaluated by a simple change of variable in conjunction with eq. (3.1.9), keeping in mind that $\delta(t)$ is an even function. Thus

$$\int_{-\infty}^{+\infty} f(\tau)\,\delta(\tau^2 - a^2)\,d\tau = \frac{1}{2a}\,f(-a) + \frac{1}{2a}\,f(a)$$

Hence both left and right hand sides of eq. (3.1.10) give the same results when substituted in the sifting integral and the equation is verified.

(c) We check, using the sifting relation, that the integral of the product of either side of the equation with $\phi(t)$ gives the same result.

$$\int_{-\infty}^{+\infty} \phi(t)\,dt \int_{-\infty}^{+\infty} \delta(\tau - a)\,\delta(t - \tau - b)\,d\tau =$$

$$= \int_{-\infty}^{+\infty} \delta(\tau - a)\,d\tau \int_{-\infty}^{+\infty} \delta(t - \tau - b)\,\phi(t)\,dt =$$

$$= \int_{-\infty}^{+\infty} \delta(\tau - a)\phi(\tau + b)\,d\tau =$$

$$= \phi(a + b) = \int_{-\infty}^{+\infty} \delta(t - a - b)\,\phi(\tau)\,d\tau$$

□

3.2 The Fourier Integral Theorem

The generalized functions, $\delta(t)$, $\mathrm{u}(t)$, can be represented in a variety of ways using ordinary functions in conjunction with the appropriate (weak) limiting process. Thus eqs. (3.1.2), (3.1.3) show $\delta(t)$ as the limit of a rectangular pulse whose height becomes infinite while its width approaches zero, but whose area remains constant at unity. Here we introduce some other approximating sequences, one of which is fundamental for proving the Fourier Integral Theorem.

Example 3.2.1
Verify the following representations for $\mathrm{u}(t)$ and $\delta(t)$.

(a)

$$\mathrm{u}(t) = \lim_{\alpha \to 0} \left(\frac{1}{2} + \frac{1}{\pi} \arctan \frac{t}{\alpha} \right) \tag{3.2.1}$$

(b)

$$\delta(t) = \frac{1}{\pi} \lim_{\alpha \to 0} \frac{\alpha}{t^2 + \alpha^2} \tag{3.2.2}$$

Solution

(a) Let $t > 0$ in eq. (3.2.1). Then as $\alpha \to 0$, the arctan function has infinite argument corresponding to an angle of $\pi/2$. Thus the right hand side is 1. The arctan function is odd, so for $t < 0$ in the limit, the right hand side is $1/2 - 1/2 = 0$. Finally, when $t = 0$, the arctan is zero so that, in this case, the right hand side is $1/2$. We, therefore, verify our definition of $\mathrm{u}(t)$ as well as the "natural" value of $\mathrm{u}(0)$ given by eq. (3.1.7).[5]

(b) To check eq. (3.2.2) use the derivative representation (3.1.8). Then

$$\delta(t) = \lim_{\alpha \to 0} \frac{d}{dt}\left(\frac{1}{2} + \frac{1}{\pi}\arctan\frac{t}{\alpha}\right) = \frac{1}{\pi}\lim_{\alpha \to 0}\frac{\alpha}{t^2 + \alpha^2}$$

The equation can also be checked directly: with $t \neq 0$, the right side of eq. (3.2.2) is zero when $\alpha = 0$. The integral of eq. (3.2.2) before going to the limit $\alpha = 0$ is

$$\frac{1}{\pi}\int_{-\infty}^{+\infty}\frac{\alpha}{\tau^2 + \alpha^2}\,d\tau = \frac{1}{\pi}\arctan\frac{\tau}{\alpha}\bigg|_{-\infty}^{+\infty} = 1$$

valid $\forall\alpha$, so the result is verified. It can also be easily shown that the limit of the sequence in eq. (3.2.2) satisfies the Definition 3.1.1, eq. (3.1.1). Also note that in the limit $\alpha \to 0$ each term of the sequence is pointwise zero for $t \neq 0$; this is by no means necessary, in general, since the only requirement is that the limit function be zero in the weak sense for $t \neq 0$.

□

A function family not even possessing a pointwise limit plays an important role in the Dirichlet theory of Fourier series and integrals. That function family is defined as

$$D(t, \alpha) = \frac{\sin \alpha t}{\pi t}$$

The generalized function defined by $d_0(t) = \lim_{\alpha \to +\infty} D(t, \alpha)$ will be shown to coincide with the impulse function, though it is not even pointwise defined for $t \neq 0$.[6] Thus we wish to prove that

$$\lim_{\alpha \to +\infty}\int_{-\infty}^{+\infty} D(t, \alpha)\, f(t)\, dt = \lim_{\alpha \to +\infty}\int_{-\infty}^{+\infty}\frac{\sin \alpha t}{\pi t} f(t)\, dt = f(0) \qquad (3.2.3)$$

[5] See also the discussion at the end of Chapter 1 in connection with Example 1.5.1.

[6] B. Van der Pol and H. Bremmer, *Operational Calculus*, Cambridge, Cambridge University Press, 1950, pp. 56-66. Included is a fascinating historical discussion of the origin of $\delta(t)$.

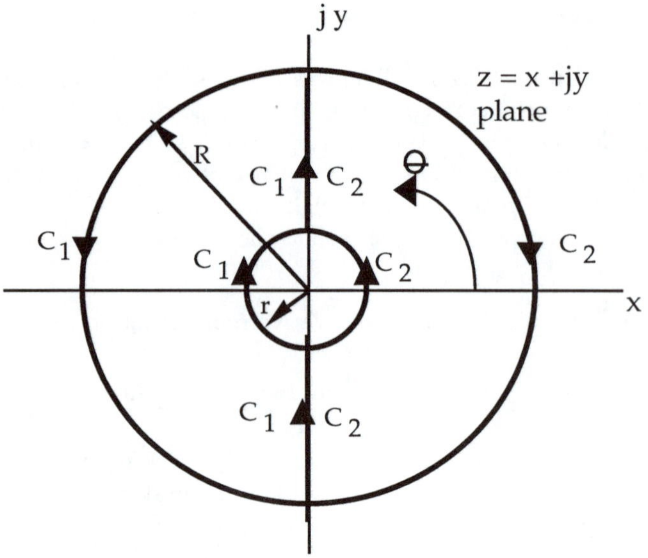

FIGURE 3.2.1
Contours for integration of *sin y/y*.

As a first step in establishing eq. (3.2.3), we show that

$$\int_{-\infty}^{+\infty} \frac{\sin y}{y}\, dy = \pi \tag{3.2.4}$$

We use contour integration (see Section A.13). The *Cauchy Integral Theorem* tells us that if $f(z)$, $z = x + jy$, is analytic in a closed region G, i.e., has a unique derivative everywhere in G and on its boundary C, then the line integral around the contour C satisfies

$$\oint_C f(z)\, dz = 0 \tag{3.2.5}$$

To apply eq. (3.2.5) use Euler's formula, sin $y = (\epsilon^{jy} - \epsilon^{-jy})/2j$ and set $z = jy$ to continue the function into the complex z plane. Thus

$$\frac{\sin y}{y}\, dy = \frac{1}{2j}\frac{\epsilon^{z}}{z} dz - \frac{1}{2j}\frac{\epsilon^{-z}}{z} dz \tag{3.2.6}$$

Integrate $\epsilon^z/2jz$ over the counterclockwise contour C_1, which, as shown in Fig. 3.2.1, is indented to the left around the small semicircle of radius $r \to 0$ to avoid the pole at $z = 0$ and is closed around the large semicircle of radius $R \to +\infty$. The second term will be integrated around C_2. For the path C_1, we allow $R \to +\infty$ and $r \to 0$.

$$\frac{1}{2j}\oint_{C_1} \frac{\epsilon^z}{z}\,dz = \frac{1}{2j}\int_{-j\infty}^{+j\infty} \frac{\epsilon^z}{z}\,dz + \frac{1}{2j}\int_{r\to 0} \frac{\epsilon^z}{z}\,dz + \frac{1}{2j}\int_{R\to +\infty} \frac{\epsilon^z}{z}\,dz = 0 \tag{3.2.7}$$

since the integrand is analytic within the region bounded by the contour C_1 and continuous in the closure.

The integral around the large semicircle has an integrand of the form $g(z)\epsilon^z$ with $g(z) = 1/z$. As $R \to +\infty$, $g(z)$ approaches zero uniformly on the large semicircle. The exponential ϵ^z goes to zero because $\Re z < 0$ in the left half of the z plane. We therefore expect this portion of the contour integral to be zero, and Jordan's Lemma[7] rigorously proves the assertion. On the small semicircle $z = r\epsilon^{j\theta}$ and $dz = jr\epsilon^{j\theta}\,d\theta$. Thus

$$\frac{1}{2j}\int_{r\to 0} \frac{\epsilon^z}{z}\,dz = \frac{1}{2j}\lim_{r\to 0}\int_{\frac{-\pi}{2}}^{\frac{-3\pi}{2}} \frac{\epsilon^z jr\epsilon^{j\theta}}{r\epsilon^{j\theta}}\,d\theta = \frac{1}{2}\int_{\frac{-\pi}{2}}^{\frac{-3\pi}{2}} d\theta = -\frac{\pi}{2}$$

The final step above follows, since as $r \to 0$, $\epsilon^z \to 1$.

Returning to eq. (3.2.7)

$$\frac{1}{2j}\int_{-j\infty}^{+j\infty} \frac{\epsilon^z}{z}\,dz = \frac{\pi}{2}$$

The second term of eq. (3.2.6) is integrated around C_2, the procedure being the same as that just completed. Note that now ϵ^{-z}/z satisfies the Jordan Lemma because $\Re(-z) < 0$ in the right half plane so that the integral on the large semicircle of C_2 is zero as $R \to +\infty$. As for the integration around the small semicircle of C_2,

$$\frac{1}{2j}\lim_{r\to 0}\int_{\frac{-\pi}{2}}^{\frac{\pi}{2}} \frac{\epsilon^{-z} jr\epsilon^{j\theta}}{r\epsilon^{j\theta}}\,d\theta = \frac{1}{2}\int_{\frac{-\pi}{2}}^{\frac{\pi}{2}} d\theta = \frac{\pi}{2}$$

Therefore,

$$\frac{1}{2j}\int_{-j\infty}^{+j\infty} \frac{\epsilon^{-z}}{z}\,dz = -\frac{\pi}{2}$$

Now integrate eq. (3.2.6)

$$\int_{-\infty}^{+\infty} \frac{\sin y}{y}\,dy = \frac{1}{2j}\int_{-j\infty}^{+j\infty} \frac{\epsilon^z}{z}\,dz - \frac{1}{2j}\int_{-j\infty}^{+j\infty} \frac{\epsilon^{-z}}{z}\,dz = \frac{\pi}{2} - \left(-\frac{\pi}{2}\right) = \pi$$

[7] For example, E. T. Whittaker and G. N. Watson, *A Course of Modern Analysis*, Cambridge, Cambridge University Press, 1940, p. 115.

We infer several immediate consequences of the above result which will be useful in discussing the sifting property. Since $\sin y/y$ is an even function

$$\int_0^{+\infty} \frac{\sin y}{y}\, dy = \frac{\pi}{2}$$

Using $y = \alpha x$, $\alpha > 0$

$$\lim_{\alpha\to+\infty} \int_0^{A} \frac{\sin \alpha x}{x}\, dx = \int_0^{+\infty} \frac{\sin y}{y}\, dy = \frac{\pi}{2} \tag{3.2.8}$$

and with $0 < A < B$

$$\lim_{\alpha\to+\infty} \int_A^{B} \frac{\sin \alpha x}{x}\, dx = \lim_{\alpha\to+\infty} \int_{\alpha A}^{\alpha B} \frac{\sin y}{y}\, dy = 0 \tag{3.2.9}$$

We can now verify that

$$\lim_{\alpha\to+\infty} \frac{\sin \alpha t}{\pi t} = \delta(t) \tag{3.2.10}$$

We give a shortened demonstration of the result emphasizing the main points of the proof. A complete derivation is in Carslaw.[8]

The proof is based on the *Second Mean Value Theorem for Integrals* which can be stated as follows. Let $f(t)$ be bounded and monotonic in the interval $(a, b) \Rightarrow a \le t \le b$. Let $\phi(t)$ be bounded and integrable in (a, b). Then (a, b) can always be divided into two intervals $(a, \zeta), (\zeta, b)$ so that

$$\int_a^b f(t)\, \phi(t) dt = f(a) \int_a^{\zeta} \phi(t)\, dt + f(b) \int_{\zeta}^{b} \phi(t)\, dt \tag{3.2.11}$$

The theorem can be qualitatively understood if we suppose $f(t)$ is monotone increasing in (a, b). Then the first integral of eq. (3.2.11) is smaller than the true value over the first interval, and the second is larger than the true value. By choosing a proper division of the interval, the deficiency and excess can cancel and the correct value obtained. A similar argument applies if $f(t)$ is monotone decreasing in the interval (a, b).

To apply the theorem, let $\phi(t) = \sin \alpha t/\pi t$, and assume that $f(t)$ is such that the infinite interval $(0, +\infty)$ can be divided into sub-intervals over each of which $f(t)$ is bounded and monotonic. Consider one such interval (a, b), $0 < a < b$.

$$\lim_{\alpha\to+\infty} \int_a^b f(t) \frac{\sin\ \alpha t}{\pi t}\, dt = f(a) \lim_{\alpha\to+\infty} \int_a^{\zeta} \frac{\sin\ \alpha t}{\pi t}\, dt + $$

$$+ f(b) \lim_{\alpha\to+\infty} \int_{\zeta}^{b} \frac{\sin\ \alpha t}{\pi t}\, dt$$

[8] H. S. Carslaw, *An Introduction to the Theory of Fourier's Series and Integrals*, New York, Dover Publications, pp. 219-227.

But, referring to eq. (3.2.9), each of the integrals on the right of the above equation is zero so that for all intervals other than the one that includes the origin

$$\lim_{\alpha \to 0} \int_a^b f(t) \frac{\sin\ \alpha t}{\pi t}\, dt = 0 \tag{3.2.12}$$

Now consider the interval $(0, c)$, $0 < c$

$$\lim_{\alpha \to +\infty} \int_0^c f(t) \frac{\sin\ \alpha t}{\pi t}\, dt = f(0) \lim_{\alpha \to +\infty} \int_0^\zeta \frac{\sin\ \alpha t}{\pi t}\, dt + \\ + f(c) \lim_{\alpha \to +\infty} \int_\zeta^c \frac{\sin\ \alpha t}{\pi t}\, dt$$

Referring to eqs. (3.2.8), (3.2.9), the first of the integrals on the right side of the above equation is 1/2, the other is zero. Thus

$$\lim_{\alpha \to +\infty} \int_0^c f(t) \frac{\sin\ \alpha t}{\pi t}\, dt = \frac{1}{2} f(0)$$

Combining this result with (3.2.12)

$$\lim_{\alpha \to +\infty} \int_0^{+\infty} f(t) \frac{\sin\ \alpha t}{\pi t}\, dt = \frac{1}{2} f(0) \tag{3.2.13}$$

The same procedure can be carried out for the interval $(-\infty, 0)$, and the integral will also be $f(0)/2$ as in eq. (3.2.13). The final result is, therefore,

$$\lim_{\alpha \to +\infty} \int_{-\infty}^{+\infty} f(t) \frac{\sin\ \alpha t}{\pi t}\, dt = f(0) = \int_{-\infty}^{+\infty} f(t)\, \delta(t)\, dt \tag{3.2.14}$$

so that the Dirichlet function in the limit, eq. (3.2.10), represents $\delta(t)$. It follows that the sifting property holds

$$\int_{-\infty}^{+\infty} f(\tau)\, \delta(t - \tau)\, d\tau = \lim_{\alpha \to +\infty} \int_{-\infty}^{+\infty} f(\tau) \frac{\sin\ \alpha(t - \tau)}{\pi(t - \tau)}\, d\tau = f(t) \tag{3.2.15}$$

The conditions for which eqs. (3.2.14), (3.2.15) are valid can be generalized to the case where $f(t)$ has a finite number of infinite discontinuities over any interval (a, b) provided (i) $f(t)$ is bounded over each such interval except for arbitrarily small neighborhoods of the discontinuity, (ii) (a, b) (excluding the discontinuities) can be divided into partial intervals in each of which $f(t)$ is monotonic, and (iii) $f(t)$ is absolutely integrable over each complete interval (a, b). These are the *Dirichlet Conditions.* Since eq. (3.2.15) is taken over the infinite interval $(-\infty, +\infty)$, we add the proviso that $f(t)$ be absolutely integrable from $-\infty$ to $+\infty$. The reader is referred to Carslaw (footnote 8 page 114) for further discussion.

We can now readily deduce the fundamental Fourier integral representation for $f(t)$ embodied in the following *Fourier Integral Theorem.*

THEOREM 3.2.1
For any $f(t)$ satisfying Dirichlet's conditions, the following representation holds.

$$\frac{1}{2\pi}\int_{-\infty}^{+\infty} f(\tau)\int_{-\infty}^{+\infty} \epsilon^{j\omega(t-\tau)}d\omega \;\; d\tau = f(t) \tag{3.2.16}$$

PROOF Use the sifting property of $\delta(t)$ expressed by eq. (3.2.15). Consider

$$\frac{1}{2\pi}\int_{-\infty}^{+\infty} \epsilon^{j\omega(t-\tau)}\, d\omega = \lim_{\Omega\to+\infty} \frac{1}{2j\pi(t-\tau)}\epsilon^{j\omega(t-\tau)}\Big|_{-\Omega}^{\Omega}$$

or

$$\frac{1}{2\pi}\int_{-\infty}^{+\infty} \epsilon^{j\omega(t-\tau)}\, d\omega = \lim_{\Omega\to+\infty} \frac{1}{\pi}\frac{\sin\,\Omega(t-\tau)}{(t-\tau)} = \delta(t-\tau)$$

Substituting the above relation into eq. (3.2.16) and using the sifting integral of eq. (3.2.15) establishes the required result. ∎

If the order of integration in eq. (3.2.16) is reversed, i.e., $d\tau$ followed by $d\omega$, that equation can be rewritten in the form we will generally employ which specifies two separate relations, *The Fourier Transform*, $F(j\omega) = \mathcal{F}f(t)$[9] and *The Inverse Fourier Transform*, $f(t) = \mathcal{F}^{-1}F(j\omega)$. Thus, after changing the order of integration evaluate eq. (3.2.16) in two steps

$$F(j\omega) = \int_{-\infty}^{+\infty} f(t)\epsilon^{-j\omega t}\, dt \equiv \mathcal{F}f(t) \tag{3.2.17}$$

and

$$f(t) = \frac{1}{2\pi}\int_{-\infty}^{+\infty} F(j\omega)\epsilon^{j\omega t}\, d\omega \equiv \mathcal{F}^{-1}F(j\omega) \tag{3.2.18}$$

The two equations, eqs. (3.2.17) and (3.2.18), are often referred to as the *Fourier Integral and its mate* or the *Fourier Integral Pair.* As discussed in connection with eq. (3.2.15), very general sufficient requirements for the convergence of the Fourier integral pair are that the Dirichlet conditions and absolute integrability be satisfied. We state the following theorem.

[9] The Fourier Transform can be viewed either as a complex function of a real variable $\omega \to \hat{f}(\omega)$ or a complex function of an imaginary variable $\omega \to F(j\omega)$; we use the second interpretation because it is consistent with the definition of Laplace Transform to be introduced later.

THEOREM 3.2.2

The Fourier transform (3.2.17) $\mathcal{F}f(t) = F(j\omega)$ exists, is bounded a.e., and the Fourier Integral Theorem (3.2.16) is a valid representation for $f(t)$ if the Dirichlet conditions are satisfied for every finite interval and $\int_{-\infty}^{+\infty} |f(t)|\, dt < +\infty$.

The above theorem requires absolute integrability of $f(t)$. Another important sufficiency theorem given by Titchmarsh[10] is applicable when $f(t)$ is square integrable, i.e., when $f(t)$ is in L^2, whereas Theorem 3.2.2 concerns functions in L^1. The L^2 theorem is stated here without proof (but see the example on the Parseval relation in Section 3.3).

THEOREM 3.2.3

If $\int_{-\infty}^{+\infty} |f(t)|^2\, dt < +\infty$, then $\mathcal{F}f(t) = F(j\omega)$ in eq. (3.2.17) converges and also belongs to L^2.

We say that the Fourier integral pair defines a norm preserving isomorphism of L^2 on itself; this result will be used in Chapter 4.

Two points should be noted in connection with our use of the Fourier transform infinite integrals: (1) $\int_{-\infty}^{+\infty}(\ ) = \lim_{A\to+\infty} \int_{-A}^{A}(\ )$ and (2) for a finite discontinuity at $t = t_o$, the Fourier integral representation eq. (3.2.16) yields the mean value $(1/2)[f(t_o^-) + f(t_o^+)]$.

Example 3.2.2

(a) Find the transform $F(j\omega)$

$$F(j\omega) = \mathcal{F}[\epsilon^{-\alpha t}\, \mathrm{u}(t)], \quad \alpha > 0$$

(b) Find the inverse transform $f(t)$

$$f(t) = \mathcal{F}^{-1} p_a(\omega)$$

The *ideal low pass function* of bandwidth $a/2$, $p_a(\omega)$, is defined as

$$p_a(\omega) = \begin{cases} 1 & |\omega| \le a/2 \\ 0 & |\omega| > a/2 \end{cases} \tag{3.2.19}$$

In the time domain, $p_a(t)$ represents a pulse of duration a.

Solution

[10]E. C. Titchmarsh, *An Introduction to the Theory of Fourier Integrals*, Oxford, Clarendon Press, 1948, p. 69 & seq.

(a) $f(t)$ is initiated at $t = 0$, hence eq. (3.2.17) has lower limit 0, thus

$$F(j\omega) = \int_0^{+\infty} \epsilon^{-(\alpha+j\omega)t}\,dt = \frac{-1}{\alpha + j\omega}\epsilon^{-\alpha t}\epsilon^{-j\omega t}\Big|_0^{+\infty}$$

Since $\alpha > 0$, the exponential goes to zero at the upper limit, therefore,

$$F(j\omega) = \mathcal{F}[\epsilon^{-\alpha t}\,\mathrm{u}(t)] = \frac{1}{\alpha + j\omega} \tag{3.2.20}$$

We can make use of (3.2.18) from the Fourier Integral Theorem and immediately infer

$$\mathcal{F}^{-1}\left[\frac{1}{\alpha + j\omega}\right] = \frac{1}{2\pi}\int_{-\infty}^{+\infty} \frac{\epsilon^{j\omega t}}{\alpha + j\omega} d\omega = \epsilon^{-\alpha t}\,\mathrm{u}(t)$$

(b) By using eq. (3.2.18), we obtain

$$f(t) = \mathcal{F}^{-1}p_a(\omega) = \frac{1}{2\pi}\int_{-a/2}^{a/2} \epsilon^{j\omega t}d\omega = \frac{1}{j2\pi t}\,\epsilon^{j\omega t}\Big|_{\omega=-a/2}^{a/2}$$

$$= \frac{\sin\ at/2}{\pi t}$$

Invoking eq. (3.2.17) of the Fourier Integral Theorem, we also have the result

$$F(j\omega) = p_a(\omega) = \int_{-\infty}^{+\infty} \frac{\sin\ at/2}{\pi t}\,\epsilon^{-j\omega t}dt \tag{3.2.21}$$

Note that by changing the variable in eq. (3.2.21) to $y = at/2$ and setting $\omega = 0$, we return to eq. (3.2.4).

□

The symmetry of the Fourier transform in the frequency and time variables leads to a number of useful properties which are easily derived. For example, with $f(t)$ a real function and $F(j\omega) = \mathcal{F}f(t)$

$$\mathcal{F}f(-t) = \int_{-\infty}^{+\infty} f(t)\,\epsilon^{-j\omega(-t)}\,dt = F(-j\omega) = F^*(j\omega), \quad f(t) \text{ real} \tag{3.2.22}$$

where $F^*(j\omega)$ is the complex conjugate of $F(j\omega)$.

It is also immediately evident from eq. (3.2.17) that if $f(t)$ is complex, then with $\mathcal{F}^{-1}f(t) = F(j\omega)$

$$\mathcal{F}f^*(t) = F^*(-j\omega) \tag{3.2.23}$$

Example 3.2.3
Find the transform $F(j\omega) = \mathcal{F}f(t)$, where

$$f(t) = \left\{ \begin{array}{ll} \epsilon^{-\alpha t} & t > 0 \\ -\epsilon^{\alpha t} & t < 0 \end{array} \right\} \equiv \epsilon^{-\alpha|t|} \operatorname{sign} t, \quad \alpha > 0$$

where $\operatorname{sign} t$ is the sign change function, i.e., $+1$ for $t > 0$, -1 for $t < 0$. Clearly $f(t)$ is an odd function whose magnitude decays exponentially on either side of the origin.

Solution From the definition of $f(t)$ we have

$$\mathcal{F}[\epsilon^{-\alpha\,|t|} \operatorname{sign} t] = \mathcal{F}[\epsilon^{-\alpha t}\, \mathrm{u}(t)] - \mathcal{F}[\epsilon^{\alpha t}\, \mathrm{u}(-t)] = \mathcal{F}f_0(t) - \mathcal{F}f_0(-t)$$

The first term above is evaluated using eq. (3.2.20). Referring to eq. (3.2.22) the second term is the negative complex conjugate of the first. Thus

$$\mathcal{F}[\epsilon^{-\alpha|t|} \operatorname{sign} t] = \frac{1}{\alpha + j\omega} - \frac{1}{\alpha - j\omega} = \frac{-2j\omega}{\alpha^2 + \omega^2} \tag{3.2.24}$$

□

Again invoking symmetry we can get two transforms for the price of one.

Example 3.2.4
Given $\mathcal{F}f(t) = F(j\omega)$, show that a second transform immediately follows, i.e.,

$$\mathcal{F}f(t) = -j2\pi f(-j\omega) \tag{3.2.25}$$

Solution From eqs. (3.2.18) and (3.2.17) $-j2\pi f(-t) = \mathcal{F}F(j\omega)$. Interchanging $j\omega$ and t, the desired expression follows.

As an illustration of the symmetry relation consider eq. (3.2.21), $F(j\omega) = p_a(\omega) = \mathcal{F}[\sin{(at/2)}/\pi t]$. Using eq. (3.2.25) the transform of a time pulse is obtained.

$$\mathcal{F}p_a(t) = \frac{2\sin\dfrac{-a\omega}{2}}{-\omega} = \frac{a\sin\dfrac{a\omega}{2}}{\dfrac{a\omega}{2}} \tag{3.2.26}$$

□

Example 3.2.5
Derive the *Delay Rule*

$$\mathcal{F}f(t - t_0) = \epsilon^{-j\omega t_0}\, F(j\omega) \tag{3.2.27}$$

Solution Refer to the Transform in eq. (3.2.17) and change the variable to $u = t - t_0$

$$\mathcal{F}f(t-t_0) = \int_{-\infty}^{+\infty} f(u)\epsilon^{-j(t_0+u)\omega}du = \epsilon^{-j\omega t_0} F(j\omega)$$

□

A similar procedure gives the symmetric relation known as the *Modulation Rule*

$$\mathcal{F}^{-1}F(j\omega - j\omega_0) = \epsilon^{j\omega_0 t} f(t) \tag{3.2.28}$$

This section is concluded with some simple though useful further consequences of symmetry. We have seen in eq. (3.2.22) that the Fourier transform $F(j\omega)$ is generally complex when $f(t)$ is real. $F(j\omega) = R(\omega) + jX(\omega)$, where $R(\omega) = \Re F(j\omega), X(\omega) = \Im F(j\omega)$. Assume $f(t)$ is a real function and separate it into its even $f_e(t)$ and odd $f_o(t)$ parts.

$$f(t) = f_e(t) + f_o(t) \tag{3.2.29}$$

$$f_e(t) = \frac{1}{2}[f(t) + f(-t)] \tag{3.2.30}$$

$$f_o(t) = \frac{1}{2}[f(t) - f(-t)] \tag{3.2.31}$$

Now compute $F(j\omega) = \mathcal{F}[f_e(t) + f_o(t)]$. Referring to eq. (3.2.17), first with the integrand in the form $f_e(t)\cos\omega t - jf_e(t)\sin\omega t$, it is clear that, since the $f_e \sin\omega t$ term is odd in t, its contribution to the integral vanishes over the interval $(-\infty, +\infty)$ and the resulting integration gives a real function. Similarly with the odd function $f_o(t)$ in the integrand of eq. (3.2.17), it is the term $f_o(t)\cos\omega t$ which is odd and makes zero contribution to the integral, so that in this case the integration yields an imaginary function. Accordingly

$$F(j\omega) = R(\omega) + jX(\omega) = \int_{-\infty}^{+\infty} [f_e(t)\ \cos\omega t + jf_o(t)\sin\omega t]\, dt$$

Thus by equating real and imaginary parts

$$R(\omega) = \int_{-\infty}^{+\infty} f_e(t)\ \cos\omega t\, dt = R(-\omega) \tag{3.2.32}$$

and

$$X(\omega) = \int_{-\infty}^{+\infty} f_o(t)\sin\omega t\, dt = -X(-\omega) \tag{3.2.33}$$

The functions $R(\omega)$, $X(\omega)$ are even and odd as indicated, since the integrands of eqs. (3.2.32), (3.2.33) are, respectively, even and odd in ω and the integration variable is t. In summary, the Fourier transform of a real, even time function is real and even in ω, and the Transform of a real odd time function is imaginary and odd in ω.

3.3 Impulse Response and Convolution

We now consider the case in which an LTI quiescent system is excited by a delta function as the input signal. We can visualize what occurs by supposing that the system is subjected to a shock of very short duration. The response will then be a linear combination of natural modes since the forcing function is zero once the impulse has occurred. If the system is stable and has some loss, i.e., all natural frequencies with a finite negative real part, these natural modes die out and the system eventually returns to its initial quiescent state.

The *impulse response* $h(t)$ is the solution of the operator equation when the forcing function is $\delta(t)$. Thus, let S represent the operator for a given LTI system

$$\delta(t) = S\{h(t)\} \tag{3.3.1}$$

Suppose the impulse response, i.e., the solution of eq. (3.3.1), is known. We will show that if the system is LTI we can determine the response to any input function $x(t)$. In other words, we can invert the operator equation

$$x(t) = S\{y(t)\}$$

and find

$$y(t) = S^{-1}\{x(t)\} \tag{3.3.2}$$

provided S is LTI.

Represent $x(t)$ using the sifting property (3.1.6) of $\delta(t)$.

$$x(t) = \int_{-\infty}^{\infty} \delta(t-\tau)\, x(\tau)\, d\tau$$

By the time invariance of S, eq. (3.3.1) becomes $\delta(t-\tau) = S\{h(t-\tau)\}$ so that

$$x(t) = \int_{-\infty}^{+\infty} S\{h(t-\tau)\}\, x(\tau)\, d\tau = S\left\{\int_{-\infty}^{+\infty} h(t-\tau)\, x(\tau)\, d\tau\right\} \equiv S\{y(t)\} \tag{3.3.3}$$

where, regarding the integral as the limit of a *sum*, we have taken S (which operates on h considered as a function of t with the integration variable τ as a parameter) outside the integral by virtue of linearity. Then by inspection of eq. (3.3.3), with $h(t)$ prescribed, we have the solution

$$y(t) = \int_{-\infty}^{+\infty} h(t-\tau)\, x(\tau)\, d\tau = S^{-1}\{x(t)\} \tag{3.3.4}$$

The operator S^{-1} defining $y(t)$ in eq. (3.3.4) is known as a *convolution integral* and is usually symbolized as

$$f(t) * g(t) = \int_{-\infty}^{+\infty} f(\tau)g(t-\tau)\,d\tau \tag{3.3.5}$$

where g and f replace h and x, respectively. The convolution has the following properties:

(a) *commutative*

$$f * g = g * f$$

(b) *associative*

$$(f * g) * h = f * (g * h) = f * g * h \tag{3.3.6}$$

(c) *distributive*

$$(f + g) * h = f * h + g * h \tag{3.3.7}$$

PROOF

(a) The commutative property is proved as follows. In eq. (3.3.5) let $u = t - \tau$. Then

$$f(t) * g(t) = \int_{-\infty}^{+\infty} f(\tau)\,g(t-\tau)\,d\tau = \int_{+\infty}^{-\infty} f(t-u)\,g(u)\,(-du)$$

$$= \int_{-\infty}^{+\infty} g(u)f(t-u)\;du = g(t) * f(t)$$

(b) The associative property is proved by transforming the integrals involved in the left side of eq. (3.3.6) in a double integral and then reversing the order of integration. For locally integrable functions, this requires absolute integrability over the infinite interval from $-\infty$ to $+\infty$. For generalized functions, a sufficient condition is that they are zero for $t < t_0$ (or for $t > t_0$).

(c) The distributive property follows immediately from the linearity properties of the integrals. ∎

COROLLARY 3.3.1
If

$$y(t) = f(t) * g(t)$$

then

$$f(t-a) * g(t-b) = y[t-(a+b)] = f(t) * g[t-(a+b)] = f[t-(a+b)] * g(t) \tag{3.3.8}$$

that is, the convolution of two delayed functions is equal to the convolution of any one of them nondelayed with the other delayed by the sum of the delays.

PROOF Eq. (3.3.5), with $u = t - \tau - b$, yields for $f(t-a) * g(t-b)$,

$$f(t-a) * g(t-b) = \int_{-\infty}^{+\infty} f(\tau - a) g(t - \tau - b)\, d\tau =$$

$$= \int_{-\infty}^{+\infty} g(u) f[t - (a+b) - u]\, du = g(t) * f[t - (a+b)]$$

and from the commutative property the fourth expression of eq. (3.3.8) follows. Similarly, using $u = \tau - a$ the third expression is verified. The second expression follows immediately by inspection. ■

COROLLARY 3.3.2

Let $h(t)$ be given. Then for an arbitrary function $x(t)$ the operator C : $C\{x(t)\} = h(t) * x(t) = y(t)$ is an LTI operator.

PROOF Linearity is immediate, for referring to the integral (3.3.5) defining convolution and eq. (3.3.7)

$$C\{a_1 x_1 + a_2 x_2\} = a_1 C\{x_1\} + a_2 C\{x_2\}$$

Time invariance also follows directly, for by the first and second expressions of eq. (3.3.8)

$$C\{x(t-a)\} = h(t) * x(t-a) = y(t-a)$$

which is the desired result. ■

In summary, we have seen in eq. (3.3.4) that for an LTI system the response to an admissible input signal can be represented as the convolution of the signal with the system's prescribed impulse response. Conversely, Corollary 3.3.2 shows that if the system response is expressible as a convolution of the signal and the system impulse response, then the system is LTI. In fact, the following theorem can be rigorously proved.

THEOREM 3.3.1

A necessary and sufficient condition for a system S to be LTI is that its response to any admissible $x(t)$ excitation function be determined by the convolution of $h(t) * x(t)$, where $h(t)$ is the impulse response of S.

THEOREM 3.3.2
An LTI system S is causal if and only if its impulse response h(t) is a causal signal.

PROOF Necessity: Given that S is causal, it must respond causally to all causal signals. Since $\delta(t) = 0,\ t < 0$ the impulse response $h(t)$ must be causal.

Sufficiency: Given that $h(t)$ is causal, we must show that the response $y(t)$ to an arbitrary causal signal $x(t) = 0,\ t < 0$, satisfies $y(t) = 0,\ t < 0$. Since the system is LTI apply convolution

$$y(t) = \int_{-\infty}^{+\infty} h(t-\tau)\, x(\tau)\, d\tau$$

By hypothesis, both $x(t)$, $h(t)$ vanish for negative values of their argument. Hence $y(t)$ may be written

$$y(t) = \int_0^t h(t-\tau)\, x(\tau)\, d\tau$$

When $t < 0$, $h(t-\tau) = 0,\ \forall\tau > 0$, thus $y(t) = 0,\ t < 0$, i.e., it is a causal signal and may be written

$$y(t) = \mathrm{u}(t) \int_0^t h(t-\tau)\, x(\tau)\, d\tau$$

■

An important application of convolution is to the evaluation of the Fourier transform of the product of two functions. Suppose $f_1(t)$, $f_2(t)$ and the product $f_1\, f_2$ are each $\mathcal{F}$ transformable, e.g., satisfy Theorem 3.2.16 or Theorem 3.2.2. Then using eq. (3.2.18) for $f_1(t)$, compute $\mathcal{F}[f_1(t)\, f_2(t)]$ reversing the order of integration in the process

$$\begin{aligned}\int_{-\infty}^{+\infty} f_1(t)\, f_2(t)\, \epsilon^{-j\omega t}\, dt &= \frac{1}{2\pi}\int_{-\infty}^{+\infty} f_2(t)\, \epsilon^{-j\omega t} \int_{-\infty}^{+\infty} F_1(j\Omega)\, \epsilon^{j\Omega t}\, d\Omega\, dt \\ &= \frac{1}{2\pi}\int_{-\infty}^{+\infty} f_2(t)\epsilon^{j\Omega t}\, \epsilon^{-j\omega t}\, dt \int_{-\infty}^{+\infty} F_1(j\Omega)\, d\Omega\end{aligned}$$

We now apply eq. (3.2.28) to the integration in t and obtain

$$\mathcal{F}[f_1(t)\, f_2(t)] = \frac{1}{2\pi}\int_{-\infty}^{+\infty} F_1(j\Omega)\, F_2(j\omega - j\Omega)\, d\Omega = \frac{1}{2\pi} F_1(j\omega) * F_2(j\omega) \tag{3.3.9}$$

A similar procedure using the Fourier Transform (3.2.17) for $F_1(j\omega)$ and the delay rule (3.2.27) yields the symmetric relation

$$\mathcal{F}^{-1}[F_1(j\omega)\, F_2(j\omega)] = f_1(t) * f_2(t) \tag{3.3.10}$$

Example 3.3.1
Find the inverse Fourier Transform

$$f(t) = \mathcal{F}^{-1}\left[\frac{1}{a+j\omega}\,\frac{\sin\omega}{\omega}\right]$$

Solution If we let $F_1(j\omega) = 1/(a+j\omega)$, and $F_2(j\omega) = \sin\omega/\omega$, then, according to eqs. (3.3.10), (3.2.20), and (3.2.26), $f(t) = f_1(t) * f_2(t) = \exp(-at)\,\mathrm{u}(t) * p_2(t)$ or

$$f(t) = \int_{-\infty}^{+\infty} \epsilon^{-a\tau}\,\mathrm{u}(\tau)\,p_2(t-\tau)\,d\tau$$

The integration is with respect to τ and the time variable t acts as a parameter. It is often helpful to think of the function $f_2(t-\tau)$ as though it were a focal plane camera shutter sliding along. As it moves its window exposes different areas of $f_1(\tau)$. For any t the area under the product function $f_1(\tau)\,f_2(t-\tau)$, i.e., of window and the common portion of f_2 it exposes, gives the convolution integral.

Referring to Fig. 3.3.1, there are three different intervals of t to be considered as $p_2(t-\tau)$ slides along:

(a) $t < -1,\qquad f_1(t) * f_2(t) = 0$

(b) $-1 \le t \le 1,\qquad f_1(t) * f_2(t) = \int_0^{t+1} A\epsilon^{-a\tau}\,d\tau = \frac{A}{a}\left[1 - \epsilon^{-a(t+1)}\right]$

(c) $1 \le t,\qquad f_1(t) * f_2(t) = \int_{t-1}^{t+1} A\epsilon^{-a\tau}\,d\tau = \frac{2A}{a}\,\epsilon^{-at}\sinh a$

The function $\mathcal{F}^{-1}F(j\omega) = f(t) = f_1(t) * f_2(t)$ is shown in Fig. 3.3.1 for $A = 1$, $a = 0.5$. □

The following Parseval's theorem gives an important application of the convolution relation (3.3.9) for the transform of a product.

THEOREM 3.3.3
(Parseval's Theorem) *If $f(t) \in L^2$ it admits of a Fourier transform $F(j\omega) \in L^2$ and moreover*

$$\int_{-\infty}^{+\infty} |f(t)|^2\,dt = \frac{1}{2\pi}\int_{-\infty}^{\infty} |F(j\omega)|^2\,d\omega \tag{3.3.11}$$

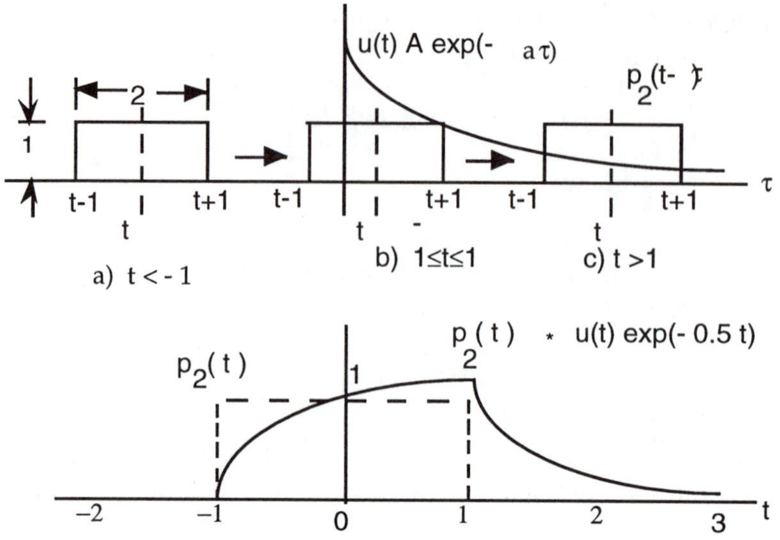

FIGURE 3.3.1
Convolution for $\mathcal{F}^{-1}F(\omega) = f(t) = A\exp(-at)\,u(t) * p_2(t)$.

PROOF We only prove eq. (3.3.11). Since for $f(t)$ real or complex $|f(t)|^2 = f^*(t)f(t)$, $\mathcal{F}[f^*(t)\,f(t)]$ may be written with the aid of eqs. (3.3.9) and (3.2.23) as

$$\int_{-\infty}^{+\infty} f^*(t)\,f(t)\,\epsilon^{-j\Omega t}\,dt = \frac{1}{2\pi}F(j\Omega) * F^*(-j\Omega)$$

The convolution integral is written

$$F(j\Omega) * F^*(-j\Omega) = \int_{-\infty}^{+\infty} F(j\omega)F^*(-j[\Omega-\omega])\,d\omega$$

and setting $\Omega = 0$, its integrand becomes $F^*(j\omega)F(j\omega) = |F(j\omega)|^2$, whereas that of the Fourier integral is $f^*(t)f(t)$, so that the Parseval relation follows. This result is clearly consistent with the L^2 Theorem 3.2.2. ∎

3.4 Real–Imaginary Part Relations; The Hilbert Transform

It is often important to determine the imaginary part of a Fourier Transform given its real part, and vice versa. In particular we shall be interested in real-imaginary part relations for the Fourier Transform of real *causal* time functions, i.e., $\mathcal{F}[g(t)\,\mathrm{u}(t)] = G(j\omega) = \Re G(j\omega) + j\Im G(j\omega)$.

The results we seek can be obtained by using the convolution-product rule (3.3.9). As will become evident a little later, it is useful (at least initially) to introduce a convergence factor $\epsilon^{-a|t|}$, $a > 0$ a real quantity. Thus define our causal function as

$$g(t)\,\mathrm{u}(t) = f(t)\ \epsilon^{-a|t|}\,\mathrm{u}(t)$$

We find even and odd parts of $g(t)\,\mathrm{u}(t)$ by using eqs. (3.2.29), (3.2.30), and (3.2.31)

$$g_e(t) = \frac{1}{2}[g(t)\,\mathrm{u}(t) + g(-t)\,\mathrm{u}(-t)]$$

$$g_o(t) = \frac{1}{2}[g(t)\,\mathrm{u}(t) - g(-t)\,\mathrm{u}(-t)]$$

Noting that $\mathrm{u}(t) = 0,\ t < 0$ and $\mathrm{u}(-t) = 0,\ t > 0$, the above equations can be written

$$g_e(t) = \begin{cases} \dfrac{g(t)}{2} & t > 0 \\ \dfrac{g(-t)}{2} & t < 0 \end{cases}$$

$$g_o(t) = \begin{cases} \dfrac{g(t)}{2} & t > 0 \\ \dfrac{-g(-t)}{2} & t < 0 \end{cases}$$

and inspecting these equations

$$g_o(t) = g_e(t)\ \mathrm{sign}(t) = f_e(t)\ \epsilon^{-a|t|}\,\mathrm{sign}(t)$$

$$g_e(t) = g_o(t)\ \mathrm{sign}(t) = f_o(t)\ \epsilon^{-a|t|}\,\mathrm{sign}(t)$$

where sign(t) is the sign change function. Let $\mathcal{F}f(t) = F(j\omega) = R(\omega) + jX(\omega)$ and referring to eqs. (3.2.32), (3.2.33) $R(\omega) = \mathcal{F}f_e(t)$, $jX(\omega) = \mathcal{F}f_o(t)$. We can now use eq. (3.3.9), the product formula, to write

$$\mathcal{F}g_o(t) = \frac{1}{2\pi}R(\omega) * \mathcal{F}[\epsilon^{-a|t|}\,\mathrm{sign}(t)]$$

$$\mathcal{F}g_e(t) = \frac{1}{2\pi}jX(\omega) * \mathcal{F}[\epsilon^{-a|t|}\,\mathrm{sign}(t)]$$

If $f(t)$ is well behaved, all the transforms employed above exist and when we substitute eq. (3.2.24)

$$\mathcal{F}[\epsilon^{-a|t|}\,\mathrm{sign}(t)] = \frac{-2j\omega}{a^2 + \omega^2}$$

The convergence factor was, in fact, introduced to secure the above relation, avoiding the convergence problems of $\mathcal{F}[\text{sign}(t)]$. The two preceding equations then become

$$j\Im G(\omega) = \frac{1}{2\pi}\int_{-\infty}^{+\infty} R(\Omega)\, \frac{-2j(\omega-\Omega)}{a^2+(\omega-\Omega)^2}\, d\Omega$$

$$\Re G(\omega) = \frac{1}{2\pi}\int_{-\infty}^{+\infty} jX(\Omega)\, \frac{-2j(\omega-\Omega)}{a^2+(\omega-\Omega)^2}\, d\Omega$$

Since the integrals exist $\forall\ a \geq 0$, we can set $a = 0$. Then $G(j\omega) = F(j\omega)$ and $\Re G(\omega) = R(\omega)$, $\Im G(\omega) = X(\omega)$. The final result is

$$X(\omega) = \frac{1}{\pi}\int_{-\infty}^{+\infty} \frac{R(\Omega)}{\Omega-\omega}\, d\Omega \tag{3.4.1}$$

and

$$R(\omega) = -\frac{1}{\pi}\int_{-\infty}^{+\infty} \frac{X(\Omega)}{\Omega-\omega}\, d\Omega \tag{3.4.2}$$

The above relations define a *Hilbert Transform* pair, yielding $R(\omega)$ and its mate $X(\omega)$. As noted earlier, these functions are real and imaginary parts of the $\mathcal{F}$ transform of a real causal time signal. The integrals are to be interpreted as *Principal Values* (designated with P), meaning that an infinitesimal neighborhood of the singularity at $\Omega = \omega$ is excluded from the integration. The principal value of the integral is defined as

$$P\int_{-\infty}^{+\infty} \frac{R(\Omega)}{\Omega-\omega}\, d\Omega = \lim_{e\to 0}\int_{-\infty}^{\omega-e} \frac{R(\Omega)}{\Omega-\omega}\, d\Omega + \int_{\omega+e}^{+\infty} \frac{R(\Omega)}{\Omega-\omega}\, d\Omega \tag{3.4.3}$$

Usually the P is understood from the context and we will often omit it.

There are a number of alternate forms of the Hilbert Transform pair which will prove useful. For example, we can make use of the fact that only the even component of the integrands in eqs. (3.4.1) and (3.4.2) contribute to the value of the integrals. Since $R(\Omega)$ is even, the even part of the eq. (3.4.1) integrand is

$$\frac{R(\Omega)}{2}\left(\frac{1}{\Omega-\omega} + \frac{1}{-\Omega-\omega}\right) = \frac{\omega R(\Omega)}{\Omega^2-\omega^2}$$

Since the integration is now symmetric about the origin, change the lower limit to 0 and double the result. Then eq. (3.4.1) becomes

$$X(\omega) = \frac{2\omega}{\pi}\int_{0}^{\infty} \frac{R(\Omega)}{\Omega^2-\omega^2}\, d\Omega \tag{3.4.4}$$

Similarly noting that $X(\Omega)$ is odd, we use the odd part of $(\Omega-\omega)^{-1}$ in eq. (3.4.2) so that

$$R(\omega)=\frac{2}{\pi}\int_0^{\infty}\frac{-\Omega X(\Omega)}{\Omega^2-\omega^2}\,d\Omega \tag{3.4.5}$$

It is often useful in determining $X(\omega)$ from $R(\omega)$ to approximate the latter by straight line segments and use numerical integration. For this purpose alternate forms for the the $\mathcal{H}$ transform of $R(\omega)$ which depend on $dR(\omega)/d\omega$ are convenient. To obtain these forms, integrate eq. (3.4.1) by parts using

$$\begin{aligned} dv &= \frac{d\Omega}{\Omega-\omega} \\ v &= \ln(\Omega-\omega) \\ u &= R(\Omega) \\ du &= \frac{dR(\Omega)}{d\Omega} \end{aligned}$$

As a principal value we then obtain

$$\begin{aligned} X(\omega) = \lim_{e\to 0}\frac{1}{\pi}\Bigg\{ & R(\Omega)\ln(\Omega-\omega)\Bigg|_{-\infty}^{\omega-e} - \int_{-\infty}^{\omega-e}\frac{dR(\Omega)}{d\Omega}\ln(\Omega-\omega)\,d\Omega \\ & + R(\Omega)\ln(\Omega-\omega)\Bigg|_{\omega+e}^{+\infty} - \int_{\omega+e}^{+\infty}\frac{dR(\Omega)}{d\Omega}\ln(\Omega-\omega)\,d\Omega\Bigg\} \end{aligned}$$

Since

$$\ln(\Omega-\omega)=\ln(-|\Omega-\omega|)=j\pi+\ln|\Omega-\omega|,\ \ \Omega<\omega$$

$$\ln(\Omega-\omega)=\ln|\Omega-\omega|,\ \ \Omega>\omega$$

the quantity $j\pi R(\Omega)$ cancels from the first two terms of the above equation for $X(\omega)$, hence the terms in $R(\Omega)\ln(\Omega-\omega)$ become

$$\lim_{e\to 0}\left[R(\Omega)\ln|\Omega-\omega|\Bigg|_{-\infty}^{\omega-e} - R(\Omega)\ln|\Omega-\omega|\Bigg|_{+\infty}^{\omega+e}\right]=0$$

and the new equation for $X(\omega)$ vs. $R(\omega)$ is

$$X(\omega)=-\frac{1}{\pi}\,P\int_{-\infty}^{+\infty}\frac{dR(\Omega)}{d\Omega}\ln|\Omega-\omega|\,d\Omega \tag{3.4.6}$$

Just as in the case of eq. (3.4.5), we can modify the form of eq. (3.4.6) by using the even part of the integrand. Since the derivative of an even function is odd, we use the odd part of $\ln|\Omega-\omega|$, i.e., $\frac{1}{2}(\ln|\Omega-\omega|-\ln|-\Omega-\omega|)$. Then

$$X(\omega) = \frac{1}{\pi} \int_0^{+\infty} \frac{dR(\Omega)}{d\Omega} \ln \frac{|\Omega + \omega|}{|\Omega - \omega|} \, d\Omega \tag{3.4.7}$$

In many applications of the Hilbert Transform, $F(j\omega)$ is an impedance function and $R(\omega)$, $X(\omega)$ its resistive and reactive components. It is evident that the impedance may include an arbitrary constant series resistance without affecting $X(\omega)$. If such a constant term does not appear in $F(j\omega)$, it is known as a *Minimum Resistance Function.* Similarly the impedance may include in series a branch made up of purely reactive elements consisting, say, of the interconnection of inductors and capacitors; such a series reactance branch introduces impedance poles on the $j\omega$-axis, and its presence has no effect on the resistance function, $R(\omega)$. If $F(j\omega)$ does not include such a series reactance term with its associated imaginary axis poles, it is known as a *Minimum Reactance Function.* When $F(j\omega)$ is a minimum resistance, minimum reactance function, $R(\omega)$ and $X(\omega)$ are uniquely related by the Hilbert Transform. It should be noted that if our causal $f(t)$ function is square integrable (hence $F(j\omega)$ as well), these arbitrary resistance and reactance terms will not be present.

Example 3.4.1
According to Example 3.2.19

$$\mathcal{F}[A_1\epsilon^{-\alpha\ t}\,\mathrm{u}(t)] = F(j\omega) = \frac{A_1}{\alpha + j\omega}$$

Changing $F(j\omega)$ slightly

$$F(j\omega) = \frac{1}{a + jb\omega} = \frac{a}{a^2 + b^2\omega^2} + \frac{-jb\omega}{a^2 + b^2\omega^2}$$

The function $f(t)$ is causal and $F(j\omega)$ is both a minimum resistance and a minimum reactance function since there are no poles on $j\omega$ and no constant term. Deduce $X(\omega)$ from $R(\omega)$ by the Hilbert Transform.

Solution Substitute $R(\omega) = a/(a^2 + b^2\omega^2)$ into eq. (3.4.4)

$$X(\omega) = \frac{2a\omega}{\pi} P \int_0^{+\infty} \frac{1}{(a^2 + b^2\Omega^2)(\Omega^2 - \omega^2)} \, d\Omega =$$

$$= \frac{2a\omega}{\pi} P \int_0^{+\infty} \left(\frac{A}{a^2 + b^2\Omega^2} + \frac{B}{\Omega^2 - \omega^2} \right) d\Omega$$

We have expanded the integrand in partial fractions and

$$A = \frac{-b^2}{a^2 + b^2\omega^2} \qquad B = \frac{1}{a^2 + b^2\omega^2}$$

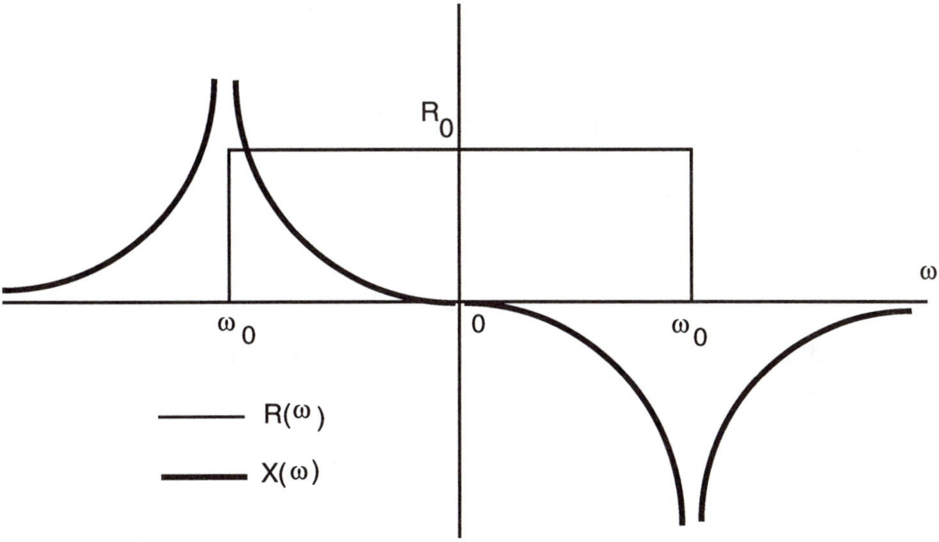

FIGURE 3.4.1
Plot of $X(\omega) = \mathcal{H}R(\omega)$, $R(\omega) = R_0 p_{2\omega_0}(\omega)$.

Integrating and letting $e \to 0$

$$P\int_0^{+\infty} \frac{B}{\Omega^2 - \omega^2}\, d\Omega = -\frac{B}{2}\ln\left|\frac{\omega+\Omega}{\omega-\Omega}\right| \Big|_0^{\omega-e} - \frac{B}{2}\ln\left|\frac{\omega+\Omega}{\omega-\Omega}\right| \Big|_{\omega+e}^{+\infty} = 0$$

$$P\int_0^{+\infty} \frac{A}{a^2 + b^2\Omega^2}\, d\Omega = \frac{A}{ab}\arctan\frac{b\Omega}{a}\Big|_0^{\omega-e} + \frac{A}{ab}\arctan\frac{b\Omega}{a}\Big|_{\omega+e}^{+\infty} = \frac{A\pi}{2ab}$$

Substituting for A, we verify $X(\omega)$

$$X(\omega) = \frac{2a\omega\pi}{2ab\pi}A = \frac{-b\omega}{a^2 + b^2\omega^2}$$

□

Example 3.4.2
Given $R(\omega) = R_0 p_{2\omega_0}(\omega)$, the rectangular resistance function of height R_0 extending from $-\omega_0 \le \omega \le \omega_0$ and zero elsewhere. Find the associated $X(\omega)$.

Solution It is convenient to use the slope form of the Hilbert Transform (3.4.4). For $0 < \omega$, $dR(\omega)/d\omega$ is zero everywhere except at $\omega = \omega_0$ where there is a discontinuous jump in $R(\omega)$ equal to $-R_0$. Referring to eq. (3.1.3), the resistance slope for $0 \le \omega$ is

$$\frac{dR(\omega)}{d\omega} = -R_0\,\delta(\omega - \omega_0) \qquad 0 \le \omega$$

Taking $F(\omega)$ as a minimum reactance function, its imaginary part corresponding to $R(\omega)$ is therefore

$$\begin{aligned} X(\omega) &= \frac{1}{\pi}\int_0^{+\infty} \frac{dR(\Omega)}{d\Omega} \ln\frac{|\Omega+\omega|}{|\Omega-\omega|}\, d\Omega \\ &= -\frac{R_0}{\pi}\int_0^{+\infty} \delta(\Omega-\omega_0) \ln\frac{|\Omega+\omega|}{|\Omega-\omega|}\, d\Omega \\ &= -\frac{R_0}{\pi} \ln\frac{|\omega_0+\omega|}{|\omega_0-\omega|} \end{aligned}$$

The Hilbert Transform $X(\omega) = \mathcal{H}R(\omega)$ is drawn on Fig. (3.4.1) and $X(-\omega) = -X(\omega)$ is also shown. There are logarithmic singularities (not poles) at $\omega = \pm\omega_0$. □

3.5 Causal Fourier Transforms

Our basic concern with LTI causal systems gives the Hilbert Transform special significance since it applies to functions $F(j\omega) = R(\omega) + jX(\omega)$ which are the Fourier transforms of causal signals $f(t)$, i.e., signals that vanish for $t < 0$. The question then arises as to the special properties of the Fourier transform of a causal signal which make its real and imaginary parts eligible for Hilbert transformation.

It will be important to employ a continuation of the Fourier Transform from the $j\omega$ axis to the plane of the complex variable $s = \sigma + j\omega$. One mode for doing this is to define the *Laplace Transform* of a function, $\mathcal{L}f(t)$, as the Fourier Transform of $f(t)\,\epsilon^{-\sigma t}$, so that

$$\mathcal{L}f(t) = \mathcal{F}[f(t)\,\epsilon^{-\sigma t}]$$

Since

$$\mathcal{F}[f(t)\,\epsilon^{-\sigma t}] \equiv \int_{-\infty}^{+\infty} f(t)\,\epsilon^{-\sigma t}\epsilon^{-j\omega t}\,dt = \int_{-\infty}^{+\infty} f(t)\,\epsilon^{-(\sigma+j\omega)t}\,dt = F(\sigma+j\omega) \tag{3.5.1}$$

we can introduce the complex variable $s = \sigma + j\omega$ and obtain

$$\mathcal{L}f(t) \equiv \int_{-\infty}^{+\infty} f(t)\,\epsilon^{-st}\,dt = F(s) \tag{3.5.2}$$

The Laplace transform $F(s)$ is defined by the integral in eq. (3.5.2) for all values of the complex variable s for which the integral converges. It is also often referred to as the *Complex Fourier Transform*.[11]

Consider first the Laplace transform of the *anti-causal* signal $f(t) = \epsilon^{at}\,\mathrm{u}(-t)$, $a > 0$. This function vanishes for $t > 0$ and decays exponentially to the left of the origin; it is clearly in L^2, and the integral (3.5.2) yields

$$\mathcal{L}[\epsilon^{at}\,\mathrm{u}(-t)] = \frac{1}{a-s} \qquad \Re s < a$$

On the other hand, evaluating the Laplace transform integral for the similar but causal function $f(t) = \epsilon^{-at}\,\mathrm{u}(t)$, $a > 0$ gives

$$\mathcal{L}[\epsilon^{-at}\,\mathrm{u}(t)] = \frac{1}{a+s} \qquad \Re s > -a$$

The basic distinction between the two transforms is evident. That of the noncausal function has a pole in the right half of the s plane at $s = a$, whereas the transform of the causal function is analytic everywhere in the right half plane. Its one pole lies to the left of the $j\omega$-axis.[12]

Let us show that, in general, the Laplace Transform of a causal signal is analytic everywhere in the right half of the s plane, i.e. in $\Re s > 0$. Suppose then that the Laplace Transform, $F(s)$ exists, $\Re s \geq 0$ and corresponds to a causal signal, $f(t) = 0$, $t < 0$. Then

$$F(s) = \int_0^{+\infty} f(t)\,\epsilon^{-st}\,dt$$

By hypothesis this is convergent for $\sigma = 0$. It is, in fact, uniformly convergent in the right half plane where $\sigma > 0$, for $|\epsilon^{-st}| = \epsilon^{-\sigma t}$ acts as a damping factor since $t > 0$. For analyticity in a region, the derivative must exist (see eq. (A.3.4)).

$$\frac{dF(s)}{ds} = \int_0^{+\infty} (-t) f(t)\,\epsilon^{-st} dt$$

This, too, is uniformly convergent in $\Re s > 0$ due to the exponential damping factor, despite the presence of the t multiplier. We conclude that $F(s)$, corresponding to a causal $f(t)$, is analytic everywhere in the open right half s plane.[13] One form of the converse result is illustrated by the following

[11]Designating $\mathcal{F}f(t) = \hat{f}(\omega)$ (see footnote 9, p. 116), the *Complex Fourier Transform* often employed by physicists is obtained by replacing ω with $z = \omega + j\tau$.

[12]Attention! The Fourier transform of certain causal time functions may not exist but their Laplace transforms, eq. (3.5.2) may be well defined. For example, $\mathrm{u}(t)$ has, strictly speaking, no Fourier transform, but its Laplace transform is $1/s$. On the other hand, $\epsilon^{-\sqrt{|t|}}$, (noncausal), has no Laplace transform, but its Fourier transform, $F(j\omega)$, exists. Using the same nomenclature for both cases can sometimes cause confusion.

[13]A detailed proof is given in D. V. Widder, *The Laplace Transform*, Princeton, Princeton University Press, 1941.

sufficiency theorem which applies to the case of L^1 Fourier transforms that correspond to causal time functions.

THEOREM 3.5.1

Let the Fourier transform $F(j\omega)$ be given and be absolutely integrable $-\infty \leq \omega \leq +\infty$, and let the Laplace transform continuation $F(s)$ have the following properties

(a) *$F(s)$ is devoid of singularities in $\Re\, s > 0$, and almost everywhere on $s = j\omega$, (i.e., except for denumerable singular points), with the proviso that $F(j\omega)$ is bounded in an arbitrarily small neighborhood of each singularity.*

(b) *$F(s)$ satisfies the Jordan conditions on the right half plane semicircle C_2 centered at the origin and of radius R. That is, $F(s)$ goes to zero uniformly on C_2 as $R \to +\infty$.*

Then $f(t)$ is causal, i.e., $f(t) = 0$, $t < 0$.

PROOF (Note that this Theorem excludes $j\omega$ poles.) We use the following argument to demonstrate the result. Consider the contour integral

$$\frac{1}{2\pi j} \oint_C F(s)\, \epsilon^{st}\, ds, \quad s = \sigma + j\omega$$

where C is the closed curve in the complex plane which follows the $j\omega$-axis from $-jR$ to jR and closes to the right on the large semicircle C_2. The portion of the path along $j\omega$ avoids any singularities by infinitesimal indentations, C_1, to the right. Thus, $F(s)\, \epsilon^{st}$ is analytic inside the closed path and continuous in the closure, and the Cauchy integral theorem tells us that the closed path integral is zero.

We are given that $F(s)$ goes to zero uniformly on the big semicircle C_2 as $R \to \infty$. Furthermore, the exponential in the transform integrand acts as a damping factor in the right half plane ($\sigma > 0$) when $t < 0$. Under these conditions, the Jordan Lemma (see Section A.13, Lemma A.13.1) applies and specifies that as $R \to +\infty$, the contribution of the line integral along C_2 is zero. The integral along the $j\omega$ portion of the path exists as a principal value, since, by hypothesis, $F(j\omega)$ is absolutely integrable $-\infty < \omega < +\infty$. Also, since $F(s)$ is bounded on $j\omega$ arbitrarily close to each singularity, (the function $\epsilon^{-\sqrt{s}}$, singular at $s = 0$, illustrates this property) and analytic in $\Re s > 0$, it is bounded on C_1. Therefore, the line integral contribution for this portion of the contour goes to zero as the radii of the indentations approach zero. Thus the zero value contour integral may be expressed as

$$\frac{1}{2\pi j} \oint_C F(s)\, \epsilon^{st}\, ds = \frac{1}{2\pi} \int_{-\infty}^{+\infty} F(j\omega)\, \epsilon^{j\omega t}\, d\omega = f(t) = 0 \quad t < 0$$

since the second integral is just the Inverse Fourier Transform. ∎

An important basic theorem is that given by Titchmarsh.[14] It applies to L^2 signals and presents a broadly applicable set of conditions, which are symmetric in t and s with respect to square integrability, for the existence of causal signals and their transforms.

THEOREM 3.5.2

(Titchmarsh Theorem) *The class of causal square-integrable functions $f(t)$ is identical with the class of functions whose Laplace Transforms $F(s)$ are both analytic in the right half of the s-plane, $\Re s = \sigma > 0$, and satisfy*

$$\int_{-\infty}^{+\infty} |F(\sigma + j\omega)|^2 \, d\omega \; < +\infty \quad \forall \sigma > 0 \tag{3.5.3}$$

The above statement of the theorem is that of Youla et al.,[15] and the following proof follows Titchmarsh.

PROOF Necessity: Given that $f(t)$ is causal and square integrable, then the Fourier transform, $F(j\omega)$, and its continuation, the Laplace transform exist. As discussed above, it follows that the Laplace transform $F(s)$ is analytic in $\Re s > 0$ since $f(t)$ is causal. Since the causal signal $f(t)$ is square integrable, so, too, is $f(t)\,\epsilon^{-\sigma t}$, $t > 0$, $\sigma > 0$, because of the exponential damping factor. Hence applying the Parseval Theorem eq. (3.3.11) to such a function and its Laplace transform given by eq. (3.5.1), we obtain

$$\frac{1}{2\pi}\int_{-\infty}^{+\infty} |F(\sigma + j\omega)|^2 \, d\omega = \int_{0}^{+\infty} |f(t)\,\epsilon^{-\sigma t}|^2 \, dt < +\infty$$

and eq. (3.5.3) follows.

Sufficiency: Given that $F(s)$ is analytic $\Re s > 0$ and that eq. (3.5.3) is satisfied in the same region, to show that $f(t)$ is causal and in L^2.

The following inverse transform exists since $F(\sigma + j\omega)$ satisfies eq. (3.5.3).

$$\phi(t,\sigma) = \frac{1}{2\pi}\int_{-\infty}^{+\infty} F(\sigma + j\omega)\,\epsilon^{j\omega t} \, d\omega \quad \sigma > 0 \tag{3.5.4}$$

[14] E. C. Titchmarsh, *An Introduction to the Theory of Fourier Integrals*, Oxford, Clarendon Press, 1948, pp. 122-132 and, in particular, Theorem 95.

[15] D. C. Youla, L. J. Castriota, and H. J. Carlin, "Bounded Real Scattering Matrices and the Foundations of Linear Passive Network Theory", *IRE Trans. on Circ. Th.*, vol. CT-6, no. 1, March 1959, pp. 102-124.

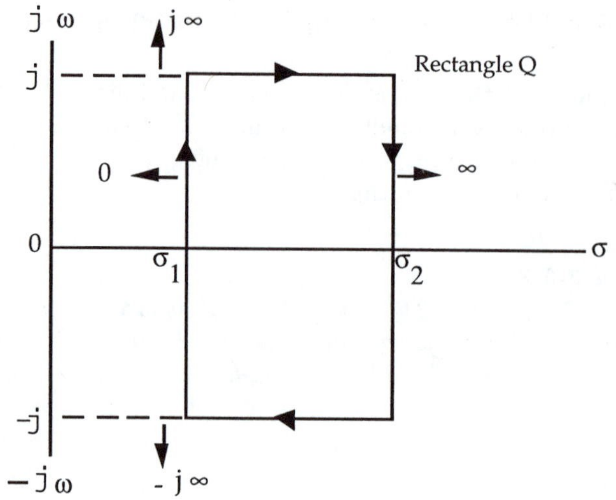

FIGURE 3.5.1
Rectangular Contour Q for Titchmarsh Theorem.

We wish to show that $\phi(t,\sigma) \equiv f(t)\,\epsilon^{-\sigma t}$, $\forall\sigma > 0$. Consider the contour integral

$$\oint_Q F(s)\,\epsilon^{st}\,ds = 0$$

which by the Cauchy Integral Theorem is zero since by hypothesis the integrand has no singularities inside and on Q, a rectangular contour in the right half plane, symmetric about σ, with one pair of sides parallel to the $j\omega$ axis, the other pair parallel to σ at a distance $\pm\omega_0$ from the real axis.

Now $F(s)$ is square integrable along any path parallel to $j\omega$ in the right half plane, so we expect it to go to zero along any such path as $\omega \to \pm\infty$.[16] Therefore, $F(s) \to 0$ along the upper and lower sides of the rectangle as $\omega_0 \to \pm\infty$. Evidently the only contribution to the contour integral around the infinite rectangle is due to the two sides parallel to $j\omega$. Since the value of the contour integral is zero, the integrals along the left and right sides of the rectangle must be equal so as to cancel over the complete contour. Thus

$$\frac{1}{2\pi}\int_{-\infty}^{+\infty} F(\sigma_1 + j\omega)\,\epsilon^{(\sigma_1+j\omega)t}\,d\omega = \frac{1}{2\pi}\int_{-\infty}^{+\infty} F(\sigma_2 + j\omega)\,\epsilon^{(\sigma_2+j\omega)t}\,d\omega$$

Remove $\epsilon^{\sigma_k t}$ to the outside of each integral and using the notation of eq. (3.5.4)

$$\epsilon^{\sigma_1 t}\phi(t,\sigma_1) = \epsilon^{\sigma_2 t}\phi(t,\sigma_2), \quad \forall\,\sigma_1,\sigma_2 > 0, \quad \forall\,t$$

[16] E. C. Titchmarsh, *An Introduction to the Theory of Fourier Integrals*, Oxford, Clarendon Press, 1948, Lemma p. 125.

Let $\sigma_1 = \sigma > 0$ and let $\sigma_2 = 1$. Then

$$\epsilon^{\sigma t}\phi(t,\sigma) = \epsilon^{t}\phi(t,1)$$

The right hand side is independent of σ, so that we can let $\epsilon^{t}\phi(t,1) = f(t)$. We therefore obtain $\phi(t,\sigma) = f(t)\,\epsilon^{-\sigma t}$. Substituting into eq. (3.5.4)

$$f(t)\,\epsilon^{-\sigma t} = \phi(t,\sigma) = \frac{1}{2\pi}\int_{-\infty}^{+\infty} F(\sigma + j\omega)\,\epsilon^{j\omega t}\,d\omega \quad \sigma > 0, \forall t$$

Thus we have the inverse transform of $F(s)$. Applying the Parseval equation

$$\frac{1}{2\pi}\int_{-\infty}^{+\infty} |F(\sigma + j\omega)|^2\,d\omega = \int_{-\infty}^{+\infty} |f(t)\,\epsilon^{-\sigma t}|^2\,dt < \infty \tag{3.5.5}$$

We can now show that $f(t)$ is causal. Consider

$$\int_{-\infty}^{-e} |f(t)|^2\,dt \le \int_{-\infty}^{-e} |f(t)\,\epsilon^{-\sigma(t+e)}|^2\,dt = \epsilon^{-2\sigma e}\int_{-\infty}^{-e} |f(t)\,\epsilon^{-\sigma t}|^2\,dt, \quad e > 0$$

The above inequality follows since $-\sigma(t+e) \ge 0$ for any (negative) t in the range of integration. But from eq. (3.5.5) the integral on the right is bounded, say by M, $\forall\sigma > 0$. Therefore, $\int_{-\infty}^{-e} |f(t)|^2\,dt \le \epsilon^{-\sigma e}M$. Taking the limit as $\sigma \to +\infty$

$$\int_{-\infty}^{-e} |f(t)|^2\,dt = 0$$

whence it follows that, since $|f(t)|^2 \ge 0$, $f(t) = 0,\ t < 0$. $f(t)$ is causal.

Finally, expression (3.5.5) is now bounded as $\sigma \to 0$, so that the causal $f(t)$ itself belongs to L^2, and the Theorem is proved. ■

An important corollary given by Titchmarsh is the following.

COROLLARY 3.5.1

If the conditions of the Titchmarsh Theorem are satisfied, then

$$\lim_{\sigma \to 0^+} F(\sigma + j\omega) \to F(j\omega), \quad \text{for almost all } \omega$$

i.e., the Laplace transform converges pointwise a.e. to the Fourier transform when σ tends to zero.

Example 3.5.1

Illustrate the Titchmarsh theorem with the causal signal $f(t) = t\epsilon^{-at}\,\mathrm{u}(t)$, $a > 0$.

Solution By direct integration we verify that the signal is in L^2

$$\int_0^\infty |f(t)|^2 \, dt = \frac{2}{a^3}$$

We find $F(j\omega)$ by a simple device. Let $\mathcal{F}f_1(t) = F_1(j\omega)$, then

$$\frac{dF_1}{d\omega} = \int_{-\infty}^{+\infty} (-jt) f_1(t) \, \epsilon^{-j\omega t} dt$$

or

$$\mathcal{F}[tf_1(t)] = F(j\omega) = j\frac{dF_1}{d\omega}$$

Apply this relation to the example with $F_1 = \mathcal{F}[\epsilon^{-at} \, \mathrm{u}(t)] = 1/(j\omega + a)$. Then

$$F(j\omega) = j\frac{d}{d\omega}\frac{1}{j\omega + a} = \frac{1}{(j\omega + a)^2}$$

The Laplace transform is

$$\mathcal{L}[t\epsilon^{-at} \, \mathrm{u}(t)] = F(s) = \frac{1}{(s + a)^2}$$

$F(s)$ has only one singularity, a pole in the left half plane at $s = -a$, so that, as expected, it is analytic $\forall \, \Re \, s > 0$. The other properties predicted by Theorem 3.5.3 are also evident; for example, $\lim_{\omega \to \infty} F(\sigma + j\omega) = 0$, $\forall \sigma > 0$.

□

Example 3.5.2

Illustrate the Titchmarsh Theorem with the causal time limited function $f(t) = \epsilon^{jt}$, $0 \leq t \leq a$, and 0 otherwise. What happens if $f(t)$ is nonzero over the interval $-a \leq t \leq a$, and zero elsewhere, i.e., is noncausal.

Solution The causal signal is clearly square integrable. By straightforward integration

$$F(s) = \int_0^a \epsilon^{jt} \epsilon^{-st} \, dt = \frac{1 - \epsilon^{-(s-j)a}}{s - j}$$

$F(s)$ is certainly analytic for all finite s in the right half plane. What about the case where $\omega \to \infty$? Under this condition $F(s)$ approaches zero uniformly, since its numerator is bounded $\forall \sigma > 0$, and the magnitude of the denominator becomes infinite. Accordingly $F(s)$ is analytic *everywhere* in $\sigma > 0$ as demanded by the Titchmarsh Theorem.

At first glance it appears that there is a pole in $F(s)$ on the $j\omega$ axis at $s = j$. This would be embarrassing since $F(j\omega)$ would then not be square integrable, contrary to the Parseval Theorem. However, the pole is only apparent, for note that both numerator and denominator approach zero as $s \to j$. If we take the derivative of numerator and denominator and apply L'Hospital's rule, then $F(j) = a$, i.e., no pole exists at $s = j$.

The noncausal signal is still in L^2, but with Fourier integral limits $-a$ to a, the result is

$$F(s) = \frac{1}{s-j}\{\epsilon^{(s-j)a} - \epsilon^{-(s-j)a}\}$$

As in the causal case, the function is analytic in the *finite* right half plane, and there is no pole at $s = j$. However, with $s = \sigma + j\omega$, $0 < \sigma \to +\infty$, the numerator becomes *exponentially* infinite, and $F(s)$ is no longer bounded, hence not analytic $\forall \Re s > 0$, despite the fact that in the denominator $|s - j| \to +\infty$. The complex transform of the noncausal function is thus neither analytic nor bounded everywhere in the right half plane. □

We conclude with another version of the Hilbert transform based on the Titchmarsh theorem.

THEOREM 3.5.3

Let $f(t)$ be causal and square integrable with Laplace transform $F(s)$. Let $s = \sigma + j\omega$, $\sigma \neq 0$ be a point in the interior of the complex plane. Then $F(s)$ is given by its boundary behavior

$$F(s) = \begin{cases} \dfrac{1}{2\pi}\displaystyle\int_{-\infty}^{+\infty} \frac{F(j\omega)}{s - j\Omega}\, d\Omega & \sigma > 0 \\ 0 & \sigma < 0 \end{cases}$$

PROOF According to the Titchmarsh Theorem, $F(s)$ is analytic in the right half plane, so that by the Cauchy Integral Formula (see eq. (A.3.1))

$$F(s) = \frac{1}{2\pi j}\oint_Q \frac{F(\zeta)}{s - \zeta}\, d\zeta \tag{3.5.6}$$

where the contour Q is the rectangle of Fig. 3.5.1 in the right half plane; s is located inside Q, and ζ is the complex variable $\Sigma + j\Omega$.

Now allow the upper and lower sides of Q to approach $\pm j\infty$, the right hand side to move to $+\infty$, and the left hand side to approach the $j\Omega$ axis. As in the proof of the Titchmarsh Theorem, $F(\Sigma + j\Omega) \to 0$ as $\Omega \to \pm\infty$ so that the contribution to the integral of the upper and lower sides of Q is 0. Since $F(\zeta)$ is bounded everywhere in $\Sigma > 0$, the integrand of eq. (3.5.6) goes to zero as $\Sigma \to +\infty$ because in the denominator $|s - \zeta| \to +\infty$. Thus the integral along the right side of Q is also zero.

In the limit, therefore, as the left side of Q approaches $j\Omega$, the only contribution to the contour integral is the upper equation of Theorem 3.5.3.

If $\Re s < 0$, the integrand of eq. (3.5.6) has no pole inside Q and the contour integral is 0 by the Cauchy Integral Theorem so that the second equation of the theorem follows. ∎

By this same argument, it is immediately evident that

$$\int_{-\infty}^{+\infty} \frac{F(j\omega)}{s+j\Omega}\, d\Omega = 0, \quad \Re s > 0 \tag{3.5.7}$$

3.6 Minimum Immittance Functions

It is evident from the preceding Section 3.5 that, starting with causal signals, the real-imaginary part Hilbert transform relations give functions $R(\omega)$, $X(\omega)$ which can be continued from the $j\omega$-axis to form a complex function analytic everywhere in the right half of the s plane. As a consequence it is possible to obtain the Laplace transform $F(s)$ as an analytic function directly from the boundary values of either component. Suppose then we proceed from $R(\omega)$ and generate the complex transform $F(s)$.

THEOREM 3.6.1

Let $R(\omega)$ be even, real, and square integrable. Then it is always possible to find a Laplace transform $F(s)$ analytic in $\Re s > 0$, so that $\Re F(j\omega) = R(\omega)$.

PROOF Since $R(\omega)$ is even, real, and in L^2, its inverse transform $f_e(t)$ exists and is even, real, and in L^2. As in Section 3.4 we can choose the odd function

$$f_o(t) = f_e(t)\, \mathrm{sign}(t)$$

Now define

$$f(t) = f_e(t) + f_o(t) = 2f_e(t)\, \mathrm{u}(t)$$

Clearly $f(t)$ is causal and in L^2, hence by the Titchmarsh Theorem 3.5.3, its complex transform is $F(s)$, analytic in $\Re s > 0$. Furthermore, $\mathcal{F} f_o(t)$ is imaginary so that

$$\Re F(j\omega) = \mathcal{F} f_e(t) = R(\omega)$$

and the Theorem is proved. ∎

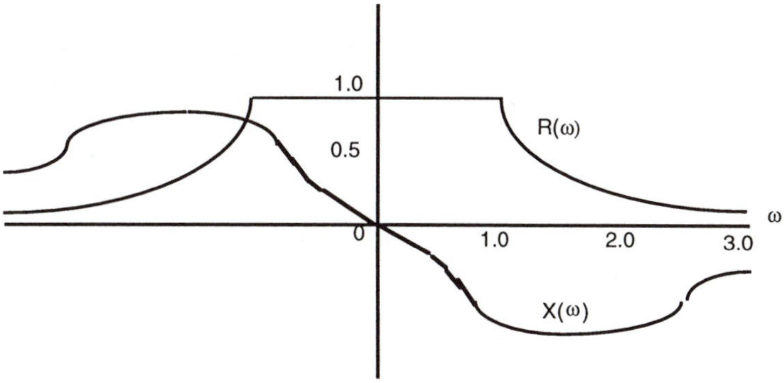

FIGURE 3.6.1
***R*, *X* from Hilbert relation.**

Since $F(j\omega)$ is square integrable it can have no poles on $j\omega$. Thus the complex function $F(s)$ generated by $R(\omega)$ is a *minimum reactance* function. This term specifically refers to an impedance function without $j\omega$ poles. For an admittance without $j\omega$ poles we would use *minimum susceptance* function, and the generic term that includes either one is *minimum immittance* function.

Having shown that $R(\omega)$ can generate a Laplace transform analytic in the right half plane, we now deduce a formula which yields $F(s)$ directly from $R(\omega)$.

Based on Theorem 3.5.3 we can write

$$F(s) = \frac{1}{2\pi}\left(\int_{-\infty}^{+\infty} \frac{F(j\Omega)}{s - j\Omega}\, d\Omega + \int_{-\infty}^{+\infty} \frac{F(j\Omega)}{s + j\Omega}\, d\Omega\right)$$

where $\Re(s) > 0$, and hence the second integral is zero by eq. (3.5.7). Combining terms

$$F(s) = \frac{s}{\pi}\int_{-\infty}^{+\infty} \frac{F(j\Omega)}{s^2 + \Omega^2}\, d\Omega, \quad \Re\, s > 0$$

Along $j\Omega$, $F(j\Omega) = R(\Omega) + jX(\Omega)$ so that

$$F(s) = \frac{s}{\pi}\int_{-\infty}^{+\infty} \frac{R(\Omega) + jX(\Omega)}{s^2 + \Omega^2}\, d\Omega = \frac{2s}{\pi}\int_{0}^{+\infty} \frac{R(\Omega)}{s^2 + \Omega^2}\, d\Omega, \quad \Re s > 0 \tag{3.6.1}$$

since the term $X(\Omega)/(s^2 + \Omega^2)$ is odd in Ω, its contributions to the integral at $+\Omega$ and $-\Omega$ cancel; the other term is even, hence the factor of 2 for limits 0 and ∞. Thus eq. (3.6.1) gives an expression for determining the analytic continuation $F(s)$ in $\Re s > 0$ directly from a real, even, square integrable $R(\omega)$.

Example 3.6.1

Let $R(\omega) = 1$, $1 \geq |\omega|$ and $R(\omega) = 1/\omega^2$, $1 \leq |\omega|$. Determine the corresponding Laplace transform $F(s)$, analytic in $\Re s > 0$.

Solution Figure 3.6.1 shows $R(\omega)$, the given square integrable resistance function. Apply eq. (3.6.1) to $R(\omega)$.

$$\begin{aligned} F(s) &= \frac{2s}{\pi}\int_0^1 \frac{1}{s^2+\Omega^2}\,d\Omega + \frac{2s}{\pi}\int_1^{+\infty} \frac{1}{\Omega^2(s^2+\Omega^2)}\,d\Omega = \\ &= \frac{2s}{\pi}\left[\frac{1}{s}\arctan\frac{\Omega}{s}\bigg|_0^1 - \left(\frac{1}{s^2\Omega} + \frac{1}{s^3}\arctan\frac{\Omega}{s}\bigg|_1^\infty\right)\right] = \\ &= \frac{2}{\pi}\left[\frac{1}{s} + (1+\frac{1}{s^2})\arctan\frac{1}{s} - \frac{\pi}{2s^2}\right] \end{aligned}$$

The final equation is simplified if the identity $\arctan 1/s = \pi/2 - \arctan s$ is used. Thus

$$F(s) = 1 + \frac{2}{\pi s} - \frac{2}{\pi}(1+\frac{1}{s^2})\arctan s \tag{3.6.2}$$

It appears that somehow a pole has been introduced at $s = 0$. However, this is only a mirage. As $s \to 0$, $\arctan s \to s$, therefore,

$$\lim_{s\to 0} F(s) = 1 - \frac{2s}{\pi}$$

and there is no pole at the origin or, for that matter, anywhere else.

If $s \to j\omega$, then according to Theorem 3.6.1 the functions $R(\omega)$ and $X(\omega)$ can be determined. Use the identity

$$\arctan s = \frac{j}{2}\ln\frac{j+s}{j-s}\bigg|_{s=j\omega} = \frac{j}{2}\ln\frac{1+\omega}{1-\omega} \tag{3.6.3}$$

In other words, $F(s)$ has a *logarithmic singularity* or *branch point* at $s = j$. Substitute eq. (3.6.3) into eq. (3.6.2) with $0 \leq \omega < 1$. Then,

$$F(j\omega) = R(\omega) + jX(\omega)$$

and

$$R(\omega) = 1 \qquad\qquad 0 \leq \omega < 1$$

$$X(\omega) = -\frac{1}{\pi}\left[\frac{2}{\omega} + \left(1 - \frac{1}{\omega^2}\right)\ln\left|\frac{1+\omega}{1-\omega}\right|\right] \qquad\qquad 0 \leq \omega < 1$$

It might be pointed out that since $x \ln x \to 0$, as $x \to 0$, it follows that $X(1) = -2/\pi$. Real and imaginary parts can also be determined when $1 \leq \omega$. In that case from eq. (3.6.3)

$$\arctan j\omega = \frac{j}{2}\ln(-1)\frac{\omega+1}{\omega-1}$$

and since we can take $\ln(-1) = jk\pi$, where $k = \pm 1$

$$\arctan j\omega = -\frac{k\pi}{2} + \frac{j}{2}\ln\left|\frac{1+\omega}{1-\omega}\right| \quad 1 \leq \omega$$

If this is substituted into eq. (3.6.2)

$$F(j\omega) = 1 + k(1 - \frac{1}{\omega^2}) - \frac{j}{\pi}\left[\frac{2}{\omega} + (1 - \frac{1}{\omega^2})\ln\left|\frac{1+\omega}{1-\omega}\right|\right] \quad 1 \leq \omega$$

It is evident that to retrieve the correct $R(\omega)$ we must choose $k = -1$ and then

$$R(\omega) = \frac{1}{\omega^2} \quad 1 \leq \omega$$

which is the prescribed resistance function for $1 \leq \omega$. The reactance is

$$X(\omega) = -\frac{j}{\pi}\left[\frac{2}{\omega} + (1 - \frac{1}{\omega^2})\ln\left|\frac{1+\omega}{1-\omega}\right|\right] \quad 1 \leq \omega$$

Figure 3.6.1 shows a plot of $R(\omega)$ and $X(\omega)$. □

3.7 Amplitude-Phase Relations

In preceding sections we have examined relations between real and imaginary parts of *causal* Fourier transforms $F(j\omega)$ whose continuation $F(s)$ is analytic everywhere in the right half plane. In this section we consider relations between the amplitude and phase of $F(j\omega)$.

Formally the basic results are easily established. Take the logarithm of $F(j\omega)$

$$\ln F(j\omega) = \ln|F(j\omega)| + j\theta(\omega)$$

Thus $\Re \ln F(j\omega) = \ln|F(j\omega)| \equiv a(\omega)$ and $\Im F(j\omega) = \arg F(j\omega) \equiv \theta(\omega)$. Referring to eq. (3.4.4), the Hilbert transform relating (log) amplitude and phase is therefore

$$\theta(\omega) = \frac{2\omega}{\pi}\int_0^{+\infty} \frac{a(\omega)}{\Omega^2 - \omega^2}\, d\Omega \tag{3.7.1}$$

And the slope version from eq. (3.4.7) is

$$\theta(\omega) = \frac{1}{\pi}\int_0^{+\infty} \frac{da(\Omega)}{d\Omega}\ln\left|\frac{\Omega+\omega}{\Omega-\omega}\right| d\Omega \tag{3.7.2}$$

The representation of the analytic continuation $\ln F(s)$ follows directly from eq. (3.6.1), which becomes

$$\ln F(s) = \frac{2s}{\pi}\int_0^{+\infty} \frac{\ln|F(j\omega)|}{s^2 + \Omega^2}\, d\Omega, \quad \Re s > 0 \tag{3.7.3}$$

The above transform relations deal with the logarithm function. Evidently, if they are to be valid, it is the function $\ln F(s)$, i.e., the continuation of $\ln F(j\omega)$, which must be analytic in $\Re s > 0$. This demands that $F(s)$ (a) be analytic in the right half plane and (b) *have no zeros in the right half plane.* The second constraint is imposed to avoid logarithmic singularities in $\Re s > 0$. A function $F(s)$ which satisfies requirements (a) and (b) is known as a *Minimum Phase Function.*

This designation stems from the following consideration. Let

$$F_0(s) = (s + \alpha)\, g(s), \quad \alpha > 0 \text{ and real}$$

where $g(s)$ is analytic and without zeros in $\Re s > 0$.

$$|F_0(j\omega)| = |(j\omega + \alpha)\, g(j\omega)|$$

Modify $F_0(s)$ by transferring the left half plane zero at $s = -\alpha$ to $s = \alpha$.

$$F(s) = (s - \alpha)\, g(s) = \frac{s - \alpha}{s + \alpha} F_0(s)$$

The amplitude of $F(j\omega)$ is

$$|F(j\omega)| = \left| \frac{j\omega - \alpha}{j\omega + \alpha} \right| |F_0(j\omega)| = |F_0(j\omega)| = a(\omega)$$

The two functions have the same amplitude on $s = j\omega$. On the other hand

$$\theta(\omega) = \theta_0(\omega) + \arg \frac{j\omega - \alpha}{j\omega + \alpha}$$

The function $B(s) = (s - \alpha)/(s + \alpha)$ is an example of a *Blaschke Fraction* (more generally, several such fractions each with unit amplitude on $s = j\omega$ are multiplied together forming a *Blaschke Product*) and its $j\omega$ phase is

$$\arg B(j\omega) = \pi - 2 \arctan \frac{\omega}{\alpha}$$

If this is substituted in the preceding equation

$$\theta(\omega) = \theta_0(\omega) + \pi - 2 \arctan \frac{\omega}{\alpha}$$

In other words, the phase $\theta(\omega)$ associated with $F(s)$ (a *nonminimum phase function*) always exceeds $\theta_0(\omega)$ associated with $F_0(s)$, though both $F(j\omega)$ and $F_0(j\omega)$ have the same amplitude on the imaginary axis. Thus for the given $a(\omega)$ the phase function is not unique, but $\theta_0(\omega)$ is the minimum possible phase associated with the given amplitude. The result clearly extends to any set of zeros not on $j\omega$, be they real and/or in complex conjugate pairs.

Evidently, when the integral converges, eq. (3.7.3) gives a minimum phase function ($F(s)$ will have no right half-plane zeros), and eqs. (3.7.1), (3.7.2) give the phase of such a function. Note that in the special case $|F(j\omega)|$ is a real constant the transform equations give a phase which is zero everywhere. The alternate solution of a phase of $j\pi$ ($F(s)$ a negative constant) is not obtained and can be included under the rubric of non-minimum phase.

In Theorem 3.6.1 it was shown that it was always possible, given $R(\omega)$ in L^2, to determine $X(\omega)$, so that $F(s)$ is analytic in $\Re s > 0$ with the corresponding $f(t)$ causal. The question now arises as to the restrictions on the amplitude function $|F(j\omega)| \equiv a(\omega)$ to insure that a phase function $\theta(\omega)$ exists so that the complex continuation $F(s)$, $F(j\omega) = |F(j\omega)| \exp j\theta(\omega)$, is analytic in the right half plane ($f(t)$ causal). The answer is given by the *Paley-Wiener Theorem.*[17]

THEOREM 3.7.1

(Paley-Wiener Theorem) *The necessary and sufficient conditions that $a(\omega)$ be the amplitude of the $\mathcal{F}$-transform of a causal L^2 function $f(t)$, nonzero a.e., are*

$$\begin{aligned} &(a) \quad a(\omega) > 0 \ \text{(a.e.)}; \\ &(b) \quad a(\omega) \text{ be in } L^2; \\ &(c) \quad \int_{-\infty}^{+\infty} \frac{|\ln a(\omega)|}{1+\omega^2}\, d\omega < +\infty. \end{aligned}$$

COROLLARY 3.7.1

If $a(\omega)$ satisfies the Paley-Wiener Theorem, then let $a(\omega) \equiv |F(j\omega)|$ and a phase function $\theta(\omega)$ can be determined such that $F(j\omega) = |F(j\omega)|\epsilon^{j\theta(\omega)}$ has a continuation $F(s)$ which is analytic in $\Re s > 0$.

The Paley Wiener Theorem can be used as a necessity test for the physical realizability of the amplitude spectrum of a transfer function. Suppose in an LTI system the impulse response $h(t)$ has Fourier transform $H(j\omega)$, defined as the system or transfer function. Let $X(j\omega)$ be the transform of an arbitrary causal input signal $x(t)$, and let the output signal $y(t)$ have transform $Y(j\omega)$. The output can be obtained by convolving the input with

[17] R.E.A.C. Paley and N. Wiener, "Fourier Transforms in the Complex Domain", *Amer. Math. Soc. Colloquium Publication 19*, New York, 1934. A simplified proof is given in: A. Papoulis, *The Fourier Integral and its Applications*, New York, McGraw Hill Book Co., 1962.

the impulse response function $y(t) = h(t) * x(t)$, and the product rule gives

$$Y(j\omega) = H(j\omega)\,X(j\omega) \tag{3.7.4}$$

Since $x(t)$ is causal, $X(j\omega)$ continues analytically into $\Re s > 0$. Since the system is causal if and only if its response to *any* causal input is also causal, realizability demands that the output $y(t)$ be causal, and its complex transform be analytic in $\Re s > 0$. Referring to eq. (3.7.4), this can only occur if the system function has an analytic continuation into the right half of the s plane. (We cannot count on cancellations by $X(s)$ of pole factors in $H(s)$, since the latter is fixed and the former is arbitrary.) It is evident, therefore, that a necessary condition for the physical realizability of a transducer is that the amplitude of its transfer function, $|H(j\omega)|$, satisfies the Paley-Wiener Theorem. One other point: a causal function which corresponds to an amplitude function satisfying the Paley-Wiener Theorem may be a complex time signal. Since the impulse response of a physical system is a real signal, this requires that $|H(j\omega)|$ be an even function (see discussion of eq. (3.2.32)). In summary:

COROLLARY 3.7.2

For a square integrable function $a(\omega)$ to be realizable as the amplitude of the transfer function of a physical LTI system, it is necessary that it be even in ω and satisfy the Paley-Wiener Theorem.

Example 3.7.1

Consider the ideal rectangular low-pass amplitude spectrum

$$|H(j\omega)| = p_2(\omega) \equiv a(\omega) = \begin{cases} 1, & |\omega| \leq 1 \\ 0, & |\omega| > 1 \end{cases}$$

Determine whether there is an associated phase function so that $H(j\omega)$ can be continued analytically into $\Re s > 0$, to define a causal system.

Solution The given amplitude is square integrable but requirement (a) of the Paley-Wiener criterion, Theorem 3.7.1 is not satisfied since $a(\omega)$ is zero over an interval. Therefore no continuation, associated with the given amplitude, exists which is analytic in the right half plane. In other words the ideal rectangular low-pass frequency response cannot be realized by a causal structure. □

Example 3.7.2

Repeat Example 3.7.1 for the low-pass amplitude spectrum

$$|H(j\omega)|^2 \equiv a^2(\omega) = \frac{1}{1+\omega^{2n}}, \quad 0 < n < +\infty \text{ and integer}$$

This is known as a Butterworth gain function.

Solution The amplitude $a(\omega)$ is square integrable, even, and positive and the integrand in (c) of Theorem 3.7.1 is $\ln(1+\omega^{2n})/[2(1+\omega^2)]$. If the integration is from 0 to ω_c and then to $+\infty$, where $\omega_c \gg 1$, the contribution up to ω_c is finite. We therefore consider the integral from ω_c to $+\infty$. Clearly we may disregard the 1 in numerator and denominator so that we need only evaluate

$$\int_{\omega_c}^{+\infty} \frac{\ln\ \omega^{2n}}{\omega^2}\, d\omega = -2n \left(\frac{\ln\ \omega}{\omega} + \frac{1}{\omega} \right) \Bigg|_{\omega_c}^{+\infty}$$

Since as $\omega \to +\infty$,

$$\left(\frac{\ln\ \omega}{\omega} + \frac{1}{\omega} \right) \to 0$$

the integral is clearly finite and the given transfer function can be continued analytically as a causal transfer function.

Note that as n becomes very large (but not infinite), the response becomes arbitrarily close to the ideal lowpass gain function $p_2(\omega)$. For any finite value of n there is a causal continuation, but not in the limit. In general we term any function which satisfies the Paley-Wiener criterion *a causal amplitude.* Evidently it is sufficient for a nonnegative function to be a causal amplitude if its square is *rational* and even, and it is square integrable. □

Example 3.7.3

Given $|H(j\omega)| = \epsilon^{-\omega^2}$ (the Gaussian function), determine if there is a continuation, analytic in $\Re\, s > 0$, which establishes this function as a causal amplitude.

Solution Since

$$\int_{-\infty}^{+\infty} \epsilon^{-2\omega^2}\, d\omega = \sqrt{\pi/2}$$

$|H|$ is square integrable. Also, it is nonnegative. The Paley Wiener integral is

$$2\int_0^{+\infty} \frac{\omega^2}{1+\omega^2}\, d\omega$$

For $\omega \gg 1$ the integrand approaches unity, so the integral is clearly divergent. Hence given the Gaussian amplitude there is no continuation which is analytic in $\Re\, s > 0$, and the transfer function is not causal.

As a verification of this result use as $\mathcal{F}$-transform $H(j\omega) = |H(j\omega)|\, \epsilon^{-j\omega t_o}$, which has the prescribed amplitude. Referring to the delay rule, eq. (3.2.27), the time domain response is $h(t - t_o)$, i.e. $h(t) =$

$\mathcal{F}^{-1}|H(j\omega)|$ delayed by t_o. The transform of the Gaussian is well known.[18] Thus

$$\mathcal{F}[\epsilon^{-a^2t^2}] = \frac{\sqrt{\pi}}{a}\,\epsilon^{-(\omega/2a)^2}$$

In other words, taking $a = 1/2$ we have

$$h(t) = \mathcal{F}^{-1}|H(j\omega)| = \frac{1}{2\sqrt{\pi}}\,\epsilon^{-(t/2)^2}$$

so that the time domain response is itself a Gaussian function. The Gaussian curve has the familiar bell shape centered about the origin, with infinitely long tails for positive and negative values of the abscissa. Thus, as expected, the final delayed response $h(t - t_o)$ is noncausal $\forall t_o$. As an alternate verification it is not difficult to show that the continuation $F(s) = \epsilon^{s^2}\,\epsilon^{-st_o}$ fails the Titchmarsh Theorem for causal transforms (Theorem 3.5.3), since eq. (3.5.3) is not satisfied. □

3.8 Numerical Evaluation of Hilbert Transforms

A numerical procedure for evaluating the Hilbert transform is often necessary since in many problems the integrals are not determinable in closed form. Furthermore the real or imaginary parts are frequently given as data rather than as mathematical functions. The numerical methods discussed by Bode[19] form the basis for the following presentation and are easily adapted as programs on a personal computer.

Suppose then that the real part $R(\omega)$, $0 \le \omega < +\infty$ is given, and the minimum reactance imaginary part $X(\omega)$ is to be computed at some specified angular frequency ω. The calculation is carried out by first approximating $R(\omega)$ as a set of connected straight line segments. Since each segment has a constant slope, its contribution to $X(\omega)$ is easily found by using the slope form of the Hilbert transform (3.4.7). The contributions of all the segments are summed and the result is X at the prescribed frequency. A computer program would then simply use a loop to calculate the imaginary part in a similar fashion at each of a succession of frequencies in a given band, and the final result would be an approximation of $X(\omega)$ over the band.

To see how this is carried out refer to Fig. 3.8.1. Part (a) shows a resistance curve approximated by n line segments. Note that the approximating

[18] For example, see Example 2-10 in Papoulis (footnote 17 page 145).

[19] H. W. Bode, *Network Analysis and Feedback Amplifier Design*, New York, D. Van Nostrand Co., 1945.

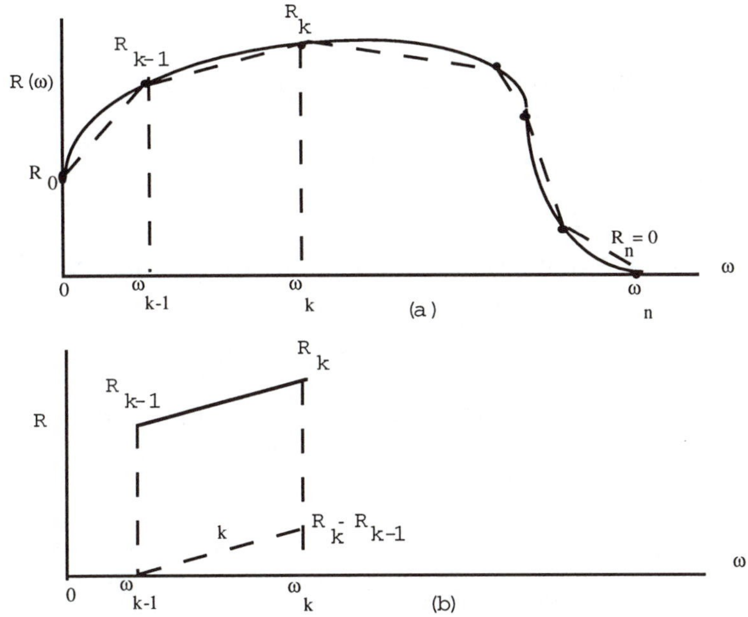

FIGURE 3.8.1
Line segment approximation of real part for Hilbert transform computation.

resistance is set to zero beyond the final line segment. This is a practical choice consistent with the assumption of a square integrable $R(\omega)$. A typical approximating line segment is shown in Fig. 3.8.1 (b). The original $R(\omega)$ is approximated by adding all the line segments typified by the line connecting R_{k-1} to R_k, to the level at $\omega = 0$, R_0. To find an approximation to the minimum reactance corresponding to $R(\omega)$ at a given value of ω, the slope form of the Hilbert transform, eq. (3.4.7), is used to find the imaginary part $X_k(\omega)$ for each line segment and $X(\omega) = \sum_k X_k(\omega)$. Thus $X_k(\omega)$ is given by

$$X_k(\omega) = \frac{M_k}{\pi} \int_{\omega_{k-1}}^{\omega_k} \ln \frac{|\Omega + \omega|}{|\Omega - \omega|} \, d\Omega \tag{3.8.1}$$

In this equation the constant slope of the k-th line segment is

$$M_k = \frac{R_k - R_{k-1}}{\omega_k - \omega_{k-1}} \tag{3.8.2}$$

The integral in eq. (3.8.1) is readily evaluated and the imaginary part $X_k(\omega)$ is

$$X_k(\omega) = \frac{M_k}{\pi} [F_k(\omega) - F_{k-1}(\omega)] \tag{3.8.3}$$

where

$$F_i(\omega) \equiv F(\omega, \omega_i) = (\omega + \omega_i) \ln |\omega + \omega_i| + (\omega - \omega_i) \ln |\omega - \omega_i| \qquad (3.8.4)$$

and

$$X(\omega) = \sum_{k=1}^{n} X_k(\omega) \qquad (3.8.5)$$

Especially when writing a general Hilbert transform program based on eqs. (3.8.3), (3.8.4), (3.8.5), it is useful to note that

$$F(\omega, 0) = 2\omega \ln \omega, \quad F(\omega_k, \omega_k) = 2\omega_k \ln \omega_k, \quad F(0, \omega_k) = 0 \qquad (3.8.6)$$

As an illustration of the numerical evaluation of Hilbert transforms we consider the resistance and reactance of a given impedance, and check approximate against exact values.

Example 3.8.1
Given a parallel RC circuit, with $R = 1$, $C = 2$, determine the reactance by using a line segment approximation of the resistance function and numerically evaluating the Hilbert transform. As notation, let the exact impedance be $z(j\omega) = r(\omega) + jx(\omega)$, and the approximation be $Z(j\omega) = R(\omega) + jX(\omega)$.

Solution In order not to complicate the numerical work, we make a rather simple line segment approximation of the resistance function which nevertheless will be seen to yield a reasonable result. The actual impedance of the RC load is

$$z(j\omega) = \frac{1}{1 + 2j\omega} = \frac{1}{1 + 4\omega^2} - j\frac{2\omega}{1 + 4\omega^2}$$

The curves of $r(\omega)$, $x(\omega)$ are shown as solid lines on Fig. 3.8.2. The resistance $r(\omega)$ is approximated by 3 line segments as shown by the dashed lines, $R(\omega)$, on Fig. 3.8.2. The tableau below gives the numerical values at the break points of the line segments:

$$\omega \Rightarrow [0 \quad .8 \quad 1.6 \quad 4]$$

$$R(\omega) \Rightarrow [1.0 \quad .281 \quad .089 \quad 0]$$

As a typical calculation we use eq. (3.8.5) to find the reactance $X(1)$ from $R(\omega)$

$$X(1) = \sum_{k=1}^{3} X_k(1)$$

Thus, referring to the tableau and eqs. (3.8.2), (3.8.3), (3.8.4),

$$M_1 = \frac{.8 - 0}{.281 - 1} = -1.1127 \quad M_2 = -.2400 \quad M_3 = -.0371$$

also

$$F_1(1) \equiv F(1, .8) = (1 + .8) \ln|1 + .8| + (1 - .8) \ln|1 - .8| = .73613$$

Similarly,

$$F_0(1) \equiv F(1, 0) = 0,\ F_2(1) \equiv F(1, 1.6) = 2.79183,\ F_3(1) \equiv F(1, 4) = 4.75135$$

and

$$X_1(1) = \frac{1}{\pi} M_1(F_1 - F_0) = \frac{1}{\pi}(-.66163)$$

$$X_2(1) = \frac{1}{\pi} M_2(F_2 - F_1) = \frac{1}{\pi}(-.49303)$$

$$X_3(1) = \frac{1}{\pi} M_3(F_3 - F_2) = \frac{1}{\pi}(-.07274)$$

so that the final result $X(1)$ compared with the true value $x(1)$ is

$$X(1) = X_1 + X_2 + X_3 = -.39073 \quad x(1) = -.4$$

A calculation at $\omega = .5$ gives

$$X(.5) = -.4676 \quad x(.5) = -.500$$

A plot showing $X(\omega)$, and $x(\omega)$ computed by a program which repeats the calculations given above over a range of frequencies is given in Fig. 3.8.2. As can be seen, even employing a small number of line segments can lead to a reasonable approximation. Obviously, a characteristic with many fluctuations requires a large number of approximating line segments. A useful point to keep in mind for rough calculations is that the contributions of a segment remote from the frequency in question is small (compare $X_1(1)$ with $X_3(1)$ in the above calculations), and line segments with large slopes make large contributions. □

The procedure just described is, of course, applicable to amplitude-phase calculations. Given a minimum phase function, $Z(j\omega) = |Z(j\omega)|\, \epsilon^{j\theta(\omega)}$, line segments are used to approximate $\ln|Z(j\omega)|$ as the real part, and the result is a Hilbert transform approximation to the minimum phase imaginary part $\theta(\omega)$.

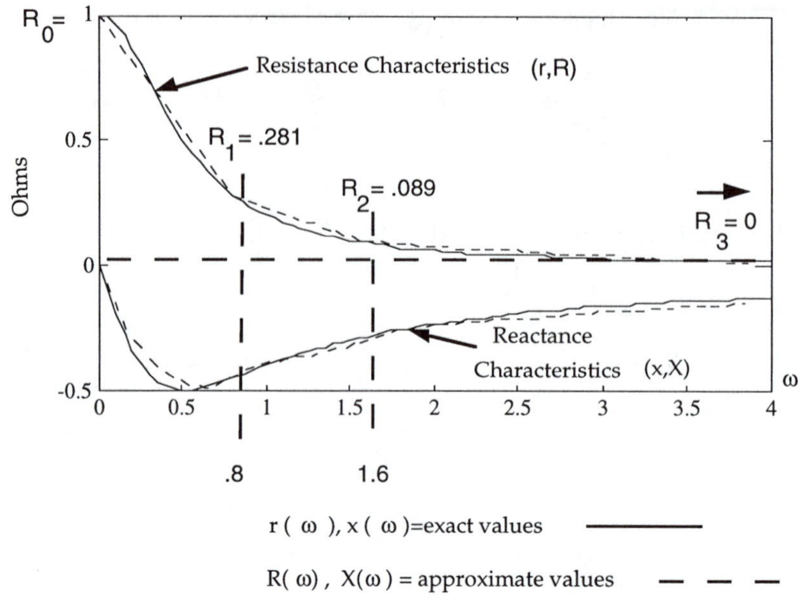

FIGURE 3.8.2
Hilbert transform approximation for parallel RC load, $R = 1$, $C = 2$.

3.9 Operational Rules and Generalized Fourier Transforms

In previous sections various properties of L^2 Fourier transforms have been obtained which can be used to find the Fourier transform of a given function, based on the known transform of a related function, without going through an explicit integration process. These properties are known as Operational Rules and in many cases simplify transform evaluation. These rules also lead to useful formal means for defining a generalization of Fourier Transforms when applied to functions that are not square integrable.

Rule (2) (see Table 3.9) is a direct consequence of the Definition (1), and Rules (3), (4) follow immediately by a change of variable in (1). The following examples give derivations of the differentiation and integration rules shown in the table.

Example 3.9.1
Verify the time and frequency derivative rules (10), (11) of the table, and find $\mathcal{F}[\frac{d^2}{dt^2}(t^2\epsilon^{-at}\,\mathrm{u}(t))]$.

Solution Differentiate the definition integral with respect to t under the

Table 3.1 Operational Rules for Fourier Transforms.

The functions $f(t) = \mathcal{F}_t^{-1}[F(j\omega)]$ and $F(j\omega) = \mathcal{F}f(t)$ are given.

Rule	Modified $f(t)$	Modified $F(j\omega)$	Reference
1. Definition	$\frac{1}{2\pi}\int_{-\infty}^{+\infty} \epsilon^{j\omega t} F(j\omega)\, d\omega$	$\int_{-\infty}^{+\infty} \epsilon^{-j\omega t} f(t) dt$	(3.2.18) (3.2.17)
2. Linearity	$a_1 f_1(t) + a_2 f_2(t)$	$a_1 F_1(j\omega) + a_2 F_2(j\omega)$	
3. Reflection	$f(-t)$	$F(-j\omega)$	
4. Scaling	$f(at)$	$F(j\omega/a)$	
5. Delay	$f(t-t_0)$	$\epsilon^{-j\omega t_0} F(j\omega)$	(3.2.27)
6. Modulation	$\epsilon^{j\omega_0 t} f(t)$	$F(j(\omega-\omega_0))$	(3.2.28)
7. Symmetry	$F(t)$	$2\pi f(-j\omega)$	(3.2.25)
8. Multiplication	$f_1(t)\, f_2(t)$	$\frac{1}{2\pi} F_1(j\omega) * F_2(j\omega)$	(3.3.9)
9. Convolution	$f_1(t) * f_2(t)$	$F_1(j\omega)\, F_2(j\omega)$	(3.3.10)
10. Derivative (t)	$f^{(n)}(t)$	$(j\omega)^n F(j\omega)$	Example 3.9.1
11. Derivative (ω)	$(-jt)^n f(t)$	$F^{(n)}(j\omega)$	Example 3.9.1
12. Integration	$\int_{-\infty}^{t} f(t)\, dt$	$\frac{F(j\omega)}{j\omega}$	Example 3.9.2

integral sign, $df(t)/dt = \mathcal{F}^{-1}F(j\omega)\}$.

$$f^{(1)}(t) = \frac{1}{2\pi}\int_{-\infty}^{+\infty} j\omega F(j\omega)\, \epsilon^{j\omega t}\, d\omega = \mathcal{F}^{-1}[j\omega F(j\omega)]$$

or

$$\mathcal{F}f^{(1)}(t) = j\omega F(j\omega)$$

It is evident that we simply repeat the process n times to obtain Rule (10). Rule (11) is similarly obtained except the starting point is $d/d\omega\, F(j\omega) = \mathcal{F}f(t)$.

To find the Fourier transform $\mathcal{F}\{\frac{d^2}{dt^2}[t^2\epsilon^{-at}\, \mathrm{u}(t)]\}$ we note that

$$t^2\epsilon^{-at}\, \mathrm{u}(t) = -(-jt)^2\epsilon^{-at}\, \mathrm{u}(t)$$

and from eq. (3.2.20) $\mathcal{F}[\epsilon^{-at}\,\mathrm{u}(t)] = 1/(a+j\omega)$. Thus using Rule (11)

$$\mathcal{F}[t^2\epsilon^{-at}\,\mathrm{u}(t)] = -\frac{d^2}{d\omega^2}\frac{1}{a+j\omega} = -\frac{2}{(a+j\omega)^3}$$

and now applying Rule (10)

$$\mathcal{F}\frac{d^2}{dt^2}[t^2\epsilon^{-at}\,\mathrm{u}(t)] = \omega^2\frac{2}{(a+j\omega)^3}$$

□

Example 3.9.2
Verify the integration Rule (12) of the table and use it to evaluate $\mathcal{F}\int_{-\infty}^{t}(-t\,\epsilon^{-a|t|})\,dt$, $a > 0$.

Solution Let

$$\mathcal{F}\int_{-\infty}^{t} f(t)\,dt = G(j\omega)$$

We then apply Rule (10), noting that the derivative of an integral with respect to a variable upper limit is the integrand

$$j\omega G(j\omega) = \mathcal{F}\frac{d}{dt}\int_{-\infty}^{t} f(t)\,dt = \mathcal{F}f(t) = F(j\omega)$$

and Rule (12) follows.

It is evident that the integration rule cannot be applied unless $F(j\omega)/j\omega$ exists so that it is necessary that $F(0) = 0$ to avoid a singularity at the origin. This leads to an additional constraint on $f(t)$:

$$F(0) = \int_{-\infty}^{+\infty} f(t)\,\epsilon^{-j\omega t}\,dt\,|_{\omega=0} = \int_{-\infty}^{+\infty} f(t)\,dt = 0$$

The example calls for the evaluation

$$G(j\omega) = \mathcal{F}\int_{-\infty}^{t} f(t)\,dt; \quad f(t) = -t\,\epsilon^{-a|t|}$$

The function $f(t)$ is odd, and the integral exists for limits $(0,\ +\infty)$, so that the integral constraint is satisfied, and we may therefore use the integration rule. First note that applying the reflection Rule (3)

$$\mathcal{F}\epsilon^{-a|t|} = \mathcal{F}[\epsilon^{at}\,\mathrm{u}(-t)] + \mathcal{F}[\epsilon^{-at}\,\mathrm{u}(t)] = \mathcal{F}h(-t) + \mathcal{F}h(t) = H(-j\omega) + H(j\omega)$$

Thus, since $H(j\omega) = \mathcal{F}[\epsilon^{-at}\,\mathrm{u}(t)] = 1/(a+j\omega)$

$$\mathcal{F}\epsilon^{-a|t|} = \frac{2a}{a^2+\omega^2}$$

Next use the differentiation Rule (11)

$$\mathcal{F}f(t) = \mathcal{F}[(-j)(-jt)\epsilon^{-a|t|}] = -j\frac{d}{d\omega}\frac{2a}{a^2+\omega^2} = \frac{4aj\omega}{(a^2+\omega^2)^2}$$

Finally applying the integration Rule (12)

$$\mathcal{F}\int_{-\infty}^{t}(-t\,\epsilon^{-a|t|})\,dt = \frac{1}{j\omega}\mathcal{F}f(t) = \frac{4a}{(a^2+\omega^2)^2}$$

□

Fourier transforms can be defined in a useful manner for singularity functions such as steps, impulses, and other functions which are not square integrable by first generalizing the Fourier Transform to include the delta function and then applying the operational rules given in the preceding table. The sifting property (3.1.6) immediately yields the transform of unit impulse

$$\mathcal{F}\delta(t) = \int_{-\infty}^{+\infty}\delta(t)\,\epsilon^{-j\omega t}\,dt = 1 \tag{3.9.1}$$

If eq. (3.9.1) is interpreted as $\delta(t) = \mathcal{F}^{-1}1$ and written out as in eq. (3.2.18), an interesting representation of the delta function is obtained

$$\delta(t) = \frac{1}{2\pi}\int_{-\infty}^{+\infty}\epsilon^{j\omega t}d\omega \tag{3.9.2}$$

This representation expresses the remarkable fact that $\delta(t)$ can be visualized as possessing a spectrum made up of all the harmonics, each with unit amplitude and zero phase. Note that the integral (3.9.2) does not converge in the strong sense (i.e., as a pointwise function), but in the weak sense (see footnote at p. 106).

$$\lim_{\Omega\to+\infty}\int_{-\infty}^{+\infty}\phi(t)\,dt\int_{-\Omega}^{+\Omega}\epsilon^{j\omega t}\,d\omega = \phi(0)$$

(In effect the second integral acts as $\delta_n(t)$.)[20]

We can also obtain the transform of a constant by using symmetry (Rule (7) of the table). Recalling that $\delta(\omega)$ is even

$$\mathcal{F}1 = 2\pi\delta(\omega) \tag{3.9.3}$$

The Fourier Transform of the unit step does not exist in the ordinary sense, but a generalized meaning can be given to the integral by considering

[20] L. Schwartz, *Théorie des distributions*, 2 vols., Paris, Hermann & Cie., 1950-1951. For a more accessible presentation see: M. J. Lighthill, *Fourier Analysis and Generalised Functions*, Cambridge, Cambridge University Press, 1978.

the complex Fourier transform. As in eq. (3.5.2) this is obtained by inserting a convergence factor $\epsilon^{-\sigma t}$, $\sigma > 0$ into the integrand. Thus

$$\mathcal{L}\,\mathrm{u}(t) = \int_{-\infty}^{+\infty} \mathrm{u}(t)\,\epsilon^{-(\sigma+j\omega)t}\,dt = \int_{0}^{+\infty} \mathrm{u}(t)\,\epsilon^{-(\sigma+j\omega)t}\,dt$$

Let $s = \sigma + j\omega$, then

$$\mathcal{L}\,\mathrm{u}(t) = \int_{0}^{+\infty} \mathrm{u}(t)\,\epsilon^{-st}\,dt = \frac{1}{s} \tag{3.9.4}$$

To see how eq. (3.9.4) is related to the generalized transform of $\mathrm{u}(t)$, consider the integral with respect to ω of the transform of $\mathrm{u}(t)\,\epsilon^{-\sigma t}$ as $\sigma \to 0$

$$\int_{-\infty}^{+\infty} \mathcal{F}[\mathrm{u}(t)\,\epsilon^{-\sigma t}]\,d\omega = \int_{-\infty}^{+\infty} \frac{1}{\sigma + j\omega}\,d\omega$$

Since $1/(\sigma + j\omega) \equiv \sigma/(\sigma^2 + \omega^2) - j\omega/(\sigma^2 + \omega^2)$, it follows that

$$\int_{-\infty}^{+\infty} \mathcal{F}[\mathrm{u}(t)\,\epsilon^{-\sigma t}]\,d\omega = \int_{-\infty}^{+\infty} \frac{\sigma}{\sigma^2 + \omega^2}\,d\omega - j\int_{-\infty}^{+\infty} \frac{\omega}{\sigma^2 + \omega^2}\,d\omega$$

Now let $\sigma \to 0$. Referring to the delta function representation of eq. (3.2.2), the integrand of the first term in the above equation is

$$\lim_{\sigma \to 0} \frac{\sigma}{\sigma^2 + \omega^2} = \pi\delta(\omega)$$

Accordingly

$$\int_{-\infty}^{+\infty} \mathcal{F}\,\mathrm{u}(t)\,dt = \int_{-\infty}^{+\infty} \pi\delta(\omega)\,d\omega + P\int_{-\infty}^{+\infty} \frac{1}{j\omega}\,d\omega$$

where, as in eq. (3.4.3) because of the singularity at the origin, the second integral must be evaluated as a principal value.

In view of the last equation, $1/s$, the Laplace transform of $\mathrm{u}(t)$, is taken to have as its limit on $j\omega$ the generalized Fourier Transform

$$\mathcal{F}\,\mathrm{u}(t) = \pi\delta(\omega) + PF\frac{1}{j\omega} \tag{3.9.5}$$

The letters PF stand for *pseudofunction* to emphasize that the argument is not integrable in the ordinary sense, but only as a principal value.

Example 3.9.3

Employ the operational rules to deduce the following generalized Fourier transforms:

(a) $\mathcal{F}[\text{sign}(t)]$;

(b) $\mathcal{F}[\epsilon^{j\omega_0 t}]$;

(c) $\mathcal{F}[\cos\,\omega_0 t]$, $\mathcal{F}[\sin\,\omega_0 t]$.

Solution

(a) Apply Rules (2), (3), and (3.9.5)

$$\begin{aligned}\mathcal{F}[\text{sign}(t)] &= \mathcal{F}[\text{u}(t) - \text{u}(-t)] \\ &= \pi\delta(\omega) + PF\frac{1}{j\omega} - \left[\pi\delta(-\omega) + PF\frac{1}{-j\omega}\right]\end{aligned}$$

The result is

$$\mathcal{F}[\text{sign}(t)] = PF\frac{2}{j\omega}$$

(b) Apply Rule (6) and (3.9.3)

$$\mathcal{F}[\epsilon^{j\omega_0 t}] = 2\pi\delta(\omega - \omega_0) \tag{3.9.6}$$

(c) Using Euler's formula for $\sin\omega_0 t$ and $\cos\omega_0 t$ together with Rule (3.9.6) gives

$$\mathcal{F}[\cos\,\omega_0 t] = \pi\delta(\omega - \omega_0) + \pi\delta(\omega + \omega_0)]$$

and

$$\mathcal{F}[\sin\,\omega_0 t] = \frac{1}{j}[\pi\delta(\omega - \omega_0) - \pi\delta(\omega + \omega_0)]$$

□

Example 3.9.4

Use the generalized Fourier transform to verify the L^2 transform of the pulse function $p_a(t)$.

Solution The pulse function is defined by eq. (3.2.19). It is readily expressed in terms of unit step functions.

$$p_a(t) = \text{u}\left(t + \frac{a}{2}\right) - \text{u}\left(t - \frac{a}{2}\right)$$

Thus, using eq. (3.9.5) together with Rule (5)

$$\mathcal{F}[p_a(t)] = \epsilon^{j\omega a/2}\left[\pi\delta(\omega) + PF\frac{1}{j\omega}\right] - \epsilon^{-j\omega a/2}\left[\pi\delta(\omega) + PF\frac{1}{j\omega}\right]$$

Combining terms

$$\mathcal{F}[p_a(t)] = 2j\sin\frac{\omega a}{2}\pi\delta(\omega) + 2j\sin\frac{\omega a}{2}PF\frac{1}{j\omega}$$

Referring to eq. (3.1.4), the first term has the form $f(0)\,\delta(x) = 0$. The second term no longer has the singularity at the origin so that PF can be deleted. Thus

$$\mathcal{F}[\,p_a(t)] = \frac{a \sin \dfrac{\omega a}{2}}{\dfrac{\omega a}{2}}$$

which agrees with eq. (3.2.26). □

In the following example the elementary theory of amplitude modulation and demodulation is presented in terms of generalized Fourier transforms.

Example 3.9.5

Given a signal $f(t)$ whose Fourier transform $F(j\omega)$ is band limited, i.e., the *frequency spectrum* is zero outside the cutoff points $\pm\omega_c$. $f(t)$ is then *amplitude modulated* (multiplied by) $\cos \omega_0 t$ where $\omega_c < \omega_0$, the *carrier frequency*.

(a) Find the Fourier transform (spectrum) of the modulated signal $g(t) = f(t) \cos \omega_0 t$.

(b) Show how to demodulate $g(t)$, i.e., retrieve $f(t)$.

Solution

(a) Since

$$f(t) \cos \omega_0 t = \frac{f(t)}{2}\left(\epsilon^{j\omega_0 t} + \epsilon^{-j\omega_0 t}\right)$$

apply the modulation Rule (6).

$$G(j\omega) = \mathcal{F}[f(t) \cos \omega_0 t] = \frac{1}{2}\{F[j(\omega - \omega_0) + F(j(\omega + \omega_0)]\} \quad (3.9.7)$$

The spectrum of the modulated signal consists of an *upper sideband* $F[j(\omega - \omega_0)]$, $\omega > \omega_0$, and a *lower sideband* $F[j(\omega - \omega_0)]$, $\omega < \omega_0$, whose bandwidth is that of $F(j\omega)$, ω_c, but centered about the carrier frequency ω_0; these are the images of $F(j\omega)$, $\omega > 0$ and $F(-j\omega)$, $\omega < 0$. The energy of the modulated signal is concentrated around the high carrier frequency, and it is this radio frequency signal which is transmitted as a radio wave. Due to eqs. (3.2.32) and (3.2.33), we have $F(-j\omega) = F^*(j\omega)$, so that the spectrum at negative frequencies is completely determined by its amplitude and phase at positive (physical) frequencies. Thus the lower sideband is completely determined by the upper one, and vice versa; it is therefore possible to transmit only one sideband, thus reducing the required channel bandwidth by one half.

(b) One method of demodulating the radio signal, $g(t)$, is to multiply it by a periodic function of frequency ω_0, and then extract the original signal $f(t)$. Thus, as in eq. (3.9.7),

$$M(j\omega) = \mathcal{F}[g(t)\, 2\cos\omega_0 t] = G[j(\omega - \omega_0)] + G[j(\omega + \omega_0)]$$

Substituting eq. (3.9.7)

$$M(j\omega) = F(j\omega) + \frac{1}{2}\{F[j(\omega - 2\omega_0)] + F[j(\omega + 2\omega_0)]\}$$

The resultant spectrum of the received radio signal has a band (composed of an upper and a lower sideband) centered at $2\omega_0$ and its reflected image for negative frequencies centered at $-2\omega_0$. In addition the original signal spectrum $F(j\omega)$ centered at zero frequency and with bandwidth ω_c is also present. To retain $F(j\omega)$ and eliminate upper and lower sidebands (and their images for negative frequencies), which are well separated from $\pm\omega_c$, $m(t)$ is passed through an ideal low pass filter whose cutoff is $\pm\omega_c$. Therefore, the filter has the frequency response $p_{2\omega_c}$. The spectrum of the signal delivered at the filter output is therefore

$$M(j\omega)\, p_{2\omega_c} = F(j\omega)$$

In other words, the upper and lower high frequency sidebands have been eliminated, and the original spectrum and associated signal, $f(t)$, have been retrieved.

□

3.10 Laplace Transforms and Eigenfunction Response

The Laplace transform, defined by eq. (3.5.2) here repeated

$$\mathcal{L}[f(t)] \equiv F(s) = \int_{-\infty}^{+\infty} f(t)\, \epsilon^{-st}\, dt, \quad s = \sigma + j\omega \tag{3.10.1}$$

converges in a vertical strip parallel to $j\omega$ in the s plane, though this is not noted in eq. (3.10.1). For example, if for $t > 0$, $f(t) = \exp(-\alpha t)$, and for $t < 0$, $f(t) = \exp(\beta t)$, with $0 < \alpha$, $0 < \beta$, then the strip of convergence for $F(s)$ has its lower limit at $\sigma_0 = -\alpha$, and its upper limit at $\sigma_1 = \beta$. For causal signals, eq. (3.10.1) reduces to[21]

$$\mathcal{L}[f(t)] = F(s) = \int_{0-}^{+\infty} f(t)\,\epsilon^{-st}\,dt, \quad s = \sigma + j\omega, \quad \sigma > \sigma_0 \qquad (3.10.2)$$

where the lower bound is taken as 0− to include in the integration interval the origin as the support of the impulse function and its derivatives, possibly present in $f(t)$.

In the Fourier integral, the convergence factor $\epsilon^{-\sigma t}$ is allowed to approach unity. In the Laplace transform the convergence factor is retained with $\sigma > \sigma_0$ set at a level, depending on $f(t)$, to produce convergence of the integral. Since this takes place in a strip parallel to the $j\omega$-axis, there exist two values of σ, say σ_0 and σ_1, which coincide with the lower and the upper bounds of the σ-set for which the integral converges; they are called the *lower* and the *upper abscissas of convergence.* When $f(t)$ is causal, $\sigma_1 \to +\infty$; hence the integral converges in the half-plane to the right of σ_0, which is simply called the *abscissa of convergence.*

For some $f(t)$ functions, e.g., $f(t) = \exp(-t^2)$, the integral eq. (3.10.1) converges for all values of σ; in such a case, $\sigma_0 = -\infty$ and $\sigma_1 = +\infty$. For other functions, like $f(t) = \exp(t^2)$, simple exponential damping is not sufficient to produce convergence whatever σ is and in such cases the Laplace transform does not exist; thus $\sigma_0 = +\infty$. Essentially it is the exponential convergence factor which differentiates the Laplace from the Fourier transform.

The definition of the *inverse Laplace transform* is obtained by employing the same change of variable in the *inverse* Fourier transform, eq. (3.2.18), that was used to obtain $F(s)$ by continuing the Fourier transform $F(j\omega)$ into the complex plane. Namely, $j\omega \Rightarrow s = \sigma + j\omega$ and $d\omega = ds/j$ (σ is constant during the integration) is substituted in the integrand of eq. (3.2.18). The corresponding change in integration limits to $\sigma \pm j\infty$ is also introduced. The abscissas of convergence σ_0 and σ_1 delimit the strip where the parallel to the $j\omega$-axis lies, along which integration is carried out. Thus the inverse transform may be written as

$$\mathcal{L}^{-1}F(s) = f(t) = \frac{1}{2\pi j}\int_{\sigma-j\infty}^{\sigma+j\infty} F(s)\,\epsilon^{st}\,ds, \quad \sigma_0 < \sigma < \sigma_1 \qquad (3.10.3)$$

It is useful to see how the Laplace transform arises naturally in the eigenfunction analysis of LTI causal systems. The remainder of this section will

[21]The integral in eq. (3.10.1) is often called *two sided* or *bilateral* Laplace transform; that in eq. (3.10.2) the *one sided* or *unilateral* Laplace transform. Such a terminology is awkward, since it suggests that for a given function, $f(t)$, one has the option of applying either transform, whereas the second is nothing but the first applied to a causal function. Hereafter we simply use the term "Laplace Transform".

be devoted to this topic, as well as its consequences for general LTI system analysis. The relation between impulse response, the Laplace transform, and phasor analysis will first be considered.

Consider a single input, single output, LTI, stable (see Definition 2.5.1 and Theorem 2.5.1 in Section 2.5) causal system excited by the eigenfunction $x(t) = \epsilon^{st}$. The response is $y(t)$. As in Chapters 1 and 2, the system equation is written in operator form as an inhomogeneous equation.

$$x(t) = \epsilon^{st} = S\{y(t)\}, \quad t > -\infty \tag{3.10.4}$$

Choose s so that the response is dominated by the natural modes (see Section 2.5). Therefore, $y(t)$ will be an eigenfunction, and the solution is just the forced response. Thus

$$y(t) = Y(s)\,\epsilon^{st}, \quad \Re s > \sigma_0 = \Re p_k \tag{3.10.5}$$

In eq. (3.10.5), the condition for dominance and eigenfunction response is related to the natural modes of the system. As in Theorem 2.5.2, Section 2.5, p_k is that natural mode located farthest to the right in the complex s plane.

As an alternate to the procedures of eigenfunction analysis of Chapter 2, convolution as in eqs. (3.3.2) and (3.3.4) can be employed for the LTI system, and the operator equation (3.10.4) inverted so that the solution $y(t) = S^{-1}\{x(t)\}$ is found as the convolution of $x(t)$ with the impulse response $h(t)$. Thus

$$y(t) = \int_{-\infty}^{+\infty} h(\tau)\,x(t-\tau)\,d\tau, \quad x(t-\tau) = \epsilon^{s(t-\tau)}$$

Since the system is causal $h(t) = 0$, $t < 0$ (Theorem 3.3.2). Therefore,

$$y(t) = \int_{0}^{+\infty} h(\tau)\,\epsilon^{s(t-\tau)}d\tau = \epsilon^{st}\int_{0}^{+\infty} h(\tau)\,\epsilon^{-s\tau}d\tau = Y(s)\,\epsilon^{st} \tag{3.10.6}$$

Comparing eqs. (3.10.6) and (3.10.5), it is clear that

$$Y(s) = H(s), \quad H(s) = \int_{0}^{+\infty} h(\tau)\,\epsilon^{-s\tau}d\tau \quad \Re s > \sigma_0$$

$H(s)$ in eq. (3.10.6) is simply the Laplace transform (as defined by eq. (3.10.2)) of the impulse response, and hence the phasor transfer function $Y(s)$ as determined by eigenfunction analysis (Chapter 2) is identical with the Laplace transform of the impulse response. Furthermore, the abscissa of convergence for the Laplace transform, σ_0, is determined by the conditions (3.10.5), which follow from dominance and insure a pure eigenfunction response.

THEOREM 3.10.1
The phasor transfer function $Y(s)$ of an LTI stable, causal, system as determined by eigenfunction (phasor) analysis is identical to the Laplace transform of the impulse response $h(t)$, with abscissa of convergence σ_0,

$$Y(s) \equiv H(s) = \mathcal{L}[h(t)] = \int_0^{+\infty} h(t)\,\epsilon^{-st}\,dt, \quad \Re s > \sigma_0 = \Re p_k \qquad (3.10.7)$$

where p_k is the natural mode located farthest to the right in the complex s plane.

The impulse response of Theorem 3.10.1 starts at $t = 0-$ and thus includes the response to a possible delta function at the origin.

The procedures for evaluation of Laplace transforms are essentially the same as those used for Fourier transforms. It is evident that, when the abscissa of convergence $\sigma_0 < 0$, the Fourier and Laplace transforms give identical results, since the $j\omega$ axis will be to the right of σ_0. But in the case of the unit step function, for example, $\sigma_0 = 0$, and as indicated by eqs. (3.9.4) and (3.9.5) the Fourier and Laplace transforms are not the same. The Laplace transform, however, exists also in those cases in which the integral (3.10.2) has a valid strip of convergence that does not contain the $j\omega$-axis.

The Fourier Transform operational rules as listed in the Table of Section 3.9 apply equally well to Laplace transforms with s replacing $j\omega$. So, for example, we have from formulae 10 and 11 of the Table of Section 3.9,

$$\mathcal{L}[\,(-t)^n f(t)\,] = \frac{d^{(n)}F(s)}{ds^{(n)}}$$

and

$$\mathcal{L}f'(t) = sF(s) \qquad (3.10.8)$$

Eq. (3.10.8) yields the Laplace transform of the *generalized* derivative of a function $f(t)$. It is also possible to find the Laplace transform of its *classical* derivative that applies only to its absolutely continuous part. Let $f(t) = g(t)\,\mathrm{u}(t)$ where $f(t)$ is causal and $g(t)$ is absolutely continuous but in general noncausal. Thus we have

$$g'(t)\,\mathrm{u}(t) = [g(t)\,\mathrm{u}(t)]' - g(t)\,\mathrm{u}'(t)$$

Transposing and taking into account eqs. (3.1.4) and (3.1.8)

$$g'(t)\,\mathrm{u}(t) = [g(t)\,\mathrm{u}(t)]' - g(0)\,\delta(t)$$

Hence from eqs. (3.10.8) and (3.9.1) with $j\omega$ replaced by s

$$\mathcal{L}[g'(t)\,\mathrm{u}(t)] = sF(s) - g(0) \qquad (3.10.9)$$

If the procedure is repeated n times

$$\mathcal{L}[g^{(n)}(t)\,\mathrm{u}(t)] = s^n F(s) - \sum_{k=1}^{n} s^{n-k} g^{(k-1)}(0)\,, \quad g^{(0)}(0) \equiv g(0) \qquad (3.10.10)$$

Theorem 3.10.1 is the basis for transient analysis of LTI circuits. The system function is found by conventional phasor analysis, which in turn permits the calculation of the impulse response by inversion of the Laplace transform. The time domain response to an arbitrary excitation can then be found by convolution. Suppose the arbitrary excitation is $x(t)$, a causal function, then with $X(s) = \mathcal{L}x(t)$, the output $y(t)$ is given by the application of convolution Rule 9 of the Table in Section 3.9

$$y(t) = h(t) * x(t), \quad Y(s) = H(s)X(s) \qquad (3.10.11)$$

It is evident that the response can equally well be given as

$$y(t) = \mathcal{L}^{-1}[H(s)X(s)]$$

and, in effect, initial conditions are automatically included by the use of eq. (3.10.10).

Particularly in the case of rational system functions, where the only singularities are poles, it is useful in implementing the above analysis procedure to use contour integration in the complex plane and the method of residues (Section A.13) to evaluate the inverse Laplace transform (3.10.3). In the discussion that follows we use contour integration to infer some general properties of inverse Laplace transforms for use in the study of an important class of systems.

Suppose that in a causal system we wish to find the time domain response (inverse Laplace transform) corresponding to a system function $F(s)$, analytic in the complex plane except for poles. If such an $F(s)$ has only a finite number of poles and has a pole or is analytic at ∞, it is a *rational function*; otherwise, it is a *meromorphic function* (essential or non-isolated singularity at ∞, depending, respectively, on whether the poles are finite or infinite in number). Further assume that $F(s)$ satisfies the Jordan Lemma (see Section A.13) which requires that as $|s| \to \infty$, $F(s)$ approaches zero uniformly (i) on the left semicircle, (center at $s = 0$ of radius $R = |s|$) when a solution for $t > 0$ is sought, or (ii) on the similar right semicircle, when a solution for $t < 0$ is sought. This means, for example, that if $F(s)$ is rational and its denominator degree exceeds that of the numerator, it approaches zero at least as fast as $1/s$ and therefore satisfies the Jordan Lemma.[22] Theorem 3.10.1 tells us that, for a causal system, the abscissa of

[22]If the denominator is not higher in degree than the numerator, the time domain response contains impulses and their derivatives, and this case can be handled separately.

convergence σ_0 is equal to the real part of the farthest right pole of $F(s)$. We can, therefore, write the inverse transform of $F(s)$ as

$$\mathcal{L}^{-1}F(s) = f(t) = \frac{1}{2\pi j}\int_{\sigma-j\infty}^{\sigma+j\infty} F(s)\,\epsilon^{st}\,ds \equiv \frac{1}{2\pi j}\int_{\mathrm{Br}} F(s)\,\epsilon^{st}\,ds \quad (3.10.12)$$

where Br stands for *Bromwich path*, that is a line parallel to $j\omega$, and of infinite extent between the points at $\sigma \pm j\infty$, $\sigma > \sigma_0$. We now consider the evaluation of the Bromwich integral (3.10.12). Initially assume that there are no poles in the right half plane or on $j\omega$. Let the closed contour C consist of the two sections C1 and C2. The portion C1 is the $j\omega$-axis between $\pm jR$. C2 is a semicircle whose radius is R, and either closes C1 to the right (C2=CR) or to the left (C2=CL) forming the closed contour C. There are no singularities within the right contour C=C1+CR, and only a finite number of poles within the left contour C=C1+CL. In either case, Cauchy's Integral formula applies to the closed contour C

$$\oint_{\mathrm{C}} F(s)\epsilon^{st}\,ds = \int_{\mathrm{C1}} F(s)\epsilon^{st}\,ds + \int_{\mathrm{C2}} F(s)\epsilon^{st}\,ds = \pm 2\pi j \sum_k r_k \quad (3.10.13)$$

where the sum is taken over the residues r_k of the poles p_k inside the closed contour C, and the sign is (+) if the direction of C is counterclockwise (CL), (−) if clockwise (CR) (Section A.1).

It now follows that, since $F(s) \to 0$ on C2 as $R \to \infty$, and if $\Re\, st < 0$ on C2, (i.e., $\epsilon^{st} \to 0$ on C2) then the conditions for the Jordan Lemma are satisfied and accordingly the contribution to the contour integral along C2 as $R \to \infty$ in eq. (3.10.13) approaches zero.

If the restriction that poles be confined to the left half plane is removed, the contour C1, now moved to the right of all the poles, becomes the Bromwich path. Again in this case, the residue equation still remains satisfied provided $\Re\, st < 0$ along that portion of the semicircle C2 entirely within the right or left half planes, regardless of the fact that a finite portion of C2 may lie in the right half plane when the contour C is closed to the left, i.e., if C1 is the Bromwich path located to the right of the $j\omega$ axis. This follows because the portion of C2=CL in the right half plane where $\Re\, st \geq 0$ is only of finite extent and $F(s) \to 0$ on this portion as $R \to +\infty$.

We consider the response given by eq. (3.10.12), in view of the above discussion. First let $t < 0$. In this case, keeping in mind that our aim is to evaluate the integral on the Bromwich path, C1, we will have $\Re\, st < 0$ provided $\Re\, s > 0$, and the result is exponential *damping* on C2=CR. In other words, when $t < 0$ the contour C must be closed to the right by C2=CR. Then as $R \to \infty$, the contribution of the integral goes to zero on C2. Referring to eqs. (3.10.12), (3.10.13), it becomes

$$\mathcal{L}^{-1}F(s) = f(t)|_{t<0} = \frac{1}{2\pi j}\int_{Br} F(s)\epsilon^{st}\,ds = -\sum_k r_k = 0 \quad (3.10.14)$$

since there are no poles within C=C1+CR. Causality is therefore verified.

For $t > 0$, C2 is taken to the left so that $\Re\, st < 0$ on that part of C2 in the left half plane, and this time we have exponential damping on C2=CL. Under these conditions apply eq. (3.10.13) with the radius $R \to \infty$ and obtain

$$f(t)|_{t>0} = \frac{1}{2\pi j} \int_{\mathrm{Br}} F(s)\, \epsilon^{st}\, ds = \sum_k r_k \qquad (3.10.15)$$

In this case the residue sum is not zero, since all the poles lie inside C1+CL. The final result, $-\infty < t < +\infty$, is a causal response, and using the unit step function, u(t), we may write

$$f(t) \equiv f(t)\,\mathrm{u}(t)$$

The question still remaining is the matter of residue evaluation at the poles. A pole is precisely defined in terms of a Laurent series (Section A.4), but for the purpose of the following discussion it suffices to note that $F(s)$ has a pole of order n_k at $s = p_k$ if in the neighborhood of p_k $F(s) \approx K/(s-p_k)^{n_k}$, where K is a constant. For example, consider

(a)

$$F(s) = \frac{\epsilon^{st}}{(s+2)^3(s+3)}$$

(b)

$$F(s) = \frac{s}{\sin^2 s}$$

In (a), F(s) has a pole of order 3 at $s = -2$, and a simple pole at $s = -3$. In (b), $F(s)$ has a simple pole at $s = 0$ (the numerator cancels an s factor from the denominator as $s \to 0$), and double poles at $s = \pm k\pi$, $k = 1, 2, 3, \ldots$ The residue at a simple pole at $s = p_0$ of $F(s)$, is given by

$$\text{Res @ } p_0 = \lim_{s \to p_0} (s - p_0)\, F(s) \quad \text{(simple pole)}$$

When the pole at $s = p_0$ is of order n, the residue is given by

$$\text{Res @ } p_0 = \frac{1}{(n-1)!} \lim_{s \to p_0} \frac{d^{n-1}[\,(s-p_0)^n\, F(s)\,]}{ds^{n-1}} \quad \text{(pole of order } n\text{)}$$

In general, a rational function can be expanded in partial fractions (see Section A.11, eq. (A.11.2)). Assuming that the degree of the numerator is less than that of the denominator, we obtain

$$F(s) = \sum_{k=1}^{n} \sum_{r=1}^{m_k} \frac{c_{-r}^{(k)}}{(s - s_k)^r}$$

By using eq. (3.10.12) with the $F(s)$ above, we obtain

$$f(t) = \begin{cases} \sum_{k=1}^{n} \sum_{r=1}^{m_k} \dfrac{c_{-r}^{(k)}}{(r-1)!} t^{r-1} \epsilon^{s_r t} & t > 0 \\ 0 & t < 0 \end{cases}$$

It is readily seen that poles in the RHP yield instability of exponential order and poles on the $j\omega$-axis yield instability of polynomial order, unless they are simple. This conclusion was already reached in Section 2.5 on the basis of eigenfunction analysis.

Example 3.10.1

(a) For the function

$$F(s) = \frac{\epsilon^{st}}{(s+2)^3(s+3)}$$

find the residues at the simple pole $s = -3$ and at the third order pole $s = -2$.

(b) For the function

$$F(s) = \frac{s}{\sin^2 s}$$

find the residue at the simple pole $s = 0$ and at the double pole $s = \pi$.

Solution

(a) Using the technique above, we obtain

$$@ \text{ simple pole } s = -3, \text{ Res} = r_{a1} = \left.\frac{\epsilon^{st}}{(s+2)^3}\right|_{s=-3} = -\epsilon^{-3t}$$

$$@ \text{ third order pole } s = -2, \text{ Res} = r_{a2} = \left[\frac{d^2}{ds^2}\frac{\epsilon^{st}}{s+3}\right]_{s=-2}$$

or

$$\begin{aligned} r_{a2} &= \left[\frac{t^2\epsilon^{st} + \epsilon^{st}}{s+3} + \frac{2\epsilon^{st}}{(s+3)^3} - \frac{t\,\epsilon^{st}}{(s+3)^2}\right]_{s=-2} \\ &= t^2\epsilon^{-2t} - t\,\epsilon^{-2t} + 3\epsilon^{-2t} \end{aligned}$$

(b) Use the expansions of $\sin^2 s$ about the points $s = 0$, $s = \pi$.

$$\left.\sin^2 s\right|_{s=0} = s^2 - \frac{1}{3}s^4 + \cdots$$

$$\sin^2 s\Big|_{s=\pi} = (s-\pi)^2 - \frac{1}{3}(s-\pi)^4$$

For our purpose, the first term of each series suffices.

$$@ \text{ simple pole } s = 0, \quad \text{Res} = \lim_{s\to 0} s\frac{s}{\sin^2 s} = 1$$

$$@ \text{ double pole } s = \pi, \quad \text{Res} = \lim_{s\to \pi} \frac{d}{ds}(s-\pi)^2\frac{s}{\sin^2 s} = 1$$

□

The application of the Laplace transform to transient analysis is demonstrated in the following examples.

Example 3.10.2

Consider the LCR two-mesh circuit (L is in mesh 1, C in 2, and R is common to the two meshes; E_1, and E_2 are the mesh excitations) of Fig. 2.2.1 under quiescent initial conditions. Find the impulse response for the currents $i_1(t)$, $i_2(t)$.

Solution Straightforward phasor mesh analysis gives eq. (2.2.9), $\boldsymbol{E}(s) = \boldsymbol{Z}(s)\boldsymbol{I}(s)$ where $\boldsymbol{E}(s)$, and $\boldsymbol{I}(s)$ are two-element phasor column vectors of mesh voltage excitation and mesh current response, respectively, and $\boldsymbol{Z}(s)$ is the impedance matrix. Let element values be $L = C = 1$, $R = 1/2$. Then the mesh equations become

$$\begin{pmatrix} E_1(s) \\ E_2(s) \end{pmatrix} = \begin{pmatrix} s+\frac{1}{2} & \frac{1}{2} \\ \frac{1}{2} & \frac{1}{s}+\frac{1}{2} \end{pmatrix} \begin{pmatrix} I_1(s) \\ I_2(s) \end{pmatrix} \tag{3.10.16}$$

The above equation is the starting point for a transient analysis of the circuit. According to Theorem 3.10.1, all the phasor entries are now to be interpreted as Laplace transforms. Thus $I_k(s) = \mathcal{L}i_k(t)$. Assume no excitation in mesh 2, and let the voltage excitation for mesh 1 be the unit impulse. Then $E_1(s) = \mathcal{L}\delta(t) = 1$, and $E_2(s) = 0$. To find the impulse response invert the mesh impedance equations, solve for the Laplace transforms of current, and then determine the time domain response using the inverse Laplace transform. Inverting the impedance matrix of eq. (3.10.16)

$$\begin{pmatrix} I_1(s) \\ I_2(s) \end{pmatrix} = \frac{1}{(s+1)^2} \begin{pmatrix} s+2 & -s \\ -s & s(2s+1) \end{pmatrix} \begin{pmatrix} 1 \\ 0 \end{pmatrix}$$

Then $i_k(t) = \mathcal{L}^{-1}I_k(s)$. Thus, noting the double pole at $s = -1$ and applying eqs. (3.10.15), (3.10.16)

$$i_1(t) = \text{Res}\,\frac{(s+2)\,\epsilon^{st}}{(s+1)^2}\bigg|_{s=-1} = \epsilon^{-t} + t\epsilon^{-t}, \quad t > 0$$

Similarly for $i_2(t)$

$$i_2(t) = \text{Res}\left.\frac{-s\,\epsilon^{st}}{(s+1)^2}\right|_{s=-1} = -\epsilon^{-t} + t\epsilon^{-t}, \quad t > 0$$

Alternately we could proceed by expanding $I_1(s)$, $I_2(s)$ in partial fractions (Section A.11) and then taking the inverse transforms term by term. The time domain response for $i_1(t)$, $i_2(t)$ shown above has exactly the same form as the homogeneous solution, since the impulse forcing function is zero $t > 0$. Referring to the equations at the end of Section 2.3, the general homogeneous solution is $i_1(t) = b\epsilon^{-t} + at\epsilon^{-t}$ and $i_2(t) = (b - 2a)\,\epsilon^{-t} + at\epsilon^{-t}$, so it is clear that our zero initial state impulse response, $t > 0$, is the homogeneous solution with $a = 1$, $b = 1$. □

The introduction of initial conditions is readily accomplished as indicated in the following example.

Example 3.10.3

For the same circuit treated in Example 3.10.2, suppose the port excitations are $e_1(t) = e_0\,\mathrm{u}(t)$, $e_2(t) = 0$. Let the initial circulating current in the coil be $i_1(0^-)$, and the initial capacitor voltage be $v_c(0^-)$. Find the Laplace transforms for $I_1(s)$, $I_2(s)$, and the associated responses in the time domain.

Solution It is important to insert the initial conditions into the dynamic equations of the system at the outset, so that the derivative (s) and integral ($1/s$, i.e., Rule 12 of Table in Section 3.9) terms are explicitly identified. Thus in eq. (3.10.16) any term of the form $LsI_k(s)$ corresponds to the Laplace transform of coil voltage Ldi_k/dt with $i_k(0^-) = 0$. When the initial value of coil current at $t = 0^-$ is $i_k(0^-)$, then referring to eq. (3.10.9), $sI_k(s)$ is replaced by $sI_k(s) - i_k(0^-)$. Furthermore, a term of the form $1/CsI_c(s)$ is the Laplace transform of the capacitor voltage $\mathcal{L}v_c(t)$ when the initial charge on the capacitor is zero. Evidently when the initial capacitor voltage is $v_c(0^-)$, the initial value acts like an additional d.c. voltage switched on at $t = 0$, i.e., $v_c(0^-)\,\mathrm{u}(t)$. Thus, including initial conditions, and since $\mathcal{L}\,\mathrm{u}(t) = 1/s$, $V_c(s) = 1/(sC)I_c(s) + v_c(0^-)/s$.

We can now introduce the initial conditions into the problem by modifying eq. (3.10.16), which is, in fact, the Laplace transform of the dynamic integro-differential equations of the circuit. In conformity with the above discussion

$$\begin{pmatrix} E_1(s) \\ E_2(s) \end{pmatrix} = \begin{pmatrix} \dfrac{e_0}{s} \\ 0 \end{pmatrix} = \begin{pmatrix} s + \dfrac{1}{2} & \dfrac{1}{2} \\ \dfrac{1}{2} & \dfrac{1}{s} + \dfrac{1}{2} \end{pmatrix} \begin{pmatrix} I_1(s) \\ I_2(s) \end{pmatrix} + \begin{pmatrix} -i_1(0^-) \\ \dfrac{v_c(0^-)}{s} \end{pmatrix}$$

The two equations are solved by transposing the initial condition vector to the left hand side and inverting the impedance matrix. Thus,

$$\begin{pmatrix} I_1(s) \\ I_2(s) \end{pmatrix} = \frac{1}{(s+1)^2} \begin{pmatrix} s+2 & -s \\ -s & s(2s+1) \end{pmatrix} \begin{pmatrix} \dfrac{e_0}{s} + i_1(0^-) \\ \dfrac{-v_c(0^-)}{s} \end{pmatrix}$$

$$I_1(s) = \frac{i_1(0^-)s^2 + [2i_1(0^-) + e_0 + v_c(0^-)]\, s + 2e_0}{s(s+1)^2}$$

$$I_2(s) = -\frac{[i_1(0^-) + 2v_c(0^-)]\, s + [\, e_0 + v_c(0^-)\,]}{(s+1)^2}$$

Residue evaluation can again be employed to obtain $i_1(t)$, $i_2(t)$. $I_1(s)$ has a simple pole at $s = 0$, and a double pole at $s = -1$. The residues of $I_1(s)\,\epsilon^{st}$ are readily calculated and the result is

$$i_1(t) = 2e_0\, \mathrm{u}(t) + \left\{ \left[i_1(0^-) - 2e_0\right] \epsilon^{-t} + \left[i_1(0^-) + v_c(0^-) - e_0\,\right] t\,\epsilon^{-t} \right\} \mathrm{u}(t)$$

The first term of this expression is the residue at $s = 0$, and $\mathrm{u}(t)$ is present since the response is zero $t < 0$. In the case of $I_2(s)$ there is only the double pole at $s = -1$.

$$i_2(t) = \left\{ -\left[i_1(0^-) + 2v_c(0^-)\,\right] \epsilon^{-t} + \left[\,(i_1(0^-) + v_c(0^-) - e_0\,\right] t\epsilon^{-t} \,\right\} \mathrm{u}(t)$$

Note that $i_1(t)$ has a d.c. component due to the unit step excitation but because of the series capacitor in mesh 2, $i_2(t)$ does not. A couple of other points worth mentioning are, first, that the natural mode portion of the above response satisfies the form of a homogeneous solution with the coefficient of $t\,\epsilon^{-t}$ in i_1 given by $= i_1(0^-)+v_c(0^-)-e_0$, and the i_1 coefficient of ϵ^{-t} given by $b = i_1(0^-)-2e_0$. Then as the homogeneous solution requires, the corresponding coefficients in i_2 are a and $b-2a$. Second, by a judicious choice of initial conditions, it is possible to suppress the natural modes entirely. Thus if $i_1(0^-) = 2e_0$, and $v_c(0^-) = -e_0$, all mode coefficients vanish except that of the d.c. term, i.e., in this case, $i_1(t) = 2e_0\,\mathrm{u}(t)$ and $i_2(t) = 0$. □

4

The Scattering Matrix and Realizability Theory

4.1 Physical Properties of n-Ports

The concept of n-port has been presented in Definition 2.12.1 as that of a circuit structure whose interaction with the outside world takes place through n *terminal pairs* or *ports*. In this section we present the general physical properties of such structures. This material supplements the discussion in Section 2.14.

For the sake of definiteness, we deal with n-ports composed of a finite number of ideal resistors, inductors, capacitors, gyrators, transformers (ideal or coupled coils), independent and dependent voltage and current generators.

All such elements, except independent voltage and current generators, are *linear* and *time-invariant*. Resistors, inductors, capacitors, gyrators, and transformers are also *passive* and, with the exception of the gyrator, *reciprocal*. Ideal inductors, transformers, capacitors, and gyrators are *lossless*.

The question arises whether the n-port relations at the ports share with the internal branches the properties of linearity, time invariance, passivity, losslessness, and reciprocity. The answer is provided by the following

THEOREM 4.1.1

An n-port whose internal branches consist of a finite number of ideal resistors, inductors, capacitors, transformers, gyrators, and dependent *voltage and current generators is* linear time-invariant *(LTI); it is* passive *if the dependent generators are excluded; if gyrators are excluded as well, it is* reciprocal; *if only inductors, capacitors, transformers, and gyrators are present, it is* lossless. *By the preceding statement we mean that the port voltage, port current relations satisfy the definitions of linearity, time invariance,* *passivity, reciprocity, and losslessness, as the case may be.*

PROOF The network is described in the time domain by a set of linear algebraic and differential equations with constant (i.e., time-independent) coefficients involving Kirchhoff's voltage and current laws (KVL and KCL) applied to the branches. The pertinent variables are the branch voltages and currents. These network equations in the branch variables are, therefore, linear and time-invariant and, as a consequence, the Laplace transform may be utilized to convert the time-domain into a frequency-domain description. The new variables are then the Laplace transforms of the branch voltages and currents. Notice that this procedure differs from that of eigenfunction and phasor analysis as discussed in Chapter 2, but, as we saw in Section 3.10, the results are the same. The n-port equations are arrived at by eliminating algebraically all *internal* branch variables and placing into evidence only the *port* voltages and currents, since the ports can be considered a subset of the total set of branches. Such a procedure involves only linear and time independent transformations which evidently preserve linearity and time-invariance. Hence the port voltage-current description is LTI.

Let the n-port internal branches consist only of ideal resistors, inductors, transformers, capacitors, and gyrators. These are individually passive, i.e., the energy absorbed by each and, of course, the total energy absorbed in the branches, is nonnegative. Direct application of Tellegen's Theorem 2.14.1 shows that the total energy measured at the ports must, therefore, be similarly nonnegative. With passive branches, the port quantities $\boldsymbol{v}(t)$, $\boldsymbol{i}(t)$ also satisfy passivity. Thus $w(t)$, the total energy absorbed by the n-port, is

$$w(t) = \Re \int_{-\infty}^{t} \boldsymbol{i}^{\dagger}(\tau)\, \boldsymbol{v}(\tau)\, d\tau \geq 0$$

Next, the satisfaction of the lossless condition for the port variables can be proved using Tellegen's Theorem along the same lines as passivity, but now we let $t \to +\infty$, and replace the "$\geq$" symbol with an "$=$" sign. That is, the total energy entering the structure through the ports over the infinite time epoch $-\infty < t < +\infty$, is zero when the internal elements are lossless. (See Chapter 1 for a discussion of the lossless concept.)

Finally, the Reciprocity Theorem 2.14.3 states that any n-port internally composed of LTI branches with symmetric mutual couplings satisfies reciprocity in terms of the port variables, validating this portion of the Theorem. ∎

Theorem 4.1.1 focuses on the physical concepts of *linearity, time-invariance, passivity, reciprocity,* and *losslessness* with reference to n-ports composed of a finite number of lumped circuit elements individually characterized by these properties. More general classes of n-ports can be defined by assuming internal structures consisting of distributed (rather than lumped) elements or a denumerable or even a nondenumerable infinity of lumped circuit elements; such element descriptions are useful in modeling the dynamics of the electromagnetic field associated with the esoteric materials of modern technology. It is therefore important to stress the role of abstract physical concepts, viewed apart from the specific structural details that make up the interior of an n-port. This point of view leads to the *axiomatic approach* to Network Theory[1], which will be discussed in the following sections.

4.2 General Representations of n-Ports

The equations of an n-port are obtained from those of the complete network on which it is based by first omitting the equations of the terminations and then eliminating the internal variables. The details are discussed below.

The initial step starts with a set of $2b$ equations (as many as the branch variables of the complete network, including the port branches) relating $2b$ variables (b branch voltages and b branch currents) to n external sources across the port branches. After the equations of the sources have been omitted, we are left with $2b-n$ equations in $2b$ variables (the *complete n-port equations describing all branches of the structure and their interconnection*); $2n$ of the variables are port voltages and currents, while the remaining $2(b-n)$ are internal variables. The second step consists in the elimination of the internal variables and of a corresponding number of equations, so that the result is n equations in $2n$ variables.

In order to obtain a representation corresponding to a set of ordinary differential equations with constant coefficients in the time domain, the original equations must have coefficients which are polynomials in s (see Sec. 2.13), and this character must be preserved in any general representation of the n-port. Moreover, the elimination of internal variables must be done in a manner which does not introduce new nonphysical modes as mathematical artifacts.

One basic form of general representation that satisfies the above requirements is the n-port (A, B) polynomial description.[2] This description is obtained by eliminating the internal variables from the complete n-port equations, that is, from the set of independent KVL, KCL, and branch

[1]H. J. Carlin, "Network Theory without Circuit Elements", *Proc. IEEE*, vol. 55, no. 4, pp. 482-497 , April 1967.

[2]V. Belevitch, *Classical Network Theory*, S. Francisco, Holden-Day, 1968, chap. 3.

equations (all in polynomial form) which describe the total structure, to arrive at the equations which hold for the port variables. The process of elimination is carried out by *elementary row operations*[3] to retain the polynomial form.

In all cases of practical interest, the number of equations for the n-port is n, equal to the number of rows in $\boldsymbol{A}$ and $\boldsymbol{B}$. The equations have the form

$$\boldsymbol{A}\boldsymbol{v} = \boldsymbol{B}\boldsymbol{i} \tag{4.2.1}$$

or in more compact form

$$(\boldsymbol{A}, -\boldsymbol{B}) \begin{pmatrix} \boldsymbol{v} \\ \\ \boldsymbol{i} \end{pmatrix} = \boldsymbol{0} \tag{4.2.2}$$

where $\boldsymbol{A} = \boldsymbol{A}(s)$ and $\boldsymbol{B} = \boldsymbol{B}(s)$ are $(n \times n)$-polynomial matrices.[4]

DEFINITION 4.2.1 *An n-port is said to be* well-behaved *if there is at least one way of applying voltage sources at some k ports and current sources at the remaining $n-k$ ports so that, for any given set of independent initial conditions, the port voltages and currents are uniquely defined.*

An obvious consequence of the definition above is the following.

THEOREM 4.2.1
A necessary condition for an n-port to be well-behaved is that the normal rank of the matrix $(\boldsymbol{A}, -\boldsymbol{B})$ be equal to n.

Indeed, in the description of any physical structure, the normal rank of $(\boldsymbol{A}, -\boldsymbol{B})$ is always n.

Every $(n \times n)$-submatrix of $(\boldsymbol{A}, -\boldsymbol{B})$ of normal rank n, composed of, say, $n-k$ voltage and k current columns, has as its determinant a polynomial of degree $\leq n$. Its roots are the natural frequencies of the n-port, when the ports corresponding to the $n-k$ voltage columns are open-circuited and those corresponding to the k current columns are short-circuited. Such submatrices are, at most, equal to the combinations of n objects taken k at a time; the number is the binomial coefficient C_k^n.

[3]Elementary row operations are: (a) exchange of rows i and j; (b) multiplication of row i by a real constant k; (c) addition to row i of row j multiplied by a polynomial.
[4]We use capital letters to denote matrices, lower case letters to denote vectors. Whether such quantities are functions of time t or the complex frequency s is either explicitly indicated or made clear from the context.

A specific rational representation relating the excitations to the responses is obtained as follows. Let $\boldsymbol{M}$ be the $(n \times n)$-submatrix of $(\boldsymbol{A}, -\boldsymbol{B})$ corresponding to $n-k$ voltage columns and k current columns, with $\boldsymbol{y}$ the vector of voltages and currents at these ports. Then $\boldsymbol{N}$, the complementary submatrix, has $n-k$ columns corresponding to currents and k columns corresponding to voltages, with the vector of these port variables designated as $\boldsymbol{x}$. Let $\boldsymbol{y}$ be the responses, $\boldsymbol{x}$ the excitations. Then eq. (4.2.1) can be rewritten as

$$\boldsymbol{M}\boldsymbol{y} = \boldsymbol{N}\boldsymbol{x}$$

or

$$\boldsymbol{y} = \boldsymbol{H}\boldsymbol{x} \tag{4.2.3}$$

with

$$\boldsymbol{H} = \boldsymbol{M}^{-1}\boldsymbol{N} \tag{4.2.4}$$

The matrix $\boldsymbol{H}$ is an n-port *hybrid* matrix.[5] If vector $\boldsymbol{x}$ consists entirely of currents, $\boldsymbol{H}$ becomes the *impedance* matrix and is denoted as $\boldsymbol{Z}$. If vector $\boldsymbol{x}$ consists entirely of voltages, $\boldsymbol{H}$ becomes the *admittance* matrix and is denoted as $\boldsymbol{Y}$. Note however that a specific rational representation exists if and only if the pertinent $\boldsymbol{M}$ has normal rank n.

Even for well-behaved n-ports not all representations expressing a group of n variables in terms of the remaining n qualify as hybrid matrices; in fact, among all such representations, whose number is at most C_n^{2n}, there are those for which both the voltage and the current at the same port appear in vector $\boldsymbol{x}$. An $\boldsymbol{x}$ defined in this fashion cannot be interpreted as an excitation vector. As an instance of this, the transmission matrix of a two-port, expressing the input quantities $(v_1,\ i_1)'$ in terms of the output variables $(v_2,\ -i_2)'$ does not belong to the hybrid class, for it is impossible to feed port 2 with an independent voltage source v_2 and an independent current source $-i_2$ at the same time. Nevertheless such nonhybrid representations are often useful in the analysis of cascaded structures.

Example 4.2.1

Discuss the two-port representation of the following equation

$$\begin{pmatrix} d & -d \\ 0 & 0 \end{pmatrix} \begin{pmatrix} v_1 \\ v_2 \end{pmatrix} = \begin{pmatrix} n & 0 \\ 1 & 1 \end{pmatrix} \begin{pmatrix} i_1 \\ i_2 \end{pmatrix}$$

where n and d are polynomials in s.

[5] In eq. (4.2.4) cancellations of common factors may occur between numerator and denominator; when these common factors occur, they are termed *hidden* modes.

Solution It is easy to see that the two-port simply consists of two wires connecting ports 1 and 2, in one of which a one-port of impedance $z = n/d$ is series inserted. The voltage polarities are so chosen that plus signs both belong to the wire where the one-port is inserted; the current is polarized to enter at the positive terminals.

For the above equation the normal rank of $\boldsymbol{A}$ is 1, that of $\boldsymbol{B}$ is 2. Thus the admittance representation, $\boldsymbol{i} = \boldsymbol{Y}\boldsymbol{v}$, exists, but not the impedance representation. If $z = 0$, corresponding to a through connection, then both $\boldsymbol{A}$ and $\boldsymbol{B}$ are rank 1 (thus singular) and neither $\boldsymbol{Z}$ nor $\boldsymbol{Y}$ exist. However, even in this case one can readily verify that both hybrid matrices (having as inputs $\boldsymbol{x} = (v_1, i_2)'$ and $\boldsymbol{x} = (v_2, i_1)'$) exist. □

The condition for the well-behaved property of Theorem 4.2.1 is necessary, but not sufficient; in certain special cases, the matrix $(\boldsymbol{A}, -\boldsymbol{B})$ may have rank n, still an excitation of the various ports with either voltage or current sources may be impossible.

Example 4.2.2

Let a two-port be described by the equation

$$\begin{pmatrix} 1 & 0 \\ 0 & 0 \end{pmatrix} \begin{pmatrix} v_1 \\ v_2 \end{pmatrix} = \begin{pmatrix} 0 & 0 \\ 1 & 0 \end{pmatrix} \begin{pmatrix} i_1 \\ i_2 \end{pmatrix}$$

Show that it is not well-behaved, although of full rank.

Solution The matrix $(\boldsymbol{A}, -\boldsymbol{B})$ has rank 2 (consider the submatrix formed by the first columns of $\boldsymbol{A}$ and $\boldsymbol{B}$), but still it is impossible to apply at port 1 (which is simultaneously an open and a short circuit) either a voltage or a current source.

The two-port above is, nevertheless, a useful model for the ideal linear *operational amplifier.* □

Another general formalism[6] for representing an LTI n-port can be derived directly from the complete n-port equations, or deduced from the (A, B) formalism. The representation is

$$\begin{pmatrix} \boldsymbol{v} \\ \boldsymbol{i} \end{pmatrix} = \begin{pmatrix} \boldsymbol{Q}_1 \\ \boldsymbol{Q}_2 \end{pmatrix} \boldsymbol{z} = \boldsymbol{Q}\boldsymbol{z} \tag{4.2.5}$$

[6]D. C. Youla, L. J. Castriota, and H. J. Carlin, "Bounded Real Scattering Matrices and the Foundations of Linear Passive Network Theory", *IRE Transactions on Circuit Theory,* Vol. CT-6, no. 3, pp. 102-124, March 1959.

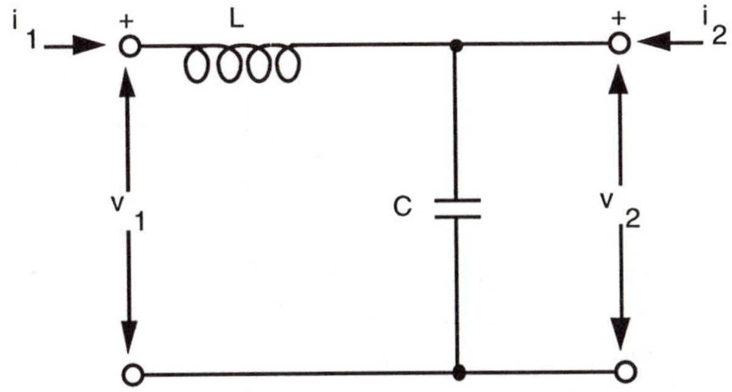

FIGURE 4.2.1
A two-port described in the text by the general representation.

where $\boldsymbol{z}(s)$ is an arbitrary n-vector, and $\boldsymbol{Q}$ can always be put into the form of a polynomial matrix by elementary row and column operations on matrix $(\boldsymbol{A}, -\boldsymbol{B})$. The representation, eq. (4.2.5), is termed the Q formalism. The (A, B) description presented n homogeneous equations relating the $2n$ port variables. The Q formalism expresses the $2n$ voltage-current port variables in terms of n arbitrary independent variables.

The general representations form the basis for obtaining all particular descriptions of an n-port, which, in principle, are equal in number to C_n^{2n}, but actually often reduce in number because the pertinent submatrices extracted from eqs. (4.2.1) are singular. Thus for a two-port the maximum number of representations is $C_2^4 = 6$. However, in the case of two-ports consisting of only a series element or of a shunt element (even if these are zero), the impedance description or admittance description, respectively, is lacking, so the actual number of admissible representations is four.

It is easy to verify that the impedance matrix $\boldsymbol{Z}$ and the admittance matrix $\boldsymbol{Y}$ can be expressed as $\boldsymbol{Z} = \boldsymbol{A}^{-1}\boldsymbol{B} = \boldsymbol{Q}_1\boldsymbol{Q}_2^{-1}$ and $\boldsymbol{Y} = \boldsymbol{B}^{-1}\boldsymbol{A} = \boldsymbol{Q}_2\boldsymbol{Q}_1^{-1}$, whenever the inverse matrices do actually exist.

As an illustration of the concepts above, consider the following.

Example 4.2.3

An LC two-port is represented in Fig. 4.2.1. Find its general description using both the (A, B) and Q formalisms.

Solution We have $v_1 - v_2 = Lsi_1$ and $i_1 + i_2 = sCv_2$, or

$$\begin{pmatrix} 1 & -1 \\ 0 & Cs \end{pmatrix} \begin{pmatrix} v_1 \\ v_2 \end{pmatrix} = \begin{pmatrix} Ls & 0 \\ 1 & 1 \end{pmatrix} \begin{pmatrix} i_1 \\ i_2 \end{pmatrix}$$

This is one choice for the (A, B) formalism. For convenience change the signs of the second rows of A and B. Then the above equation can be written

$$\begin{pmatrix} 1 & -1 & -Ls & 0 \\ 0 & -Cs & 1 & 1 \end{pmatrix} \begin{pmatrix} v_1 \\ v_2 \\ i_1 \\ i_2 \end{pmatrix} = \begin{pmatrix} 0 \\ 0 \\ 0 \\ 0 \end{pmatrix}$$

Since columns 1 and 4 form the identity matrix this is a simple choice of nonsingular minor. Then the variables $v_2 = z_1$, $i_1 = z_2$ can be set arbitrarily and, when transposed to the right hand side of the equation, a representation in the Q formalism immediately follows

$$\begin{pmatrix} v_1 \\ v_2 \\ i_1 \\ i_1 \end{pmatrix} = \begin{pmatrix} 1 & Ls \\ 1 & 0 \\ 0 & 1 \\ Cs & -1 \end{pmatrix} \begin{pmatrix} z_1 \\ z_2 \end{pmatrix}$$

□

In the next section we shall introduce still another representation derivable from the general description. This involves the so-called *scattering matrix* which always exists whenever the n-port is passive and well defined.

As a concluding remark we note that an important extension of the previous analysis arises when the n-port is composed of a nondenumerable infinity of circuit elements. If linearity and time-invariance can still be assumed, general descriptions like the (A, B) and Q formalisms may still be valid, but the the matrix functions will be nonpolynomial in character.

As a simple illustration, consider the case when matrices $\boldsymbol{A}$ and $\boldsymbol{B}$ are transcendental entire functions of the complex variable s. This arises, for example, for a lossless coaxial transmission line[7] which can be described by the equations

$$\begin{pmatrix} 1 & -\cosh sT \\ 0 & Y_0 \sinh sT \end{pmatrix} \begin{pmatrix} v_1 \\ v_2 \end{pmatrix} = \begin{pmatrix} 0 & -Z_0 \sinh sT \\ 1 & \cosh sT \end{pmatrix} \begin{pmatrix} i_1 \\ i_2 \end{pmatrix}$$

Here subscripts 1 and 2 represent input and output ports of a length of the line. Also Z_0 and $Y_0 = 1/Z_0$ are the characteristic impedance and admittance, respectively, of the transmission line, and T is the delay from the the input to the output port. For this example all particular representations

[7] It is well known from elementary transmission line theory that such a structure can be described as a ladder network consisting of the cascade of a nondenumerable infinity of L, C sections. (See Section 7.1.)

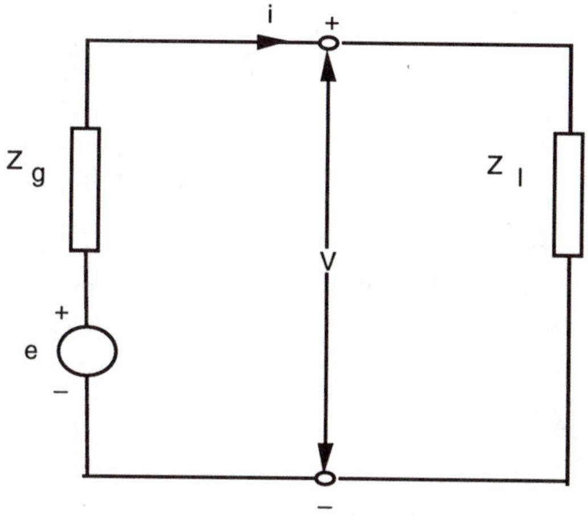

FIGURE 4.3.1
Power transfer from a resistive generator to a resistive load.

(e.g., immittance) derived from the general A, B representation above are meromorphic matrix functions of s.

4.3 The Scattering Matrix Normalized to Positive Resistors

As an introduction to the *scattering* formalism, we consider the problem of maximum power transfer from a generator to a terminating load. The solution is presented in the following theorem.

THEOREM 4.3.1

Given an independent voltage source whose phasor value is e and whose internal source impedance is $z_g = r_g + jx_g$*, terminated in an adjustable load* $z_l = r_l + jx_l$ *as in Fig. 4.3.1. Then maximum active power is transmitted to the termination when the load impedance is a* conjugate match *to the prescribed source impedance, i.e.,* $z_l = z_g^* = r_g - jx_g$*. The maximum or* available *generator power is* $P_{av} = |e|^2/4r_g$.

PROOF The real power delivered to the load is

$$p = r_l|i|^2 = \frac{r_l|e|^2}{|z_g + z_l|^2}$$

so one requirement for a maximum is to cancel the system reactance by setting $x_l = -x_g$. Under these conditions

$$p = \frac{r_l|e|^2}{(r_g + r_l)^2}$$

Setting the derivative with respect to r_l equal to zero yields the condition $r_l = r_g$, for maximum p. Accordingly, the available active source power, a property of the generator alone, is $P_{av} = |e|^2/4r_g$, occurring under conjugate match conditions, $z_l = z_g^*$. If the source impedance and termination are prescribed as pure resistors, and the source is a time domain signal, $e(t)$, the Theorem applies to maximum instantaneous power delivered to the load. ∎

A much more demanding problem is that of matching a given generator to a frequency dependent load *over a finite frequency band.* If the band is narrow compared to its central frequency, the Theorem may provide a rough and ready answer whose advantage is simplicity, by specifying a conjugate match at the midband frequency, resulting in some degree of mismatch over the rest of the band. This solution is only acceptable as an approximation if the load reactance to resistance ratio ("Q" factor) is small over the band, and even then the result is not the best available. In general, the achievement of an optimum match over the entire band is a difficult matter. At the outset one must understand that an exact match for transmission of available power over a finite frequency band to a load that is not a pure resistance is not an option, for it would require interposing a physical matching transducer between generator and load such that the latter sees an analytic transducer impedance function z_t, which is the conjugate of the load impedance over a finite interval of frequency. This is not mathematically possible; the load impedance being analytic satisfies the Cauchy-Riemann differential equations (A.1.1) and (A.1.2). Changing the sign of the imaginary part to achieve a matching transducer output *over the band* results in an impedance function $z_t(s) = z_g^*(s)$ which no longer satisfies the Cauchy-Riemann equations and, thus, is not analytic, hence not physically realizable. The *gain-bandwidth problem* is therefore that of determining and implementing the best *approximation* to available power transfer over a given frequency band when the load is specified. The treatment of methods to do this forms a rich body of theoretical concepts and practical techniques which will be discussed in subsequent chapters. The scattering matrix is central to the solution of the broadband matching problem, and the conjugate match principle is a basic aspect of the definition of the scattering formalism under real (or complex) normalization.

Start with an n-port $\mathcal{N}$ to be described in scattering terms. The n-port is *augmented* by placing in series with each i-th port a resistance r_i as in

Section 1.4. The *augmented* n-port is then excited with the independent voltage source vector $\boldsymbol{e} = (e_i)$. Let

$$\boldsymbol{R}_0 = \text{diag}\{r_1, r_2, ..., r_i, ..., r_n\}$$

be the diagonal matrix of the augmenting resistances. We now introduce the *wave* or *scattering* variables

$$\begin{aligned} 2\boldsymbol{R}_0^{\frac{1}{2}}\boldsymbol{a} &= \boldsymbol{v} + \boldsymbol{R}_0\boldsymbol{i} \\ 2\boldsymbol{R}_0^{\frac{1}{2}}\boldsymbol{b} &= \boldsymbol{v} - \boldsymbol{R}_0\boldsymbol{i} \end{aligned} \tag{4.3.1}$$

where $\boldsymbol{v}$ and $\boldsymbol{i}$ are the voltage and current vectors, respectively, measured at the ports of $\mathcal{N}$. The wave variables have an interesting physical interpretation. Vector $\boldsymbol{a}$, except for the constant factor $2\boldsymbol{R}_0^{\frac{1}{2}}$, can be identified with the source vector $\boldsymbol{e}$ and hence represents a signal applied to the augmented n-port. Vector $\boldsymbol{b}$, on the other hand, would be zero if all ports were matched to the augmenting resistors, since then $\boldsymbol{v} = \boldsymbol{R}_0\boldsymbol{i}$; thus it represents the deviation of the performance of the actual n-port from the matching condition, or the *mismatch* with respect to the given augmenting network. Vector $\boldsymbol{a}$ is called the *incident* (vector) wave, vector $\boldsymbol{b}$ the *reflected* (vector) wave. The scattering terminology stems from transmission line theory and can be interpreted in that context if we assume that the n-port is connected to the independent voltage generators through transmission lines whose characteristic impedances are the r_i's; however, such an interpretation is not necessary, since we can understand the incident wave as the signal applied to the augmented network and the reflected wave as the response to the incident signal. The definition has been formulated so that the responses are zero when there is a perfect match between the ports of $\mathcal{N}$ and its surrounding augmenting network. The wave variables can be taken as real constants (d.c. analysis), or complex constants (single frequency a.c. analysis), or analytic functions of the complex variable s (frequency domain or eigenfunction analysis), or signals in time (time domain analysis). The context will indicate which domain is under consideration.

The reflected wave in the frequency domain can be linearly related to the incident wave by using eqs. (4.2.2) ((A, B) formalism) or (4.2.5) (Q formalism) and eqs. (4.3.1) , where $\boldsymbol{a}$, $\boldsymbol{b}$ are analytic functions of s (constants for d.c. or single frequency a.c. analysis). Add eqs. (4.3.1) to solve for $\boldsymbol{v}$ in terms of $\boldsymbol{a}$, $\boldsymbol{b}$. Next subtract the equations to obtain $\boldsymbol{i}$. When the voltage and current variables are then substituted in the (A, B) formalism, and terms in $\boldsymbol{b}$ collected on the left side, terms in $\boldsymbol{a}$ on the right side, it is simple to obtain the form

$$\boldsymbol{b} = \boldsymbol{S}\boldsymbol{a} \tag{4.3.2}$$

where

$$\boldsymbol{S} = (\boldsymbol{B}\boldsymbol{R}_0^{-\frac{1}{2}} + \boldsymbol{A}\boldsymbol{R}_0^{\frac{1}{2}})^{-1}(\boldsymbol{B}\boldsymbol{R}_0^{-\frac{1}{2}} - \boldsymbol{A}\boldsymbol{R}_0^{\frac{1}{2}}) \tag{4.3.3}$$

An analogous procedure can be employed in the Q formalism to give an alternate expression for $\boldsymbol{S}$

$$\boldsymbol{S} = \boldsymbol{R}_0^{-\frac{1}{2}}(\boldsymbol{Q}_1 - \boldsymbol{R}_0\boldsymbol{Q}_2)(\boldsymbol{Q}_1 + \boldsymbol{R}_0\boldsymbol{Q}_2)^{-1}\boldsymbol{R}_0^{\frac{1}{2}} \tag{4.3.4}$$

When $n = 1$, the scattering matrix reduces to a scalar that is termed the *reflectance* or *reflection factor.*

When $n > 1$, the matrix element S_{ii} is called the *reflectance* at the port i; the element S_{ij} is called the *transmittance* from the port j to the port i. Such definitions are substantiated by the physical interpretations of the quantities above

$$\begin{aligned} S_{ii} &= \left(\frac{b_i}{a_i}\right)_{a_h=0} \qquad (h \neq i) \\ S_{ij} &= \left(\frac{b_i}{a_j}\right)_{a_h=0} \qquad (h \neq j) \end{aligned} \tag{4.3.5}$$

which show that S_{ii} is the ratio of the reflected to the incident wave at port i, when the ideal generator emf's at all other ports are replaced by short circuits, and S_{ij} is the ratio of the reflected wave at port i to the incident wave at port j, when the ideal generator emf's, e_i at all ports but port i are replaced by short circuits; all ports h at which the independent sources are short circuited are evidently terminated in generator resistances r_h.

The formalism above is simplified if one introduces the *normalized* variables $\underline{\boldsymbol{v}}$ and $\underline{\boldsymbol{i}}$ (see also eqs. (1.4.7) and (1.4.8))

$$\begin{aligned} \underline{\boldsymbol{v}} &= \boldsymbol{R}_0{}^{-\frac{1}{2}}\boldsymbol{v} \\ \underline{\boldsymbol{i}} &= \boldsymbol{R}_0{}^{\frac{1}{2}}\boldsymbol{i} \end{aligned} \tag{4.3.6}$$

in terms of which eqs. (4.3.1) are rewritten as

$$\begin{aligned} 2\boldsymbol{a} &= \underline{\boldsymbol{v}} + \underline{\boldsymbol{i}} = \underline{\boldsymbol{e}} \\ 2\boldsymbol{b} &= \underline{\boldsymbol{v}} - \underline{\boldsymbol{i}} \end{aligned} \tag{4.3.7}$$

As a consequence of eqs. (4.3.7) $\boldsymbol{a} = \underline{\boldsymbol{e}}/2$ (see also eq. (1.4.9)), therefore, $2\boldsymbol{a} = \boldsymbol{R}_0^{-1/2}\boldsymbol{e}$. We can thus view the normalized description as representing the emf vector $\underline{\boldsymbol{e}}$ applied to a normalized n-port through unit augmenting resistors.

The normalized port variables correspond to a normalization of the (A, B) and Q formalisms for the n-port, and of any of the particular representations derived from them. In particular, whenever the impedance and/or the admittance matrices exist, the equations for the normalized immittance matrices are as follows.[8]

$$\begin{aligned} \underline{v} &= \boldsymbol{z}\underline{i} \\ \underline{i} &= \boldsymbol{y}\underline{v} \end{aligned} \tag{4.3.8}$$

with

$$\boldsymbol{z} = \boldsymbol{R}_0^{-\frac{1}{2}}\boldsymbol{Z}\boldsymbol{R}_0^{-\frac{1}{2}}$$

$$\boldsymbol{y} = \boldsymbol{R}_0^{\frac{1}{2}}\boldsymbol{Y}\boldsymbol{R}_0^{\frac{1}{2}}$$

The impedance and admittance normalized matrices are related to the scattering matrix by the following equations, which easily follow from eqs. (4.3.2), (4.3.7), and (4.3.8)

$$\begin{aligned} \boldsymbol{S} &= (\boldsymbol{z}+\boldsymbol{I}_n)^{-1}(\boldsymbol{z}-\boldsymbol{I}_n) \\ \boldsymbol{z} &= (\boldsymbol{I}_n+\boldsymbol{S})(\boldsymbol{I}_n-\boldsymbol{S})^{-1} \end{aligned} \tag{4.3.9}$$

and

$$\begin{aligned} \boldsymbol{S} &= (\boldsymbol{I}_n+\boldsymbol{y})^{-1}(\boldsymbol{I}_n-\boldsymbol{y}) \\ \boldsymbol{y} &= (\boldsymbol{I}_n-\boldsymbol{S})(\boldsymbol{I}_n+\boldsymbol{S})^{-1} \end{aligned} \tag{4.3.10}$$

In the equations above $\boldsymbol{I}_n$ is the unit matrix of order n.

The manner in which the scattering matrix $\boldsymbol{S}$ was derived from the general representations insures that it is *real and rational.* Rationality is a consequence of eq. (4.3.3) and of the fact that matrices $\boldsymbol{A}$ and $\boldsymbol{B}$ are polynomial in s. Reality means that $\boldsymbol{S}$ takes on real values on the real axis of the s-plane, which is a consequence of the rational function coefficients being all real.

However, no definite conclusion can be stated about the *existence* of matrix $\boldsymbol{S}$ in any specific case; that is, there is no general guarantee that the matrix

$$\boldsymbol{B}\boldsymbol{R}_0^{-\frac{1}{2}} + \boldsymbol{A}\boldsymbol{R}_0^{\frac{1}{2}}$$

is nonsingular a.e., so that its inverse, which enters in the definition of eq. (4.3.3) of $\boldsymbol{S}$, is defined. As an example of a situation in which the scattering description fails to exist, consider a voltage source with unit internal resistance feeding a one-port consisting of a unit negative resistance. The equation of the one-port is $v = -i$ and therefore $A = 1$ and $B = -1$; since on the other hand $R_0 = 1$, we have $B + A = 0$, and $S = (B+A)^{-1}(B-A)$ does not exist, (or note that $(z+1) = 0$ in the scalar version of eq. (4.3.9)).

A simple observation clarifies the problem of existence of $\boldsymbol{S}$. We suppose the normalized n-port is excited by $\underline{e}$. If the first of eqs. (4.3.7) admits of a (generally) unique solution for $\underline{v}$ and $\underline{i}$ (in which case the n-port is said to

[8]To simplify notation we denote the normalized impedance and admittance matrices as $\mathbf{z}$ and $\mathbf{y}$ instead of $\underline{\mathbf{Z}}$ and $\underline{\mathbf{Y}}$.

be *solvable*), then from the second of eqs. (4.3.7), $\boldsymbol{b}$ is uniquely determined. As a consequence of the linear relation between $\boldsymbol{b}$ and $\boldsymbol{a}$, it follows that $\boldsymbol{S}$ will then exist. In the negative resistor example $\underline{e}$ is terminated by an impedance $1 + (-1) = 0$, so there is no way of determining the current or the voltage across the negative resistance and S is not defined.

Note moreover that the existence of the scattering matrix $\boldsymbol{S}$ does not imply that of the impedance and/or the admittance matrices $\boldsymbol{z}$ and $\boldsymbol{y}$. In fact, according to eqs. (4.3.9) and (4.3.10), they fail to exist whenever the normal ranks of $\boldsymbol{I}_n - \boldsymbol{S}$ and/or $\boldsymbol{I}_n + \boldsymbol{S}$ are less than n in the complex s-plane.

Example 4.3.1

Determine the normalized impedance, admittance, and scattering matrices for the LC circuit of Fig. 4.2.1.

Solution Let the normalization resistances be $R_1 = R_2 = R$. Based on the normalized impedances corresponding to sL and $1/sC$ we can define the normalized parameters $l = L/R$ and $c = 1/RC$ and so form the normalized two-port. Then by the methods of Section 2.12, or simply by inspection, we have the normalized open circuit impedance matrix $\boldsymbol{z}$ and the short circuit admittance matrix $\boldsymbol{y}$ of the normalized two-port.

$$\boldsymbol{z} = \begin{pmatrix} ls + \dfrac{1}{cs} & \dfrac{1}{cs} \\[2ex] \dfrac{1}{cs} & \dfrac{1}{cs} \end{pmatrix}$$

and

$$\boldsymbol{y} = \begin{pmatrix} \dfrac{1}{ls} & -\dfrac{1}{ls} \\[2ex] -\dfrac{1}{ls} & cs + \dfrac{1}{ls} \end{pmatrix}$$

When $\boldsymbol{z}$ or $\boldsymbol{y}$ is substituted in the first of either eqs. (4.3.9) or (4.3.10), respectively, the result is

$$\boldsymbol{S} = \begin{pmatrix} \dfrac{lcs^2 + (l-c)s}{lcs^2 + (l+c)s + 2} & \dfrac{2}{lcs^2 + (l+c)s + 2} \\[2ex] \dfrac{2}{lcs^2 + (l+c)s + 2} & \dfrac{-lcs^2 + (l-c)s}{lcs^2 + (l+c)s + 2} \end{pmatrix}$$

□

Another approach which often simplifies the calculation of $\boldsymbol{S}$ proceeds directly from the normalized admittance matrix of the *augmented* n-port.

This matrix is $\boldsymbol{y}_A = \boldsymbol{R}_0^{\frac{1}{2}} \boldsymbol{Y}_A \boldsymbol{R}_0^{\frac{1}{2}}$, and we have

$$\underline{i} = y_A \underline{e}$$

Substitute the wave variables using eqs. (4.3.7) (subtract them to obtain $\boldsymbol{i}$ in terms of $\boldsymbol{a}$, $\boldsymbol{b}$)

$$\boldsymbol{a} - \boldsymbol{b} = 2\boldsymbol{y}_A \boldsymbol{a}$$

so that

$$\boldsymbol{b} = (\boldsymbol{I}_n - 2\boldsymbol{y}_A)\boldsymbol{a}, \quad \boldsymbol{S} = \boldsymbol{I}_n - 2\boldsymbol{y}_A \tag{4.3.11}$$

Based on the above equation, a useful formula for transmittance is

$$S_{ji} = -2y_{jiA}$$

When the normalized n-port impedance matrix exists, $\boldsymbol{y}_A = (\boldsymbol{z} + \boldsymbol{I}_n)^{-1}$, but it is not necessary that the immittance matrices of the given n-port exist for eq. (4.3.11) to be applicable. Thus consider the following example.

Example 4.3.2

Determine the scattering matrix of a 2-port consisting of an $n:1$ ideal transformer, taking normalization resistors as R_1, R_2. Show how $\boldsymbol{S}$ measures the matching properties of the transformer.

Solution The augmenting resistors R_1, R_2 have unit values in the normalized structure, and are in series with the primary and secondary of the transformer, whose normalized turns ratio is $n':1$. The ampere turns equation for the transformer is $0 = ni_1 + i_2 = n\underline{i}_1/\sqrt{R_1} + \underline{i}_2/\sqrt{R_2}$. Thus $n\underline{i}_1\sqrt{R_2/R_1} + \underline{i}_2 = 0$ and $n' = n\sqrt{R_2/R_1}$. The elements of $\boldsymbol{y}_A$, a symmetric matrix, immediately follow by successively short circuiting the ports of the normalized augmented network and calculating the appropriate current-voltage ratios

$$y_{11A} = (1 + n'^2)^{-1}, \; y_{22A} = \left(1 + \frac{1}{n'^2}\right)^{-1}, \; y_{21A} = y_{12A} = -2n'(1 + n'^2)^{-1}$$

The scattering matrix is then given by eq. (4.3.11)

$$\boldsymbol{S} = \begin{pmatrix} \dfrac{n^2R_2 - R_1}{n^2R_2 + R_1} & \dfrac{2n\sqrt{R_1R_2}}{2n^2R_2 + R_1} \\ \dfrac{2n\sqrt{R_1R_2}}{n^2R_2 + R_1} & -\dfrac{n^2R_2 - R_1}{n^2R_2 + R_1} \end{pmatrix} \tag{4.3.12}$$

For available power transfer at port 1, port 2 terminated in R_2, b_1 should be zero, or $S_{11} = b_1/a_1 = 0$. Referring to eq. (4.3.12), this requires $R_1 = n^2R_2$, verifying that the generator impedance at port 1 be matched. When n is so chosen the transformer is known as a *matching transformer*. Under these conditions the transmittance $S_{21} = 1$, i.e., total available power is also transmitted to the load R_2, an expected result since the ideal matching transformer is lossless. The diagonal elements of eq. (4.3.12) are easily checked by a calculation of reflectance using the scalar version of the first of eqs. (4.3.9). □

4.4 Scattering Relations for Energy and Power

Given an LTI n-port operating in an eigenmode, $s = j\omega$, with column vector phasors $\boldsymbol{i}, \boldsymbol{v}$ at the ports, then the average power absorbed is $P = \Re\, \boldsymbol{i}^\dagger \boldsymbol{v}$, eq. (1.4.1). In terms of the normalized variables, eq. (4.3.6),

$$P = \Re\, \underline{\boldsymbol{i}}^\dagger \boldsymbol{R}_0^{-\frac{1}{2}} \boldsymbol{R}_0^{\frac{1}{2}} \underline{\boldsymbol{v}} = \Re\, \underline{\boldsymbol{i}}^\dagger \underline{\boldsymbol{v}}$$

since $\boldsymbol{R}_0$ is a real diagonal matrix. Thus the power expression is invariant to resistor normalization. Substituting the wave variables obtained from eqs. (4.3.6) and (4.3.7)

$$P = \Re\,(\boldsymbol{a}^\dagger - \boldsymbol{b}^\dagger)(\boldsymbol{a} + \boldsymbol{b}) = |\boldsymbol{a}|^2 - |\boldsymbol{b}|^2 = \sum_{i=1}^{n}(|a_i|^2 - |b_i|^2) = \sum_{i=1}^{n} P_i \quad (4.4.1)$$

since $\boldsymbol{a}^\dagger \boldsymbol{b} - \boldsymbol{b}^\dagger \boldsymbol{a}$, as the difference of a complex scalar and its conjugate, is purely imaginary.[9]

Using $\boldsymbol{b}(j\omega) = \boldsymbol{S}(j\omega)\boldsymbol{a}(j\omega)$ (a.c. operation), we can rewrite the last equation as

$$P = \boldsymbol{a}^\dagger \boldsymbol{a} - \boldsymbol{b}^\dagger \boldsymbol{b} = \boldsymbol{a}^\dagger (\boldsymbol{I}_n - \boldsymbol{S}^\dagger \boldsymbol{S})\, \boldsymbol{a} \quad (4.4.2)$$

The physical meaning of the above equations can be more easily understood by assuming that port i is fed by a normalized generator emf, e_i, of unit normalized internal resistance (actual value R_i), while all other ports are also terminated in unit normalized resistors (actual values R_j). This is the configuration discussed in the previous section when the physical interpretation of the scattering matrix elements was discussed. The ratio of the power delivered to the resistor at port j to the available power of the generator connected at port i is the *transducer power gain* G_{ji} from port i

[9] We denote by $|\mathbf{u}|$ the euclidean norm of vector $\mathbf{u}$, i.e., $\sqrt{\mathbf{u}^\dagger \mathbf{u}}$.

to port j, and is given by

$$G_{ji} = \frac{\dfrac{|v_j|^2}{r_j}}{\dfrac{|e_i|^2}{4r_g}} = \frac{|\underline{v}_j|^2}{\dfrac{|\underline{e}_i|^2}{4}} = \frac{|b_j|^2}{|a_i|^2} = |S_{ji}|^2$$

Moreover, the power delivered to port i,

$$P_i = (|a_i|^2 - |b_i|^2) = (1 - |S_{ii}|^2)\,|a_i|^2$$

is the difference between the available or *incident* power $|a_i|^2$ of the i-th generator and the *reflected* power at port i, $|b_i|^2 = |S_{ii}^2||a_i|^2$, the latter a consequence of the mismatch. Thus under unit normalized terminations, normalized source at port i only, the ratio of the reflected power to the incident power at port i is the squared modulus of the reflectance at port i; the ratio of the reflected power at port j to the incident power at port i is the squared amplitude of the transmittance as defined in eqs. (4.3.5).

In the general situation in which all ports are fed by generators with internal resistances R_i (the normalizing values), eq. (4.4.1) shows that the actual power delivered to the n-port is the total *incident* or available power $|\boldsymbol{a}|^2$ which is one-quarter the sum of the normalized generator emf's (eq. (4.3.7)), less the total *reflected* power $|\boldsymbol{b}|^2$ summed over all n-ports.

When time domain signals are employed, the energy absorbed by the n-port over the interval $-\infty$ to t is given by

$$w(t) = \int_{-\infty}^{t} (|\boldsymbol{a}(\tau)|^2 - |\boldsymbol{b}(\tau)|^2)\, d\tau \tag{4.4.3}$$

Under eigenmode excitation of an LTI system the incident wave vector is $\boldsymbol{a}(s)\,\epsilon^{st}$, with $\boldsymbol{b}(s) = \boldsymbol{S}(s)\,\boldsymbol{a}(s)$ and $s = \sigma + j\omega$. The average power relation, eq. (4.4.2), is then replaced by the instantaneous power

$$p(t) = \boldsymbol{a}^{\dagger}(\boldsymbol{I}_n - \boldsymbol{S}^{\dagger}\boldsymbol{S})\,\boldsymbol{a}\,\epsilon^{2\sigma t} \tag{4.4.4}$$

and eq. (4.4.3) for dissipated energy gives

$$w(t) = \frac{1}{2\sigma}\boldsymbol{a}^{\dagger}(\boldsymbol{I}_n - \boldsymbol{S}^{\dagger}\boldsymbol{S})\,\boldsymbol{a}\,\epsilon^{2\sigma t} \qquad \forall \sigma > 0$$

The integral in eq. (4.4.3) does not converge for $\sigma \leq 0$ because the excitation will not approach zero for $t \to -\infty$. Also, the average power of an exponential signal, $\sigma > 0$, depends on time over *any* interval, since the amplitude is exponentially increasing.

In the limit $\sigma \to 0+$, equation (4.4.4) gives the average power over any integer number of periods in the a.c. steady state, eq. (4.4.2),

$$P = \boldsymbol{a}^{\dagger}(j\omega)\,[\boldsymbol{I}_n - \boldsymbol{S}^{\dagger}(j\omega)\boldsymbol{S}(j\omega)]\,\boldsymbol{a}(j\omega)$$

The power and energy exchanged by the n-port can also be expressed in terms of its immittance matrices. In the a.c. case, we have

$$P = \Re \boldsymbol{i}^\dagger \boldsymbol{v} = \frac{1}{2} (\boldsymbol{i}^\dagger \boldsymbol{v} + \boldsymbol{v}^\dagger \boldsymbol{i})$$

If, say, the impedance exists, then using normalized quantities $\underline{\boldsymbol{v}} = \boldsymbol{z}\underline{\boldsymbol{i}}$ and substituting in the above equation

$$P = \frac{1}{2} \underline{\boldsymbol{i}}^\dagger(j\omega) \left[\boldsymbol{z}(j\omega) + \boldsymbol{z}^\dagger(j\omega)\right] \underline{\boldsymbol{i}}(j\omega)$$

with a similar expression in terms of $\underline{\boldsymbol{v}}(j\omega)$ and $\boldsymbol{y}(j\omega)$.

Analogously, under eigenfunction excitation proportional to ϵ^{st}, we have,

$$p(t) = \frac{1}{2} \underline{\boldsymbol{i}}^\dagger(s) \left[\boldsymbol{z}(s) + \boldsymbol{z}^\dagger(s)\right] \underline{\boldsymbol{i}}(s)\, \epsilon^{2\sigma t} \tag{4.4.5}$$

and a similar form employing $\boldsymbol{y}(s)$.

4.5 Bounded Real Scattering Matrices

Properties of linearity, time-invariance, passivity, losslessness, and reciprocity have been shown to be preserved under interconnection of a *finite* number of circuit elements forming an n-port. On the other hand, we have stressed the importance of extending the concept of n-port to structures composed of a nondenumerable infinity of interconnected elements, for example, to *distributed* structures. Since their internal detailed analysis usually requires the solution of partial differential equations under complicated spatial boundary conditions, the best one can do, in general, is to set up a purely external description of the port performance, imbedding the restrictions imposed by the general internal properties mentioned above. Thus we introduce a set of postulates formally expressing the physical constraints due to the nature of the medium or media inside the structure, and derive from these postulates comprehensive external properties of the n-port, mainly in terms of the scattering and immittance matrices.

Here are the postulates we shall employ:

REALIZABILITY POSTULATES

1. • The n-port is linear.

2. • The n-port is time-invariant.

3. • The n-port is solvable.

4. • The n-port is passive.
5. • The n-port is real.
6. The n-port is lossless.
7. The n-port is reciprocal.
8. The n-port is finite lumped, i.e. contains a finite number of lumped elements.

The first five postulates describe the broad physical assumptions which underlie the theory of linear, time-invariant, passive, real n-ports, and from these we can derive a complete characterization of realizable structures. Postulates 6, 7, and 8 can be added in any combination to the first five to obtain the description of more specialized networks.

We introduce the concept of a *Bounded Real Matrix*, which plays a basic role in physical realizability theory, and some related restrictions.

DEFINITION 4.5.1 *An $n \times n$ scattering matrix, $\boldsymbol{S} = \boldsymbol{S}(s)$, is Bounded Real (BR) if it satisfies:*

(1) $\boldsymbol{S}$ is analytic in the open RHP.

(2) $\boldsymbol{I}_n - \boldsymbol{S}^\dagger \boldsymbol{S}$ is nonnegative definite $\forall s : \Re s > 0$ and $\boldsymbol{I}_n - \boldsymbol{S}^\dagger(j\omega)\boldsymbol{S}(j\omega)$ is nonnegative definite a.e. on $j\omega$.

(3) $\boldsymbol{S}$ is real on the positive real σ-axis of the complex s plane.

At this point we augment the n-port with positive resistors and assume that the independent voltage source vector for the augmented structure is $\boldsymbol{e} \in L_n^2$. Then the postulate of solvability, (3), will have precise applicability, and we will be able to show that all responses are restricted to the space of square integrable functions, a result which will enable us to employ L_n^2 Fourier Transforms in the subsequent theory.

The fundamental Theorem of passive physical Realizability Theory is then

THEOREM 4.5.1

The necessary and sufficient condition for an n-port to satisfy the realizability postulates of (1) linearity, (2) time invariance, (3) solvability, (4) passivity, (5) reality, is that its scattering matrix $\boldsymbol{S}$ exists and is Bounded Real (BR).

PROOF Necessity: To prove necessity we assume the n-port satisfies the five postulates and show that the BR conditions follow.

The solvability postulate (3) means that normalized time domain vectors $\underline{v}(t)$ and $\underline{i}(t)$ are uniquely determined for any $\underline{e} \in L_n^2$. Thus a vector $\boldsymbol{b}(t)$ is uniquely determined from a vector $\boldsymbol{a}(t)$ whose norm is square integrable. That is, we have a mapping of the L_n^2 vector space of incident waves into a space B corresponding to the reflected wave vectors $\boldsymbol{b}$. From postulate (1) we know that such a mapping is *linear*; from postulate (2) it is time-invariant. Thus the structure is LTI. From the passivity postulate (4) it follows that the energy delivered to the LTI system is nonnegative (see Definition 1.4.1) so that the energy relation eq. (4.4.3) may be written[10]

$$w(t) = \int_{-\infty}^{t} (|\boldsymbol{a}(\tau)|^2 - |\boldsymbol{b}(\tau)|^2)\, d\tau \geq 0 \quad \forall t$$

Then, for $t \to +\infty$, we have

$$\int_{-\infty}^{+\infty} (|\boldsymbol{a}(\tau)|^2 - |\boldsymbol{b}(\tau)|^2)\, d\tau \geq 0 \qquad (4.5.1)$$

The first integral is finite since $\boldsymbol{a} \in L_n^2$; thus, to satisfy the above inequality, the reflected wave integral must also be finite. Therefore, $\boldsymbol{b} \in L_n^2$ or B coincides with L_n^2. The result is simply a slightly modified form of Theorem 1.4.3.

The Parseval Theorem[11] 3.3.11 shows that a square integrable function has a square integrable Fourier Transform and vice versa, and that for such functions the integral of the squared amplitude in both the frequency and time domains is the same within a $1/2\pi$ multiplier, eq. (3.3.11). The result is readily extended to the squared norm of n component vectors. Summarizing, we say that the Fourier transform on L_n^2 is an *isometry*, i.e., a norm preserving isomorphism of the space onto itself. Therefore, applying the Parseval Theorem for vectors to eq. (4.5.1) yields a similar result in the frequency domain

$$\int_{-\infty}^{+\infty} |\boldsymbol{b}(j\omega)|^2 d\omega \leq \int_{-\infty}^{+\infty} |\boldsymbol{a}(j\omega)|^2 d\omega < +\infty$$

As a consequence of the first three postulates we have now shown that the vectors $\boldsymbol{b}(t)$ are an LTI mapping of the incident wave vectors. Bochner's L^2-Theorem[12] states that a linear mapping of L_n^2 into L_n^2, $\boldsymbol{b}(t) = S\{\boldsymbol{a}(t)\}$,

[10] Once again for the sake of notational simplicity we will deviate from our earlier practice and use lower case letters for *both* time functions or vectors and their Laplace or Fourier Transforms. This only poses some slight risk of ambiguity which can be resolved in the context.

[11] Also known as Plancherel's Theorem.

[12] D. C. Youla, L. J. Castriota, and H. J. Carlin, "Bounded Real Scattering Matrices and the Foundations of Linear Passive Network Theory", *IEEE Trans. on Circ. Th.*, vol. CT-6, no. 1, pp. 102-124, March 1959.

is time-invariant if and only if there exists an $(n \times n)$-matrix $\boldsymbol{S}(j\omega)$ invariant to the excitation, whose elements are measurable and a.e. uniformly bounded on $j\omega$, such that $\boldsymbol{b}(j\omega) = \boldsymbol{S}(j\omega)\,\boldsymbol{a}(j\omega)$ where $\boldsymbol{b}(j\omega)$ and $\boldsymbol{a}(j\omega)$ are the L_n^2-Fourier Transforms of $\boldsymbol{b}(t)$ and $\boldsymbol{a}(t)$, respectively.[13] Thus the scattering matrix $\boldsymbol{S}$ exists a.e. on the $j\omega$-axis for all LTI passive n-ports, and now we show that it can be continued from the $j\omega$-axis into the RHP and that it is analytic there. We do this by employing a set of simple test functions. Consider the collection of incident column vectors $\boldsymbol{a}_i(t)$, $i = 1, 2, ..., n$, any one of which can be applied to the given n-port. All components of each vector are chosen to be zero except the i-th which equals $\epsilon^{-t}\,\mathrm{u}(t)$. The assemblage of these vectors arranged in order forms the $n \times n$ matrix of simple, L_n^2, *causal* test signal $\boldsymbol{A}_n(t) = \epsilon^{-t}\,\mathrm{u}(t)\,\boldsymbol{I}_n$, with Laplace Transform $\boldsymbol{A}_n(s) = 1/(s+1)\,\boldsymbol{I}_n$. By examining the $\boldsymbol{a}(s)$, $\boldsymbol{b}(s)$ relation for all n incident vectors, we can display the entire $\boldsymbol{S}$ matrix (when it exists). Using eq. (4.3.2) in the form $\boldsymbol{SA} = \boldsymbol{B}$ and taking the $s+1$ scalar factor to the right hand side, we obtain

$$\boldsymbol{S}(s) = (s+1)\boldsymbol{B}_n(s)$$

Theorem 1.5.2 shows that an augmented passive structure is causal. Thus the responses $\boldsymbol{B}(t)$ to the causal incident waves are themselves causal signals and, since they are also square integrable, the hypotheses of the Titchmarsh Theorem 3.5.3 are satisfied. Hence all components of $\boldsymbol{B}(s)$ must be analytic in the right half plane. Therefore, by the above equation, $\boldsymbol{S}$ must also be analytic in $\Re\, s > 0$. Finally, Corollary 3.5.1 to the Titchmarsh Theorem guarantees that $\boldsymbol{S}$ is continuous from the RHP to the $j\omega$-axis a.e.

The precise bound on $\boldsymbol{S}$ for LTI n-ports stems from the passivity postulate involving dissipated energy $w(t)$. This energy under eigenfunction excitation can be expressed as a hermitian form in terms of incident waves and the $\boldsymbol{S}$ matrix, according to eq. (4.4.4). Since the condition for passivity is that $w(t) \geq 0$, $\forall t$ and $\epsilon^{2\sigma t}/\sigma > 0$, $\sigma > 0$, we have as a consequence

$$\boldsymbol{a}^\dagger(\boldsymbol{I}_n - \boldsymbol{S}^\dagger\boldsymbol{S})\,\boldsymbol{a} \geq 0 \quad \Re s > 0$$

In other words, since $\boldsymbol{a}$ is arbitrary, $\boldsymbol{Q}(\sigma, \omega) = \boldsymbol{I}_n - \boldsymbol{S}^\dagger\boldsymbol{S}$ is the matrix of a nonnegative definite hermitian form, notationally represented with the inequality sign below

$$\boldsymbol{Q} = \boldsymbol{I}_n - \boldsymbol{S}^\dagger\boldsymbol{S} \geq 0 \quad \Re s > 0 \tag{4.5.2}$$

Since $\boldsymbol{S}$ is continuous from the RHP to the $j\omega$-axis a.e., eq. (4.5.2) yields in the limit as $\sigma \to 0+$

$$\boldsymbol{I}_n - \boldsymbol{S}^\dagger(j\omega)\boldsymbol{S}(j\omega) \geq 0 \quad \text{a.e. on } j\omega \tag{4.5.3}$$

[13] The earlier equivalent result for transfer functions, eq. (3.10.11), was based on convolution with the impulse response.

The preceding result demonstrates that $\boldsymbol{S}$ has no poles on the $j\omega$-axis, for if it did, the boundedness property on the imaginary axis expressed by eq. (4.5.3) would be contradicted.

Consequently, the necessity of the BR properties (1) and (2) is proved.

For a real n-port, a real excitation causes a real response. Thus if we choose $\boldsymbol{a}(t) = \boldsymbol{a}(\sigma)\,\epsilon^{\sigma t}$ with $\sigma > 0$ and $\boldsymbol{a}(\sigma)$ real, we must obtain $\boldsymbol{b}(t) = \boldsymbol{b}(\sigma)\,\epsilon^{\sigma t}$ with $\boldsymbol{b}(\sigma)$ real. This in turn implies that $\boldsymbol{S}(\sigma)$ is real and the necessity of "reality", BR property (3), is proved.

As a useful result employed in subsequent discussions, we note that the Schwarz Reflection Principle A.12.1, when applied to $\boldsymbol{S}$ by taking into account the reality of $\boldsymbol{S}(\sigma)$, yields

$$\boldsymbol{S}^{\dagger}(s) = \boldsymbol{S}'(s^{*}) \quad \Re s \geq 0 \tag{4.5.4}$$

Sufficiency: To prove sufficiency we assume a BR scattering matrix and show that the first five postulates are then satisfied.

First note that for $\boldsymbol{a}(t)$ in L_n^2 it follows that so, too, is $\boldsymbol{b}(t)$. This is because by hypothesis the hermitian form $\boldsymbol{Q}(0,\omega) \geq 0$, eq. (4.5.3), and the total energy integral therefore satisfies

$$\int_{-\infty}^{+\infty} \boldsymbol{a}^{\dagger}(j\omega)\,\boldsymbol{Q}(0,\omega)\,\boldsymbol{a}(j\omega)\,d\omega \geq 0$$

From $\boldsymbol{b}(j\omega) = \boldsymbol{S}(j\omega)\,\boldsymbol{a}(j\omega)$ and the Parseval Theorem, we find that, with $\boldsymbol{a}(t) \in L_n^2$, then $\boldsymbol{b}(t) \in L_n^2$. Linearity and solvability are clearly satisfied since $\boldsymbol{S}(j\omega)$ is a linear operator and this, coupled with the L_n^2 mapping property, satisfies the hypotheses of the Bochner Theorem, whose conditions are necessary and sufficient for time invariance. Therefore, the n-port is LTI.

For σ real, $\boldsymbol{S}(\sigma)$ is real. Consequently, if a real signal $\boldsymbol{a}(t)$ is applied to the n-port, $\boldsymbol{b}(j\omega) = \boldsymbol{S}(j\omega)\,\boldsymbol{a}(j\omega)$ satisfies the conjugacy property of eq. (4.5.4). Therefore, when the inverse transform for $\boldsymbol{b}(t) = \mathcal{F}^{-1}\boldsymbol{b}(j\omega)$ is evaluated, the imaginary part of the integrand cancels over the path from $-\infty$ to $+\infty$, hence $\boldsymbol{b}(t)$ is real and the reality postulate is verified. As discussed above, the energy integral in eq. (4.4.3) is clearly nonnegative when $t = +\infty$ in the upper limit of the integral, as a consequence of the nonnegative definite character of $\boldsymbol{Q}(\sigma,\omega)$, $\sigma \geq 0$, but, for passivity, we must prove this property $\forall t$.

As a preliminary step in the proof, we demonstrate that the n-port is causal. Let the causal incident wave $\boldsymbol{a}(t) = 0$, $t < 0$ be in L_n^2. Then by the Titchmarsh Theorem $\boldsymbol{a}$ is analytic in $\Re s > 0$, and its norm is square integrable along any path Ω, defined as $-\infty < \omega < +\infty$ in the RHP parallel to $j\omega$. Since the scattering matrix is BR, $\boldsymbol{b}(s)$ must be analytic in the RHP. It is also evident, using the nonnegative definite character of $Q(\sigma,\omega)$, $\sigma > 0$, that $|\boldsymbol{b}(\sigma+j\omega)|^2$ when integrated along Ω, gives a nonnegative result. From the sufficiency form of the Titchmarsh Theorem we have $\boldsymbol{b}(t) = 0,\ t < 0$, so the n-port satisfies causality, Definition 1.3.1.

We complete the sufficiency proof by demonstrating that the energy integral is nonnegative $\forall t$, eq. (4.4.3). Choose any L_n^2 incident signal $\boldsymbol{a}(\tau)$ with response $\boldsymbol{b}(\tau)$. Now construct the signal

$$\hat{\boldsymbol{a}}(\tau) = \begin{cases} \boldsymbol{a}(\tau) & \tau \leq t \\ 0 & \tau > t \end{cases}$$

Since the n-port is causal, the response to $\hat{\boldsymbol{a}}(\tau)$ satisfies

$$\hat{\boldsymbol{b}}(\tau) = \boldsymbol{b}(\tau) \qquad \tau \leq t$$

Since both $\hat{\boldsymbol{a}}(\tau)$ and $\hat{\boldsymbol{b}}(\tau)$ are in L_n^2, we can apply the Parseval Theorem to

$$\hat{I} = \int_{-\infty}^{+\infty} (|\hat{\boldsymbol{a}}(\tau)|^2 - |\hat{\boldsymbol{b}}(\tau)|^2)\, d\tau$$

and obtain

$$\hat{I} = \int_{-\infty}^{+\infty} (|\hat{\boldsymbol{a}}(j\omega)|^2 - |\hat{\boldsymbol{b}}(j\omega)|^2)\, d\omega = \int_{-\infty}^{+\infty} [\hat{\boldsymbol{a}}^\dagger(j\omega)\, Q(0,\omega)\, \hat{\boldsymbol{a}}(j\omega)]\, d\omega \geq 0$$

where the inequality results since by hypothesis $Q(0,\sigma) = \boldsymbol{I}_n - \boldsymbol{S}^\dagger(j\omega)\boldsymbol{S}(j\omega)$ is nonnegative definite.

Finally, use the properties of $\hat{\boldsymbol{a}}(\tau)$, $\boldsymbol{a}(\tau)$, $\hat{\boldsymbol{b}}(\tau)$, $\boldsymbol{b}(\tau)$, and revise the infinite upper integral limit on $\hat{I}$ accordingly.

$$0 \leq \hat{I} = \int_{-\infty}^{t} (|\boldsymbol{a}(\tau)|^2 - |\boldsymbol{b}(\tau)|^2)\, d\tau - \int_{t}^{+\infty} |\hat{\boldsymbol{b}}(\tau)|^2\, d\tau$$

In the above equation the integral from t to $+\infty$, as that of a squared modulus, is always nonnegative. Consequently to satisfy the inequality, the first integral must be nonnegative and passivity is established. Thus the first five postulates follow from the BR property. Hence a Bounded Real scattering matrix describes a *linear, time-invariant, passive* n-port and vice versa. ■

An n-port whose scattering matrix $\boldsymbol{S}$ is BR is said to be *realizable.* Note that this concept of realizability does not imply the existence of a physical structure implementing the n-port, but simply refers to the fact that its port description satisfies a set of constraints imposed by the physical nature of the internal structure.

If Postulates (6), (7), and (8) are assumed, they introduce additional constraints on the scattering matrix. We start with the following definitions.

***DEFINITION* 4.5.2** *Given a Bounded Real Matrix,* $\boldsymbol{S}$, *then*

(4) $\boldsymbol{S}$ *is Lossless Bounded Real (LBR) if* $\boldsymbol{I}_n - \boldsymbol{S}^\dagger(j\omega)\boldsymbol{S}(j\omega) = \boldsymbol{0}$ *a.e., i.e.,* $\boldsymbol{S}$ *is* unitary *a.e. on the* $j\omega$*-axis.*

(5) $\boldsymbol{S}$ *is Reciprocal Bounded Real if* $\boldsymbol{S} = \boldsymbol{S}'$, $\Re s > 0$, *i.e.,* $\boldsymbol{S}$ *is symmetric in the RHP.*

(6) $\boldsymbol{S}$ *is Rational Bounded Real if its elements are rational functions of* s.

Thus we have

THEOREM 4.5.2
Necessary and sufficient for a passive LTI n-port to satisfy Postulate (6), losslessness, is that its scattering matrix $\boldsymbol{S}$ *be Lossless Bounded Real (LBR).*

PROOF Necessity: According to Definition 1.4.3 for a lossless n-port, the total energy dissipated over the infinite epoch $-\infty < t < +\infty$ is zero. Thus using eq. (4.4.3)

$$\int_{-\infty}^{+\infty} (|\boldsymbol{a}(\tau)|^2 - |\boldsymbol{b}(\tau)|^2)\, d\tau = 0$$

From Parseval's theorem, we have

$$\int_{-\infty}^{+\infty} (|\boldsymbol{a}(\tau)|^2 - |\boldsymbol{b}(\tau)|^2)\, d\tau = \int_{-\infty}^{+\infty} (|\boldsymbol{a}(j\omega)|^2 - |\boldsymbol{b}(j\omega)|^2)\, d\omega = 0$$

Since $\boldsymbol{b} = \boldsymbol{S}\boldsymbol{a}$ and with $\boldsymbol{Q}(0,\omega) = \boldsymbol{I}_n - \boldsymbol{S}^\dagger(j\omega)\boldsymbol{S}(j\omega)$,

$$\int_{-\infty}^{+\infty} \boldsymbol{a}^\dagger(j\omega)\, \boldsymbol{Q}(0,\omega)\, \boldsymbol{a}(j\omega)\, d\omega = 0$$

The vector $\boldsymbol{a}$ is arbitrary and the hermitian matrix $\boldsymbol{Q}(0,\omega)$ is nonnegative definite for all passive systems, eq. (4.5.3). It follows that

$$\boldsymbol{Q}(0,\omega) = \boldsymbol{I}_n - \boldsymbol{S}^\dagger(j\omega)\boldsymbol{S}(j\omega) = \boldsymbol{0} \quad \text{a.e.} \tag{4.5.5}$$

That is,

$$\boldsymbol{S}^\dagger(j\omega)\boldsymbol{S}(j\omega) = \boldsymbol{I}_n \ \text{ a.e.}$$

The matrix $\boldsymbol{S}(j\omega)$ is *unitary.* Thus, the necessity of property (6) is proved.
Sufficiency: The proof of sufficiency only requires the reversal of the arguments leading to necessity. ∎

THEOREM 4.5.3
Given an n-port $\mathcal{N}$ described by a BR scattering matrix, $\boldsymbol{S}$. The necessary and sufficient conditions that $\mathcal{N}$ satisfy the Postulates (i) of reciprocity, 7, and/or (ii) lumped finiteness, 8, are that $\boldsymbol{S}$ be Reciprocal Bounded Real and/or Rational Bounded Real.

PROOF Necessity: We first consider the reciprocity property, Postulate (7). Corollary 2.14.3 is concerned with the reciprocity property at the ports of an n-port whose internal branches satisfy reciprocity. However we wish to consider the postulational approach, where the internal construction of the n-port is not specifically delineated. The cited Corollary suggests the global form for the *Lorenz reciprocity postulate* at the ports. Consider two different states (α, β) of voltage-current vector time signals at the ports, then the Lorenz reciprocity postulate for the ports states

$$\boldsymbol{i}'_{\beta}(t)\,\boldsymbol{v}_{\alpha}(t) = \boldsymbol{i}'_{\alpha}(t)\,\boldsymbol{v}_{\beta}(t) \tag{4.5.6}$$

A passive electromagnetic multiport cavity all of whose internal media are isotropic would satisfy this postulate. Or one might imagine a complicated system modeled by a large number (perhaps not even denumerable) of branches, each satisfying reciprocity. The time domain variables of eq. (4.5.6) can be transformed into the frequency domain and wave variables substituted by employing eqs. (4.3.6) and (4.3.7). Using commutativity of an inner product we get

$$\boldsymbol{a}'_{\alpha}(s)\,\boldsymbol{b}_{\beta}(s) - \boldsymbol{a}'_{\beta}(s)\,\boldsymbol{b}_{\alpha}(s) = 0$$

Again, employ inner product commutativity and substitute $\boldsymbol{b} = \boldsymbol{S}\boldsymbol{a}$, $\forall \boldsymbol{a}$,

$$\boldsymbol{a}'_{\alpha}\,(\boldsymbol{S} - \boldsymbol{S}')\,\boldsymbol{a}_{\beta} = 0$$

Therefore,

$$\boldsymbol{S} = \boldsymbol{S}'$$

Thus the necessity of postulate (7) is proved.

If the n-port is composed of a finite number of lumped elements, Postulate (8), its $\boldsymbol{S}$ matrix can be computed using the general representation matrices, (A, B), Q, eqs. (4.3.3) and (4.3.4). These are polynomial matrices in the lumped case and the resulting scattering matrix $\boldsymbol{S}$ is therefore rational.

Sufficiency: Reciprocity sufficiency follows by reversing the necessity argument. To prove the sufficiency of rationality, one must show that any Rational Bounded Real matrix can be realized as the scattering matrix of an n-port containing a finite number of lumped elements. This result, in its most complete form, was the landmark contribution of Y. Oono and

K. Yasuura who showed that any rational BR matrix may be realized by the interconnection of a finite number of resistors, inductors, capacitors, transformers, and gyrators.[14] ■

We conclude that (i) if and only if the scattering matrix is unitary on the $j\omega$-axis, the n-port is *lossless*; (ii) if and only if it is symmetric, the n-port is *reciprocal*; and (iii) if and only if it is rational, the n-port is *finite lumped.*

A useful concept for extending the domain of definition of the scattering matrix $\boldsymbol{S}$ of a lossless n-port is that of a *paraconjugate matrix.*

DEFINITION 4.5.3 *The* paraconjugate *of the matrix $\boldsymbol{S}$ is denoted as $\boldsymbol{S}_{\dagger}$, and defined by*

$$\boldsymbol{S}_{\dagger}(s) = \boldsymbol{S}'(-s)$$

We prove the following.

THEOREM 4.5.4

If an n-port is lossless, its scattering matrix $\boldsymbol{S}$ can be continued to the entire complex s-plane, except, at most, a set of zero measure of the $j\omega$-axis. Such a continuation has the property that $\boldsymbol{S}$ is continuous across the $j\omega$-axis a.e., in the sense that

$$\lim_{\sigma \to 0-} \boldsymbol{S}(\sigma + j\omega) = \lim_{\sigma \to 0+} \boldsymbol{S}(\sigma + j\omega) = \boldsymbol{S}(j\omega)$$

and is said to be a pseudomeromorphic *matrix.*

PROOF The matrix function $\boldsymbol{S}(s)$ is continuous to the $j\omega$-axis a.e. from the RHP and $[\boldsymbol{S}'(-s)]^{-1}$ defines a function continuous to the $j\omega$-axis a.e. from the LHP plane. The right limit is $\boldsymbol{S}(j\omega)$, the left limit is $[\boldsymbol{S}'(-j\omega)]^{-1}$. As a consequence of the conjugacy property eq. (4.5.4), we have $[\boldsymbol{S}'(-j\omega)]^{-1} = [\boldsymbol{S}^{\dagger}(j\omega)]^{-1}$. The unitary condition on the $j\omega$-axis thus yields $[\boldsymbol{S}'(-j\omega)]^{-1} = \boldsymbol{S}(j\omega)$, which proves that the right and the left limits on $j\omega$ are equal a.e. The function $[\boldsymbol{S}'(-s)]^{-1}$ has, at most, LHP poles, corresponding to zeros of $\det \boldsymbol{S}$ in the RHP. Now a single function can be defined on the entire complex s-plane, equal to $\boldsymbol{S}(s)$ in the RHP and to $\boldsymbol{S}'(s)^{-1}$ in the LHP, continuous a.e. across the $j\omega$-axis; such a function will still be denoted as $\boldsymbol{S}$. ■

[14] Y. Oono and K. Yasuura, "Synthesis of Finite Passive $2n$-Terminal Networks with Prescribed Scattering Matrices", *Mem. Fac. Eng. Kyushu Univ.*, vol. 14, no. 2, 1954. Also see V. Belevitch, *Classical Network Theory*, S. Francisco, Holden-Day, 1968, chap. 10.

If the singularities on the finite part of the $j\omega$-axis are only poles, the matrix $\boldsymbol{S}$ is *meromorphic.* If moreover they are finite in number, it is *rational.*

Theorem 4.5.4 allows the extension of the unitary condition from a.e. on the $j\omega$-axis to the entire complex s-plane. We introduce the following definition.

DEFINITION 4.5.4 *A BR matrix $\boldsymbol{S}$ is said to be* paraunitary *if it satisfies*

$$\boldsymbol{I} - \boldsymbol{S}_{\dagger}\boldsymbol{S} = \boldsymbol{0} \tag{4.5.7}$$

for all s except at most a set of measure zero on the $j\omega$-axis.

Then we have

COROLLARY 4.5.1
The scattering matrix of a lossless n-port is paraunitary.

PROOF As a consequence of Theorem 4.5.4 we can write $[\boldsymbol{S}'(-s)]^{-1} = \boldsymbol{S}(s)$ or using paraconjugate notation $\boldsymbol{S}_{\dagger}^{-1} = \boldsymbol{S}$ for all s except for at most a set of points of zero measure on the $j\omega$-axis. Eq. (4.5.7) immediately follows. ∎

In the rational case the BR conditions are simplified, because the poles in the LHP are finite in number and therefore the $j\omega$-axis belongs entirely to the region of analyticity. Consequently, the Maximum Modulus Theorem A.7.2, applied to the RHP, shows that the test for bounded amplitude need only be confined to $s = j\omega$.

THEOREM 4.5.5
A rational *scattering matrix, $\boldsymbol{S}$, is Bounded Real (BR) if and only if it satisfies:*

1. *$\boldsymbol{S}$ is analytic in the open RHP.*
2. *$\boldsymbol{I}_n - \boldsymbol{S}'(j\omega)\boldsymbol{S}(j\omega)$ is nonnegative definite $\forall\omega$.*
3. (Reality) *With $S_{ij} = N_{ij}/D_{ij}$, the coefficients of the polynomials N_{ij}, D_{ij} are real.*

COROLLARY 4.5.2
A rational BR scattering matrix represents a lossless n-port if and only if it is paraunitary, i.e., $\boldsymbol{I}_n - \boldsymbol{S}'(-j\omega)\boldsymbol{S}(j\omega) = \boldsymbol{0}$, $\forall\omega$.

We illustrate the preceding results with several examples.

Example 4.5.1

Discuss the BR character of the following reflectances.

1.
$$S = \frac{1}{s^3 + 1}$$

2.
$$S = \frac{1}{d(s)} = \frac{1}{as^2 + bs + c}, \quad a, b, c \text{ real and } > 0$$

3.
$$S = \epsilon^s, \quad S = \epsilon^{-s}, \quad S = \epsilon^{-1/s}$$

Solution

1. The denominator roots of S are the cubic roots of -1, so two of these must correspond to poles located in the RHP. Thus, the reflectance is not BR.

2. The denominator is a quadratic polynomial with positive real coefficients, so $S(\sigma)$ is real, and the usual formula for the roots shows they are strictly in the LHP. Thus S is analytic for $\Re\, s > 0$ and devoid of poles on the $j\omega$-axis. We must now establish whether the reflectance has a squared amplitude of upper bound unity in the closed RHP. The analytic behavior in this region permits us to simply explore the amplitude criterion on $j\omega$, Theorem 4.5.5.

   At $s = 0$, the amplitude is $1/c$, hence $c \geq 1$ is necessary. The squared amplitude is $|d(j\omega)|^2 = d(j\omega)d(-j\omega)$, so the modulus restriction is

   $$M = |d|^2 - 1 = a^2x^2 + (b^2 - 2ac)x + c^2 - 1 \geq 0 \quad \forall \omega^2 = x$$

   By direct inspection, a sufficient condition for the inequality is that $b^2 \geq 2ac$. To check in the case that $(b^2 - 2ac) < 0$ we note that if M has a simple real positive root x_0, then M will change sign at a real ω_0, and the restriction will be violated. Thus calculate the two roots of M

   $$2ap_{1,2} = -(b^2 - 2ac) \pm \sqrt{(b^2 - 2ac)^2 - 4a^2(c^2 - 1)}$$

   The first term is assumed positive; therefore, the only guarantee for nonreal roots is that the radical be imaginary which requires that

   $$b^2 > 2a(c - \sqrt{c^2 - 1}) \text{ and real} \tag{4.5.8}$$

Note that if $b^2 > 2ac$ the above inequality is automatically valid. Therefore for S to be BR we need only the single restriction of eq. (4.5.8).

3. The first function is paraunitary. It is analytic in the open RHP, but its magnitude there is $\epsilon^{\sigma} > 1$. Accordingly it is not BR and not physically realizable. The second function is also paraunitary and analytic in the RHP. Its amplitude $\epsilon^{-\sigma}$ is less than unity for $\sigma > 0$. It is therefore a BR lossless function. It is also an entire function, since it is devoid of singularities in the finite part of the s-plane. In fact, it represents the input reflectance of a length of transmission line terminated in an open circuit. The third function is paraunitary and analytic in the RHP. Its amplitude is $\exp[-\sigma/(\sigma^2 + \omega^2)] < 1$ for $\sigma > 0$ so it, too, is BR and lossless. The continuity across the $j\omega$-axis holds a.e., except at $s = 0$, where the there is an essential singularity. Thus the function is pseudomeromorphic. It is the reflectance of a terminated distributed structure with series capacitance per unit length and shunt inductance per unit length existing only as an abstract realization. □

Example 4.5.2
Determine whether the following matrices are BR.

1.
$$\boldsymbol{S} = \frac{1}{5}\begin{pmatrix} 3 & 4 \\ 4 & -3 \end{pmatrix}$$

2.
$$\boldsymbol{S} = \frac{1}{2}\begin{pmatrix} 1 & \dfrac{s^2 - 3s + 1}{s^2 + 3s + 1} \\ \dfrac{s^2 - 3s + 1}{s^2 + 3s + 1} & -1 \end{pmatrix}$$

Solution

1. $\boldsymbol{S}$ is symmetric and constant. We need only check

$$\boldsymbol{S}\boldsymbol{S} = \frac{1}{25}\begin{pmatrix} 3^2 + 4^2 & 3 \cdot 4 + 4 \cdot (-3) \\ 4 \cdot 3 + (-3) \cdot 4 & 4^2 + 3^2 \end{pmatrix} = \begin{pmatrix} 1 & 0 \\ 0 & 1 \end{pmatrix}$$

So $\boldsymbol{S}$ is real, constant, and unitary (*orthogonal*). Thus it is the scattering matrix of a lossless two-port. In fact, it is the matrix of an ideal transformer, see eq. (4.3.12).

2. $\boldsymbol{S}$ is real, rational, and symmetric. Since the denominator polynomial coefficients are positive, it is analytic in $\Re s \geq 0$, including $s = \infty$. From

Theorem 4.5.5, there remains the task of checking the matrix $\boldsymbol{Q}(0,\omega) = \boldsymbol{I}_2 - \boldsymbol{S}(-j\omega)\boldsymbol{S}(j\omega)$. We note that

$$S_{12}(j\omega) = \frac{1}{2}\frac{1-\omega^2-3j\omega}{1-\omega^2+3j\omega} = \epsilon^{j\phi}, \;\; \phi(\omega) = -\phi(-\omega) = -2\arctan\frac{3\omega}{1-\omega^2}$$

or

$$\boldsymbol{Q} = \boldsymbol{I}_2 - \boldsymbol{S}(-j\omega)\boldsymbol{S}(j\omega) = \frac{1}{2}\begin{pmatrix} 1 & -j\sin\phi \\ j\sin\phi & 1 \end{pmatrix}$$

$\boldsymbol{Q}$ is an *hermitian matrix*, $\boldsymbol{Q}^\dagger = \boldsymbol{Q}$. Its diagonal elements are positive, and $\det\boldsymbol{Q} = [1-\sin^2\phi(\omega)]/4 \geq 0$, $\forall\omega$. Therefore, $\boldsymbol{Q}$ is nonnegative definite $\forall\omega$ and $\boldsymbol{S}$ is BR, hence realizable. □

4.6 Positive Real Immittance Matrices

In many applications it is important to deal directly with immittance matrices rather than with the scattering matrix. This is particularly the case in problems of concrete realizability. Since $\boldsymbol{S}$ exists for all passive n-ports, we start with a BR scattering matrix and then investigate the properties of the corresponding immittance representation. This immittance, when it exists, evidently describes an n-port which satisfies the same physical postulates associated with the BR scattering description. As a matter of convenience, we treat the normalized impedance matrix $\boldsymbol{z} = \boldsymbol{R}_0^{-\frac{1}{2}}\boldsymbol{Z}\boldsymbol{R}_0^{-\frac{1}{2}}$; the properties of the admittance matrix readily follow.

The basic concept for impedance realizability is that of a Positive Real Matrix.

DEFINITION 4.6.1 *An $n\times n$ matrix function $\boldsymbol{z}$ is Positive Real (PR) if it satisfies:*

(1) $\boldsymbol{z}$ is analytic, $\Re s > 0$.

(2) $2\boldsymbol{r}(\sigma,\omega) = \boldsymbol{z}+\boldsymbol{z}^\dagger$ is nonnegative definite, $\forall\Re s > 0$, and a.e. on $s=j\omega$, $\boldsymbol{r}(\sigma,\omega) = \Re\,\boldsymbol{z}(\sigma+j\omega)$.

(3) $\boldsymbol{z}(s)$ is real on the positive real axis, or $\boldsymbol{z}(s^*) = \boldsymbol{z}^*(s)$, $\forall\Re s>0$, and a.e. on $s=j\omega$.

The fundamental immittance realizability theorem for passive n-ports is then as follows.

THEOREM 4.6.1
The impedance matrix of an n-port, when it exists, is Positive Real (PR), if and only if the associated scattering matrix is Bounded Real (BR).

PROOF Necessity: Given that $\boldsymbol{S}$ is BR, we wish to prove that the associated $\boldsymbol{z}$ is PR. The impedance-scattering relation is given by the second of eqs. (4.3.9). The reality property of $\boldsymbol{z}$ follows immediately from the reality of $\boldsymbol{S}$. It is also evident that since the BR $\boldsymbol{S}$ is analytic in the open RHP, so is $\boldsymbol{z}$, except at points where $\det(\boldsymbol{I}_n - \boldsymbol{S})$ is zero. We now show that such points are not present. Let $\det(\boldsymbol{I}_n - \boldsymbol{S})$ be singular at the point $s = p$, $\Re p > 0$. Then the homogeneous equations $[\boldsymbol{I}_n - \boldsymbol{S}(p)]\boldsymbol{a} = 0$ are of rank $\rho < n$, and hence have nonzero solutions for the vector $\boldsymbol{a}$. As a consequence $\boldsymbol{a} = \boldsymbol{S}(p)\,\boldsymbol{a}$. At $s = p$ the hermitian form

$$\boldsymbol{a}^\dagger[\boldsymbol{I}_n - \boldsymbol{S}^\dagger(p)\boldsymbol{S}(p)]\,\boldsymbol{a} = \boldsymbol{a}^\dagger\boldsymbol{a} - \boldsymbol{a}^\dagger\boldsymbol{a} = 0$$

Since $\boldsymbol{S}\boldsymbol{a}$ is a vector analytic in the RHP, its maximum norm occurs on $j\omega$. Referring to the nonnegative definite character of the hermitian form $\boldsymbol{a}^\dagger(\boldsymbol{I}_n - \boldsymbol{S}^\dagger\boldsymbol{S})\,\boldsymbol{a}$, this maximum occurs where the matrix of the form is zero, so the maximum norm is equal to $\boldsymbol{a}^\dagger\boldsymbol{a}$. But we have seen that we get the same value at the RHP interior point p. Thus, by the Maximum Modulus Theorem, the vector $\boldsymbol{S}\boldsymbol{a} = \boldsymbol{a}$ *everywhere* in the RHP. That is, $\det(\boldsymbol{I}_n - \boldsymbol{S})$ is singular everywhere and, contrary to hypothesis, $\boldsymbol{z}$ exists nowhere. The analyticity of $\boldsymbol{z}$ in $\Re s > 0$ is therefore established. Based on the equality of power and energy expressions in terms of $\boldsymbol{S}$ and $\boldsymbol{z}$, eqs. (4.4.4) and (4.4.5), we have

$$\boldsymbol{a}^\dagger(\boldsymbol{I}_n - \boldsymbol{S}^\dagger\boldsymbol{S})\,\boldsymbol{a} = \frac{1}{2}\boldsymbol{i}^\dagger(\boldsymbol{z} + \boldsymbol{z}^\dagger)\,\boldsymbol{i} \tag{4.6.1}$$

By the BR property, the hermitian form in $\boldsymbol{S}$ is nonnegative definite in $\Re s > 0$, and a.e. on $j\omega$. Therefore the form in $\boldsymbol{z}$ is nonnegative definite in the same region, and conclusion 2 of the Theorem is validated. The necessity proof is, therefore, complete.

Sufficiency: We are given a PR impedance matrix and must show that $\boldsymbol{S}$ is BR. The scattering matrix $\boldsymbol{S}$ is related to the impedance $\boldsymbol{z}$ by the first of eqs. (4.3.9). Consequently the reality property of $\boldsymbol{S}$ follows from the reality of $\boldsymbol{z}$. The analyticity of $\boldsymbol{S}$ in the RHP is evident, except if $\det(\boldsymbol{z} + \boldsymbol{I}_n)$ is zero at a point $s = p$ in the RHP. At such a point there is a nonzero vector, $\boldsymbol{i}$, which satisfies $[\boldsymbol{z}(p) + \boldsymbol{I}_n]\boldsymbol{i} = 0$, or $\boldsymbol{i} = -\boldsymbol{z}(p)\,\boldsymbol{i}$. But then the quadratic energy form in $\boldsymbol{z}$ at $s = p$ is

$$\frac{1}{2}\boldsymbol{i}^\dagger[\boldsymbol{z}(p) + \boldsymbol{z}^\dagger(p)]\boldsymbol{i} = -\boldsymbol{i}^\dagger\boldsymbol{i} < 0$$

which contradicts the nonnegative definite hypothesis for PR matrices. Therefore $\boldsymbol{S}$ must be analytic $\forall\,\Re s > 0$. Finally, starting with the nonnegative definite character of the energy form in terms of $\boldsymbol{z}$, the nonnegative definite property of the equivalent form in terms of $\boldsymbol{S}$ follows, so the sufficiency proof is complete. ∎

Note that the theorem and proof could equally well have been expressed in terms of the admittance matrix $\boldsymbol{y}$.

COROLLARY 4.6.1

With $\boldsymbol{S}$ BR, and $\boldsymbol{z}$ PR, the ranks of matrix functions $\boldsymbol{I}_n \pm \boldsymbol{S}$, $\boldsymbol{I}_n \pm \boldsymbol{z}$, $\boldsymbol{I}_n - \boldsymbol{S}^\dagger \boldsymbol{S}$, and $\boldsymbol{z}^\dagger + \boldsymbol{z}$ are invariant in the open RHP.

PROOF In each instance, because the hermitian energy form associated with either $\boldsymbol{S}$ or $\boldsymbol{z}$, as the case may be, is nonnegative definite in $\Re s > 0$, it follows that if the matrix is singular at a point in the open RHP, it is singular everywhere in the half-plane. Moreover, with $\boldsymbol{H}$ in any one of the matrices, the solution space of the homogeneous equations $\boldsymbol{Hx} = \boldsymbol{0}$, i.e., the space spanned by the m independent vector solutions $\boldsymbol{x}$ is invariant over the open RHP. Therefore the rank of each of the $n \times n$ matrices, $\rho = n - m$, is invariant in the open RHP. ∎

The following corollary needs no special comment.

COROLLARY 4.6.2

(1) An n-port satisfies the first five postulates (linearity, time invariance, solvability, passivity, reality) if and only if its impedance (admittance) matrix exists and is PR.

(2) If an LTI n-port satisfies Lorentz reciprocity, Postulate 7, eq. (4.5.6), its immittance matrices, if they exist, are symmetric.

(3) If an LTI n-port is composed of a finite number of lumped elements, postulate 8, its immittance matrices, if they exist, are rational.

A lossless structure is an extremely useful abstraction of a physical system which has finite but small dissipation. The PR concept is readily extended to this case.

COROLLARY 4.6.3

A passive, lossless n-port, postulate 6, has an impedance matrix, when it exists, that is PR and satisfies

$$\boldsymbol{z}^\dagger + \boldsymbol{z} = \boldsymbol{0} \quad \text{a.e.} \quad s = j\omega$$

z is LPR.

PROOF When the lossless postulate holds, it means that energy dissipated in the n-port over the infinite epoch from $-\infty < t < +\infty$ is always zero. From this we have shown in eq. (4.5.5) that a BR lossless $\boldsymbol{S}$ satisfies $\boldsymbol{I}_n - \boldsymbol{S}^\dagger \boldsymbol{S} = \boldsymbol{0}$ a.e. $s = j\omega$. But the hermitian energy forms in $\boldsymbol{S}$, and $\boldsymbol{z}$ are equal, eq. (4.6.1), so that

$$\frac{1}{2}\boldsymbol{i}^\dagger(\boldsymbol{z}^\dagger + \boldsymbol{z})\boldsymbol{i} = 0 \quad \text{a.e.} \quad s = j\omega, \quad \forall \boldsymbol{i}$$

and the result follows. ∎

The concept of a PR matrix function of a matrix function that is itself PR often plays a useful role in network analysis and synthesis.

THEOREM 4.6.2

Let $f(s)$ and $g(s)$ be two scalar PR (LPR) functions, then $f[g(s)]$ is also a scalar PR (LPR) function.

PROOF Function $g(s)$ maps the open RHP into a portion of itself, since $\Re f(s) > 0$, $\Re s > 0$, and we have used the strict inequality because the minimum resistance (zero) must occur on $j\omega$. Function $f(s)$ maps the image of the RHP through $g(s)$ into itself again. Moreover, $g(s)$ also maps the real axis into itself and so does $f(s)$; hence, $f[g(s)]$ maps the real axis into itself. Thus $f[g(s)]$ maps the RHP into itself and the real axis into itself, so it is a PR function. If $f(s)$ and $g(s)$ are both LPR, the same reasoning applies except that all mappings take the entire open RHP onto itself. ∎

A simple application is the following.

COROLLARY 4.6.4

The reciprocal of a scalar PR (LPR) function is a scalar PR (LPR) function.

PROOF Assume the PR function $f(s) = \dfrac{1}{s}$, and apply Theorem 4.6.2. ∎

We now prove the following.

THEOREM 4.6.3

Let $\boldsymbol{z}$ be a PR (LPR) immittance matrix. If its inverse, $\boldsymbol{y}$ exists a.e., then *it, too, is a PR (LPR) immittance matrix.*

PROOF The inverse $\boldsymbol{y}$ evidently satisfies reality. Moreover, $\boldsymbol{i}^\dagger(\boldsymbol{z}+\boldsymbol{z}^\dagger)\boldsymbol{i} > 0$ in the open RHP, for arbitrary vectors $\boldsymbol{i}$. Then choose $\boldsymbol{i} = \boldsymbol{y}\boldsymbol{v}$, so that $\boldsymbol{v}^\dagger\boldsymbol{y}^\dagger(\boldsymbol{z}+\boldsymbol{z}^\dagger)\boldsymbol{y}\boldsymbol{v} = \boldsymbol{v}^\dagger(\boldsymbol{y}^\dagger+\boldsymbol{y})\boldsymbol{v} > 0$ in the open RHP. Next, since $\boldsymbol{z}$ is analytic in $\Re s > 0$, so is $\boldsymbol{y}$, except possibly at a point $s = p$ in the RHP where $\det \boldsymbol{z} = 0$. In that case, $\boldsymbol{z}(p)\boldsymbol{x} = \boldsymbol{0}$ for some nonzero vector $\boldsymbol{x}$, and the hermitian form $\boldsymbol{x}^\dagger[\boldsymbol{z}^\dagger(p)+\boldsymbol{z}(p)]\boldsymbol{x} = 0$, or the analytic vector $\boldsymbol{z}\boldsymbol{x}$ takes on its minimum real part (i.e., zero) in the interior of the RHP so by the Minimum Real Part Theorem for vectors, eq. (A.7.4), it must be constant and $\det \boldsymbol{z}$ singular in the entire open RHP, contradicting the hypothesis. This proves analyticity and, hence, the PR character of the inverse. If $\boldsymbol{z}$ is LPR, the hermitian energy form for $\boldsymbol{z}$ is zero on $j\omega$, and so is the transformed expression in $\boldsymbol{y}$. Furthermore all the above reasoning still holds in the RHP. Thus, the inverse of an LPR matrix is LPR. ∎

If a matrix is PR, this does not preclude the possibility that it may have poles on $j\omega$, an analytic consequence of the physical property of resonance. The following Theorem addresses the restrictions on imaginary axis poles.

THEOREM 4.6.4
A PR matrix may have poles in conjugate pairs on the $j\omega$-axis, provided they are simple with hermitian nonnegative definite residue matrices which take on conjugate values at the conjugate poles.

PROOF A PR matrix has an hermitian part which is nonnegative definite a.e. on the $j\omega$-axis, or $\boldsymbol{z}(j\omega)+\boldsymbol{z}^\dagger(j\omega) \geq \boldsymbol{0}$. Suppose the matrix $\boldsymbol{z}$ has a pole of order n on $j\omega$. In the neighborhood of the pole, the matrix can be approximated by the principal term of its Laurent's expansion

$$\boldsymbol{z} = \frac{\boldsymbol{Z}_{-n}}{(s-j\omega_0)^n}$$

with $\boldsymbol{Z}_{-n}$ a constant matrix. Its elements are obtained from the behavior of the scalar functions, $\boldsymbol{z}_{(ij)}$, at the pole. Thus the scalar form $\boldsymbol{x}^\dagger\boldsymbol{z}\boldsymbol{x}$ where x is an arbitrary complex vector, can be written as $h_n/(s-j\omega_0)^n$, valid in the proximity of the pole. The constant $h_n = \boldsymbol{x}^\dagger\boldsymbol{Z}_{-n}\boldsymbol{x}$. If we put $s-j\omega_0 = \rho\epsilon^{j\theta}$ and denote by ϕ the argument of the complex number h_n, the result is

$$\frac{h_n}{(s-j\omega_0)^n} = \frac{|h_n|\epsilon^{-j(n\theta-\phi)}}{\rho^n}$$

If a small semicircle of radius ρ, centered on $j\omega_0$, is described in the RHP by allowing θ to vary between $-\pi/2$ and $\pi/2$, then the real part of the above expression changes its sign $n-1$ times and will be negative in regions of the RHP. The hermitian matrix $\boldsymbol{z}(s)+\boldsymbol{z}^{\dagger}(s)$ will fail the nonnegative definite requirement in those regions. The nonnegative definite condition can be satisfied if and only if $n=1$ (i.e., the pole is simple) and $\phi=0$, for this implies that h_1 is real and positive so that the residue matrix at the pole, $\boldsymbol{Z}_{-1}$ will be hermitian and nonnegative definite. The same result follows at the pole $(-j\omega_0)$, except that, since $\boldsymbol{z}(s^*)=\boldsymbol{z}^*(s)$, the residue matrix becomes $\boldsymbol{Z}^*_{-1}=\boldsymbol{Z}'_{-1}$. ∎

An LPR matrix, Corollary 4.6.3, has an energy form which is zero a.e. on $s=j\omega$. This property can be stated so that its applicability extends over the entire s plane, analogous to the use of a paraunitary matrix for representing a lossless BR scattering matrix, Corollary 4.5.7.

COROLLARY 4.6.5

Using paraconjugate notation, an LPR matrix, $\boldsymbol{z}$, satisfies

$$\boldsymbol{z}+\boldsymbol{z}_{\dagger}=\boldsymbol{0} \tag{4.6.2}$$

$\forall s$, except possibly a set of points of measure zero on the $j\omega$-axis. $\boldsymbol{z}$ is called a paraskew *matrix. The singularities of $\boldsymbol{z}$ are confined to the $j\omega$ axis.*

PROOF Referring to Corollary 4.6.3 for LPR matrices, the lossless criterion is

$$\boldsymbol{z}(j\omega)+\boldsymbol{z}'(-j\omega)=\boldsymbol{0} \quad \text{a.e. on } j\omega$$

Substituting $s=j\omega$, eq. (4.6.2) follows. The paraskew condition implies that a singularity in the open LHP must be accompanied by the image singularity in the open RHP, violating the RHP analyticity requirement. Thus for LPR matrices all singularities must be confined to $j\omega$. ∎

As pointed out earlier, concrete realizability, i.e., synthesis employing physical network elements, makes extensive use of rational functions. Accordingly, some of the general theorems and postulates of abstract realizability for PR and LPR matrices will be restated here for the rational matrix function case. The primary result is the following.

THEOREM 4.6.5

A rational immittance matrix $\boldsymbol{z}$ is PR if and only if:

(1) The rational functions composing the matrix array have real coefficients.

(2) The matrix is analytic in the open right half plane.

(3) Poles that occur on $j\omega$ (included the point at ∞) must be simple with hermitian nonnegative definite residue matrices.

(4) The hermitian form

$$z^{\dagger}(j\omega) + z(j\omega) \geq 0 \quad \text{a.e. on } j\omega$$

PROOF The important point to note is that for the rational case and according to the Theorem, the nonnegative definite test need be carried out only on the $j\omega$-axis. Since the matrix is rational, its only singularities are poles. The residue requirement at simple poles given in item (3) guarantees that the hermitian form is nonnegative definite in the RHP near a $j\omega$ pole, and also establishes the nonnegative definite property on the infinite radius semicircle in the RHP. Then including the provisos of item (2) and item (4) permits the use of the Minimum Real Part Theorem on the boundary consisting of the $j\omega$-axis with tiny RHP indentations at the imaginary axis poles. Therefore, the hermitian form is nonnegative definite $\Re s > 0$ and a.e. on the $j\omega$-axis. Since item (1) implies reality, the PR character is demonstrated. The converse readily follows. ∎

The LPR rational case takes on a particularly simple form since, according to Corollary 4.6.5, all poles are confined to the $j\omega$-axis. Moreover, all singularities of a rational matrix must be poles.

COROLLARY 4.6.6

A rational matrix is LPR if and only if all its poles are simple, and all are confined to $j\omega$ with nonnegative definite residue matrices.

The previous discussion of rational LPR matrices when applied to one-ports yields a statement of the classic Foster Reactance Theorem.[15] The Theorem is probably the earliest significant result of analytic network theory, and still finds extensive application.

THEOREM 4.6.6

((**Foster Theorem**) *A rational function, z, is LPR if and only if the following conditions hold. The poles of z are simple and occur in conjugate pairs, except that if a pole occurs at zero or infinity, it is unpaired. Moreover all poles are confined to the $j\omega$-axis, have positive real residues, and residues at a conjugate pole pair are equal.*

[15] R. M. Foster, "A Reactance Theorem", *Bell Sys. Tech. J.*, Vol. 3, pp. 259-267, April 1924.

Example 4.6.1
Determine whether the following are PR:

1. $z = \dfrac{as^2 + bs + c}{ds^2 + es + f} = \dfrac{N}{D}, \quad a, b, c, d, e, f \geq 0$

2. $z_1 = \sqrt{1 + s^2}, \quad z_2 = \sqrt{1 - s^2}$

3. $z = \tanh s$

4. $z = \exp(s^2)$

5. $\boldsymbol{z} = \begin{pmatrix} \dfrac{2s}{s^2+1} & \dfrac{s-1}{s^2+1} \\ \dfrac{s+1}{s^2+1} & \dfrac{2s}{s^2+1} \end{pmatrix}$

Solution

1. First, assume all coefficients are real and positive. Consequently reality is satisfied and all denominator roots will be in the LHP, so that z is analytic in $\Re\, s \geq 0$. Thus applying Corollary 4.6.5, we need only check whether $r(\omega) = \Re\, z(j\omega) \geq 0$. Or

$$r(\omega) = \frac{1}{2}[z(j\omega) + z{-}j\omega)] = \frac{ad\omega^4 + (be - af - cd)\omega^2 + cf}{2|D(j\omega)|^2} \geq 0$$

Only the positive character of the numerator need be tested. Differentiate with respect to ω^2 and set equal to zero to find the minimum point at $\omega^2 = (af + cd - be)/2ad$, and then substitute back to obtain the minimum value, $M = cf - (af + cd - be)^2/4ad$, and for $M \geq 0$

$$be \geq (\sqrt{af} - \sqrt{cd})^2 \tag{4.6.3}$$

This is the condition that z be PR. It is valid even if we allow nonnegative rather than strictly positive coefficients. For example, suppose $e = 0$. Then there is a pair of simple poles at $\pm j\sqrt{f/d}$ with residue $b/2 - j[(c - af/d)\,/\,2\sqrt{f/d}]$. For this to be real and positive, $af = cd$, which satisfies eq. (4.6.3). The other coefficient possibilities are also limited by the same inequality.

2. On $j\omega$, taking the resistances as $r_1(\omega)$, $r_2(\omega)$

$$r_1(\omega)|_{|\omega|<1} = \sqrt{1 - \omega^2} > 0, \quad r_1(\omega)|_{|\omega|>1} = 0,$$
$$r_2(\omega) = \sqrt{1 + \omega^2} > 0 \quad \forall\omega$$

So both are well behaved, i.e., nonnegative, on $j\omega$, especially r_2. But ... z_2 has branch points at $s = \pm 1$ so it fails the RHP analyticity requirement and is *not* PR. On the other hand, z_1 satisfies reality (z_2 does not) and is analytic in the RHP with a pair of boundary branch points at $s = \pm j$. We must test the nonnegative resistance requirement in the RHP for z_1. Write

$$z_1 = \sqrt{(s+j)(s-j)} = |\sqrt{1+s^2}|\epsilon^{j(\phi_a+\phi_b)}, \quad -\frac{\pi}{2} < \arg s < \frac{\pi}{2}$$

The restriction on $\arg s$ is to confine z_1 to a single branch (see Section A.9) where its PR character is to be investigated. The phase of the two factors in the RHP are $\phi_a = \arctan[(\omega+1)/\sigma]$ and $\phi_b = \arctan[(\omega-1)/\sigma]$. It is clear that for $\sigma > 0$, both $|\phi_a|$ and $|\phi_b| \le \pi/2$. Therefore in the RHP $\arg z_1 = \theta = (\phi_a + \phi_b)/2 \le \pi/2$. The result is that $r_1(\sigma,\omega) = |z_1|\cos\theta \ge 0$, $\sigma > 0$, and z_1 is PR. It is, in fact, the input (characteristic) impedance of an infinite length of lossless wave guide operating in a TM mode.[16]

3. A simple means for investigating the PR character of $z = \tanh s = (1-\epsilon^{-2s})/(1+\epsilon^{-2s})$ is to look at the corresponding reflectance. Thus

$$S = \frac{\tanh s - 1}{\tanh s + 1} = -\epsilon^{-2s}$$

We have already seen in Example 4.5.1 that ϵ^{-s} is BR and lossless; clearly then, the same is true for the reflectance S. Accordingly $z = \tanh s$ is LPR. It is, in fact, the input impedance of a length of short circuited transmission line (see Section 7.1).

4. On the boundary $z = \epsilon^{-\omega^2}$, so the real part on $j\omega$ is nonnegative. Furthermore, $z(s)$ is analytic in the open RHP and is real for real s. In the RHP

$$\Re z = \Re[\exp(\sigma^2 - \omega^2 + j2\sigma\omega)] = \epsilon^{\sigma^2-\omega^2}\cos 2\sigma\omega, \ \sigma > 0$$

Evidently $\cos 2\sigma\omega < 0$, $\pi/2 < 2\sigma\omega < \pi$, thus $\Re z$ is negative in the RHP and, despite a promising start, z is *not* PR.

5. A moment's inspection reveals that $\boldsymbol{z}$ is rational, satisfies reality and is analytic in the open RHP with poles all confined to $j\omega$. Furthermore $\boldsymbol{z}_\dagger = -\boldsymbol{z}$, so that the matrix is paraskew and therefore qualifies as an LPR immittance matrix. We have only to examine the poles on the

[16] The input impedance is not LPR since energy crossing the input terminal plane propagates to infinity and is not returned to the source. In the case of a lossless TEM transmission line of infinite length, the input impedance is a constant resistance.

boundary. The one pair of poles is at $s = \pm j$. The residue matrix at $s = j$ is $\boldsymbol{A}_{(j)}$. That at $s = -j$ is $\boldsymbol{A}_{(-j)}$ with

$$\boldsymbol{A}_{(j)} = \begin{pmatrix} 1 & \dfrac{j-1}{2j} \\ \dfrac{-j-1}{-2j} & 1 \end{pmatrix} = \boldsymbol{A}^*_{(-j)}$$

Since $\boldsymbol{A}^\dagger_{(j)} = \boldsymbol{A}_{(j)}$, the residue matrices are hermitian. Also the leading principal minor is $A_{11} = 1 > 0$ and $\det \boldsymbol{A}_{(j)} = 1 - [(j-1)(-j-1)/(2j)(-2j)] = 1/2 > 0$, so the residue matrices are nonnegative definite. The immittance matrix $\boldsymbol{z}$ is LPR. □

We conclude this section with two important results concerning stored energy in an LTI lossless reciprocal network.

THEOREM 4.6.7

Given a lossless, reciprocal *n-port described by an LPR impedance matrix* $\boldsymbol{z} = \boldsymbol{r} + j\boldsymbol{x}$, *then, under eigenmode excitation, with port currents and voltages given as* $\boldsymbol{i}(j\omega)\,\epsilon^{j\omega t}$, $\boldsymbol{v}(j\omega)\,\epsilon^{j\omega t}$, *it follows that:*

(1) The difference of the magnetic ($\overline{W}_m$) and electric ($\overline{W}_e$) average stored energies satisfies

$$\boldsymbol{i}^\dagger(j\omega)\,\boldsymbol{x}(\omega)\,\boldsymbol{i}(j\omega) = 2\omega\,(\overline{W}_m - \overline{W}_e) \tag{4.6.4}$$

(2) The sum of the stored energies satisfies

$$\boldsymbol{i}^\dagger(j\omega)\,\frac{d\boldsymbol{x}}{d\omega}\,\boldsymbol{i}(j\omega) = 2\omega\,(\overline{W}_m + \overline{W}_e) \tag{4.6.5}$$

PROOF

(1) Referring to Corollary 4.6.3, the impedance matrix of a lossless n-port satisfies $\boldsymbol{z}(j\omega) + \boldsymbol{z}^\dagger(j\omega) = \boldsymbol{0}$. Under reciprocity $\boldsymbol{z}^\dagger(j\omega) = \boldsymbol{z}^*(j\omega)$, so that

$$2\Re\,\boldsymbol{z}(j\omega) = \boldsymbol{z}(j\omega) + \boldsymbol{z}^*(j\omega) = \boldsymbol{0} \qquad \boldsymbol{z}(j\omega) = j\boldsymbol{x}(\omega) \tag{4.6.6}$$

i.e., the impedance matrix is purely imaginary on $j\omega$. As a consequence of Boucherot's Theorem, the reactive power in an LTI passive system under a.c. operation, eq. (2.14.2), is $Q = 2\omega(\overline{W}_m - \overline{W}_e)$. Expressing the result in terms of port quantities

$$\Im[\boldsymbol{i}^\dagger(j\omega)\,\boldsymbol{v}(j\omega)] = \Im[\boldsymbol{i}^\dagger(j\omega)\,\boldsymbol{z}(j\omega)\boldsymbol{i}(j\omega)] = \boldsymbol{i}^\dagger(j\omega)\,\boldsymbol{x}(\omega)\,\boldsymbol{i}(j\omega) = Q$$

which is the required result.

(2) Consider the variational expression (all arguments are $s = j\omega$ and $d\boldsymbol{v}/d\omega = (d\boldsymbol{z}/d\omega)\boldsymbol{i} + \boldsymbol{z}(d\boldsymbol{i}/d\omega)$);

$$\boldsymbol{i}^\dagger \frac{d\boldsymbol{v}}{d\omega} + \boldsymbol{v}^\dagger \frac{d\boldsymbol{i}}{d\omega} = \boldsymbol{i}^\dagger \frac{d\boldsymbol{z}}{d\omega}\boldsymbol{i} + \boldsymbol{i}^\dagger(\boldsymbol{z} + \boldsymbol{z}^\dagger)\frac{d\boldsymbol{i}}{d\omega} = \boldsymbol{i}^\dagger \frac{d\boldsymbol{z}}{d\omega}\boldsymbol{i}$$

since by losslessness $\boldsymbol{z} + \boldsymbol{z}^\dagger = \boldsymbol{0}$.

We can repeat all the above steps for the internal branches of the lossless reciprocal n-port and obtain

$$\boldsymbol{i}_b^\dagger \frac{d\boldsymbol{v}_b}{d\omega} + \boldsymbol{v}_b^\dagger \frac{d\boldsymbol{i}_b}{d\omega} = \boldsymbol{i}_b^\dagger \frac{d\boldsymbol{z}_b}{d\omega}\boldsymbol{i}_b$$

taken over all the inductors, capacitors, and ideal transformers of the lossless, reciprocal structure. The contribution of the frequency independent ideal transformers is readily shown to be zero. Over the coils with branch impedance matrix $j\omega\boldsymbol{L}$

$$\boldsymbol{i}_b^\dagger \frac{d\boldsymbol{z}_{bl}}{d\omega}\boldsymbol{i}_b = j\boldsymbol{i}_b^\dagger \frac{d\omega\boldsymbol{L}}{d\omega}\boldsymbol{i}_b = j\boldsymbol{i}_b^\dagger \boldsymbol{L}\boldsymbol{i}_b = j2\overline{W}_m$$

Over the capacitors with branch impedance matrix $\boldsymbol{z}_{bc} = -j\boldsymbol{C}^{-1}/\omega$, the result is

$$\boldsymbol{i}_b^\dagger \frac{d\boldsymbol{z}_{bc}}{d\omega}\boldsymbol{i}_b = \frac{j}{\omega^2}\boldsymbol{i}_b^\dagger \boldsymbol{C}^{-1}\boldsymbol{i}_b = j2\overline{W}_e$$

According to Tellegen's Theorem we can equate the port and branch expressions

$$\boldsymbol{i}^\dagger \frac{d\boldsymbol{z}}{d\omega}\boldsymbol{i} = \boldsymbol{i}_b^\dagger \frac{d\boldsymbol{z}_{bl}}{d\omega}\boldsymbol{i}_b + \boldsymbol{i}_b^\dagger \frac{d\boldsymbol{z}_{bc}}{d\omega}\boldsymbol{i}_b = j2(\overline{W}_m + \overline{W}_e)$$

and since $\boldsymbol{z} = j\boldsymbol{x}$, the required result eq. (4.6.5) follows.[17] ∎

4.7 The Degree of a One-Port

A finite passive one-port is characterized by a rational immittance, whose poles lie in the closed LHP (see Section 4.6). Thus its impedance must have the form

$$z = \frac{n}{d} = \frac{n_0\alpha}{d_0\beta}$$

[17]The results of this theorem can be deduced for more general structures by electromagnetic considerations outside the domain of circuit theory, including the use of variational expressions based on Maxwell's equations and the computation of field integrals over the n-ports of access. See, for example, C. G. Montgomery, R. H. Dicke, E. M. Purcell, *Principles of Microwave Circuits*, New York, McGraw Hill, 1948, chap. 5.

where polynomials n_0 and d_0 have all their zeros in $\Re s < 0$ and polynomials α and β have all their zeros on the imaginary axis.

DEFINITION 4.7.1 *A polynomial, all of whose zeros lie in the open LHP, $\Re s < 0$ is said to be a* Hurwitz polynomial.

Thus n_0 and d_0 are Hurwitz polynomials.

Hurwitz polynomials have the necessary, but not sufficient, property that all their coefficients are positive (if the leading coefficient is chosen to be positive).

On the other hand, a polynomial all of whose zeros lie on the imaginary axis must necessarily be a product of factors of the form $s^2 + \omega_i^2$ (because all zeros appear in conjugate imaginary pairs), as well as the factor s^q, $q = 0$ or 1

$$p(s) = \prod_i s^q(s^2 + \omega_i^2)$$

$p(s)$ is even or odd depending on whether $q = 0$ or $q = 1$.

Thus α and β in eq. (4.7.1) have the above form, obviously with different values for the ω_i's.

DEFINITION 4.7.2 *The degree of a rational function is the greater of the degrees of the numerator and denominator polynomials.*

(See also Section A.8.) The definition above is a natural extension to rational functions of the concept of degree of a polynomial; it reduces to the polynomial case for rational functions whose numerators are arbitrary polynomials and denominators constant. In the following, the degree of a function f will be denoted as $\delta[f]$.

The *degree of a one-port* is the degree of its immittance. The following theorem holds.

THEOREM 4.7.1
For a finite passive one-port, the degrees of the immittance and of the reflectance are equal.

PROOF Let the impedance z be expressed as

$$z = \frac{n}{d}$$

Then,

$$\delta[z] = \max(\delta[n], \delta[d])$$

The reflectance S is given by

$$S = \frac{n-d}{n+d}$$

The degree of $n-d$ cannot exceed that of $n+d$ because otherwise S would have a pole at infinity; thus $\delta[S] = \delta[n+d]$. But $\delta[n+d]$ can neither be greater nor less than the greater of the degrees of n and d, for in polynomial $n+d$ no cancellations can occur between terms of n and d. The reason for this is that the polynomials are Hurwitz (all coefficients must be positive) with possible additional $s^2+\omega_i^2$ root factors whose presence cannot change the positive character of the coefficients. The degree of $n+d$ must therefore be equal to the greater of the degrees of n and d. ∎

Theorem 4.7.1 shows that the admittance, impedance, and reflectance functions of a finite passive one-port all have the same degree, which we then define as the degree of the one-port. The physical meaning of this concept will be clarified in the subsequent discussion of one-port synthesis.

5

One-Port Synthesis

5.1 Introduction

Realizability, as discussed in the previous chapter, expressed the frequency domain properties of a scattering matrix which describes a system satisfying a specific set of time domain postulates. If the postulates were satisfied, the system was "realizable", but there was no reference to the internal structure of the n-port in terms of its circuit element building blocks. The emphasis was purely on the consequences of LTI system performance as measured at the access ports. In this sense we can speak of *abstract* realizability. If, on the other hand, an abstractly realizable scattering matrix is shown to be constructively realizable as an n-port composed of a finite or infinite number of elements, we can say that it has the property of *concrete* realizability.[1]

While abstract realizability results are available for relatively wide classes of analytic matrices, concrete realizability has been completely investigated only for finite n-ports, i.e., for the class of rational matrices, and for certain other cases transformable to the rational domain.

One of the main results of rational matrix concrete realization is Darlington's procedure of synthesis, a cornerstone for the design of passive networks. Although originally motivated to provide a method of filter design as an alternate to image parameter theory, its scope broadens to the design of all structures composed of a lossless two-port inserted between given generator and load with the aim of realizing a specified input-output frequency response. The procedure has further been generalized to n-ports, with $n > 2$ (admittedly employing rather unwieldy algebraic methods and complicated building blocks), but the two-port case is most important since it includes most practical passive filters and equalizers. These applications

[1]The concepts of abstract and concrete realizability were introduced in V. Belevitch, *Classical Network Theory*, S. Francisco, Holden-Day, 1968, chap. 3.

will be discussed more thoroughly in subsequent chapters. An important extension of the Darlington idea in the sense of abstract realizability of non-rational structures is also available. However, the discussion which follows refers exclusively to the rational case.

We first tackle the synthesis of lossless one-ports and show that it can always be accomplished by using two-terminal networks including only inductors and capacitors (LC networks); such networks can always be realized in specific forms, called the Foster and Cauer canonic forms, that involve the minimum number of circuit elements. We then extend the synthesis to one-ports realizable on two-terminal networks only including resistors and inductors (RL networks) or resistors and capacitors (RC networks) by a simple dimensional transformation of a basic LC network.

As a second step, we introduce the *canonic form* of the scattering matrix of a lossless two-port, the related open-circuit impedance and short-circuit admittance matrices, and define the concept of *transmission zeros.* Then we present Darlington's cascade synthesis; an arbitrary PR function is shown to be concretely realizable as the input immittance of a lossless two-port loaded by a unit resistor. The lossless two-port (which can always be chosen to be reciprocal) can be realized either in global form, generally involving the series and/or parallel interconnection of component two-ports, or in cascade form. The latter realization, involving the use of relatively simple sections arranged in a ladder-type structure is the only one we consider in detail. The global forms are generally impractical to build.

Although lossless one-ports can be considered within the purview of resistively terminated lossless two-ports by including zero or infinite resistors as loads, practically a lossless one-port is synthesized in Foster or Cauer canonic forms, or possibly by more complicated purely lossless structures.

In our subsequent discussions, unless otherwise stated, we will generally omit the complex frequency variable in the function statement, using f instead of $f(s)$. Also (scalar and/or matrix) immittances will be systematically considered as normalized and denoted by lower case letters as stipulated in footnote 8 at page 183.

5.2 Lossless One-Port Synthesis

The properties satisfied by a rational one-port immittance to be LPR were given in the Foster Theorem 4.6.6. A function with these characteristics is often termed a Foster function. The crucial conditions are that all poles of the function be simple, all occur in conjugate pairs (except at zero or infinity) and be confined to $j\omega$, and all have positive real residues.

We now explore the consequences of these conditions. Since an LPR z is paraskew, so that it satisfies eq. (4.6.2), it is an odd function. Therefore, it must be the ratio of either an even and an odd polynomial or an odd and an even polynomial; all polynomial zeros lie on the $j\omega$-axis, are simple and, except when a zero is at the origin, occur in conjugate complex pairs. These conditions, for the denominator polynomial d, stem from the constraints on the poles of the function z; for the numerator polynomial n, they derive from the constraints on the poles of the function $y = 1/z$. The degrees of the two polynomials must differ by at least 1 since one polynomial is even, the other odd; they differ by precisely 1, otherwise there would be a pole of higher order than 1 at infinity in either z or y.

Now observe that the LBR reflectance deduced from LPR $z = n/d$ is

$$S = \frac{z-1}{z+1} = \frac{n-d}{n+d}$$

It follows from the BR property of S, which is analytic in the RHP, that $n+d$ is Hurwitz, i.e., devoid of zeros in the RHP, and hence its coefficients must be strictly positive. Therefore the coefficients of both n and d, as the even and odd parts of the Hurwitz polynomial $n+d$, must all be strictly real and positive. Finally, since all residues of z are real and conjugates of each other at the conjugate pole pairs, the paired residues must also be equal to each other.

We now apply the partial fraction expansion described in eq. (A.11.2) to the impedance z

$$z = h_\infty s + \sum_{i=-m}^{m} \frac{h_i}{s - j\omega_i} + \frac{h_0}{s}$$

where in the sum index i skips the value $i = 0$, and $2m$ is the number of poles at finite nonzero frequencies ω_i (*internal* poles). Conjugate complex poles are grouped as follows

$$\frac{h_i}{s - j\omega_i} + \frac{h_i}{s + j\omega_i} = \frac{2h_i s}{s^2 + \omega_i^2}$$

so that the expansion can be rewritten as

$$z = h_\infty s + \sum_{i=1}^{m} \frac{2h_i s}{s^2 + \omega_i^2} + \frac{h_0}{s} \tag{5.2.1}$$

Note the following:

(i) $h_\infty s$ is the impedance of an h_∞ inductance;

(ii) each term in the sum can be written as

$$\frac{2h_i s}{s^2 + \omega_i^2} = \frac{1}{\dfrac{1}{2h_i}s + \dfrac{\omega_i^2}{2h_i s}}$$

which represents the impedance of a parallel resonator consisting of capacitance $c_i = 1/2h_i$ in shunt with inductance $l_i = 2h_i/\omega_i^2$. This circuit resonates at angular frequency ω_i;

(iii) h_0/s is the impedance of capacitance $1/h_0$.

We conclude that the impedance z is realizable by series connecting the inductor mentioned in (i), the capacitor mentioned in (ii), and the m parallel resonators mentioned in (iii).

We note that eq. (5.2.1) refers to the case in which the numerator is an even polynomial, the denominator is an odd polynomial, and the degree of the numerator exceeds by 1 that of the denominator. This is easily verified as follows: (a) the presence of term h_0/s implies the existence of a pole at the origin, hence of a zero at $s = 0$ of the denominator polynomial; evidently the zero is present in the odd polynomial, and not in the even one, otherwise the two zeros would cancel, eliminating the pole; (b) the presence of term $h_\infty s$ implies the existence of a (simple) pole at infinity, which is only possible if the degree of the numerator is one more than that of the denominator, and since there are $2m$ finite nonzero poles, the denominator will be of degree $2m + 1$, the numerator of degree $2m + 2$.

There are four possibilities that arise from combining the parity of the polynomials (numerator even and denominator odd, or vice versa) with the difference between their degrees ($+1$ or -1). These possibilities correspond to the four choices for h_∞ and h_0 as zero or nonzero. The specific problem dictates these coefficient values, and the form of the partial fractions expansion.

Alternatively, the synthesis procedure can be applied to the admittance y. Its partial fraction expansion has the same form as that of the impedance, because both functions satisfy the same set of rules (Theorem 4.6.1)

$$y = k_\infty s + \sum_{i=1}^{m'} \frac{2k_i s}{s^2 + \omega_i'^2} + \frac{k_0}{s} \tag{5.2.2}$$

Here $2m'$ is the number of internal poles of y (coinciding with the internal zeros of z). Eq. (5.2.2) can be interpreted as the parallel connection of

(i) capacitance k_∞;

(ii) m' series resonant circuits each composed of inductance $l_i = 1/2k_i$ in series with capacitance $c_i = 2k_i/\omega_i'^2$;

(iii) inductance $1/k_0$.

The above results are summarized in the following.

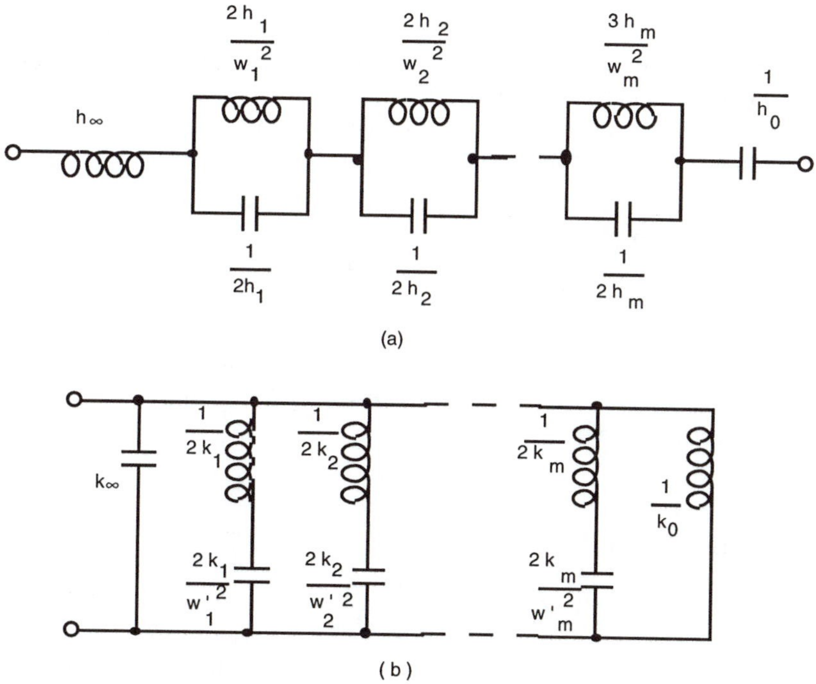

FIGURE 5.2.1
Foster realization of an LPR function: (a) series realization; (b) parallel realization.

THEOREM 5.2.1
(**Foster Synthesis**) *An LPR function can be realized as the impedance z of a lossless one-port by series connecting:*

i) an inductance h_∞ ;

ii) m parallel resonant circuits each composed of a capacitance $1/2h_i$ and an inductance $2h_i/\omega_i^2$;

iii) a capacitance $1/h_0$.

m is the number of internal poles of z; the h_i's are the residues of z at these poles; h_∞ and h_0 are the residues at poles, which occur at infinity and/or the origin, respectively (external poles).

The dual procedure of parallel interconnection of series resonators leads to the equivalent result for admittances.

The two concrete realizations are depicted in Fig. 5.2.1.

It is a straightforward matter to count the number of circuit elements required for the Foster synthesis by inspection of the partial fraction expansion. The result for either the impedance or admittance representation is the same and is given as twice the number of internal poles plus the number of external poles (the latter at zero and/or infinity). Alternately, the number of elements is equal to the higher of the degrees of n and d or, even more simply put, equal to the degree of the polynomial $n + d$, which in turn is the degree of the rational function z or y. Thus

THEOREM 5.2.2
The minimum number of lossless elements necessary to realize a lossless immittance is equal to its degree.

The two Foster structures are *canonical* since they allow the realization of any LPR function; they are *minimal* since they require the minimum number of circuit elements.

Example 5.2.1
Consider the function

$$f = \frac{n}{d} = \frac{30s^4 + 23s^2 + 1}{30s^3 + 2s} \tag{5.2.3}$$

Realize it as the impedance of a lossless one-port both in series and in parallel canonical forms.

Solution

(a) We represent the function (5.2.3) as an impedance

$$z = \frac{n}{d} = \frac{30s^4 + 23s^2 + 1}{30s^3 + 2s}$$

The higher degree (4) is that of n. Thus four reactances will be required. Decompose the impedance into partial fractions. The poles of z lie at infinity and at the roots of the denominator d

$$s_0 = 0; \qquad s_{1,2} = \pm j\frac{1}{\sqrt{15}}$$

The residue at infinity is calculated as

$$h_\infty = \lim_{s\to\infty} \frac{z}{s} = 1$$

The residues, h_i at the finite poles, according to eq. (A.11.5), are found by using the expressions below, where d' is the denominator derivative with respect to s

$$g = \frac{n}{d'} = \frac{30s^4 + 23s^2 + 1}{90s^2 + 2}, \qquad h_i = g(s_i) \tag{5.2.4}$$

The results, since residues at conjugate poles are equal, are

$$h_0 = g(s_0) = \frac{1}{2}, \qquad h_{1,2} = g(s_1) = g(s_2) = \frac{1}{10}$$

The partial fraction expansion of z is therefore

$$z = s + \frac{1}{2s} + \frac{\frac{1}{5}s}{s^2 + \frac{1}{15}}$$

According to the Foster Synthesis Theorem 5.2.1 the impedance above is realized by series connecting:

(i) an inductance $l_\infty = h_\infty = 1$;

(ii) a parallel resonant circuit consisting of capacitance $c_1 = 1/2h_{1,2} = 5$ and inductance $l_1 = 2h_{1,2}/\omega_1^2 = 3$;

(iii) a capacitance $c_0 = 1/h_0 = 2$.

The realization has the configuration of Fig. 5.2.1 (a).

(b) The second Foster form is realized by expanding the admittance $y = 1/z$

$$y = \frac{d}{n} = \frac{30s^3 + 2s}{30s^4 + 23s^2 + 1}$$

Its four poles are internal and located at the roots of n

$$s_{1,2} = \pm j0.21511\,; \qquad s_{3,4} = \pm j0.84876$$

For the residues, use eq. (5.2.4)

$$e = \frac{d}{n'} = \frac{15s^2 + 1}{60s^2 + 23},$$
$$h_{1,2} = e(s_1) = 0.015128, \quad h_{3,4} = e(s_3) = 0.48487$$

The partial fraction expansion of y is therefore

$$y = \frac{2 \cdot 0.015128s}{s^2 + 0.21511^2} + \frac{2 \cdot 0.48487s}{s^2 + 0.84876^2}$$

The Foster Synthesis now yields a realization as the parallel connection of two series resonators:

i) a first series resonator composed of an inductance $l_1 = 1/2h_{1,2} = 33.052$ and a capacitance $c_1 = 2h_{1,2}/\omega_1^2 = 0.65387$;

ii) a second series resonator composed of an inductance $l_3 = 1/2h_{3,4} = 1.0312$ and a capacitance $c_3 = 2h_{3,4}/\omega_3^2 = 1.3461$.

The realization has the configuration of Fig. 5.2.1 (b), but the individual pure capacitance and inductance at input and output, respectively, are not present.

Both realizations require four elements, the degree of z or y. □

The Foster realization requires the factorization of a polynomial equal to degree m, the number of internal poles of the prescribed function. However, the synthesis can be carried out with purely rational operations as described by Cauer, leading to either one of two ladder structures, another classic result of network synthesis.[2]

The basic idea is the following. Let z be a function to be realized as the impedance of a lossless one-port; since the difference between the degrees of the numerator and the denominator equals 1 or -1, either z or $y = 1/z$ has a simple pole at infinity. Suppose first that the pole appears in z with residue a_1. If it is extracted from the function, the remainder

$$z_1 = z - a_1 s$$

is still an LPR function, as is immediately evident by inspection of the partial fraction expansion (5.2.1). The remainder no longer has a pole at infinity, but must have a zero there. (Remember that the difference between the numerator and denominator degrees can only be 1 or -1, thus any Foster function must have either a pole or a zero at infinity.) The admittance $y_1 = 1/z_1$ now has a pole at infinity, which can, in turn, be extracted so as to obtain

$$y_2 = y_1 - a_2 s$$

where y_2 is again LPR and has a simple zero at infinity. The procedure continues with a degree reduction of one at each step, corresponding to the elimination of a term proportional to s (only this term need be computed) from each new partial fraction expansion. The process terminates when the degree reduces to zero. Note that removing a pole at infinity from an impedance means extracting a series inductance; removing a pole at infinity from an admittance means extracting a shunt capacitance.

Thus, starting from z, a series inductance is extracted leaving z_1 as remainder; next, a shunt capacitance is extracted from y_1 leaving y_2 as remainder; then a series inductance is extracted from z_2, and so on, to completion.

If the given impedance z has a zero at infinity, the above procedure starts with y (which has the pole at infinity) instead of z, and the first element is a shunt capacitance. In either case, the resulting structure is a

[2] W. Cauer, "Ein Reaktanztheorem", *Sitzber. preuss. Akad. Wiss.*, vol. 30-32, pp. 673-681, 1931.

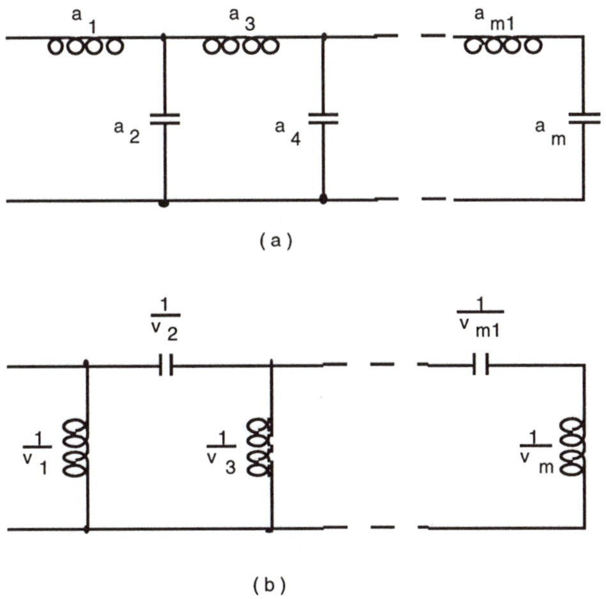

FIGURE 5.2.2
Cauer realization of an LPR function: (a) lowpass realization; b) highpass realization.

ladder network with inductive series branches and capacitive shunt branches (*lowpass ladder*) and is shown in Fig. 5.2.2 (a). The first element is either a series inductor or a shunt capacitor, depending on whether z or y has the pole at infinity. The last element is a shunt capacitor or series inductor depending on whether z or y has a pole at the origin.

Algebraically, the realization just discussed (*first Cauer structure*) corresponds to a continuous fraction expansion about the point at infinity

$$z = a_1 s + \cfrac{1}{a_2 s + \cfrac{1}{a_3 s + \cfrac{1}{a_4 s + \cdots}}} \tag{5.2.5}$$

The a_i are always residues at infinity, and can be determined by an iterative long division procedure, in which at each step the remainder is divided into the divisor of the previous step; the details of this iteration are readily worked out and can be found elsewhere.[3]

As an alternate choice, one can extract poles at the origin, resulting in a ladder structure in which series capacitors alternate with shunt inductors (*highpass ladder*) as shown in Fig. 5.2.2 (b).

[3]H. Baher, *Synthesis of Electrical Networks*, New York, Wiley, 1984, p. 53.

Algebraically, the alternate realization (*second Cauer structure*) corresponds to a continuous fraction expansion about the origin

$$z = \frac{b_1}{s} + \cfrac{1}{\cfrac{b_2}{s} + \cfrac{1}{\cfrac{b_3}{s} + \cfrac{1}{\cfrac{b_4}{s} + \cdots}}}$$

The second Cauer expansion does not require essentially new ideas nor new techniques, since it is the dual of the previous case; at each step of the extraction process the term proportional to $1/s$ (rather than proportional to s) is removed corresponding to the pole at zero.

Cauer realizations are both canonical and minimal.

Example 5.2.2

Realize the impedance of eq. (5.2.3) (a) as a lowpass ladder, and (b) as a highpass ladder.

Solution

(a) The extraction of the pole at infinity from z in eq. (5.2.3) gives

$$z_1 = z - a_1 s = \frac{30s^4 + 23s^2 + 1}{30s^3 + 2s} - s = \frac{21s^2 + 1}{30s^3 + 2s}$$

The series inductor has inductance $l_1 = a_1 = 1$. Then we extract the pole at infinity from $y_1(s)$

$$y_2 = y_1 - a_2 s = \frac{30s^3 + 2s}{21s^2 + 1} - \frac{10}{7}s = \frac{\frac{12}{21}s}{21s^2 + 1}$$

The shunt capacitor has capacitance $c_2 = a_2 = 10/7$. With a further step, we obtain

$$z_3 = z_2 - a_3 s = \frac{21s^2 + 1}{\frac{12}{21}s} - \frac{441}{12}s = \frac{21}{12s}$$

The series inductor has inductance $l_3 = a_3 = 147/4$. Finally, we have

$$y_4 = y_3 - a_4 s = \frac{12}{21}s - \frac{12}{21}s = 0$$

The shunt capacitor has capacitance $c_4 = a_4 = 4/7$. After its extraction the degree is zero and the remainder is a zero admittance, i.e., an open circuit. The resultant circuit has the lowpass configuration of Fig. 5.2.2 (a).

b) The extraction of the pole at the origin from z in eq. (5.2.3) (rewritten as a function of $1/s$), gives

$$z_1 = z - \frac{b_1}{s} = \frac{\frac{1}{s^4} + \frac{1}{s^2} + 30}{\frac{2}{s^3} + \frac{30}{s}} - \frac{1}{2s} = \frac{\frac{8}{s^2} + 30}{\frac{2}{s^3} + \frac{30}{s}}$$

The first element is series capacitance $c_1 = 1/b_1 = 2$. Then we extract the pole at the origin from y_1.

$$y_2 = y_1 - \frac{b_2}{s} = \frac{\frac{2}{s^3} + \frac{30}{s}}{\frac{8}{s^2} + 30} - \frac{1}{4s} = \frac{\frac{45}{2s}}{\frac{8}{s^2} + 30}$$

The shunt inductance $l_2 = 1/b_2 = 4$. Similarly the next step gives

$$z_3 = z_2 - \frac{b_3}{s} = \frac{\frac{8}{s^2} + 30}{\frac{45}{2s}} - \frac{16}{45s} = \frac{30}{\frac{45}{2s}}$$

and the series capacitance $c_3 = 1/b_3 = 45/16$. Finally,

$$y_4 = y_3 - \frac{b_4}{s} = \frac{45}{60s} - \frac{45}{60s}$$

The last element is shunt inductance $l_4 = 1/b_4 = 4/3$. After its extraction the degree is zero, and the remainder is a zero admittance i.e., an open circuit. The resultant circuit has the highpass configuration of Fig. 5.2.2 (b). The two circuits are equivalent at their input ports. As a simple check note that at zero and infinite frequencies both ladders present an open circuit at the input.

□

Minimal realizations are by no means limited to the previous four cases. As matter of fact one can extract in any order poles at infinity, at the origin and possibly at finite frequencies, thus arriving at a variety of structures, all containing the same (minimum) number of elements.

If the condition that the realization be minimal is removed, poles can be extracted *partially*, i.e., only a portion of the residue removed; this obviously implies that the pole remains in the function and no reduction in degree is achieved until it is completely extracted. This clearly produces realizations with superfluous elements.

The Foster Theorem provides an elegant test for the Hurwitz character of a polynomial.

THEOREM 5.2.3
Let a polynomial in s be represented as $p = e + o$, where e is the polynomial of even powers and o contains the odd powers. Then p is a strict Hurwitz polynomial (no roots in the closed RHP) if and only if $z = e/o$ is a Foster function.

PROOF Necessity: Given that e/o is Foster, then the corresponding reflectance $S = (e - o)/(e + o)$ is LBR. But then $p = e + o$ has no roots in the closed RHP and is therefore strictly Hurwitz.

Sufficiency: Given p Hurwitz, then all its coefficients are positive. Consider the reflectance $S = (e - o)/(e + o)$. Along $s = j\omega$ the amplitudes of numerator and denominator af $S(j\omega)$ are equal since the odd power terms (imaginary terms) of $e - o$ are the negatives of those of $e + o$, whereas the even powers (real terms) are equal. Thus $|S(j\omega)| = 1$ and, since its denominator is strictly Hurwitz, S is LBR. But then the corresponding impedance $z = e/o$ is LPR, i.e., a Foster function. ■

Note that applying a Cauer expansion for z to verify its Foster character by determining whether all the continued fraction coefficients are positive, gives a rational procedure for performing the Hurwitz test.

A fundamental property of lossless impedance functions is presented in the next Theorem. Since an LPR function is odd, it must be purely imaginary on $j\omega$, so we may write $z(j\omega) = jx(\omega)$, where $x(\omega)$ is the reactance function.

THEOREM 5.2.4
Reactance Slope Theorem *Let the reactance corresponding to a lossless LPR rational function be $x(\omega)$. Then (excluding the trivial case of $x \equiv 0$)*

(1) The reactance slope must satisfy

$$\frac{dx(\omega)}{d\omega} > 0, \quad \forall\omega \tag{5.2.6}$$

(2) Except for the case that the input impedance to the one-port is that of a pure inductance or capacitance, the reactance function satisfies

$$\frac{dx(\omega)}{d\omega} > \left|\frac{x(\omega)}{\omega}\right|, \quad \forall\omega \tag{5.2.7}$$

(3) If the input to the one port is a pure inductance or capacitance, then the reactance satisfies

$$\frac{dx(\omega)}{d\omega} = \left|\frac{x(\omega)}{\omega}\right|, \quad \forall\omega \tag{5.2.8}$$

Thus the minimum attainable reactance slope for a Foster function which interpolates to reactance value x_0 at ω_0, is

$$\min \frac{dx(\omega_0)}{d\omega} = \left|\frac{x_0}{\omega_0}\right|$$

realizable by a pure inductance one-port ($\omega_0 L = x_0$, when $x_0 > 0$), or a pure capacitance one-port ($1/\omega_0 C = -x_0$, when $x_0 < 0$).

PROOF The slope theorem can be demonstrated as an immediate consequence of the average stored energy vs. immittance relations of Theorem 4.6.7, which apply to a finite n-port under a.c. operation. Rewriting the matrix results of eqs. (4.6.4), (4.6.5) for a a rational LPR scalar impedance rather than an n-port

$$\frac{dx(\omega)}{d\omega}|i(j\omega)|^2 = 2(W_m + W_e), \quad \frac{x(\omega)}{\omega}|i(j\omega)|^2 = 2(W_m - W_e) \qquad (5.2.9)$$

where $i(j\omega)$ is the port current and $W_m \geq 0$, $W_e \geq 0$ are the average stored magnetic and electric energies, respectively.

For items (1) and (2) of the Theorem, the network is not purely inductive or capacitive so that neither W_m nor W_e is identically zero. Inspection of eq. (5.2.9) then immediately confirms the slope constraints of eqs. (5.2.6), (5.2.7).

In case the system is purely inductive, $W_e \equiv 0$, and the equality eq. (5.2.8) is established. The same result follows in the purely capacitive case with $W_m \equiv 0$. This proves item (3) of the Theorem. ∎

The Slope Theorem is also valid for susceptance, $b(\omega)$. Thus a Foster admittance function is given as $y(j\omega) = jb(\omega)$ and, with minor changes in wording, Theorem 5.2.4 becomes the "Susceptance Slope Theorem" by replacing $x(\omega)$ with $b(\omega)$.

As a consequence of the Slope Theorem, the reactance and susceptance are increasing functions of angular frequency $\forall\omega$, so that the following Corollary is evident.

COROLLARY 5.2.1

Separation Property *The zeros and poles of a rational Foster (LPR) function must alternate on the $j\omega$-axis, as indicated in the typical reactance plot of Fig. 5.2.3.*

The positive slope property can be generalized to lossless n-ports. In fact, the structure need not be finite. The following theorem encompasses these possibilities.

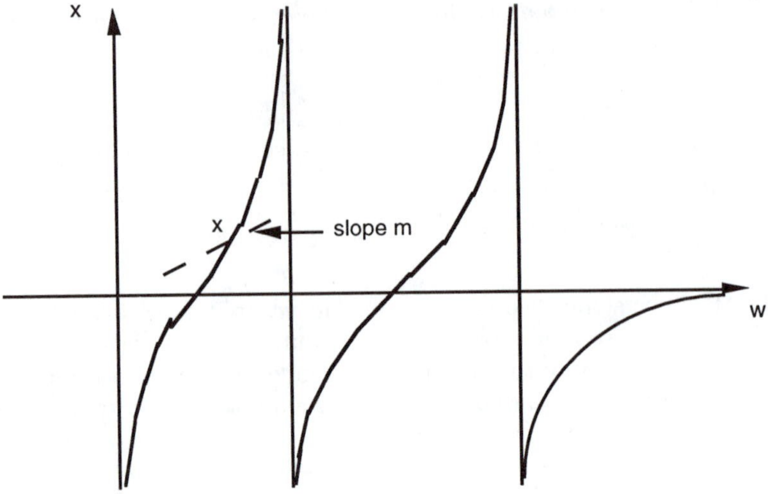

FIGURE 5.2.3
Typical reactance vs. frequency graph. $x(\omega)$ with pole at origin, zero at infinity. Slope of dashed straight line, $m = x/\omega < dx/d\omega$.

***THEOREM* 5.2.5**

Let a reciprocal n-port be described by an LPR immittance matrix $\boldsymbol{z}$. Then

$$\boldsymbol{z}(j\omega) = j\boldsymbol{x}(\omega)$$

and

$$\frac{d\boldsymbol{x}(\omega)}{d\omega} \text{ is positive definite a.e. on } j\omega \tag{5.2.10}$$

PROOF Since by hypothesis the n-port is reciprocal, $\boldsymbol{z} = \boldsymbol{z}'$. Furthermore, as a consequence of the LPR property, $\boldsymbol{z}$ is skew hermitian on $j\omega$. Therefore $\boldsymbol{z}^*(j\omega) = \boldsymbol{z}(-j\omega) = -\boldsymbol{z}(j\omega)$, and thus $\Re\,\boldsymbol{z}(j\omega) = \boldsymbol{0}$, or $\boldsymbol{z}(j\omega) = j\boldsymbol{x}(\omega)$.

Now consider the quadratic form (a scalar function) $f(s) = \boldsymbol{i}^\dagger \boldsymbol{z}\boldsymbol{i}$, for an arbitrary complex vector $\boldsymbol{i}$. Since $\boldsymbol{z}$ is symmetric, f is real when s is real. Also f satisfies all the remaining LPR requirements since $\boldsymbol{z}$ is an LPR matrix. Thus f is an LPR function, and $\Re f(s) = 0$, a.e. on $j\omega$.

Consider $r(\sigma, \omega) = \Re\, f(\sigma + j\omega)$ where σ is real and positive. Then no matter how small we take σ, $r(\sigma, \omega) > 0$, by the LPR property (except for the trivial case that $\boldsymbol{z} \equiv \boldsymbol{0}$). Thus the derivative $\partial r/\partial\sigma > 0$. Apply the Cauchy-Riemann equations for an analytic function (A.1.1) and (A.1.2). Taking $\Im f = x(\sigma, \omega)$

$$\frac{\partial r}{\partial \sigma} = \frac{\partial x}{\partial \omega} > 0$$

In other words, the quadratic form

$$\boldsymbol{i}^{\dagger}\frac{dx(\omega)}{d\omega}\boldsymbol{i} > 0, \quad \text{a. e. } \forall \boldsymbol{i}$$

The positive character of the slope matrix, eq. (5.2.10), follows.[4] ∎

Because of the positive slope property, the reactance plots for nonrational LPR immittances look very much like the curves shown on Fig. (5.2.3), except that as ω increases the patterns do not terminate but repeat indefinitely. However, a glance at the diagram verifies the relation between slope, $dx(\omega)/d(\omega)$, and $x(\omega)/\omega$, eq. (5.2.7), even in the case where the curve patterns continue to infinity. In essence then, the properties of immittances delineated by the reactance slope theorem remain valid for structures that are not finite.

5.3 RC and RL One-Port Synthesis

The synthesis of one-ports including exclusively resistors and capacitors (RC one-ports) or resistors and inductors (RL one-ports) can be derived by a *dimensional transformation* from that of lossless one-ports (that being realizable as seen in the previous Section by only inductors and capacitors may be properly called LC one-ports).

Thus the synthesis of two-element one-ports can be accomplished by elementary means, contrasted with that of RLC one-ports, to which the rest of this chapter is devoted.

THEOREM 5.3.1
A rational one-port impedance z is realizable with only positive real Rs and Cs if and only if

(1) all poles are simple and confined to the negative real axis with positive residues;

(2) there is no pole at infinity.

PROOF Necessity: Let an RC one-port be given. We define a new frequency variable p as follows. Let every impedance in the one-port be

[4] The present theorem can be regarded as an alternate derivation of the Reactance Slope Theorem 5.2.4. However, the earlier derivation is important because it emphasizes the physical concepts associated with stored energy.

multiplied by p, and transform the frequency by $s = p^2$. Each resistance transforms according to $r_h \to r_h p$, that is, becomes an inductance r_h in the p domain. Each capacitive impedance transforms according to $1/c_h s \to p(1/c_h p^2)$, i.e., stays a capacitor. Then the structure is an LC in p. The new LC impedance, in the p domain by Foster's Theorem, is

$$Z = d_\infty p + \sum_{h=1}^{m} \frac{d_h p}{p^2 + \omega_h^2} + \frac{d_0}{p}$$

We now return to $z = (1/p)Z$

$$z = d_\infty + \sum_{h=1}^{m} \frac{d_h}{p^2 + \omega_h^2} + \frac{d_0}{p^2}$$

or with $s = p^2$ and $\sigma_h = \omega_h^2$

$$z = d_\infty + \sum_{h} \frac{d_h}{s + \sigma_h} + \frac{d_0}{s} \tag{5.3.1}$$

and this is the canonic form for an RC impedance. Inspection of the form shows that items (1) and (2) are validated.

Sufficiency: If conditions (1) and (2) are satisfied, the partial fraction expansion has the form above. Thus z can be realized by series connecting a resistance d_∞, m parallel RC circuits each composed by a capacitance $1/d_h$ and a resistance d_h/σ_h, and a capacitance $1/d_0$. ■

Note that the admittance y of the RC one-port can be similarly calculated from the admittance Y of the LC one-port

$$Y = d'_\infty p + \sum_{h=1}^{m} \frac{d'_h p}{p^2 + \omega_h'^2} + \frac{d'_0}{p}$$

by multiplying times p and replacing p^2 with s.

$$y = d'_\infty s + \sum_{h=1}^{m} \frac{d'_h s}{s + \sigma'_h} + \frac{d'_0}{s} \tag{5.3.2}$$

where $\sigma'_h = \omega_h'^2$ is obtained.

The circuits described by eqs. (5.3.1) and (5.3.2) are described in Fig. 5.3.1 (a) and (b).

Example 5.3.1

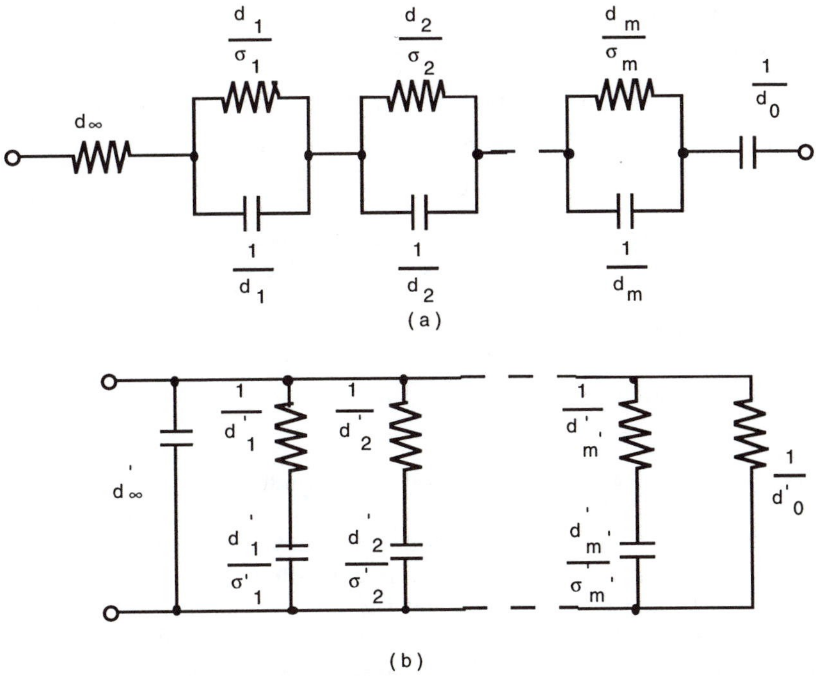

FIGURE 5.3.1
Fosterlike realization of an RC impedance: (a) series realization; (b) shunt realization.

(a) Realize the function

$$z = \frac{30s^2 + 23s + 1}{30s^2 + 2s}$$

as the impedance of an RC series one-port.

(b) Realize the reciprocal of the function above as the admittance of an RC shunt one-port.

Solution

This example illustrates the realization of an RC one-port impedance by *dual* series and parallel canonic forms.

(a) Realization by a series structure. By decomposing z into partial fractions, we obtain

$$z = 1 + \frac{3}{15s + 1} + \frac{1}{2}$$

According to Theorem 5.3.1 the impedance above is realized by series connecting a resistance $r_\infty = h_\infty = 1$, a parallel RC circuit composed of a resistance $r_1 = h_1'/\sigma_1 = 3$ and a capacitance $c_1 = 1/h_1' = 5$, and a capacitance $c_0 = 1/h'_0 = 2$. The resulting one-port has the structure shown in Fig. 5.3.1 (a).

(b) Shunt realization. The partial fraction expansion of $y = 1/z$ is

$$y = \frac{0.030255s}{s + 0.046271} + \frac{0.96974s}{s + 0.72040}$$

Again, eq. (5.3.2) shows that the given impedance is realized as the parallel connection of a first series circuit composed of a resistance $r_1 = 1/k'_1 = 33.052$ and a capacitance $c_1 = k'/\sigma_1 = 0.65387$, and a second series circuit composed of an inductance $r_2 = 1/k'_1 = 1.0312$ and a capacitance $c_1 = k'/\sigma_1 = 1.3461$. The resulting one-port has the structure shown in Fig. 5.3.1 (b). □

The lowpass Cauer form is obtained by alternately extracting at infinity the finite value of the impedance and the pole of the admittance, thus obtaining a ladder with series resistors and shunt capacitors; the highpass Cauer form, on the other hand, is obtained by alternately extracting at the origin the pole of the impedance and the finite value of the admittance, thus obtaining a ladder with series capacitors and shunt resistors.

The resulting continued fraction expansions have the form

$$z = e_1 + \cfrac{1}{e_2 s + \cfrac{1}{e_3 + \cfrac{1}{e_4 s + \cdots}}}$$

and

$$z = \frac{f_1}{s} + \cfrac{1}{f_2 + \cfrac{1}{\frac{f_3}{s} + \cfrac{1}{f_4 + \cdots}}}$$

The circuit structures are illustrated in Fig. 5.3.2 (a) and (b).

Example 5.3.2

Realize the impedance z of Example 5.3.1:

(a) as a lowpass ladder;

(b) as an highpass ladder.

Solution By successive extractions at infinity (a) and at the origin (b) we find:

(a) $r_1 = a_1 = 1$, $c_2 = a_2 = 10/7$, $r_3 = a_3 = 147/4$, $c_4 = a_4 = 4/7$;

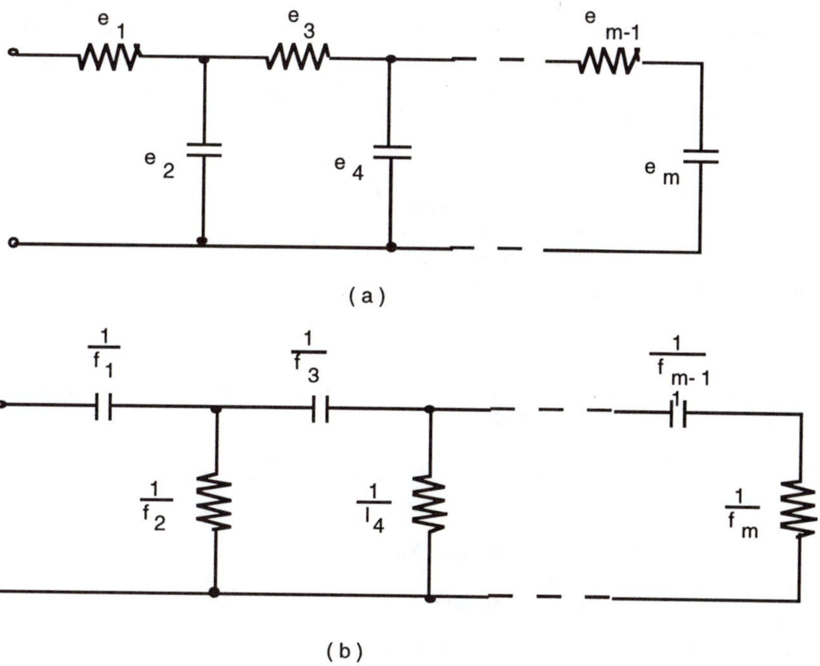

FIGURE 5.3.2
Cauer realization of an RC impedance: (a) low-pass realization; (b) high-pass realization.

(b) $c_1 = 1/b_1 = 2$, $r_2 = 1/b_2 = 4$, $c_3 = 1/b_3 = 45/16$, $r_4 = 1/b_4 = 3/4$.

The lowpass and highpass Cauer realizations have the form shown in Fig. 5.3.2 (a) and (b). □

The case of RL one-ports is readily deduced from the above synthesis by observing that an RL impedance is dual to an RC admittance and vice versa. Thus an RL admittance has the form of eq. (5.3.1) and an RL impedance that of eq. (5.3.2). The synthesis, both in series-parallel and in ladder form, may be carried out along dual lines of those described above.

5.4 The Canonic Form of the Scattering Matrix of a Lossless Two-Port

Let $\boldsymbol{S}$ be a rational (2×2)-matrix function of the complex variable s. Thus each of its entries is a ratio of polynomials in s with real coefficients; we represent each matrix element in the form $S_{ij} = N_{ij}/D$ where D is their least common denominator. Passivity requires that the matrix $\boldsymbol{S}$ be BR; thus cancellations of common factors between some numerators and the denominator can occur only in the open LHP, because the denominator D must be strictly Hurwitz. The additional restriction of losslessness requires that $\boldsymbol{S}$ be para-unitary, that is

$$\boldsymbol{S}_{\dagger}\boldsymbol{S} = \boldsymbol{I}_2 \tag{5.4.1}$$

Since the right and left inverses of square matrices are equal, eq. (5.4.1) can be rewritten in the equivalent form

$$\boldsymbol{S}\boldsymbol{S}_{\dagger} = \boldsymbol{I}_2 \tag{5.4.2}$$

Expanding eq. (5.4.1) and multiplying both sides by D_*D, we obtain the following polynomial equalities

$$N_{11*}N_{11} + N_{21*}N_{21} = D_*D \tag{5.4.3}$$

$$N_{11*}N_{12} + N_{21*}N_{22} = 0 \tag{5.4.4}$$

$$N_{12*}N_{12} + N_{22*}N_{22} = D_*D \tag{5.4.5}$$

Expanding eq. (5.4.2) in the same way, we get

$$N_{11}N_{11*} + N_{12}N_{12*} = DD_* \tag{5.4.6}$$

$$N_{11}N_{21*} + N_{12}N_{22*} = 0 \tag{5.4.7}$$

$$N_{22}N_{22*} + N_{21}N_{21*} = DD_* \tag{5.4.8}$$

Comparison of eqs. (5.4.3) and (5.4.8) yields

$$N_{11*}N_{11} = N_{22*}N_{22} \tag{5.4.9}$$

while from eqs. (5.4.3) and (5.4.6), one finds

$$N_{21*}N_{21} = N_{12*}N_{12} \tag{5.4.10}$$

Eqs. (5.4.9) and (5.4.10) show that both N_{11}/N_{22*} and N_{12}/N_{21*} are all-passes, i.e., they have amplitude unity on $s = j\omega$. Eq. (5.4.7) shows that they can be represented as

$$\frac{N_{11}}{N_{22*}} = \mp\frac{\theta_*}{\theta}$$

$$\frac{N_{12}}{N_{21*}} = \pm\frac{\theta_*}{\theta}$$

where θ is a polynomial and θ_*/θ is assumed to be irreducible (no common factors in numerator and denominator). Thus

$$N_{11} = h_0\theta_* \qquad N_{12} = f_0\theta_*$$

$$N_{21} = \pm f_{0*}\theta_* \qquad N_{22} = \mp h_{0*}\theta_*$$

where h_0 is a polynomial consisting of the common factors of N_{11} and N_{22*} and f_0 is a polynomial consisting of the common factors of N_{12} and N_{21*}. Eq. (5.4.3) shows that D_*D must contain the factor $\theta_*\theta$. Since D is by assumption the least common denominator of all entries of the matrix $\boldsymbol{S}$, it must be a Hurwitz polynomial and therefore is representable as $g_0\theta$, where both g_0 and θ are strictly Hurwitz. Thus the scattering matrix of a lossless two-port has the representation

$$\boldsymbol{S} = \begin{pmatrix} \dfrac{h_0\theta_*}{g_0\theta} & \dfrac{f_0\theta_*}{g_0\theta} \\ \pm\dfrac{f_{0*}\theta_*}{g_0\theta} & \mp\dfrac{h_{0*}\theta_*}{g_0\theta} \end{pmatrix} \tag{5.4.11}$$

where, from eq. (5.4.3),

$$h_{0*}h_0 + f_{0*}f_0 = g_{0*}g_0 \tag{5.4.12}$$

and upper or lower signs are chosen together.

The form can be further simplified if we add additional extra factors by multiplying all entries of the matrix $\boldsymbol{S}$ in eq. (5.4.11) by θ/θ, so that the Hurwitzian character of the common denominator is not impaired. Clearly one can recover the irreducible form if all common factors (including those already present in eq. (5.4.11) are cancelled. Thus we have

THEOREM 5.4.1

The Belevitch canonic form of the scattering matrix of a finite lossless two-port *is*[5]

$$\boldsymbol{S} = \begin{pmatrix} \dfrac{h}{g} & \dfrac{f}{g} \\ \pm\dfrac{f_*}{g} & \mp\dfrac{h_*}{g} \end{pmatrix} \tag{5.4.13}$$

where $h = h_0\theta_*\theta$, $f = f_0\theta_*\theta$, *and* $g = g_0\theta^2$, *and, as a consequence of eq.* (5.4.12),

$$h_*h + f_*f = g_*g \tag{5.4.14}$$

Eq. (5.4.13) describes the structure of a (2×2) rational bounded real matrix satisfying the para-unitary conditions; i.e., the structure of the scattering matrix of a LTI, finite, passive, and lossless two-port. As we have seen, to attain its simplified form may require the use of common numerator and denominator factors.

[5] V. Belevitch, *Classical Network Theory*, S. Francisco, Holden-Day, 1968, p. 276.

If the two-port is *reciprocal*, then the matrix $\boldsymbol{S}$ is symmetric and

$$\pm f_* = f \tag{5.4.15}$$

Thus in the reciprocal case, f is even or odd, according to whether the $+$ or the $-$ sign is required in eq. (5.4.15). Thus

COROLLARY 5.4.1
The canonic form for the scattering matrix of a finite, lossless, reciprocal two-port *is*

$$\boldsymbol{S} = \begin{pmatrix} \dfrac{h}{g} & \dfrac{f}{g} \\ \dfrac{f}{g} & \mp\dfrac{h_*}{g} \end{pmatrix}, \quad (-) \text{ if } f = f_*, \quad (+) \text{ if } f = -f_* \tag{5.4.16}$$

where f is either an even or an odd polynomial and signs $+$ and $-$ are chosen accordingly.

DEFINITION 5.4.1 *A reciprocal two-port is said to be* symmetric *if the permutation of its ports changes the two-port into itself. A reciprocal two-port is said to be* antimetric *if the permutation of its ports changes the two-port into its dual.*

If the two-port is *symmetric*, both reflectances are obviously equal and therefore $h = \mp h_*$; h is odd or even depending on whether f is even or odd.

If the two-port is antimetric, input and output reflectances are negatives of each other (the input and output impedances of the resistively terminated two-port are reciprocals of each other). In this case $h = \pm h_*$ and the parity for h is the same as for f.

Hence

COROLLARY 5.4.2
Symmetric and antimetric two-ports satisfy constraints

$$h = \mp h_* \tag{5.4.17}$$

$$h = \pm h_* \tag{5.4.18}$$

where upper signs correspond to f even, lower signs to f odd.

As earlier noted, some (but not all) of the entries S_{ij} in eq. (5.4.13) can contain surplus factors, beyond those contained in θ, which cancel between the numerator and the denominator; they are necessarily LHP common zeros of g_0 and of one or more (but not all) of the polynomials h_0, h_{0*}, f_0, and f_{0*}.

However, the determinant

$$|\boldsymbol{S}| = \mp\frac{g_*}{g}$$

is an irreducible expression, since g is Hurwitz. The degree of g can be shown to be the *degree of the scattering matrix*, that is, the minimum number of reactive elements necessary to realize it.[6]

The representation of Theorem 5.4.1 is said to be *canonic* because the scattering matrix of any finite lossless two-port can be put in that form, also possibly imbedding the further constraints of reciprocity, symmetry, or antimetry.

By assuming g to be *monic* (coefficient of highest power is unity), the form for the canonic representation is unique.

The fact that the representation is canonic, unique, and of minimal degree, though not irreducible, makes it most convenient for describing the constraints imposed by passivity and losslessness (phrased by eq. (5.4.14)), and possibly by reciprocity, symmetry, or antimetry (phrased by eqs. (5.4.15), (5.4.17), (5.4.18) on reflectances and transmittances.

THEOREM 5.4.2

The matrix $\boldsymbol{S}$ is completely and uniquely defined by the characteristic function

$$\psi = \frac{h}{f}$$

PROOF Polynomials h and f are defined from ψ within a common factor; their knowledge allows us to calculate from eq. (5.4.14) the product $g_* g$ which still contains the common factor and is uniquely factorized into g and g_* under the requirement that g is Hurwitz. We can use the constant imbedded in the common factor to make g monic and ψ unique. ∎

Note that *any* rational function ψ, including a polynomial, qualifies to be the characteristic function of a lossless two-port. Examples of the use of the characteristic function will be presented later.

Since the four entries of the matrix $\boldsymbol{S}$ depend only on the two polynomials h and f, and there may be cancellations in the LHP between numerators and the denominator, the distribution of their zeros is subject to some restrictions.

[6]V. Belevitch, *Classical Network Theory*, S. Francisco, Holden-Day, 1968, p. 277.

A zero of a function is *paired* if it is not on the $j\omega$-axis and is accompanied by a zero symmetric with respect to $j\omega$. It is *unpaired* if it is not on the $j\omega$-axis and appears unaccompanied by its symmetric zero.

Since all functions we are dealing with are real, complex zeros always occur in conjugate pairs. Thus unpaired zeros appear either as a single real zero or as a complex conjugate pair; paired zeros appear either as a symmetric pair of real zeros or as a symmetric pair of complex zero pairs (symmetric pairs of real zeros and symmetric quadruplets of complex zeros are said to exhibit *quadrantal symmetry*).

Eq. (5.4.13) shows that the zeros of one of the transmittances (reflectances) lying in the LHP necessarily appear in the other one in the symmetric position of the RHP. The converse may or may not be true or because cancellations may occur in the LHP (but not in the RHP since all denominators must be Hurwitz polynomials). Thus the *unpaired* zeros of one of the transmittances (reflectances) lying in the LHP necessarily appear in the other function as unpaired zeros located at the RHP image positions, whereas the unpaired RHP zeros of one transmittance (reflectance) may appear in the other as unpaired zeros in the LHP image position or may not (in the event of cancellations). When cancellation of the LHP image occurs, the original transmittance (reflectance) clearly contains an allpass function analytic in the RHP. Finally, paired zeros in the numerators may be shared by both transmittances (reflectances) or LHP cancellations may occur, so that unpaired RHP zeros in one function can be accompanied by unpaired RHP zeros in the other.

The finite zeros on the $j\omega$-axis are a limit case of paired zeros collapsing on the axis from both half-planes, and thus must appear as a pair of conjugate pure imaginary zeros common to both transmittances (reflectances). A zero at infinity has the same order for both transmittances (reflectances) because the differences between the degree of the denominator and the degrees of numerators are the same.

If the two-port is *reciprocal*, the distribution of zeros for the *transmittances* have special restrictions since the two off-diagonal functions are now equal. The Belevitch canonic form calls for the zeros of the transmittance function to always be paired, since according to eq. (5.4.15) the polynomial f is either purely even or odd. However, we must examine the irreducible form of S to determine under what conditions a realizable Belevitch form can be constructed. If the irreducible form has paired transmittance zeros, these appear as well in the Belevitch form. If the irreducible form has unpaired RHP transmittance zeros, surplus factors introduced in numerator and denominator yield the paired zeros of the Belevitch form, and the denominator is still a Hurwitz polynomial. If the irreducible form has unpaired LHP transmittance zeros, surplus RHP factors to obtain paired zeros for the Belevitch form are ruled out since these would force the denominator to become non-Hurwitz. Thus unpaired zeros in the LHP are not permissible for the reciprocal case. We have the surprising result that *a transmittance function containing an unpaired LHP zero is not realizable by a lossless reciprocal two-port terminated in a resistor.*

If the two-port is *symmetric* or *antimetric*, according to eqs. (5.4.17) and (5.4.18), the polynomial h is odd or even, respectively, and the zeros of the reflectances are either paired or unpaired in the RHP. However, exactly as discussed for reciprocal transmittances, in the symmetric or antimetric cases unpaired reflectance zeros in the LHP are not realizable for lossless two-ports.

Finally from eq. (5.4.13) one obtains

$$\frac{S_{11}}{S_{22*}} = \mp\frac{g_*}{g} \qquad \frac{S_{12}}{S_{21*}} = \pm\frac{g_*}{g}$$

Thus we have the following.

THEOREM 5.4.3

At real frequencies the amplitudes of the reflectances, as well as those of the transmittances, are equal.

$$|S_{11}(j\omega)| = |S_{22}(j\omega)|, \quad |S_{12}(j\omega)| = |S_{21}(j\omega)|$$

and the phases are constrained as follows:

$$\phi_{11} + \phi_{22} - \phi_{12} - \phi_{21} \pm \pi = 0$$

which reduces to

$$\phi_{11} + \phi_{22} - 2\phi_{21} \pm \pi = 0$$

in the reciprocal case.

The input reflectance S and the input impedance z of the lossless two-port terminated on a unit resistor, from eqs. (4.3.9), are expressed in terms of the polynomials h and g as follows:

$$\begin{aligned} S &= \frac{z-1}{z+1} = \frac{h}{g} \\ z &= \frac{1+S}{1-S} = \frac{g+h}{g-h} \end{aligned} \tag{5.4.19}$$

From the second of eqs. (5.4.19), it turns out that both $g+h$ and $g-h$ are the product of Hurwitz polynomials, with polynomials all of whose zeros lie on the $j\omega$-axis.

5.5 The Immittance Matrices of a Lossless Two-Port

The open-circuit impedance matrix $\boldsymbol{z}$ and the short-circuit admittance matrix $\boldsymbol{y}$, which play a fundamental role in Darlington's synthesis, are derived from the canonic form of the $\boldsymbol{S}$ matrix, eq. (5.4.11), by using eqs. (4.3.9) and (4.3.10).

As a matter of notation, the even and odd parts of a polynomial are denoted by the subscripts e and o, respectively. Thus $\boldsymbol{z}$ has either the form

$$\boldsymbol{z} = \begin{pmatrix} \dfrac{g_e + h_e}{g_o - h_o} & \dfrac{f}{g_o - h_o} \\ \dfrac{f_*}{g_o - h_o} & \dfrac{g_e - h_e}{g_o - h_o} \end{pmatrix} \tag{5.5.1}$$

or the form

$$\boldsymbol{z} = \begin{pmatrix} \dfrac{g_o + h_o}{g_e - h_e} & \dfrac{f}{g_e - h_e} \\ \dfrac{-f_*}{g_e - h_e} & \dfrac{g_o - h_o}{g_e - h_e} \end{pmatrix} \tag{5.5.2}$$

according to whether we choose the upper or lower signs in eq. (5.4.11).

$\boldsymbol{y}$, in turn, has either the form

$$\boldsymbol{y} = \begin{pmatrix} \dfrac{g_e - h_e}{g_o + h_o} & \dfrac{-f}{g_o + h_o} \\ \dfrac{-f_*}{g_o + h_o} & \dfrac{g_e + h_e}{g_o + h_o} \end{pmatrix} \tag{5.5.3}$$

or the form

$$\boldsymbol{y} = \begin{pmatrix} \dfrac{g_o - h_o}{g_e + h_e} & \dfrac{-f}{g_e + h_e} \\ \dfrac{f_*}{g_e + h_e} & \dfrac{g_o + h_o}{g_e + h_e} \end{pmatrix} \tag{5.5.4}$$

according to the preceding rule.

We introduce the following notation for the even and odd parts of the polynomials $g + h$ and $g - h$

$$\begin{aligned} m_1 &= g_e + h_e \qquad & n_1 &= g_o + h_o \\ m_2 &= g_e - h_e \qquad & n_2 &= g_o - h_o \end{aligned} \tag{5.5.5}$$

Thus eqs. (5.5.1) and (5.5.2) can be rewritten as

$$\boldsymbol{z} = \begin{pmatrix} \dfrac{m_1}{n_2} & \dfrac{f}{n_2} \\ \dfrac{f_*}{n_2} & \dfrac{m_2}{n_2} \end{pmatrix} \tag{5.5.6}$$

and

$$\boldsymbol{z} = \begin{pmatrix} \dfrac{n_1}{m_2} & \dfrac{f}{m_2} \\ \dfrac{-f_*}{m_2} & \dfrac{n_2}{m_2} \end{pmatrix} \tag{5.5.7}$$

Similarly, eqs. (5.5.3) and (5.5.4) can be rewritten as

$$\boldsymbol{y} = \begin{pmatrix} \dfrac{m_2}{n_1} & \dfrac{-f}{n_1} \\ \dfrac{-f_*}{n_1} & \dfrac{m_1}{n_1} \end{pmatrix} \tag{5.5.8}$$

and

$$\boldsymbol{y} = \begin{pmatrix} \dfrac{n_2}{m_1} & \dfrac{-f}{m_1} \\ \dfrac{f_*}{m_1} & \dfrac{n_1}{m_1} \end{pmatrix} \tag{5.5.9}$$

In the equations above, referring to eqs. (5.4.14) and (5.5.5), f is any solution of

$$ff_* = (g_e + g_o)(g_e - g_o) - (h_e + h_o)(h_e - h_o) = m_1 m_2 - n_1 n_2 \tag{5.5.10}$$

If the two-port is *reciprocal*, $f_* = \pm f$ so that eq. (5.5.10) becomes

$$\pm f^2 = m_1 m_2 - n_1 n_2 \tag{5.5.11}$$

It must be emphasized that the above equations are based on the $\boldsymbol{z}$ or $\boldsymbol{y}$ matrix as deduced from the Belevitch canonic form *without cancellation of surplus factors.* This forces the expression at the right side of eq. (5.5.11) or its negative to be a perfect square for any reciprocal two-port. Thus polynomial f has the form

$$f = \pm\sqrt{m_1 m_2 - n_1 n_2} \tag{5.5.12}$$

for eqs. (5.5.6) (5.5.8) and

$$f = \pm\sqrt{n_1 n_2 - m_1 m_2} \tag{5.5.13}$$

for eqs. (5.5.7) (5.5.9). Note that the choice between eqs. (5.5.12) and (5.5.13) is dictated from being the expression $m_1 m_2 - n_1 n_2$ even or odd.

The input impedance z and the input reflectance S of the lossless two-port terminated on a unit resistor are expressed in terms of the polynomials m_1, m_2, n_1, and n_2 as follows

$$\begin{aligned} z &= z_{11} - \frac{z_{12}^2}{z_{22}+1} = \frac{m_1 + n_1}{m_2 + n_2} \\ S &= \frac{z-1}{z+1} \qquad\quad = \frac{m_1 - m_2 + n_1 - n_2}{m_1 + m_2 + n_1 + n_2} \end{aligned} \tag{5.5.14}$$

From the second of eqs. (5.5.14) it turns out that $m_1 + m_2 + n_1 + n_2$ is a (strict) Hurwitz polynomial, while $m_1 - m_2 + n_1 - n_2$ can have its zeros anywhere in the complex plane.

The impedance matrix $\boldsymbol{z}$, in a sufficiently small neighborhood of one of its poles, can be represented as

$$\boldsymbol{z} = \begin{pmatrix} h_{11} & h_{12} \\ h_{21} & h_{22} \end{pmatrix} f(s)$$

where $f(s)$ has the form s, $1/s$, or $2s/(s^2 + \omega_i^2)$, if the pole is at infinity, at the origin, or at the finite frequency $s_i = j\omega_i$.

In the same neighborhood, the input impedance z of the two-port terminated on a load impedance not containing the pole is written

$$z = \left(h_{11} - \frac{h_{12}^2}{h_{22}} \right) f(s)$$

$$= \frac{h_{11}h_{22} - h_{12}^2}{h_{22}} f(s)$$

It is seen that the pole does not appear in z if and only if the determinant $|H| = h_{11}h_{22} - h_{12}^2$ is zero. In such a case, the residue matrix H is positive semidefinite, or, according to network theory jargon, *compact.*

If the pole is present in the load impedance with residue h_l, it appears in the input impedance with residue

$$h = \frac{h_{11}h_{22} - h_{12}^2 + h_{11}h_l}{h_{22} + h_l} \tag{5.5.15}$$

It is seen that the compactness condition amounts to reduce the value of the residue in the input impedance, for a given value of that in the load impedance, to its minimum. Conversely from eq. (5.5.15), we obtain

$$h_l = \frac{h_{22}h - (h_{11}h_{22} - h_{12}^2)}{h_{11} - h}$$

Thus the compactness condition causes the value of the residue in the load impedance, for a given value of that in the input impedance, to reach its maximum.

5.6 Transmission Zeros

An important concept is that of *transmission zeros* of a lossless two-port. Let the two-port be terminated by unit resistors and the port 1 be fed with an exponential excitation $a\exp(st)$ with $\Re s \geq 0$. Then the signal is not transmitted from the generator to the load whenever s is a zero of S_{21}; if the generator and the load are exchanged, the same applies to S_{12}. Thus we can define as *transmission zeros* the RHP zeros of the function $F = S_{12}S_{21}$. Taking into account the expressions of S_{12} and S_{21} appearing in eq. (5.4.13), we obtain the following.

DEFINITION 5.6.1 *The transmission zeros of a lossless two-port are the closed RHP zeros of the expression*

$$F = \pm\frac{f_*f}{g^2} \tag{5.6.1}$$

where all possible common factors between the numerator and the denominator have been cancelled and the zeros on the $j\omega$-axis are counted for half their multiplicity.

Note that f_*f is an even real polynomial, so that its zeros in the s-plane exhibit quadrantal symmetry, while those on the $j\omega$-axis (where it is real and positive) are necessarily of even order.

It follows that the number of finite transmission zeros is equal to one-half the degree of f_*f and that the number of infinite transmission zeros is equal to one-half the difference between the degree of g_*g and that of f_*f. Thus the total number of transmission zeros is equal to the degree of g, which is also the degree of matrix $\boldsymbol{S}$.

Alternatively the transmission zeros can be defined on the basis of the open-circuit impedance matrix $\boldsymbol{z}$ or the short-circuit admittance matrix $\boldsymbol{y}$. From eqs. (5.6.1), (4.3.9), and (4.3.10), we obtain

$$F = \frac{4z_{12}z_{21}}{[(z_{11}+1)(z_{22}+1) - z_{12}z_{21}]^2}$$

and

$$F = \frac{4y_{12}y_{21}}{[(y_{11}+1)(y_{22}+1) - y_{12}y_{21}]^2}$$

The function F can also be expressed in terms of the input functions z or S, when the reactance two-port is terminated in unit resistor. In the first case, by replacing in eq. (5.6.1) f_*f by its expression given by eq. (5.4.14), and using eqs. (5.4.19), we obtain

$$F = 2b\frac{z+z_*}{(z+1)^2} \tag{5.6.2}$$

where b is the analytic allpass built on the denominator of z

$$b = \pm\frac{m_2 - n_2}{m_2 + n_2} \tag{5.6.3}$$

Note that the zeros of $m_2 - n_2$, all in the RHP, do not appear in F because they cancel with poles of $z + z_*$. Thus the transmission zeros are either zeros of $z + z_*$ or poles of z (which can be zeros of $z + z_*$ or not). In the second case, using eqs. (5.6.1) and (5.4.14), we find

$$F = \pm g_*/g(1 - S_*S) \tag{5.6.4}$$

where g_*/g is the analytic allpass built on the denominator of S. Thus it is seen that the transmission zeros are the zeros of $1 - S_*S$; the RHP zeros of g_* cancel with the poles of S_*.

Finally, we can replace the impedance z in eq. (5.6.2) with its expression given by the second of eqs. (5.5.14) and we obtain

$$F = 4\frac{m_1m_2 - n_1n_2}{(m_1 + m_2 + n_1 + n_2)^2} \tag{5.6.5}$$

The same expression is obviously obtained by using eqs. (5.5.6), (5.5.7), (5.5.8), (5.5.9).

Thus the finite transmission zeros are the closed RHP zeros of the even polynomial $m_1m_2 - n_1n_2$, while the transmission zeros at infinity are one-half the difference between the degrees of the denominator and the numerator.

The fact that the transmission zeros are completely defined either in terms of the input impedance or the reflectance of the lossless two-port loaded by a unit resistor suggests that they can be considered a property of the two-port. Moreover, it is evident that they can be defined, at least formally, by using eq. (5.6.2) or (5.6.4) for any prescribed PR impedance or BR reflectance.

5.7 Darlington's Procedure of Synthesis

Darlington's synthesis consists in extracting from a given PR impedance z all of its transmission zeros, thus reducing its degree to zero and leaving a resistor (which can always be assumed of unit value) as the remainder. The structure realizing the transmission zeros is a lossless (reciprocal) two-port terminated on the resistor, exhibiting at its input port the prescribed impedance (Fig. 5.7.1).

If the impedance z is LPR, every point of the complex s-plane is a transmission zero and, therefore, in principle, the synthesis can be achieved by successive extractions at arbitrary points eventually leaving a lossless one-port of zero degree, i.e., an open or a short circuit. As presented in Section 5.1, the structure of Fig. 5.7.1 can be restored by inserting between

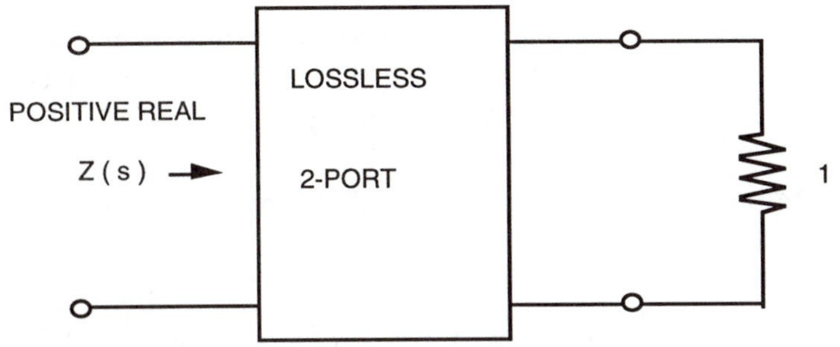

FIGURE 5.7.1
Darlington's realization of an impedance.

the open or short circuit and the unit resistor an ideal transformer of ratio equal to infinity or zero, respectively.

The extraction may be done in one or in several steps. In the first case (*global synthesis*), the lossless two-port has a cumbersome and generally impractical structure. In the second (*cascade synthesis*), the lossless two-port is obtained by cascading a number of simple sections, each of which realizes a small number (from one to four) of transmission zeros.

The extraction may be done directly on the prescribed impedance z (this is the usual technique), or on the open-circuit input impedance z_{11} (or the short-circuit input admittance y_{11}) of the lossless two-port exhibiting z as its input when terminated on a unit resistor. The two procedures yield the same final results, but the second (which we will label the z_{11} method) presents consistent advantages, as it only works with reactance functions, thus reducing by about one half the required number of operations with the result of improved numerical accuracy.[7] Furthermore, the theoretical proofs carried out in terms of the elements of the lossless two-port immittance matrix (rather than in terms of the input immittance of the resistively terminated two-port) are significantly simpler and more direct, particularly for the Brune, type C, and type D sections (as defined below) and lead to a unified technique for extracting such sections. It is surprising that this approach seems not to have been used in any of the technical literature and is probably presented here for the first time.

We first present a lemma that will be useful in subsequent discussions.

LEMMA 5.7.1
(Spectral Factorization) *Let the even nonnegative polynomial with real coefficients* $\hat{N}(\omega^2) \geq 0$, $\forall$ *real* ω^2. *Then writing* $\hat{N}(\omega^2) = \hat{N}(-s^2)$, $-s^2 =$

[7] J. Komiak and H. J. Carlin, "Improved Accuracy for Commensurate-Line Synthesis", *IEEE Trans. on Microwave Th. and Tech.*, vol. 24, no. 4, p. 212-215, April 1976.

ω^2, it is always possible to perform a factorization such that

$$\hat{N}(-s^2) = c^2 N(-s)N(s) = c^2 N_* N, \quad c \, real, \; N \; monic$$

so that the roots of N as well as those of N_* are real or occur in complex conjugate pairs.

PROOF Because of the nonnegative property of $\hat{N}(\omega^2)$, the constant c^2 is positive, $c^2 > 0$, and all the root factors of $\hat{N}(\omega^2)$ are real (even), and nonnegative $\forall\omega^2$. In terms of the roots in $\omega^2 = -s^2$, the factors can therefore only have the following forms (n, m, r are integers)

$$\begin{aligned}
\omega_k \text{ real}, \quad (\omega_k^2 - \omega^2)^{2n} &= (\omega_k^2 + s^2)^{2n} \\
\alpha_k \text{ real}, \quad (\alpha_k^2 + \omega^2)^m &= (\alpha_k^2 - s^2)^m \\
s_k^2 \text{ complex}, \quad (s_k^2 + \omega^2)^r (s_k^{*2} + \omega^2)^r &= (s_k^2 - s^2)^r (s_k^{*2} - s^2)^r
\end{aligned}$$

Each of the above terms may be split into a pair of factors in s and $-s$.

$$(\omega_k^2 + s^2)^{2n} = (\omega_k^2 + s^2)^n (\omega_k^2 + s^2)^n \tag{5.7.1}$$

$$(\alpha_k^2 - s^2)^m = (\alpha_k - s)^m (\alpha_k + s)^m \tag{5.7.2}$$

$$(s_k^2 - s^2)^r (s_k^{*2} - s^2)^r = (s_k - s)^r (s_k^* - s)^r (s_k + s)^r (s_k^* + s)^r \tag{5.7.3}$$

In order to carry out the final factorization, we simply assign the first factor on the right of eq. (5.7.1) to N and the other factor to N_*; similarly for eq. (5.7.2). In the case of eq. (5.7.3) we assign the first pair of factors to N and the remaining pair to N_*. Then all the roots in N are real or in complex conjugate pairs, i.e., $s = \pm j\omega_k, \; \pm\alpha_k, \; s_k, s_k^*$. Those of N_* have the opposite sign. ∎

It will be noted that the distribution of roots into the RHP and LHP for N, N_* is arbitrary; but when the Lemma is applied to the realization procedure, the factorization must take into account the constraints imposed by passivity, losslessness, and reciprocity upon the location of the zeros of transmittance.

The main result of Darlington's synthesis is embodied in the following.

THEOREM 5.7.1

Any positive real rational immittance can be realized as the input immittance of a lossless two-port, that can always be chosen to be reciprocal, terminated on a unit resistor (fig. 5.7.1).

PROOF It is particularly convenient (though not necessary[8]) to base the proof of the Theorem on the Belevitch canonic representation of a lossless S matrix. For definiteness, we assume that a positive real impedance z is assigned. The associated reflectance S is given by the first of eqs. (5.4.19) and identifies the polynomials h and g (except for common factors) in eq. (5.4.13). The polynomial f is determined by factorization using eq. (5.4.14) and applying Lemma 5.7.1. The polynomial f is not unique, because the zeros of the even polynomial f_*f must be assigned one-half to f, the other half to f_*, according to Lemma 5.7.1, with the possibility of exchanging any zero of f with the opposite zero of f_* (the zeros on the imaginary axis, being all of of even order, are accordingly divided each with half its order between the two polynomials). Once one distribution is chosen, the scattering matrix (lossless, but not necessarily reciprocal) of the two-port is completely defined. Thus when the two-port is terminated on a unit resistor, its reflectance at the input port is just S.

The procedure above leads, in general, to a polynomial f which is neither even nor odd, so that the matrix (5.4.13) is not symmetric and, therefore, describes a nonreciprocal lossless two-port. The remainder of the proof shows that a reciprocal realization is always possible.

Preservation of reciprocity is important conceptually, because a one-port is reciprocal and it is somewhat unnatural to have recourse to nonreciprocal devices to realize it. Moreover practical realizations of wideband passive lossless nonreciprocal circuit elements (e.g., gyrators), even at microwave frequencies, are more difficult than capacitors and inductors.

To obtain a reciprocal structure, the Belevitch form for S shows we must have $f = \pm f_*$, that is $f_*f = \pm f^2$. This means that an additional condition on f_*f is that it be the square (with the plus or the minus sign) of an even or odd polynomial f and, therefore, that all its zeros are of even order and paired. Since, in general, this is not the case, we can artificially obtain the result in the following way. Decompose f_*f as

$$f_*f = f_{1*}f_1\phi_*\phi \tag{5.7.4}$$

In eq. (5.7.4) $f_{1*}f_1$ includes all $j\omega$-axis zeros, which are of even order, and therefore is the square of a even or odd polynomial; ϕ_* includes all RHP zeros, and consequently ϕ all the symmetrically located LHP zeros. Suppose, for definiteness, that all zeros are simple. Thus we multiply the left side of eq. (5.7.4) by $\phi_*\phi$, so that all zeros in both the RHP and the LHP become double. The modified f_*f is then

$$f_*f = f_{1*}f_1\phi_*^2\phi^2$$

[8] H. J. Carlin and R. La Rosa, "On the Synthesis of Reactance 4-poles", *J. of Appl. Phys.*, vol. 24, no. 10, p. 1336-1337, Oct. 1953.

Such a modification requires that factor $\phi_*\phi$ must be appended to both products h_*h and g_*g as a consequence of eq. (5.4.14). Both products must be factorized under the conditions that the common denominator in matrix $\boldsymbol{S}$ remains Hurwitz and that $S = h/g$ remains unchanged, i.e., when surplus factors are cancelled. It is immediately seen that this can be accomplished in a unique way, and the result is the *augmented* scattering matrix

$$\boldsymbol{S} = \begin{pmatrix} \dfrac{h\phi}{g\phi} & \dfrac{f_1\phi_*\phi}{g\phi} \\ \dfrac{f_1\phi_*\phi}{g\phi} & \mp\dfrac{h_*\phi_*}{g\phi} \end{pmatrix} \tag{5.7.5}$$

The augmented matrix entries of eq. (5.7.5) can each be identified with the matrix elements of the reciprocal canonic form, eq. (5.4.16). The case in which the zeros of f_*f are of arbitrary orders is treated in much the same way and need not to be discussed here in detail. Eq. (5.7.5) shows that the adopted augmentation procedure has succeeded in producing a symmetric scattering matrix describing a reciprocal lossless two-port. It should be noted that $h\phi$ in eq. (5.7.5) is identified with h in eq. (5.4.16), and similarly $g\phi$ with g, $f_1\phi_*\phi$ with f. Hopefully, the dual meaning of some of the symbols (because of surplus factors and their cancellation) should cause no problem. This completes the proof. ∎

An important point is that the process of augmentation introduces common factors in the numerator and the denominator of both reflectance $S = h\phi/g\phi$ and impedance z.

$$z = \frac{1+S}{1-S} = \frac{(g+h)\phi}{(g-h)\phi}$$

Thus the realization (5.7.5) of the impedance z is not of minimum degree, inasmuch as the zeros of f_*f in the RHP are not originally even, so augmentation is needed. This shows that minimal synthesis is not possible in general with reciprocal structures, except in the special case where all such zeros are on the imaginary axis.

Finally, note that the original zeros of the transmittance are not altered by the process of augmentation, because the added factors cancel in the irreducible form.

The above considerations hold equally for global and cascade synthesis.

Since in the next sections we will present the latter approach in detail, based on the z_{11} method, some general considerations specific to cascade synthesis are appropriate at this point.

Brune's landmark paper[9] on the synthesis of a PR driving point impedance z is based on successive extractions of lossless or dissipative sections

[9]O. Brune, "Synthesis of a Finite Two-terminal Network Whose Driving-point Imped-

until the degree of z is reduced to zero.

In Darlington's classic paper[10] it is shown that only lossless sections associated with the zeros of transmission need be extracted from z, and the total dissipation is localized at the output, since the final extraction is a pure resistance termination. The presentation of the next sections is based on the Brune-Darlington idea of extracting two-port sections (all lossless), where, in general, each such two-port contains a nonpassive one-port which is absorbed by employing coupled coils. However, the approach here differs significantly from Brune-Darlington in that it is based on the open-circuit input impedance z_{11} of the lossless two-port derived from z. Furthermore, we do not assume that z be minimum reactance and minimum susceptance. For an alternative approach, in which lossless two-ports are extracted in successive single steps from the prescribed impedance z by employing two-port matrices, see, for example, the paper by Scanlan and Rhodes.[11]

The transmission zeros may be extracted in any order; the various Darlington's two-ports which are obtained are all *equivalent* in the sense that they present the same input impedance z when terminated on a unit resistor.

The transmission zeros can be partitioned into three sets:

(a) complex transmission zeros in the RHP;

(b) real transmission zeros in the RHP;

(c) imaginary transmission zeros.

The three sets are each invariant with respect to the order of extraction. The set (c) can be further divided into three subsets as follows:

(c1) transmission zeros which are zeros of z;

(c2) transmission zeros which are poles of z;

(c3) transmission zeros at which z has a finite nonzero (imaginary) value.

Such a division is *not* invariant with respect to the extraction order except for poles and zeros at the origin and infinity. This means that, say, a finite nonzero pole of z, if not extracted at the outset, could at a further step appear as a transmission zero at which the remainder impedance has a finite

ance is a Prescribed Function of Frequency", *J. Math. Phys.*, vol. 10, no. 3, p. 191-236, Aug. 1931.

[10] S. Darlington, "Synthesis of Reactance 4-Poles which Produce Prescribed Insertion Loss Characteristics", *J. Math. Phys.*, vol. 18, no. 4, p. 257-353, Sept. 1939.

[11] J. O. Scanlan and J. D. Rhodes, "Unified Theory of Cascade Synthesis", *Proc. IEE*, vol. 17, no. 4, p. 665-670, April 1970.

value. This would change radically the type of section to be employed. The converse is obviously true.

The synthesis procedure may terminate prematurely on an open-circuit before all transmission zeros have been extracted. This occurs when z_{22} has *private poles*, i.e., poles which do not appear in both z_{11} and z_{12}, and the transmission zeros they produce have not been completely extracted at that stage. In such a case, a series branch connected at the output will realize the above transmission zeros without affecting z_{11}.

With these premises, the procedure to be described here for synthesizing a PR impedance z is as follows:

1. As discussed in Sections 5.3 and 5.4, the zeros of transmission and the reactance two-port matrix $\boldsymbol{z} = (z_{ij})$ are computed from z.

2. The LPR function z_{11} is expanded in a manner which realizes all the zeros of transmission.

3. The transmission zeros due to private poles of z_{22} not yet extracted when an open-circuit remainder is left are all realized by a series branch at the output. Note that the extraction of transmission zeros associated with private poles of z_{22}, wherever done, simply adjusts z_{22} without affecting z_{11}.

4. The resultant two-port structure realizes z_{11} and z_{22}, and simultaneously contains *all* the transmission zeros (these include the zeros of z_{12}). Thus it realizes all the z_{ij}.[12] When the two-port is terminated in a pure resistor, the input impedance is the prescribed z.

Step 2 above is at the heart of the procedure. The zeros of transmission that must be extracted from z_{11} are of four types:

(a) those which correspond to $j\omega$ poles and zeros of z; these are extracted as *A and B sections*, respectively;

(b) those which are the $j\omega$ zeros of $ff_* = m_1m_2 - n_1n_2$ (eq. (5.5.10)); they are associated with the *Brune section*;

(c) those which are the zeros of ff_* on the real axis; they are associated with the Darlington *Type C section*;

(d) those which occur as complex quadruplet zeros of ff_*; they are associated with the Darlington *Type D section*.

The transmission zeros are all extracted in successive cycles. The final network structure is a realizable losssless two-port, $\mathcal{D}$, consisting of cascaded

[12] Actually z_{12} is realized to within a constant factor whose correct value may require an ideal transformer across the load. The series output branch associated with z_{22} is also adjusted to account for the transformer, which can finally be removed by employing a nonunit load resistance. The argument is based on the fact that the two-port, except for elements in the final output series branch, is compact at all its poles.

sections. The input to $\mathcal{D}$ realizes the open circuit impedance function z_{11}. Evidently $\mathcal{D}$ also realizes z_{12}, since the transfer function poles must be common to some or all the poles of z_{11} (realizability demands that z_{12} have no "private poles") and its zeros must be common to some or all the transmission zeros. Because $\mathcal{D}$ is lossless, its 2×2 residue matrices at its poles must all be positive definite (leading diagonal element positive and determinant exceeds zero) or semi-definite (leading diagonal element nonzero and determinant zero, termed *compact*). Thus with z_{11} and z_{12} realized, z_{22} is also (almost) automatically realized. The only adjustment which may be required to fully realize it is to add in series at the back end additional reactive elements which produce those of its private poles, if any, which have not been extracted before the remainder reduces to an open circuit. One may also need to add reactive elements producing poles already in z_{11}, but for which the residue matrices are not compact. These adjustments do not affect z_{11}, z_{12} or transmission zeros already extracted. Terminate the final $\mathcal{D}$ in an appropriate resistor and the synthesis of z is complete.

Often one initially extracts A sections from z to make it minimum reactance. If this is done, then all the poles of $\mathcal{D}$ have compact residue matrices, and z_{22} need only be checked for private poles (the residue matrix only has nonzero entries in the "22" position), a relatively simple matter.

5.8 An Example

The concepts presented in the previous sections will be illustrated by the following.

Example 5.8.1

For the impedance

$$z = \frac{8s+7}{2s+7} \tag{5.8.1}$$

construct both a minimal and a reciprocal abstract realization.

Solution From z, we derive S as

$$S = \frac{z-1}{z+1} = \frac{3s}{5s+7}$$

so that

$$h = 3s \tag{5.8.2}$$

and

$$g = 5s + 7 \tag{5.8.3}$$

From eq. (5.4.14), using eqs. (5.8.2) and (5.8.3), we obtain

$$f_* f = -16s^2 + 49 \tag{5.8.4}$$

which can be factorized as

$$f = -4s + 7 \qquad f_* = 4s + 7$$

At this point, the symbols h, g, and f really refer to the canonic form for S with no augmentation. The transmission zeros are calculated by using the formula (5.6.1)

$$F = \pm \frac{49 - 16s^2}{(5s + 7)^2}$$

We have a simple transmission zero at $s_0 = 7/4$. It is seen that the allpass factor common to all entries reduces to ± 1, and that the same holds for the common factor f_0 to N_{12} and N_{21*} (i.e., f_0 and $\pm f_0$). Thus the canonic form of the scattering matrix of the lossless two-port is

$$\boldsymbol{S} = \begin{pmatrix} \dfrac{3s}{5s + 7} & \dfrac{-4s + 7}{5s + 7} \\ \pm\dfrac{4s + 7}{5s + 7} & \mp\dfrac{-3s}{5s + 7} \end{pmatrix}$$

where the choice of both the upper or lower signs is arbitrary. It is immediately noted that the matrix is nonsymmetric and, therefore, the lossless two-port is nonreciprocal; this is an obvious consequence of the fact that the transmission zero in the transmittance S_{12} is unpaired. It is also noted that

$$|\boldsymbol{S}| = \mp \frac{-5s + 7}{5s + 7}$$

The degree of matrix $\boldsymbol{S}$ is 1, like the degree of impedance z. Thus the realization is minimal.

A reciprocal realization involves augmentation so that the zero at s_0 in the transmittance becomes paired. From eq. (5.8.4) and eq. (5.7.4), we see that we can choose $f_1 = 1$ and

$$\phi = 4s + 7$$

so that we obtain from eq. (5.7.5)

$$\boldsymbol{S} = \begin{pmatrix} \dfrac{3s(4s + 7)}{(5s + 7)(4s + 7)} & \dfrac{(-4s + 7)(4s + 7)}{(5s + 7)(4s + 7)} \\ \dfrac{(-4s + 7)(4s + 7)}{(5s + 7)(4s + 7)} & \dfrac{3s(-4s + 7)}{(5s + 7)(4s + 7)} \end{pmatrix} \tag{5.8.5}$$

The transmittance now has paired zeros at $\pm 7/4$. Finally, by multiplying the factors, we obtain

$$\boldsymbol{S} = \begin{pmatrix} \dfrac{12s^2 + 21s}{20s^2 + 63s + 49} & \dfrac{-16s^2 + 49}{20s^2 + 63s + 49} \\ \dfrac{-16s^2 + 49}{20s^2 + 63s + 49} & \dfrac{-12s^2 + 21s}{20s^2 + 63s + 49} \end{pmatrix} \tag{5.8.6}$$

Since f is even, we have chosen the upper signs in eq. (5.4.13). The impedance z based on the Belevitch form eq. (5.8.6) with surplus factors is from the second of eqs. (5.4.19)

$$z = \frac{32s^2 + 84s + 49}{8s^2 + 42s + 49} \tag{5.8.7}$$

which can be rewritten as

$$z = \frac{(8s + 7)(4s + 7)}{(2s + 7)(4s + 7)}$$

It is seen that impedance (5.8.7) contains the common factor $\phi = 4s + 7$ both in the numerator and the denominator and, therefore, although algebraically the same as the impedance (5.8.1), a hidden mode is now present which allows preservation of reciprocity based on the Belevitch form. Its appearance is best portrayed by the irreducible expression

$$\det \boldsymbol{S} = -\frac{20s^2 - 63s + 49}{20s^2 + 63s + 49}$$

whose degree is 2, equal to the degree of the augmented impedance, one more than the degree of the original impedance z. Thus the realization is noncanonic. The price we have to pay to avoid nonreciprocity is a doubling of the degree of the realization.

Note that when surplus factors are cancelled, matrix (5.8.5) has the irreducible form

$$\boldsymbol{S} = \begin{pmatrix} \dfrac{3s}{(5s + 7)} & \dfrac{(-4s + 7)}{(5s + 7)} \\ \dfrac{(-4s + 7)}{(5s + 7)} & \dfrac{3s(-4s + 7)}{(4s + 7)(5s + 7)} \end{pmatrix} \tag{5.8.8}$$

which still describes a lossless reciprocal two-port, but not in the Belevitch canonic form. However, a realization of eq. (5.8.8) would yield the same number of reactive elements as that derived from the reciprocal Belevitch form (i.e., when all zeros are paired by the use of common factors). It must be emphasized that, although the Belevitch form is a useful convenience, it does not change the physical reality. The zeros of the transmittance in eq. (5.8.8) automatically have the right location (unpaired in the RHP); an unpaired zero in the LHP, given by the factor $4s + 7$, cannot arise and, in fact, as discussed in section 5.2, would not be physically realizable.

By using the first of eqs. (5.5.14) we find from the expression of z in eq. (5.8.7).

$$m_1 = 32s^2 + 49 \qquad n_1 = 84s$$

$$m_2 = 8s^2 + 49 \qquad n_2 = 42s$$

This concludes the example. □

5.9 Cascade Synthesis: Type A and B Sections

The *Type A section* extracts a pole of z at real frequency (Fig. 5.9.1 (a), (b), (c)). A transmission zero which is a pole of z on the $j\omega$-axis is extracted by one of the sections in Fig. 5.9.1 (a), (b), (c), according to whether the pole is at infinity, at the origin, or at a finite nonzero angular frequency ω_i. The extraction of such sections from the open-circuited lossless two-port leaves a remainder whose open-circuit input impedance is still a reactance function. This is because the operation reduces to the complete extraction of one of the poles of a reactance function.

For the three cases, the one-step extraction procedure results in the following equations:

$$z_{11}^{(1)} = z_{11} - h_\infty s \tag{5.9.1}$$

$$z_{11}^{(1)} = z_{11} - \frac{h_0}{s} \tag{5.9.2}$$

$$z_{11}^{(1)} = z_{11} - \frac{2h_i s}{s^2 + \omega_i^2} \tag{5.9.3}$$

In the equations above, the parameters h_∞, h_0, h_i are the residues of the impedance z_{11} at the poles. The extraction procedure reduces the degree of z_{11} by 1 in cases (5.9.1) and (5.9.2) and by 2 in case (5.9.3) as the pole is extracted completely. A partial extraction would yield no degree reduction and therefore no progress in the synthesis.

The *Type B section* extracts a zero of z at real frequency (Fig. 5.9.1 (d), (e), (f)). This case can be treated as the dual of the previous one by replacing the impedance z by the admittance y. A transmission zero, which is a pole of y on the $j\omega$-axis, is extracted by one of the sections in Fig. 5.9.1 (d), (e), (f), according to whether the pole is at the infinity, at the

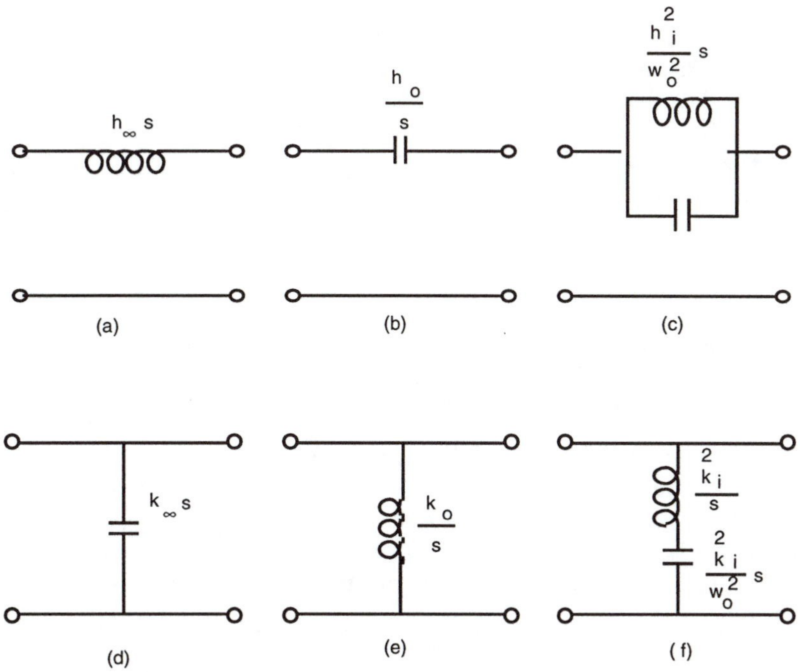

FIGURE 5.9.1
Sections for cascade synthesis: (a) A section: pole at infinity; (b) A section: pole at the origin; (c) A section: pole at real frequency; (d) B section: zero at infinity; (e) B section: zero at the origin; (f) B section: zero at real frequency.

origin, or at a finite nonzero frequency. The extraction of such sections from the short-circuited lossless two-port leaves a remainder whose short-circuit input impedance is still a reactance function. This is because the operation reduces to the complete extraction of a susceptance function pole. For the three cases, the one-step extraction procedure results in the following equations:

$$y_{11}^{(1)} = y_{11} - k_\infty s \tag{5.9.4}$$

$$y_{11}^{(1)} = y_{11} - \frac{k_0}{s} \tag{5.9.5}$$

$$y_{11} = y_{11}^{(1)} - \frac{2k_i s}{s^2 + \omega_i^2} \tag{5.9.6}$$

In the equations above, the parameters k_∞, k_0, and k_i are the residues of the admittance y at the poles. The extraction procedure reduces the degree of y by 1 in cases (5.9.4) and (5.9.5) and by 2 in case (5.9.6) as the pole is extracted completely. A partial extraction would yield no degree reduction and, therefore, no progress in the synthesis.

Let z be a prescribed impedance. Let all of its poles on the $j\omega$-axis be extracted by type A sections, leaving a remainder of admittance $y^{(1)}$. Let all poles of $y^{(1)}$ be extracted by type B sections, leaving a remainder of impedance $z^{(2)}$. Now repeat the initial procedure until a new remainder is obtained whose immittance is devoid of both poles and zeros on the $j\omega$-axis. Such an immittance is said to be *minimum reactance and minimum susceptance.*

5.10 Cascade Synthesis: Brune's Section

The Brune section (Fig. 5.10.1 (g)) produces a pair of simple imaginary conjugate transmission zeros at $\pm j\omega_i$ (*real frequency* transmission zeros). The extraction process is carried out in three successive steps, at each of which a *real* (though possibly nonpositive) immittance is extracted, thus leaving a real remainder. From now on, for the sake of brevity, this point will not be repeated.

The open-circuit input impedance of the lossless two-port z_{11} is deduced from the expression (5.5.14) of z by taking into account eqs. (5.5.6) and (5.5.7). The first or the second solution is chosen according to whether eq. (5.5.12) or eq. (5.5.13) holds. In order to have a pair of simple imaginary conjugate transmission zeros, z_{11} must be of degree at least two.

At the transmission zero $j\omega_i$, we have $z_{11}(j\omega_i) = jx_i$, while at the conjugate point, the conjugate equation holds.

x_i may be positive or negative. First, assume that $x_i < 0$.

The cycle is begun by extracting an inductance l_a such that the resulting function has a zero at the prescribed zero of transmission

$$l_a = \frac{x_i}{\omega_i}$$

Thus we have $l_a < 0$. After the extraction we are left with the lossless impedance

$$z_{11}^{(1)} = z_{11} - l_a s$$

which has *simple* zeros at $s = \pm j\omega_i$ because, since z_{11} is LPR, eq. (5.2.6) yields

$$\frac{dz_{11}^{(1)}}{ds} = \left(\frac{dz_{11}}{ds} - l_a\right)_{s=j\omega_i} = \left(\frac{dx_{11}}{d\omega} - \frac{x_{11}}{\omega}\right)_{\omega=\omega_i} > 0$$

A similar relation holds at every other point $s = j\omega_r$ at which $z_{11}^{(1)}$ has a zero.

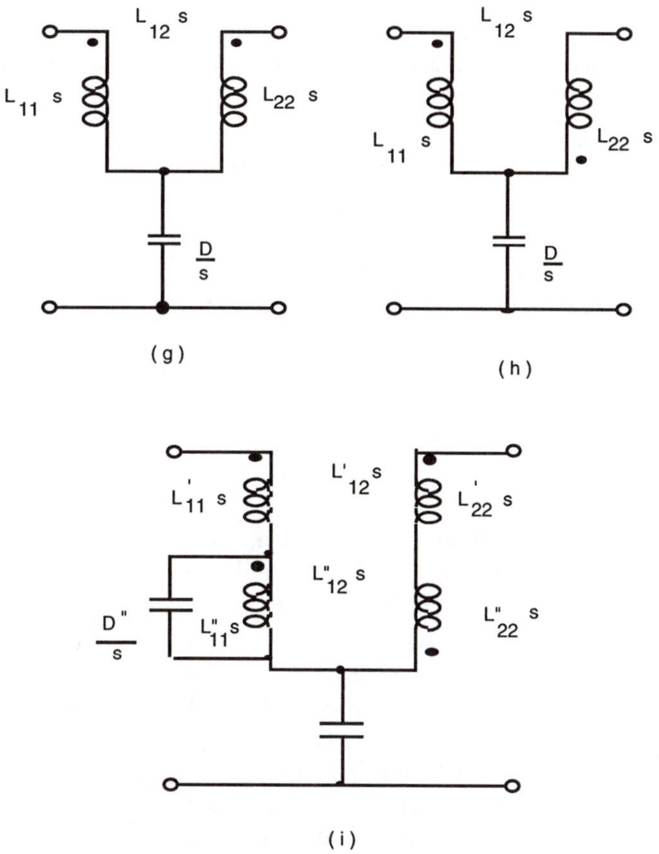

FIGURE 5.10.1
Sections for cascade synthesis (cont.): g) Brune's section; h) Darlington's C-section; i) Darlington's D-section.

The impedance $z_{11}^{(1)}$ is still LPR, because it is the sum of two LPR functions z_{11} and $-l_a s$. The function $y_{o1}^{(1)} = 1/z_{11}^{(1)}$ or

$$y_{o1}^{(1)} = \frac{1}{z_{11} - l_a s} \tag{5.10.1}$$

is LPR, as the reciprocal of the LPR function $z_{11}^{(1)}$. (See Theorem 4.6.3.) At $s = j\omega_i$ it has a pair of simple poles (because $z_{11}^{(1)}$ has simple zeros there), at which the residues are thus positive.

Now we start the second step of the cycle, consisting of the extraction from $y_{o1}^{(1)}$ of the conjugate imaginary poles at $s = j\omega_i$ (the possible other poles at $s = j\omega_r$, $r \neq i$, are extracted in successive later cycles)

$$y_{o1}^{(2)} = y_{o1}^{(1)} - \frac{2k_i s}{s^2 + \omega_i^2} \tag{5.10.2}$$

where k_i is the residue of $y_{o1}^{(1)}$ at $s = \pm j\omega_i$. This residue is always positive as shown above. Thus the extracted admittance y_b represents a series resonant LC branch which is connected in shunt, where

$$l_b = \frac{1}{2k_i} \qquad c_b = \frac{1}{l_b \omega_i^2} \tag{5.10.3}$$

Note that the resonator consists of a positive inductor and capacitor and, therefore, is always realizable. The admittance $y_{o1}^{(2)}$ is LPR because it is the remainder of an LPR function $y_{o1}^{(1)}$ from which the poles at $\pm j\omega_i$ have been completely extracted.

At this point, we claim that $y_{o1}^{(2)}$ can be represented in the form

$$y_{o1}^{(2)} = \frac{1}{z_{11}^{(3)} + l_c s} \tag{5.10.4}$$

where $z_{11}^{(3)}$ is an LPR function, and l_c is such that the entire Brune section is realizable despite the presence of negative inductance l_a within the section. A possibility at the end of this step is that the remainder impedance $z_{11}^{(2)} = 1/y_{o1}^{(2)}$ may be infinite, i.e., it represents an open circuit. This occurs if the transmission zeros to be still extracted, if any, are private poles of z_{22}. But note that in such a case a branch may always be added in series at the output of the circuit realized up to this point without affecting the open circuit input impedance function z_{11}. It will later be shown that lossless elements can always be placed in this branch which make the Brune section realizable, and, at the same time, are consistent with the z_{22} function. Leaving this special case aside for the moment, we now assume there is a finite LPR remainder function $y_{o1}^{(2)}$ after the pair of poles is extracted as a shunt element. To prove that the entire Brune section is realizable, we consider the behavior at infinity of the function $y_{o1}^{(2)}$, defined by eq. (5.10.2), taking into account eqs. (5.10.1) and (5.10.3)

$$\frac{1}{h_{2\infty}} = -\frac{1}{l_a - h_\infty} - \frac{1}{l_b} \tag{5.10.5}$$

where h_∞ and $h_{2\infty}$ are the (positive) residues at infinity of z_{11} and $z_{11}^{(2)} = 1/y_{o1}^{(2)}$. Solving eq. (5.10.5) for $h_{2\infty}$, we find

$$h_{2\infty} = \frac{-(l_a - h_\infty) l_b}{l_a + l_b - h_\infty} \tag{5.10.6}$$

Since $h_{2\infty} > 0$ because $y_{o1}^{(2)}$ is LPR, and $l_b > 0$, $l_a < 0$, it follows that in eq. (5.10.6)

$$l_a + l_b - h_\infty > 0 \tag{5.10.7}$$

Hence

$$l_a + l_b > 0 \tag{5.10.8}$$

In eq. (5.10.7) the inequality (5.10.8) remains strict even if $h_\infty = 0$, otherwise, the input impedance of the section would become a shunt capacitor and the structure would reduce to a Type B section.

From eqs. (5.10.1), (5.10.2), and (5.10.4), we find

$$\frac{1}{z_{11} - l_a s} - \frac{2k_i s}{s^2 + \omega_i^2} = \frac{1}{z_{11}^{(3)} + l_c s} \tag{5.10.9}$$

We denote as $h_{3\infty}$ the residue at infinity of $z_{11}^{(3)}$. Taking the limit for $s \to \infty$ in eq. (5.10.9), we obtain

$$\frac{1}{h_\infty - l_a} - 2k_i = \frac{1}{h_{3\infty} + l_c}$$

or, taking into account eq. (5.10.3),

$$\frac{1}{h_\infty - l_a} - \frac{1}{l_b} = \frac{1}{h_{3\infty} + l_c} \tag{5.10.10}$$

Using eq. (5.10.10) and (5.10.7), we get

$$h_{3\infty} = \frac{h_\infty(l_c + l_b) - (l_a l_b + l_b l_c + l_c l_a)}{l_a + l_b - h_\infty} \tag{5.10.11}$$

We shall see later that the physical realizability of the section requires that

$$l_a l_b + l_b l_c + l_c l_a \geq 0 \tag{5.10.12}$$

Therefore, we need to show that there exists an $l_c > 0$ together with an LPR function $z_{11}^{(3)}$, both satisfying eq. (5.10.4), such that $h_{3\infty} > 0$, and that eq. (5.10.12) is satisfied. Under the latter condition, l_c satisfies the inequality

$$l_c \geq -\frac{l_a l_b}{l_a + l_b} > 0$$

Since the denominator of eq. (5.10.11) has been shown to be positive, the requirement that $h_{3\infty} > 0$ imposed on eq. (5.10.11) results in

$$l_c \leq -\frac{(l_a - h_\infty) l_b}{l_a + l_b - h_\infty}$$

The two inequalities are mutually consistent. We have therefore shown that there exists an l_c which meets the requirements presented above, and it is bounded by two positive quantities as follows,

$$0 < -\frac{l_a l_b}{l_a + l_b} \leq l_c \leq -\frac{(l_a - h_\infty) l_b}{l_a + l_b - h_\infty} \tag{5.10.13}$$

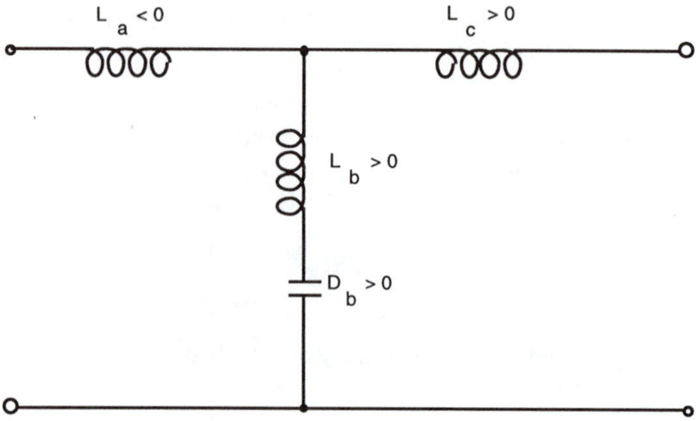

FIGURE 5.10.2
T equivalent circuit of Brune's section for $x_i < 0$.

which defines the interval of the allowable values of l_c. It has been shown in Section 5.5 that if the input impedance z is minimum reactance, its transmission zeros are extracted by compact sections. Thus the upper and lower bounds of l_c in eqs. (5.10.13) must coincide, that is, it must be $h_\infty = 0$; it is concluded that *the open circuit input impedance z_{11} of the lossless two-port has a zero at infinity whenever z is minimum reactance.*

It remains to prove that $z_{11}^{(3)}$ is LPR. We have $z_{11}^{(3)} = z_{11}^{(2)} - l_c s$ which shows that it is odd and that its finite $j\omega$ poles must coincide with those of the LPR function $z_{11}^{(2)}$, and have the same positive residues. Moreover, we have also just shown that with the proper choice of $l_c > 0$, (eq. (5.10.13)) the residue at infinity satisfies $h_{3\infty} > 0$. It follows that $z_{11}^{(3)}$ is LPR.

Note that the usual choice of l_c is that corresponding to the equality sign in eq. (5.10.12) and in the first of eqs. (5.10.13); such a choice is compulsory when z is minimum reactance and therefore h_∞ is zero.[13]

At this point the cycle is completed, leaving a PR remainder impedance $z_{11}^{(3)}$, whose degree is equal to that of z_{11} minus two. In fact, the extraction of l_a increases the degree of z_{11} by one if such a function has a zero at infinity, by zero, if it has a pole. The extraction of the shunt branch reduces the degree by two. The extraction of l_c reduces the degree by one, eliminating from $z_{11}^{(3)}$ the pole at infinity, artificially introduced by the extraction of l_a, or by zero, if the pole already existed.

The two-port realizing Brune's section is shown in Fig. 5.10.2. It contains the negative inductance l_a, so that it is not immediately realizable as a

[13]If h_∞ is not zero, we can give l_c a value greater than the minimum and leave the difference as a series inductor to be later used for practical adjustment of the circuit elements.

lossless structure. However, it is immediately seen that the matrix

$$\boldsymbol{L} = \begin{pmatrix} l_a + l_b & l_b \\ l_b & l_b + l_c \end{pmatrix} \tag{5.10.14}$$

under condition (5.10.12) represents two coupled inductors, with self inductances

$$\begin{aligned} l_{11} &= l_a + l_b \\ l_{22} &= l_c + l_b \end{aligned} \tag{5.10.15}$$

and mutual inductance

$$l_{12} = l_b \tag{5.10.16}$$

Eq. (5.10.12) is precisely the condition that the matrix $\boldsymbol{L}$ in eq. (5.10.14) be positive semidefinite or that the energy stored in the coupled inductor system be nonnegative. If the equality sign is chosen, the inductance matrix is compact and the inductors are perfectly coupled.

The special case referred to earlier, namely when $y_{01}^{(2)} = 1/z_{11}^{(2)} = 0$, is handled as follows. Eqs. (5.10.13) are replaced by

$$0 < -\frac{l_a l_b}{l_a + l_b} \leq l_c \leq h_{22\infty} \tag{5.10.17}$$

where $h_{22\infty}$ is the residue of z_{22} at infinity; in fact, the second eq. (5.10.13) breaks down because $l_a + l_b - h_\infty = 0$ as a result of having $h_{2\infty} = \infty$ in eq. (5.10.6). Thus a series branch containing any $l_c > 0$ as defined by eqs. (5.10.17) is simply added to the output port of the z_{11} circuit. This has no effect on the prescribed open circuit impedance z_{11}, nor does it affect the zero of transmission, and l_c may be absorbed so as to produce realizable coupled coils exactly as described above. If z_{22} demands more inductance, this can be simply placed in series at the output.

The final form of Brune's section is shown in Fig. 5.10.1 (g). The compactness condition (zero residue matrix) is always satisfied for the pole at the origin, while it may be satisfied or not for the pole at infinity. Note that if the latter is a pole of z, it can be used to relax the perfect coupling requirement, thus obtaining a more realistic model of a transformer.

We now turn to the case in which $x_i > 0$.

In this case we can extract from the impedance z_{11} a pole at the origin in the form of a series negative elastance $d_a = -x_i/\omega_i$ without impairing positive reality. This produces a pair of zeros at $j\omega_i$ in the remainder impedance $z_{11}^{(1)}$, which can be extracted as poles of $y_{o1}^{(1)}$ using a series resonant circuit in shunt (exactly as in the preceding case) with elastance d_b in series with inductance l_b. The remainder admittance $y_{o1}^{(2)}$, still LPR, has a pole at the origin with positive residue. The extraction of the latter in the form of a series capacitor d_c finally leaves a positive real impedance remainder

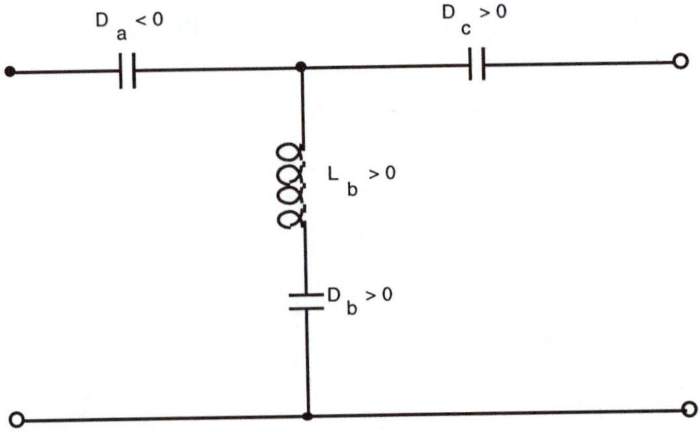

FIGURE 5.10.3
T equivalent circuit of Brune's section for $X_i > 0$.

$z_{11}^{(3)}$ and the cycle is completed. We need not enter into more detail because the procedure, including the singular case $y_{01}^{(2)} \equiv 0$, is exactly the same as described for the case $x_i < 0$ with s replaced by $1/s$.

At the end of the cycle we have the structure of Fig. 5.10.3. The Tee of capacitors is replaced by a capacitor of elastance $d_a + d_b$ shunting the input of an ideal transformer of ratio $n = (d_a + d_b)/d_b < 1$. To eliminate the ideal transformer across the capacitor, we insert another ideal transformer of ratio $1/n$ at the output port of the section and correspondingly readjust the elements in the series resonant branch. This does not change either z_{11} or the zero of transmission. The net effect is to obtain the section of Fig. (5.10.1 (g)) in which the capacitor transformer is cancelled, while a new transformer now shunts the inductor; thus the parameters of the section become

$$l_{11} = l_b \qquad l_{12} = l_b/n \qquad l_{22} = l_b/n^2 \qquad d = d_a + d_b$$

The result for the section is a coupled coil circuit in series with a capacitor, exactly the same circuit form as the Brune section for the case $x_i < 0$. The coupling turns out to be perfect, but it can be relaxed if there is some series inductance, just as in the previous case.

Note that the Tee equivalent for the coupled coil circuit would now show a negative inductance ($l_c = l_{22} - l_{12} < 0$) at the output rather than at the input as in the case $x_i < 0$. Indeed, one may simply carry out the computations by extracting a positive or negative inductor, as the case may be, at the beginning of the cycle. The output transformer is eventually absorbed by adjusting the resistive load.

The concepts above are illustrated by the following.

Example 5.10.1
Realize in Darlington cascade form the impedance

$$z = \frac{2s^3 + 2s^2 + (4 - 2\sqrt{2})s + 1}{2s^2 + s + 1}$$

Solution We calculate

$$m_1 m_2 - n_1 n_2 = (\sqrt{2}s^2 + 1)^2 \tag{5.10.18}$$

From eq. (5.10.18), we see that the impedance z has two simple transmission zeros at $s = \pm j\omega_i$ where

$$\omega_i = \frac{1}{\sqrt[4]{2}}$$

Since the right side of the expression is the square of polynomial, we must choose

$$z_{11} = \frac{m_1}{n_2} = \frac{2s^2 + 1}{s}$$

Since z has a pole at infinity with residue 1, we can extract it from z_{11}

$$z_{11}' = \frac{2s^2 + 1}{s} - s = \frac{s^2 + 1}{s}$$

We have $x_i = z_{11}'(j\omega_i)/j < 0$ and $l_a < 0$, so eq. (5.10.13) gives

$$l_a = \frac{z_{11}'(j\omega_i)}{j\omega_i} = -(\sqrt{2} - 1)$$

After extracting l_a as a series inductance from z_{11}', we find

$$z_{11}^{(1)} = \frac{s^2 + 1}{s} - [-(\sqrt{2} - 1)s] = \frac{\sqrt{2}s^2 + 1}{s}$$

$y_{o1}^{(1)} = 1/z_{11}^{(1)}$ has a pole at $s = j\omega_i$, which is extracted in the form of a parallel connected series resonator to yield $y_{o1}^{(2)}$

$$y_{o1}^{(2)} = y_{o1}^{(1)} - \frac{s}{\sqrt{2}s^2 + 1} = 0$$

The inductance and the capacity have the values

$$l_b = \sqrt{2} \; c_b = 1$$

The remainder is a zero admittance, therefore, this is the special case of an open circuit. To complete the synthesis, compute the minimum permitted value of l_c using eq. (5.10.13)

$$l_c = -\frac{l_a l_b}{l_a + l_b} = -\frac{-(\sqrt{2} - 1)\sqrt{2}}{-(\sqrt{2} - 1) + \sqrt{2}} = 2 - \sqrt{2}$$

We can now check z_{22}

$$z_{22} = \frac{m_2}{n_2} = 2s + \frac{1}{s}$$

But this is precisely the impedance seen at the back end, i.e., l_b, c_b, l_c in series, so no further adjustment for z_{22} is required. This was to be expected since initially the pole of z at infinity was extracted to make it a minimum reactance function.

Thus the parameters of the Brune section are, according to eqs. (5.10.15) and (5.10.16),

$$l_{11} = 1 \qquad l_{22} = 2 \qquad l_{12} = \sqrt{2} \qquad c_b = 1$$

Alternatively, we could have deferred the initial extraction of the pole at infinity. In that case we would have obtained the value $2 - \sqrt{2}$ for l_a, l_b would be unchanged, and for l_c any value in the interval $[1 - \sqrt{2},\ 2 - \sqrt{2}]$ could be chosen. The minimum value for l_c corresponds to a compact section, which leaves a remainder inductance in series with the perfectly coupled coils. There are obviously an infinite number of solutions since we can partially extract the pole at infinity at the outset, and then realize the remainder impedance with the proviso of determining l_c so as to satisfy the compactness condition leaving a remainder coil to contribute to z_{22}. However, a better choice would be to absorb the pole at infinity within the section, so as to realize the more realistic condition of imperfectly coupled coils. Returning to the case where the extraction of the pole at infinity is deferred, we choose complete absorption of inductance within the coupled coil section (the imperfectly coupled coil case). We then have $l_a = l_c = 2 - \sqrt{2}$, $l_b = \sqrt{2}$. Thus the inductance matrix is given by $l_{11} = l_{22} = 2$ and $l_{12} = \sqrt{2}$. Its determinant is two rather than zero, the latter value corresponding to the compact case. □

5.11 Cascade Synthesis: Darlington's C-Section

The *Darlington C-section* extracts a double transmission zero on the positive real axis (Fig. 5.10.1 (h)). Let the positive real impedance z have such a transmission zero at $s = \sigma_i$. We assume multiplicity two to conform with the Darlington procedure of Section 5.7 which calls for zeros of transmission in $\Re s > 0$ to be of even multiplicity. At $s = \sigma_i$ the open circuit input impedance of the lossless two-port is real and positive since it is PR

$$z_{11}(\sigma_i) = r_i > 0$$

We begin the cycle by extracting l_a such that

$$l_a = \frac{r_i}{\sigma_i} > 0$$

Thus

$$z_{11}^{(1)} = z_{11} - l_a s \tag{5.11.1}$$

is an odd function, analytic in the open RHP, with real positive residues at the finite poles (necessarily on the imaginary axis), (i) a simple unique RHP zero at $s = \sigma_i$, and (ii) a real negative residue at the pole at infinity.

We need only prove statements (i) and (ii). Write

$$z_{11}^{(1)} = z_q Q_1 Q_2$$

where Q_1 is the zero factor (even since $z_{11}^{(1)}$ is odd) introduced by eq. (5.11.1)

$$Q_1 = \sigma_i^2 - s^2$$

and Q_2 is a further even zero factor including any other zeros in the RHP, complex or real, some possibly coinciding with σ_i. Q_2 is readily defined to be positive on the $j\omega$-axis. z_q is LPR because (a) it is odd, (b) its residues at its finite poles on the $j\omega$-axis (these poles are the same as those of z_{11}) are positive since $Q_1Q_2 > 0$ there, and (c) it has no pole at infinity because the degree of Q_1Q_2 is at least 2, while z_{11} has, at most, a first order pole at infinity. Evidently the introduced zero at σ_i is simple, for if Q_2 were of degree m (Q_1Q_2 of degree $m+2$), z_q would have a zero of order $m+1$ at infinity ($z_{11}^{(1)}$ has a simple pole at infinity). But since z_q is an LPR function, m must be zero, i.e., $Q_2 = 1$. Thus the extraction produces a unique simple zero at $s = \sigma_i$ and (i) is proved.

As for (ii), note first that since z_q is LPR, its Taylor coefficient for $1/s$ in the expansion about infinity satisfies $c_{-1}^{(\infty)} > 0$. We can now infer that the residue of $z_{11}^{(1)}$ at its simple pole at infinity is negative. Thus

$$\lim_{s\to\infty} \frac{z_{11}^{(1)}}{s} = \lim_{s\to\infty} \frac{c_{-1}^{(\infty)}}{s^2}(\sigma_i^2 - s^2) = -c_{-1}^{(\infty)} \equiv -h_\infty < 0 \tag{5.11.2}$$

and (ii) is proved.

We now consider $y_{o1}^{(1)} = 1/z_{11}^{(1)}$. The function is odd because its reciprocal is odd; it is analytic in $\Re s > 0$ except for a simple pole at $s = \sigma_i$ because $z_{11}^{(1)}$ has a simple zero there. The finite $j\omega_h$ poles of $y_{o1}^{(1)}$ coincide with the imaginary axis zeros of $z_{11}^{(1)}$ and are simple. These poles will necessarily have positive residues if the coefficients of the first order Taylor expansion terms of $z_{11}^{(1)}$ at its corresponding zeros are positive. To prove that these coefficients are indeed positive, consider the partial fraction expansions of $z_{11}^{(1)}$ and $dz_{11}^{(1)}/ds$ using eq. (5.11.2)

$$\begin{aligned} z_{11}^{(1)} &= -h_\infty s + \frac{h_0}{s} + \sum_j \frac{2h_j s}{s^2+\omega_j^2} \\ \frac{dz_{11}^{(1)}}{ds} &= -h_\infty - \frac{h_0}{s^2} + \sum_j \frac{2h_j(\omega_j^2 - s^2)}{(s^2+\omega_j^2)^2} \end{aligned} \tag{5.11.3}$$

where all coefficients h_∞, h_0, and h_i are positive. Divide the first of eqs. (5.11.3) by s, then subtract it from the the second and evaluate the difference at the zero $s_h = j\omega_h$ of $z_{11}^{(1)}$. The result is

$$\left(\frac{dx_{11}^{(1)}}{d\omega}\right)_{\omega=\omega_h} = \frac{2h_0}{\omega_h^2} + \omega_h^2 \sum_i \frac{4h_j}{(\omega_j^2 - \omega_h^2)^2} > 0 \tag{5.11.4}$$

Now $(dx_{11}^{(1)}/d\omega)_{\omega=\omega_h}$ is just the coefficient of the first order term of the Taylor expansion of $z_{11}^{(1)}$ around its zero at $j\omega_h$, so the statement is proved.

The second step in the cycle consists in extracting the pole of $y_{o1}^{(1)} = 1/z_{11}^{(1)}$ at σ_i. We wish to show that the residue of $y_{o1}^{(1)}$ at this pole is negative or, equivalently, that the first term of the power series for $z_{11}^{(1)}$ at $s = \sigma_i$ is negative, i.e., that

$$k_i = \left(\frac{1}{dz_{11}^{(1)}/ds}\right)_{s=\sigma_i} < 0$$

To prove this, consider the equations (5.11.3) at the zero of $z_{11}^{(1)}$ at $s = \sigma_i$. By proceeding as in eq. (5.11.4)

$$\frac{1}{k_i} = \left(\frac{dz_{11}^{(1)}}{ds}\right)_{s=\sigma_i} = -\frac{2h_0}{\sigma_i^2} - \sum_j \frac{4h_j\omega_j^2}{(\omega_j^2+\sigma_i^2)^2} < 0$$

As a result of the extraction, we obtain

$$y_{o1}^{(2)} = y_{o1}^{(1)} - \frac{k_i}{s-\sigma_i} - \frac{k_i}{s+\sigma_i} = y_{o1}^{(1)} - \frac{2k_i s}{s^2-\sigma_i^2} \tag{5.11.5}$$

It is convenient to include the extra term, $-k_i/(s+\sigma_i)$. It is PR so that it does not introduce realizability problems, and serves the purpose of making the last term in eq. (5.11.5) an odd function. The admittance

$$y_b = \frac{2k_i s}{s^2-\sigma_i^2} = \frac{1}{\dfrac{s}{2k_i} - \dfrac{\sigma_i^2}{2k_i s}}$$

is realized as a series resonant LC branch in shunt with the line

$$l_b = \frac{1}{2k_i} \qquad c_b = -\frac{1}{l_b\sigma_i^2} = -\frac{2k_i}{\sigma_i^2}$$

Since k_i is negative, l_b is negative, whereas c_b is positive. Because of the negative inductance, the extracted LC branch "resonates" at $s = \sigma_i$, rather than at a point on the imaginary axis.

The admittance $y_{o1}^{(2)}$, the remainder after the resonator extraction, is an odd function because it is the difference of two odd functions, is analytic in the RHP because the only RHP pole of $y_{o1}^{(1)}$ has been extracted, and has real positive residues at its finite poles (necessarily all on the imaginary axis) because they are just the residues at the poles of $y_{o1}^{(1)}$, for which the property has already been proved. Finally, $y_{o1}^{(2)}$ has no pole at infinity because $y_{o1}^{(1)}$ is the reciprocal of $z_{11}^{(1)}$, which has a pole there, and the extracted term y_b is regular at infinity. Thus $y_{o1}^{(2)}$ is LPR.

From here on, the extraction proceeds exactly as for a Brune section with $x_i < 0$. Therefore, the inductance l_c, corresponding to the residue of the LPR function $1/y_{o1}^{(2)}$ at infinity, is positive.

The extraction of a Darlington's C-section reduces the degree of the (possibly augmented) impedance by two; the proof is carried out in the same manner as for Brune's section. If the transmission zero was originally double, no augmentation is required and the impedance degree is actually reduced by two; if, however, the transmission zero was originally simple, the impedance degree is reduced only by one, because the other zero that also disappears was artificially introduced by augmentation.

Darlington's C-section can be realized in a form strongly suggestive of Brune's section (Fig. 5.10.1 (h)). The coupled inductance parameters are

$$l_{11} = l_a + l_b$$

$$l_{22} = l_c + l_b$$

$$l_{12} = \quad l_b$$

Although the equations above are formally identical to eqs. (5.10.15) and (5.10.16), the presence of a negative, instead of a positive, mutual inductance produces a transmission zero on the positive real axis, rather than on the imaginary axis.

The concepts above are illustrated below.

Example 5.11.1

Realize the impedance already discussed in Section 5.8 in Darlington cascade form

$$z = \frac{8s + 7}{2s + 7}$$

which presents a simple transmission zero at $\sigma_0 = 7/4$.

Solution We augment it by the factor $4s + 7$ and obtain, as earlier, eq. (5.8.7)

$$z = \frac{32s^2 + 84s + 49}{8s^2 + 42s + 49}$$

We have

$$m_1 m_2 - n_1 n_2 = (16s^2 - 49)^2$$

Since $m_1 m_2 - n_1 n_2$ is the square of an even function, we compute z_{11} as

$$z_{11} = \frac{m_1}{n_2} = \frac{32s^2 + 49}{42s}$$

Calculate l_a

$$l_a = \frac{z(7/4)}{7/4} = \frac{8}{7}$$

We proceed to extract from z_{11} the impedance $l_a s$

$$z_{11}^{(1)} = \frac{32s^2 + 49}{42s} - \frac{8}{7}s = \frac{-16s^2 + 49}{42s}$$

Thus

$$y_{o1}^{(1)} = \frac{42s}{-16s^s + 49} = \frac{1}{-\frac{8}{21}s + \frac{7}{6s}}$$

We have

$$y_b = y_{o1}^{(1)}$$

and the shunt resonator consists of

$$l_b = -\frac{8}{21} \qquad c_b = \frac{6}{7}$$

We are left with an open circuit remainder. Thus, as discussed in connection with the cycle used to extract a Brune section, to determine the last branch of the Tee we examine the z_{22} element of the open circuit two-port impedance matrix

$$z_{22} = \frac{m_2}{n_2} = \frac{8s^2 + 49}{42s} = \frac{4}{21}s + \frac{7}{6s}$$

Thus $l_{22} = 4/21$ and we have

$$l_{11} = l_a + l_b = \frac{16}{21}, \qquad l_{12} = l_b = -\frac{8}{21}, \qquad l_{22} = l_c + l_b = \frac{4}{21}$$

$$c_b = \frac{6}{7}, \qquad l_c = l_{22} - l_b = \frac{12}{21}$$

The inductive coupling is perfect, i.e., $l_{11}l_{22} - l_{12}^2 = 0$, defining a compact residue matrix at infinity, a consequence of the minimum reactance property of the prescribed impedance z. □

5.12 Cascade Synthesis: Darlington's D-Section

The *Darlington's D-section* extracts a pair of double conjugate complex transmission zeros in the RHP (Fig. 5.10.1 (i)).

Let the positive real impedance z have a pair of double conjugate complex transmission zeros at $s_i = \sigma_i + j\omega_i$, $s_i^* = \sigma_i - j\omega_i$. At such points the open circuit impedance of the lossless two-port takes on the values

$$z_{11}(\sigma_i \pm j\omega_i) = r_i \pm jx_i$$

with $r_i > 0$, $\sigma_i > 0$, $\omega_i > 0$.

We begin the cycle by extracting a series resonator of impedance $z_a = l_a s + d_a/s$ such that

$$z_{11}^{(1)}(s_i) = z_{11}(s_i) - l_a s_i - \frac{d_a}{s_i} = 0 \tag{5.12.1}$$

Eq. (5.12.1) can be solved for l_a and d_a by equating real and imaginary parts. The result is

$$\begin{aligned} l_a &= \frac{\omega_i r_i + \sigma_i x_i}{2\sigma_i\omega_i} \\ d_a &= \frac{\omega_i r_i - \sigma_i x_i}{2\sigma_i\omega_i}(\sigma_i^2 + \omega_i^2) \end{aligned}$$

Consider the partial fraction expansion of z_{11} at $s = s_i$. Then for each term z_{11k} it is easy to show that since $\Re\, z_{11k}(s_i) > 0$, (z_{11} and z_{11k} are PR), it follows that $\omega_i/\sigma_i \geq |\Im\, z_{11k}|/\Re\, z_{11k}$, so that for the entire expansion

$$\frac{\omega_i}{\sigma_i} \geq \frac{|x_i|}{r_i}$$

It follows, regardless of the sign of x_i, that l_a, $d_a > 0$. Note too, that since $z_{11}^{(1)}$ is a real, odd, rational function, $z_{11}^{(1)} = 0$ at $\pm s_i$ and $\pm s_i^*$, i.e., $z_{11}^{(1)}$ has a quadruplet of complex zeros with quadrantal symmetry. The zeros in the RHP introduced by eq. (5.12.1) are simple and unique. This is proved in the same fashion as for the previously discussed C-section. We write

$$z_{11}^{(1)} = z_q Q_1 Q_2 = z_q Q \tag{5.12.2}$$

where we define Q_1 to contain the zero factor introduced by eq. (5.12.1), as well as an additional pole at the origin

$$Q_1 = \frac{(s_i^2 - s^2)(s_i^{*2} - s^2)}{-s^2} \tag{5.12.3}$$

and Q_2 is a further even factor including any other zeros in the RHP, complex or real, possibly partially or completely coinciding with s_i. Q_2 is readily defined to be positive on the $j\omega$-axis. z_q is LPR because (a) it is odd, with all its poles on $j\omega$, (b) its residues at its finite poles on the $j\omega$-axis (identical with those of z_{11}) are positive, since $Q_1(j\omega)Q_2(j\omega) > 0$, (c) it has no pole at infinity because the degree of Q_1Q_2 is at least 2 while z_{11} has at most a first order pole at infinity, and (d) it has no pole in the origin, because it is cancelled by the denominator s^2 of Q_1. Note that the introduced RHP zeros of $z_{11}^{(1)}$ are not present in z_q; they are contained in Q_1Q_2.

The introduced zeros at s_i, s_i^* are simple, for if Q_2 were of degree $m > 0$ (Q_1Q_2 of degree $m+2$), z_q would have a zero of order $m+1$ at infinity ($z_{11}^{(1)}$ has a simple pole at infinity). But since we have just proved that z_q is LPR, m must be zero, i.e., $Q_2 = 1$. Thus the extraction produces a unique pair of simple zeros in the RHP at s_i, s_i^*. Evidently $z_{11}^{(1)} = z_qQ_1$ and, since z_q is LPR and has no pole at zero or infinity, the residues for the poles in $z_{11}^{(1)}$ at these points are, respectively, h_0, h_∞ and are negative. For using eq. (5.12.3)

$$h_0 = \lim_{s\to 0} sz_qQ_1 \propto -|s_i|^4 < 0, \quad h_\infty = \lim_{s\to\infty} z_qQ_1/s \propto -1 < 0 \qquad (5.12.4)$$

We now show that the function $z_{11}^{(1)}$, as defined by eq. (5.12.2), can be represented as a nonpassive one-port producing the complex zeros of transmission, shunted by an LPR one-port. Write

$$y_{o1}^{(1)} = \frac{1}{z_{11}^{(1)}} = \frac{y_q}{Q_1} = y_{o1}^{(2)} + y_b \qquad (5.12.5)$$

where $y_q = 1/z_q$, as the reciprocal of z_q, is an LPR admittance and has positive residues at all its poles (necessarily on the $j\omega$-axis). If $y_{o1}^{(1)}$ is expanded in partial fractions, we get the sum of two sets of terms, as shown in eq. (5.12.5). The first set, $y_{o1}^{(2)}$, is a portion of the partial fractions expansion of $y_{o1}^{(1)}$ taken at the poles of y_q, say $j\omega_h$, whose residues are those of y_q each divided by $Q_1(j\omega_h) > 0$. Thus the resultant residues are all positive, and $y_{o1}^{(2)}$ is therefore an LPR function. It is important to note, for later use, that since $z_{11}^{(1)}$, as defined by eq. (5.12.1), has simple poles at zero and infinity, $y_{o1}^{(2)}$ in eq. (5.12.5) does not. Its reciprocal is LPR with simple poles and positive residues at these two points.

The second set of terms, y_b, consists of partial fractions, taken at the poles of $1/Q_1$, i.e., located at the zeros of Q_1. y_b consists of four terms, one for each of the quadruplet of zeros. It is clear that y_b as defined by eq. (5.12.5) is a real, rational, odd function. Thus its residues at the poles $\pm s_i$ are equal, as are the pair of residues at $\pm s_i^*$. Furthermore, the residues at conjugate poles are conjugates of each other. As a result, the partial fraction expansion for y_b has the form

$$y_b = \frac{k_i}{s - s_i} + \frac{k_i}{s + s_i} + \frac{k_i^*}{s - s_i^*} + \frac{k_i^*}{s + s_i^*}$$

If the terms of y_b are combined, then $z_b = 1/y_b$ can be written in the form

$$z_b = -b_0 \frac{s^4 + b_1 s^2 + b_2}{s(s^2 + b_3)}$$

The b_k are real constants.

We have proved that the residues of $z_{11}^{(1)}$ at infinity and at the origin are negative, eq. (5.12.4); referring to eq. (5.12.5) this must originate from negative residues of the poles of $z_b = 1/y_b$ at zero and infinity because $y_{o1}^{(2)}$ is LPR. It therefore follows that $b_0 > 0$ (pole at infinity), and $b_2/b_3 > 0$ (pole at zero). Finally, in order that the polynomial $s^4 + b_1 s^2 + b_2$ have a quadruplet of roots, $b_2 > (b_1/2)^2$, therefore $b_2 > 0$, hence $b_3 > 0$. The residue of z_b at the finite pole on the $j\omega$-axis at $s = \pm j\sqrt{b_3}$ is easily calculated as $h_{b3} = (b_0/2b_3)(b_3^2 - b_1 b_3 + b_2)$. The sign of the residue is that of the trinomial $b_3^2 - b_1 b_3 + b_2$, but $b_3^2 - b_1 b_3 + b_2 > b_3^2 - b_1 b_3 + (b_1/2)^2 = (b_3 - b_1/2)^2 > 0$, or $h_{b3} > 0$.

As a result of the discussion above, we can write z_b in the form

$$z_b = l_b s + \frac{d_b}{s} + \frac{2h_{b_3} s}{s^2 + b_3}$$

where $l_b < 0$, $d_b < 0$, and $h_{b_3} > 0$. Thus z_b is a branch consisting of the series connection of a negative inductor, a negative capacitor, and a lossless parallel resonant circuit.

We now extract from $z_{11}^{(2)} = 1/y_{o1}^{(2)}$ a series resonator of impedance $z_c = l_c s + d_c/s$. Since we have shown above that $1/y_{o1}^{(2)}$ is LPR with simple poles and positive residues at zero and infinity, it follows that $l_c > 0$, $d_c > 0$, and the remainder impedance $z_{11}^{(3)} = z_{11}^{(2)} - z_c$ is also LPR. Our remaining task is to show that the Tee section formed by the three inductors l_a, l_b, l_c has the property that the resulting two-port impedance matrix has a positive semidefinite residue matrix at infinity; similarly, the Tee of capacitors of elastances d_a, d_b, d_c has the same property at the origin. The proof for the inductances is carried out in exactly the same manner as for the Brune extraction in Section 5.10 for the case that z_{11} at the zero of transmission is $x_i < 0$. The only modification is that here $l_a > 0$, $l_b < 0$, whereas for the Brune section, $l_a < 0$, $l_b > 0$. For the zero at the origin, the Brune section proof for the case $x_i > 0$ is used, with the only change that here $d_a > 0$, $d_b < 0$ instead of $d_a < 0$, $d_b > 0$. The final result is that the realizability equations for the Tee of inductors and the Tee of capacitors are satisfied

$$\begin{aligned} l_a l_b + l_b l_c + l_c l_a &\geq 0 \\ d_a d_b + d_b d_c + d_c d_a &\geq 0 \end{aligned} \tag{5.12.6}$$

Again, as in the case of the Brune section, the extraction can be done with the inequality sign if the pole at infinity and the pole at the origin of z have not been completely extracted; otherwise, the equality sign must be chosen, resulting in compact residue matrices. In this latter case the input elements l_a, d_a are absorbed into the coupled coils, consistent with the assumption that z is a minimum reactance function.

The degree of the LPR impedance $z_{11}^{(3)}$ is four units less than that of z_{11}. In fact, the extraction of z_a increases the degree at most by two, the extraction of z_b decreases it by four, and the extraction of z_c decreases it exactly by the amount by which it was increased with the extraction of z_a. The same remarks on augmentation done in Section 5.11 with reference to Darlington's C-section apply here; if the transmission zero is not originally double, the reduction in degree of the impedance is really only two.

The final structure is a Tee composed of the branches z_a, z_b, and z_c. Assuming compactness, the negative elements are eliminated by replacing the Tee of inductors by two perfectly coupled inductors with negative mutual inductance and the Tee of capacitors has as equivalent a single capacitor shunted by an ideal transformer with primary and secondary wound in opposite directions so that the primary to secondary turns ratio is $n/1 = -\sqrt{(d_a + d_b)/(d_c + d_b)}/1$; the resonant circuit remains unchanged. The resulting structure is shown in Fig. 5.12.1. The ideal transformer may be eliminated exactly as was done for the Brune section extraction of a negative capacitor (for the case $x_i > 0$) by inserting between the output port and the load an ideal transformer (which can later be absorbed in the load) of ratio $n/1$. This, in turn, allows the cancellation of the ideal transformer shunting the capacitor, and requires a modification of the two other components of the shunt branch, caused by the compensating presence of $1/n$ transformers.

Thus the coupled inductors are changed according to the rule

$$l'_{11} = l_{11}, \quad l'_{12} = n l_{12}, \quad l'_{22} = n^2 l_{22} \tag{5.12.7}$$

The parallel resonant circuit (l_r, c_r) now has a $1/n$ ideal transformer across it, but this is absorbed by replacing the coil and transformer with its equivalent, a pair of perfectly coupled coils; the primary (shunted by the capacitor c_r) having self inductance l_r, the secondary with self inductance $n^2 l_r$, and the mutual inductance becomes $n l_r$. The resulting type D section has no ideal transformers and is connected across the line. It consists of three two-ports in series. The first is a pair of perfectly coupled coils. The second is

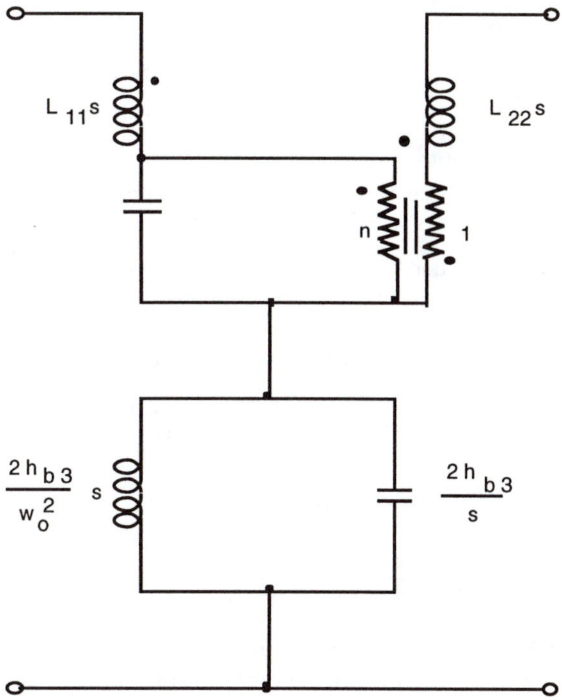

FIGURE 5.12.1
Darlington's D-section: structure with ideal transformers.

formed by a capacitor connected across the primary of another pair of coupled coils. The third two-port is just another capacitor. The structure is shown in Fig. 5.10.1 (i). It should be noted that this form for the D-section is precisely equivalent to that shown (but not proved) in Darlington's original paper.[14]

Example 5.12.1

As an example, consider the following open-circuit impedance which we assume to have been derived from a minimum reactance PR driving point impedance z

$$z_{11} = s + \frac{1}{s} + \frac{s}{s^2+1} + \frac{s}{s^2+4}$$

and assume that a quadruplet of transmission zeros at $s_i = \pm 1 \pm j$ is to be extracted. Other transmission zeros may of course be present. Realize the Darlington two-port.

[14]S. Darlington, "Synthesis of Reactance 4-Poles which Produce Prescribed Insertion Loss Characteristics", *J. Math. Phys.*, vol. 18, no. 4, p. 257-353, Sept. 1939.

Solution We have

$$z_{11}(1+j) = 2.4 + j0.4 = (1+j)l_a + \frac{1}{1+j}d_a \tag{5.12.8}$$

From eq. (5.12.8) we obtain

$$l_a = \tfrac{7}{2} \qquad d_a = 2$$

We extract the impedance $z_a = 1.4s + 2/s$ from z_{11}

$$z_{11}^{(1)} = z_{11} - 1.4s - \frac{2}{s} = -\frac{0.4s^6 + s^4 + 1.6s^2 + 4}{s(s^2+1)(s^2+4)} =$$

$$= -\frac{0.4(s^2+2.5)(s^4+4)}{s(s^2+1)(s^2+4)}$$

Then $y_{o1}^{(1)}$ is calculated and decomposed into partial fractions

$$y_{o1}^{(1)} = -\frac{2.5s(s^2+1)(s^2+4)}{(s^2+2.5)(s^4+4)} = \frac{\frac{45}{82}s}{s^2+2.5} - \frac{\frac{125}{41}s^3 + \frac{200}{41}s}{s^4+4} \tag{5.12.9}$$

The last partial fraction in eq. (5.12.9), representing y_b, can be inverted and decomposed in its turn to yield the following representation of z_b

$$z_b = \frac{s^4+4}{\frac{125}{41}s^3 + \frac{200}{41}s} = -\frac{41}{125}s - \frac{41}{50s} + \frac{\frac{1681}{1250}s}{s^2+\frac{8}{5}} \tag{5.12.10}$$

Hence eq. (5.12.9) represents an active branch whose impedance, z_b, is zero at $\pm s_i$, $\pm s_i^*$, shunted by a series resonant circuit with

$$l_0 = \frac{82}{45}, \quad d_0 = \frac{41}{9}$$

The input impedance z is assumed to be minimum reactance so that the input branch l_a, d_a must be absorbed into the Tee's of inductance and capacitance whose residue matrices are compact. We therefore use a portion of l_0 and d_0 to realize eq. (5.12.6) with equal signs. Thus

$$l_c = \frac{287}{670}, \quad d_c = \frac{82}{59}$$

The remainder in the output arm consists of an inductance l_t in series with a capacitor of elastance d_t

$$l_t = l_0 - l_c = \frac{1681}{1206}, \quad d_t = d_0 - d_c = \frac{1681}{531}$$

Now we proceed to eliminate the negative elements by replacing the Tee of inductances with a pair of perfectly coupled coils, $l_{11} = l_a + l_b$, $l_{12} = l_b$, and $l_{22} = l_c + l_b$. The Tee of capacitors is replaced by a shunt element equal to $d_{11} = d_a + d_b$ followed by an ideal transformer of ratio $n/1 = -\sqrt{d_{11}/(d_c + d_b)}/1$. For our example we have

$$l_{11} = 1.0720, \quad l_{22} = 0.1004, \quad l_{12} = -0.3280$$

and

$$d_{11} = 1.1800, \quad n = -1.4390$$

At this point we have arrived at one form of a type D section (the ideal transformer can also be eliminated as discussed below), which in turn is shunted by the LPR remainder z_t. Actually we have simply shown how to realize z_{11} so as to include the complex quadruplet of zeros of transmission. The synthesis of the lossless two-port, which entails the location of the output port, must still be completed. Under the assumption of a minimum reactance input impedance z, the output port is placed across the z_t remainder. By inspection of the resultant circuit, it is evident that when the resistive termination is in place z will be a minimum reactance function, i.e., no poles anywhere on $j\omega$. The resultant compactness property automatically yields the appropriate z_{22} and z_{12}. There is an additional zero of transmission at the series resonance of the type B section z_t. If the prescribed z were not minimum reactance and the output port placed across the d_t capacitor, the resultant structure terminated in a resistor would have two zeros of transmission at infinity, associated with the "el" consisting of the series inductor l_t, followed by the capacitor across the output. If instead, the output port were placed across l_t, this would introduce two transmission zeros at the origin. Finally, even under the minimum reactance assumption, there could be additional real frequency zeros of transmission if z_{22} had private poles on $j\omega$ with $0 < \omega < \infty$, realized by Type A sections.

Our earlier discussion showed how to eliminate the ideal transformer across the capacitor by inserting an ideal transformer of ratio $n:1$, with its primary across the active branch, z_b. The resultant structure is shown in Fig. 5.10.1 (h), with the parameters of the mutually coupled inductors as given by eq. (5.12.7)

$$l'_{11} = 1.072, \quad l'_{22} = 0.2079, \quad l'_{12} = -0.4720$$

The capacitor $d_{11} = 1.1800$ remains as a series element in the shunt branch z_b, while two perfectly coupled inductors replace the coil $l_r = 0.8405$ of the parallel resonant circuit (see eq. (5.12.10)),

$$l_{r11} = l_r = 0.8405, \quad l_{r22} = n^2 l_r = 1.7404, \quad l_{r12} = n l_r = 1.2095$$

In the final form, the impedance of the system elements beyond the D section must be multiplied by n^2 to eliminate the output transformer. □

5.13 Ladder Synthesis; Fujisawa's Theorem

In the last five sections, it has been shown that any finite passive (rational and PR) immittance can be realized as a resistively terminated cascade of lossless sections, each producing one or two or four of the transmission zeros. The resultant structure may have coupled coils if Brune, type C or type D sections are required, as well as an ideal transformer. It is, of course, desirable to avoid the complexity of these magnetically coupled elements. Thus the use of an ideal transformer can be avoided if the requirement that the load be held to a prescribed value, conventionally unity, is relaxed. On the other hand, the avoidance of the sections that require coupled coils can be achieved only in the earlier stages of design (generally where an approximation process is used) by attempting to construct an immittance function whose transmission zeros require only A and B sections. The resultant function is then realized as a resistively terminated lossless ladder structure which alternates shunt and series elements, each a pure inductance or capacitance or resonator, all without coupled coils. (See Fig. 5.13.1 (a) and (b).) Unfortunately, the general necessary and sufficient conditions to allow this transformerless ladder synthesis by a PR immittance function whose zeros of transmission are all on $j\omega$ are still not known.

The most general case that has been solved is probably that given by Fujisawa's Theorem.[15] Fujisawa's result treats the particular case of a lowpass ladder, which characterizes a structure having a direct connection between input and output at d.c. with transmission zero(s) at infinity. Specifically, the ladder of Fujisawa's theorem alternates shunt and series branches, with each shunt branch a single capacitor or an LC pair forming a series resonant circuit. The series branches each contain a single inductor or an LC pair forming a parallel resonant circuit. The initial element of the ladder is either a shunt capacitor (*mid-shunt ladder*) or a series inductor (*mid-series ladder*). The theorem is particularly useful for realizing filters which have zeros of transmission at finite frequencies in the stop band such as elliptic filters.

THEOREM 5.13.1

(Fujisawa Theorem) *Given a rational PR impedance represented in the*

[15] D. C. Youla, "A Tutorial Exposition of Some Key Network-Theoretic Ideas Underlying Classical Insertion-Loss Filter Design", *Proc. IEEE*, vol. 59, no. 5, p. 760-799, May 1971. As a matter of historical interest, Darlington gave a similar result on p. 49 of his original 1939 paper. (See footnote 10 at p. 247.)

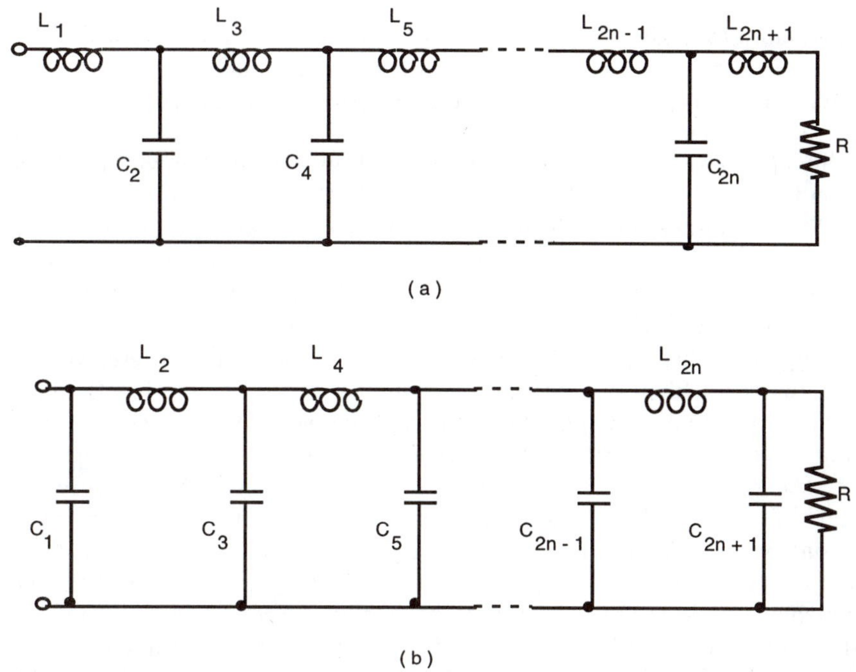

FIGURE 5.13.1
(a) Lowpass ladder mid-series structure; (b) lowpass ladder mid-shunt structure.

(irreducible) form

$$z = \frac{m_1 + n_1}{m_2 + n_2}$$

m_k *even,* n_k *odd, eq.* (5.5.5). *The necessary and sufficient conditions that* z *be realizable as a resistively terminated lossless lowpass ladder are:*

(1) $z(0) = r$*, where* $r > 0$ *is the terminating resistance;*

(2) at infinity, z *has either a zero or a pole;*

(3) either

$$M = m_1 m_2 - n_1 n_2 = k \prod_{i=1}^{l} (s^2 + \omega_i^2)^2$$

or

$$M = k$$

where $k > 0$ *and the* ω_i*'s are positive and form a nondecreasing sequence;*

(4) either $M = k$*, or for every* m*,* $m = 1, 2, 3, ...l$*, one of the polynomials* n_2 *or* n_1 *possesses at least* m *imaginary complex conjugate zero pairs whose magnitudes are not larger than* ω_m*.*

We note that, depending on whether z has a zero or pole at $s = \infty$, the ladder is mid-shunt or mid-series, respectively. Although we do not present a formal proof of Fujisawa's theorem here, we will illustrate qualitatively the ideas upon which it rests, and which are at the basis of the synthesis technique. The reader is referred to Youla's tutorial paper[16] for a complete discussion.

We first observe without proof that, if z has a pole anywhere on the $j\omega$ axis including the point at infinity, the pole can be completely extracted without impairing the properties (1) to (4) listed in the Theorem.

We can therefore assume that the prescribed impedance z has a zero at infinity (mid-shunt case), then the admittance y has a pole there. We *partially* extract a Type B section, i.e., a shunt capacitor which is less than that required to remove the pole. The effect of increasing the extracted capacitance as complete extraction is approached, is to decrease the susceptances at each transmission zero (the associated conductances are already zero and $y(j\omega_i) = jb(\omega_i)$). We stop the extraction when the susceptance, $b(\omega_i)$, at one or more of the transmission zeros becomes zero. This produces an admittance zero at the point in question, i.e., a pole of impedance. The impedance pole is then extracted by a Type A section (series-connected shunt resonator), and the cycle is repeated with a new partial capacitance extraction. The effect of this process of *partial* extraction of the immittance pole at infinity is to associate a *pole* of impedance with a transmission zero to avoid the necessity of a Brune section.

Intuitively we might expect that, to produce an admittance zero at a zero of transmission, the value of the partially extracted capacitance c_1 should be chosen to satisfy

$$c_1 = \min_i \frac{b(\omega_i)}{\omega_i}$$

where index i varies over the set of transmission zeros at which the ratio $b(\omega_i)/\omega_i$ is positive. The proof of the theorem actually shows that as we carry out this choice successively to account for all the zeros of transmission, then at each extraction the remainder impedance still satisfies the conditions of the theorem. Thus the process can be continued until the last remainder reduces to a resistor.

We have described the synthesis of a mid-shunt ladder network. Evidently the mid-series ladder can be treated in a dual fashion, using partial inductance extraction.

This process of ladder synthesis is based on repeated *partial* extractions of an immittance pole at infinity; therefore, superfluous elements will appear in the final ladder structure. Thus the realization is not minimal; this is the price one pays to avoid coupled coils.

[16]loc. cit.

As discussed in Section 5.7 in connection with the general Darlington realization of a PR immittance, it is convenient for the synthesis to be carried out on the open-circuit impedance z_{11} (or the short-circuit admittance y_{11}), rather than directly on the prescribed impedance z. To do this (mid-shunt case) partial shunt capacitance extraction (c_1) of the pole in $y_{o1} = 1/z_{11}$ at infinity is carried out to produce a pole in the remainder impedance, $z_{11}^{(1)} = z_{11} - 1/sc_1$, at one of the zeros of transmission, and the process is repeated till all the finite zeros of transmission are accounted for. As in the general Darlington synthesis, final adjustment, if necessary, of the back end of the ladder is carried out by referring to z_{22}.

We illustrate Fujisawa's theorem by the following example.

Example 5.13.1

Realize by a ladder structure the impedance

$$z = \frac{2s^2 + s + 1}{3s^3 + 2s^2 + 2s + 1}$$

Solution The function F, whose zeros determine the zeros of transmission, is by eq. (5.6.5)

$$F = 4\frac{(s^2+1)^2}{(3s^3 + 4s^2 + 3s + 2)^2}$$

Thus z possesses simple transmission zeros at $s = \infty$ and $s = \pm j1$. The open circuit input impedance of the reactive two-port is

$$z_{11} = \frac{m_1}{n_2} = \frac{2s^2 + 1}{3s^3 + 2s}$$

At the finite zero of transmission we calculate $y_{o1}(j1) = j \neq 0$; therefore, a synthesis starting with the extraction of a finite transmission zero would require a Brune section. Alternatively, we might immediately extract the transmission zero at infinity by removing the pole of y_{o1} at infinity, yielding a Type B section with a capacitance $c = 3/2$ and leaving a remainder

$$z_{11}^{(1)} = \frac{4s^2 + 2}{s}$$

which would result in $z_{11}^{(1)}(j1) = 2j$ and again would require the extraction of a Brune section.

However, the conditions of Fujisawa's theorem are satisfied by z:

(1) $z(0) = 1$;

(2) z has a zero at infinity (mid-shunt case);

(3) $m_1 m_2 - n_1 n_2 = (s^2+1)^2$;

(4) $n_2 = 3s^3 + 3s$ has a conjugate zero pair at $s = \pm j$, so that the corresponding magnitude $|s| = 1$ does not exceed the transmission zero frequency of $\omega_1 = 1$.

The impedance z is therefore realizable as a mid-shunt ladder.

Recall that $y_{o1}(j) = j$ while with the pole at infinity removed from y_{o1}, the remainder admittance satisfies $y_{o1}^{(1)}(j1) = -j/2$. Thus the complete extraction of capacitance $c = 3/2$ causes the susceptance at $s = j1$ to change its sign. This indicates that an appropriate partial extraction would force $y_{o1}^{(1)}(j)$ to be zero. In fact, if we extract from $y_{o1} = jb$ a capacitance

$$c_1 = 1$$

which is exactly $b(1)/1$, we find a remainder

$$y_{o1} - sc_1 = y_{o1}^{(1)} = \frac{s(s^2+1)}{2s^2+1}$$

whose reciprocal $z_{11}^{(1)}$ has a pole at $s = \pm j1$, with residue 1/2. We extract this pole by a Type A section (parallel resonator in series) with

$$l_2 = 1 \qquad c_2 = 1$$

and we are left with the impedance

$$z_{11}^{(1)} - \frac{s}{s^2+1} = z_{11}^{(2)} = \frac{1}{s}$$

which is realized by a shunt capacitor

$$c_3 = 1$$

Finally, the load (a unit resistor) is connected across c_3, and the input impedance of the terminated ladder is z, Fig. 5.13.2. It is seen that the third-degree impedance has a nonminimum realization with four elements; however, only three of these are independent, since the three capacitors form a mesh. Physically this means they allow the specification of only two initial voltages; the initial current in the inductor yields the third initial condition. Thus, despite the extra element, the degree of the impedance agrees with the number of independent initial conditions.

The procedure discussed above can be generalized somewhat by including in the synthesis technique the possibility of partial extraction of both a pole at infinity *or at the origin* from a given immittance so as to make it zero

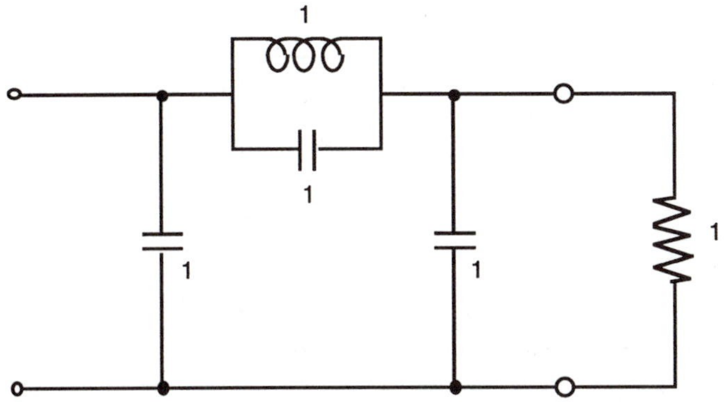

FIGURE 5.13.2
Ladder realization of the impedance of Example 5.13.1.

at a finite transmission zero, with subsequent extraction of a pole from the inverse. However, although the conditions can be precisely specified for the process to be successful at an individual extraction step, the overall necessary and sufficient conditions on the prescribed immittance are not presently known. Moreover, physical realizability of the ladder may depend on the order of the extractions. □

5.14 Transmission Zeros All Lying at Infinity and/or the Origin

An important special case arises when all transmission zeros lie at infinity and/or at the origin. In this case the PR immittance can always be realized by a ladder using only Type A and Type B sections, each consisting of a single L or C element.

In particular, for the case of a PR impedance with $z(0) = r$ and all transmission zeros at infinity, we have

$$F = \frac{4k}{(m_1 + m_2 + n_1 + n_2)^2}$$

Since $M = k$, z trivially satisfies Theorem 5.13.1 (Fujisawa) and is the input to a resistively terminated lowpass ladder. The ladder synthesis is simply a Cauer continued fraction expansion of the LPR function, z_{11}. Thus in the mid-series case the ladder is found by first removing a transmission zero by completely extracting the impedance pole of z_{11} at infinity as a series inductor. The next zero is removed by fully extracting the pole at infinity of the remainder admittance as a shunt capacitor. The procedure

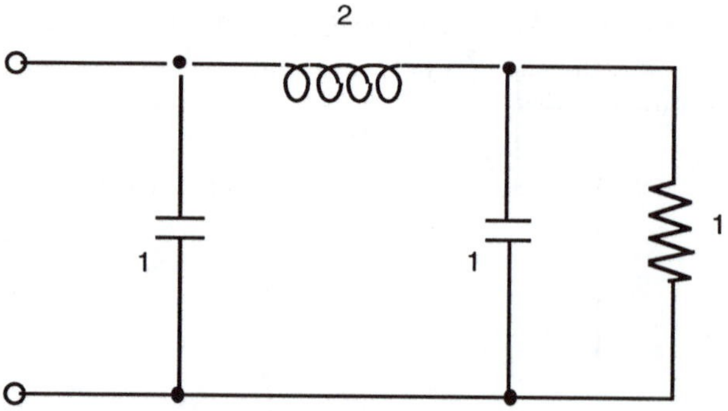

FIGURE 5.14.1
Realization of the impedance of Example 5.14.1.

continues, alternately extracting series inductors and shunt capacitors, until the remainder is zero. At this point we check z_{22} to determine whether a final series inductive branch is required. The process is the same for the mid-shunt case except we start with $y_{o1} = 1/z_{11}$ and the first extracted element is a shunt capacitor. A dual technique is employed if the short circuit admittances y_{ij} are used instead of z_{ij}. Finally, the ladder is terminated in a resistor.

We illustrate the concepts above.

Example 5.14.1
Realize the impedance

$$z = \frac{2s^2 + 2s + 1}{2s^3 + 2s^2 + 2s + 1} \tag{5.14.1}$$

Solution The open-circuit input impedance is

$$z_{11} = \frac{m_1}{n_2} = \frac{2s^2 + 1}{2s^3 + 2s}$$

$M = m_1 m_2 - n_1 n_2 = 1 = k$ (thus all transmission zeros are at infinity) and z has a zero at infinity so that we have the case of a mid-shunt lowpass ladder. Since F has a third order zero at infinity, there are three ladder elements. First extract a shunt capacitance $c_1 = 1$ from $y_{o1} = 1/z_{11}$, leaving as a remainder an admittance $y_{o1}^{(1)}$ whose reciprocal is

$$z_{11}^{(1)} = \frac{2s^2 + 1}{s}$$

Now we can extract a series inductance $l_2 = 2$ from $z_{11}^{(1)}$ leaving as remainder an impedance $z_{11}^{(2)}$, whose reciprocal is

$$y_{(o1)}^{(2)} = s$$

which corresponds to a shunt capacitor $c_3 = 1$. The impedance z is then realized by terminating the two-port in a unit resistor. The complete ladder realization is shown in Fig. 5.14.1. □

If z has all its transmission zeros at the origin (order n), the polynomial $M = m_1m_2 - n_1n_2$ has the form s^{2n}, and

$$F = \frac{s^{2n}}{(m_1 + m_2 + n_1 + n_2)^2}$$

This case does not fall directly under the rubric of Fujisawa's theorem, but the synthesis is readily carried out (it can be handled by the theorem if the frequency transformation $p = 1/s$ is used). The structure realizing z is known as a *highpass ladder* (direct connection between input and output at infinity) with shunt inductors and series capacitors. The ladder is found by extracting the pole at the origin (series capacitor) from the open-circuit impedance, then the pole from the open-circuit admittance (shunt inductor) remainder, and so forth until the remainder is zero. The procedure corresponds to a high pass Cauer continued fraction expansion.

The reader can work out an illustrative example by changing s into $1/s$ in eq. (5.14.1) and repeating step by step the procedure described above. He will find a network that alternatively could have been obtained from the lowpass configuration of Fig. 5.14.1 by replacing the series inductor with a series capacitor, and the two shunt capacitors by two shunt inductors. The new inductors and capacitors have L, C values that are each reciprocals, respectively, of the original C, L values.

The third case, that in which transmission zeros are either at the origin or at infinity, leads to a function F in the following form.

$$F = \frac{s^{2q}}{(m_1 + m_2 + n_1 + n_2)^2}$$

where q is the number of transmission zeros at the origin, and one-half the difference in degree between denominator and numerator of F is the order of the zero of transmission at infinity. Since the transmission zeros may be extracted in any sequence, the network will contain series arm inductors or shunt arm capacitors whenever transmission zeros at infinity are removed, and series arm capacitors or shunt arm inductors, when transmission zeros at the origin are removed. A resonant circuit may sometimes be formed if a series arm inductor and capacitor occur in series in the course of the synthesis; similarly, a shunt arm capacitor and inductor may form a parallel resonant circuit. However, the resonant frequencies of these circuits do not correspond to zeros of transmission, e.g., a series branch containing a series resonator is a short circuit at resonance but this has no special significance.

Of the three cases that have been examined in this section, the first is of major importance since it describes the prototype network realizing *low-pass filters*, from which *highpass*, *bandpass*, *bandstop* filters can be obtained by means of *frequency transformations.* These transformations correspond to the replacement of inductors and capacitors by suitable reactances and susceptances (i.e., single capacitors or inductors replacing inductors and capacitors, or resonators replacing inductors and capacitors) thus preserving the ladder form of the structure. Frequency transformations will be further discussed in later sections devoted to filter synthesis.

6

Insertion Loss Filters

6.1 The Concept of a Filter and the Approximation Problem

An information-bearing electric signal, $f(t)$, is generally transmitted by varying the properties of a carrier signal in accordance with the characteristics of $f(t)$ (modulation). Commonly used systems modulate the amplitude, frequency, or phase of the carrier. The effect of the modulation is to produce *sidebands* which contain the information, i.e., sinusoidal signals of suitable amplitude and phase whose frequencies cover two equal bands contiguous to the carrier frequency. The sideband whose frequencies are higher than the carrier frequency is the upper sideband; the one with lower frequencies is the lower sideband (see Example 3.9.5). The width of the sidebands depends on the modulation scheme and on the transmitted signal, particularly the amount of signal information. However, for a given modulation system and a given type of signal (phone, radio, television, etc.), and for a prescribed quality of reception, the bandwidth is more or less determined.

A large number of messages can be conveyed on a given channel by using carriers of different frequencies for each of the various signals, with the proviso that the difference in frequency between two *adjacent* carriers must be at least equal to twice the width of the sideband, if *interference* is to be avoided. Actually, it is possible to transmit only one sideband, reducing by one half the bandwidth requirement, because the other sideband is spectrally symmetric with respect to the carrier frequency, but this consideration does not enter our present discussion. It is clear that increasing the number of messages on the same physical channel requires higher and higher carrier frequencies. Currently these frequencies extend into the optical band.

Our concern here, however, is with the signal processing at the receiving end of the channel. It is obvious that the sidebands carrying the various messages must be separated from the carriers before each message is sent to the receiver. The separation is achieved by employing special two-ports, once called *wave filters* or *frequency filters*, nowadays simply referred to as *filters*. Roughly speaking, a filter has the property of being transparent to all signals whose frequencies lie in a prescribed band, the *passband*, and opaque to those frequencies which lie in the complementary band, the *stopband*. If the passband includes zero frequency and excludes infinite frequency, the filter is *lowpass*; if the passband includes infinite frequency but not zero, the filter is *highpass*; if it includes neither zero nor infinity, it is *bandpass*; if it includes both, it is *bandstop*.

Before discussing the problem of concrete realization of filters, we consider the ideal conditions under which the information contained in a filtered signal is completely preserved. It is evident that if, in traversing the filter, the signal undergoes multiplication by a constant, and experiences a fixed delay, that is if the output signal is related to the input, $f(t)$, by

$$f_o(t) = Kf(t-T)$$

there will be no signal distortion. The Fourier transform of the output is

$$\mathcal{F}f_o(t) = KF(j\omega)\exp(-j\omega T)$$

so that distortionless transmission is equivalent, in the frequency domain, to the dual requirement that all the harmonics contained in the signal be multiplied by the same real constant $K > 0$, and the phase response decrease linearly with frequency, the proportionality factor being $T \geq 0$.

Thus an *ideal* lowpass filter should have a transfer function

$$H(j\omega) = K[\mathrm{u}(\omega+1) - \mathrm{u}(\omega-1)]\exp(-j\omega T)$$

The effect of $H(j\omega)$ is that a signal whose harmonics are contained within a normalized band $-1 \leq \omega \leq 1$ is transmitted with a fixed delay, T, and an amplitude unchanged except for a multiplicative constant, K. A signal whose harmonics are outside the band is completely stopped. We are, of course, assuming that adjacent sidebands do not overlap.

The ideal lowpass filter (as well as its bandpass and highpass versions) is physically unrealizable. The application of the Paley Wiener criterion 3.7.1 given in Example 3.7.1 shows that a transducer with rectangular transfer function amplitude response is noncausal, since there is no possible phase response which can be associated with the amplitude that permits an analytic continuation into the Right Half Plane. On the other hand, Example 3.7.2 shows that one may approximate arbitrarily closely to the rectangular response. Qualitatively, however, inspection of the Hilbert Transform relating amplitude slope and phase, eq. (3.7.2), shows that in the transition region between pass and stop bands, where the amplitude slope is steep, the phase becomes very large (logarithmically infinite for a rectangular transfer function) and nonlinear. This is the situation for a minimum-phase transfer function. Matters can be improved by adding one or more equalizer allpass sections (unit amplitude response) so that the composite transfer function approximates phase response $(-j\omega T)$; if T is large enough, we can attain phase linearity to a prescribed tolerance in any frequency band $-\Omega \leq \omega \leq \Omega$ with $|\Omega| < 1$, with $T \rightarrow +\infty$, $\Omega \rightarrow 1$. The limiting case might be called a *quasi-ideal filter* characterized by a transfer function with constant amplitude and linear phase in the passband; such a structure, however, would require an infinite number of elements to be realized and, moreover, would introduce an infinite delay between input and output. It is therefore clear that the quasi-ideal filter is an abstract concept, which can be approximated as closely as desired, but the limit can never be reached with a physical design.

Physical realization of filters with a finite and reasonably small number of elements is generally performed in two steps. The first is the solution of an *approximation problem* which entails devising a transfer function $H(j\omega)$ (the ratio between the Fourier transforms of the output and the input signals of a physical two-port) satisfying a prescribed set of *specifications.* The second step is the concrete realization of the two-port by *analytic* or *numerical* methods. The approximation problem can be tackled along several different lines, some of which will be described in the following sections. At this point we only wish to stress that simultaneous amplitude and phase specifications cannot, in general, be satisfied by a minimum phase realization; the Hilbert transform uniquely relates amplitude and minimum phase. However, the use of allpass equalizer sections, which do not modify the amplitude but change the phase, allows both sets of requirements to be satisfied to any prescribed tolerance, as mentioned above in the case of the quasi-ideal filter. In the end, however, a compromise must always be worked out between the conflicting demands of amplitude selectivity, phase linearity, and design practicality.

We will consider passive, lossless, filter realizations. Negligible dissipation is usually realistically attainable, particularly at microwave frequencies and, if necessary, one can compensate for incidental dissipation. Furthermore the lossless assumption simplifies design procedures. Finally, power loss does not generally introduce any advantage in the quality of response and, if present to a significant extent, may cause difficulties due to noise generation.

The remainder of this chapter is devoted to the elements of *insertion loss filter design.* In this method, the available gain of the filter is chosen, subject only to clearly stated and not overly onerous constraints. The Darlington synthesis is then employed to design the lossless filter and, when it is inserted between real resistors, the prescribed transducer gain is realized. A major advantage of the procedure is the wide flexibility that is made available in the choice of transfer functions, and the fact that constant resistance loads are employed. The insertion loss technique supplanted an earlier approach, the method of *image parameters.* A finite number of cascaded identical lossless sections, when terminated by the correct frequency varying irrational image impedances, has a passband over which the signal propagates, and a cutoff (stop) region, where the image impedance becomes imaginary. A wide and sophisticated literature describes image parameter design, but the difficulty of taking into account the irrational loads and the lack of flexibility in selecting a transfer function eventually led Darlington to his outstanding contribution, and made the older method obsolete.

6.2 Synthesis of doubly terminated filters

The design of doubly terminated filters is one of the important applications of the Darlington theory discussed in Chapter 5. As a prerequisite for carrying out the design process, the filter is represented by means of its two-port normalized scattering matrix, and the transfer function H of the two-port is identified with the transmittance S_{21}.

The *transducer power gain function* $T(\omega^2)$ is related to the transmittance transfer function by

$$T(\omega^2) = \frac{P}{P_A} = |S_{21}(j\omega)|^2 \leq 1 \qquad 0 \leq \omega < +\infty \tag{6.2.1}$$

where P is the average power delivered to the terminating load resistor and P_A is the available power of the generator. The inequality stems from the fact that $P \leq P_A$. Eq. (6.2.1) can be continued into the s-plane as

$$T(-s^2) = S_{21*}S_{21}$$

where $S_{ij*}(s) = S_{ij}(-s)$.

The even, nonnegative, rational character of $T(\omega^2)$ expressed by eq. (6.2.1) guarantees that $T(-s^2)$ always permits a spectral factorization to separate S_{21}, as a BR function, from S_{21*}. (See Lemma 5.7.1.) Once this is done, we can determine the remaining scattering functions to form an LBR scattering matrix which describes the lossless two-port. In turn, the LBR matrix can be expressed using the Belevitch canonic representation, eq. (5.4.13), repeated here for convenience

$$\boldsymbol{S} = \begin{pmatrix} \dfrac{h}{g} & \dfrac{f}{g} \\ \pm\dfrac{f_*}{g} & \mp\dfrac{h_*}{g} \end{pmatrix}$$

where f, g, h are polynomials in s. Using Belevitch notation, the paraunitary condition, $S_{11*}S_{11} + S_{21*}S_{21} = 1$ applied to $S_{21*}S_{21}$ results in

$$S_{21*}S_{21} = \frac{f_*f}{f_*f + h_*h} = \frac{1}{1 + \dfrac{h_*h}{f_*f}} = \frac{1}{1 + \psi_*\psi} \tag{6.2.2}$$

Eq. (6.2.2) shows that the characteristic function, ψ, completely determines the transducer power gain $T(-s^2)$, whereas, given the gain, ψ is determined within an arbitrary allpass. The numerator of $\psi_*\psi$ yields the product h_*h, and the denominator of the function $S_{21*}S_{21}$ yields the product g_*g. The factorization of the denominator is unique since g must be Hurwitz, but the zeros of h_*h may be freely divided between h and h_*. The sole requirement in the allocation is that each zero of one polynomial is reflected to the image location in the other polynomial. Thus $S_{11} = h/g$ is determined within an analytic allpass η, and the general solution of the factorization problem is

$$S_{11} = \eta \hat{S}_{11}$$

where $\hat{S}_{11}$ is the minimum phase solution.

Our major interest is the realization of the transducer gain as a reciprocal structure. This realization, as discussed in Section 5.4, can be achieved by augmenting $S_{21*}S_{21}$ with common numerator and denominator factors so that all the root factors of f_*f are of even multiplicity. When the elements of $\boldsymbol{S}$ are then readjusted (the augmenting common factors are not cancelled), the resulting scattering matrix has the form given in Corollary 5.4.1

$$\boldsymbol{S} = \begin{pmatrix} \dfrac{h}{g} & \dfrac{f}{g} \\ \dfrac{f}{g} & \mp\dfrac{h_*}{g} \end{pmatrix}$$

where the (+) sign applies if f is even, the (−) if f is odd.

From S_{11} we derive

$$z = \frac{1 + S_{11}}{1 - S_{11}}$$

Finally, Darlington's cascade synthesis is used to realize z as the input impedance of a lossless, reciprocal two-port terminated at the output port by a unit resistor.

We note that the procedure is always successful, because from eq. (6.2.1) and

$$|S_{11}(j\omega)|^2 + |S_{21}(j\omega)|^2 = 1$$

we find that $|S_{11}| \leq 1$ on the $j\omega$-axis and, therefore, since S_{11} has an analytic continuation into the RHP, it is BR.

Referring to eq. (6.2.2) and noting that $|\psi(j\omega)|^2$ is nonnegative, we can summarize matters with the following theorem which is the basis for the construction of transfer functions that approximate ideal filter characteristics.

THEOREM 6.2.1
A rational, real, even, transducer gain function which satisfies

$$0 \leq T(\omega^2) \leq 1, \;\; \forall \omega$$

is always realizable by a lossless, finite, reciprocal *two-port with resistive terminations. Furthermore, the gain function satisfies the realizability requirements when the characteristic function, $\psi(s)$, is rational with real coefficients.*

We now turn our attention to phase equalization. Let

$$\boldsymbol{S} = \begin{pmatrix} S_{11} & S_{12} \\ S_{21} & S_{22} \end{pmatrix}$$

be the scattering matrix of the lossless two-port. We insert between its output port and the load resistance an allpass two-port with scattering matrix

$$\boldsymbol{\Sigma} = \begin{pmatrix} 0 & \eta_1 \\ \eta_2 & 0 \end{pmatrix}$$

with $\eta_{1*}\eta_1 = \eta_{2*}\eta_2 = 1$. Then it is easily seen that the resulting two-port has the scattering matrix

$$\boldsymbol{S} = \begin{pmatrix} S_{11} & \eta_1 S_{12} \\ \eta_2 S_{21} & \eta_1\eta_2 S_{22} \end{pmatrix} \tag{6.2.3}$$

If the original two-port is reciprocal, we may preserve the property by choosing $\eta_1 = \eta_2 = \eta$. Then the input reflectance S_{11} and hence the input impedance z are unchanged as an obvious consequence of the fact that the load resistance, seen through the allpass, remains equal to itself. Also the modulus of the transmittance S_{21} remains unchanged on the $j\omega$-axis, while its phase is changed by the phase of η. By properly designing the allpass, it is, therefore, possible to introduce phase equalization for improving phase linearity.

We have briefly summarized a procedure for synthesizing a lossless two-port realizing prescribed compatible amplitude and phase characteristics. In the following sections some basic approximation schemes will be discussed. For the sake of simplicity, we will study a lowpass *prototype* filter whose bandwidth is normalized to unity, and whose resistive terminations are normalized to the internal resistance of the generator. It will be shown that these assumptions are not real restrictions since, by appropriate scaling and frequency transformations, we can derive from the prototype, lowpass, bandpass, highpass, and bandstop filters with arbitrary finite resistive terminations and arbitrary passbands.

6.3 Impedance and Frequency Scaling, Frequency Transformations

We define as a *lowpass prototype filter* a lowpass filter with a unit passband and resistive terminations normalized to the generator resistance, realizing a prescribed transducer gain which approximates the ideal lowpass characteristic discussed in Section 6.1. The prototype filter is said to be *normalized* with respect to both frequency and impedance.

From a lowpass prototype filter, highpass, bandpass, and bandstop filters can be derived with arbitrary bandwidths and arbitrary resistive terminations. The derivation is based on the property that a PR (LPR) function of a PR (LPR) function remains PR (LPR) (see Theorem 4.6.2); also the PR (LPR) property is not affected by multiplication by a positive constant. Impedance scaling and frequency transformation can be simultaneously applied in the form of an LPR function (which occasionally may reduce to a frequency scaling) which changes the prototype element values (denoted in the following by lower case letters) to yield those of the final derived filter (denoted by capital letters); impedance and frequency scaling to obtain the final structure is called *denormalization*. An example of this process was presented in Section 5.3.

Impedance scaling of ratio R is described by the following equations

$$Z_i = Rz_i$$
$$Y_j = \frac{y_j}{R} \tag{6.3.1}$$

where indices i and j refer to the branches that are described on an impedance and an admittance basis, respectively, and R is a reference resistance, e.g., the resistance of the generator. Note that the prototype filter impedances and admittances z_i and y_j are dimensionless.

Frequency transformation is represented by an equation of the type

$$S = f(s)$$

where $S = j\Omega$ is the complex frequency of the prototype filter, $s = j\omega$ the complex frequency of the transformed filter, and f a LPR function such that S is dimensionless.

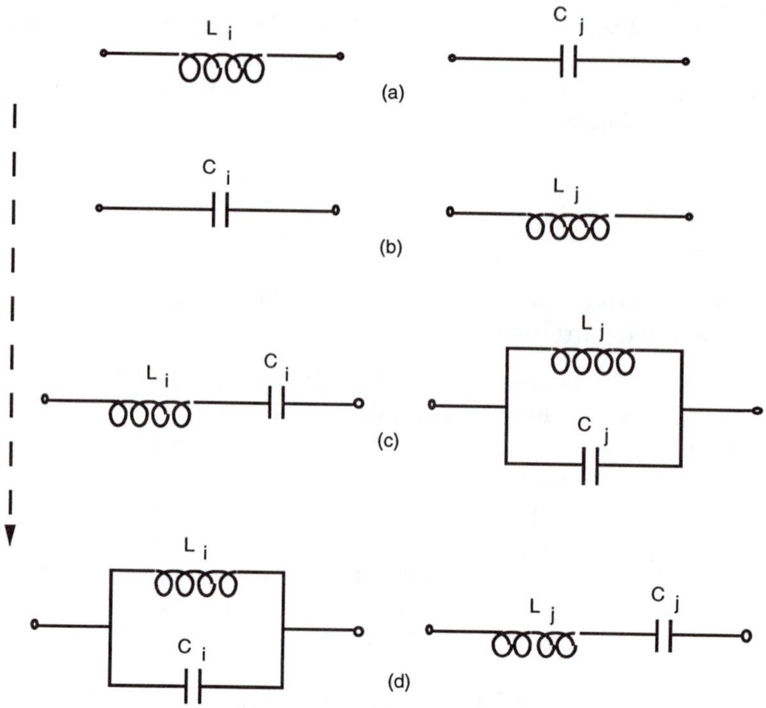

FIGURE 6.3.1
Transformation of basic elements under frequency transformation: (a) lowpass; (b) highpass; (c) bandpass; (d) bandstop.

In the prototype filter, inductor impedances and capacitor admittances are represented as $l_i S$ and $c_j S$, respectively.

The following impedance and frequency transformations are basic for designing lowpass, highpass, bandpass, and bandstop filters starting from lowpass prototypes.

(1) Lowpass to lowpass transformation, to change the impedance level and the cutoff frequency of the passband: the frequency transformation is given by the following LPR mapping

$$S = \frac{s}{\omega_0} \tag{6.3.2}$$

where ω_0 is the cutoff frequency of the transformed filter. Combining the two mappings (6.3.2) and (6.3.1) we obtain the branch inductances L_i and capacitances C_j

$$Rl_i S = \frac{Rl_i}{\omega_0} s \ = L_i s$$

$$\frac{c_j}{R} S \ = \frac{c_j}{R\omega_0} s = C_j s$$

and, therefore,

$$\begin{aligned} L_i &= \frac{Rl_i}{\omega_0} \\ C_j &= \frac{c_j}{R\omega_0} \end{aligned} \tag{6.3.3}$$

Transformation (6.3.2) maps the band $0 \leq \Omega \leq 1$ onto the band $0 \leq \omega \leq \omega_0$, thus producing a simple frequency scaling. Eq. (6.3.3) shows that all inductances are transformed into inductances and all capacitances remain capacitances (Fig. 6.3.1 (a)) with values depending on both the impedance and the frequency scaling.

(2) Lowpass to highpass transformation: the frequency transformation is given by the following LPR mapping

$$S = \frac{\omega_0}{s} \tag{6.3.4}$$

Combining the two mappings (6.3.4) and (6.3.1) we obtain the transformation from the prototype to the transformed filter

$$\begin{aligned} Rl_i S &= \frac{Rl_i\omega_0}{s} = \frac{1}{C_i s} \\ \frac{c_j}{R} S &= \frac{c_j\omega_0}{Rs} = \frac{1}{L_j s} \end{aligned}$$

and, therefore,

$$\begin{aligned} C_i &= \frac{1}{Rl_i\omega_0} \\ L_j &= \frac{R}{c_j\omega_0} \end{aligned} \tag{6.3.5}$$

Transformation (6.3.5) maps the band $-1 \leq \Omega \leq 0$ onto the band $\omega_0 \leq \omega < +\infty$, thus making a frequency inversion. (Note that the prototype response has even symmetry about $\omega = 0$.) Eq. (6.3.5) shows that all inductances are transformed into capacitances and all capacitances into inductances (Fig. 6.3.1 (b)) with values depending on both the impedance scaling and the frequency inversion.

(3) Lowpass to bandpass transformation: We seek a frequency transformation that takes the prototype passband $-1 < \Omega < 1$ into the bandpass filter (simply "filter") passband $\omega_1 < \omega < \omega_2$. To do this, the elements of the filter should take on reactance values in the transformed band equal to those of the prototype elements in the lowpass band. This may be accomplished by replacing prototype inductors by series resonant circuits, and prototype capacitors by parallel resonant

circuits. Thus an inductance of the prototype has zero reactance at the origin, and this point maps into the series resonant frequency ω_0 for the transformed filter branch. The points $\Omega = 1, -1$ should map into $\omega = \omega_2, \omega_1$, and just as the prototype inductance has reactance values which are negatives of each other at the band edges, ± 1, the same thing can be true for the transformed filter branch at ω_2, ω_1, since the reactance changes sign above and below resonance. The replacement of a coil l_i of the prototype by a circuit series resonant at ω_0 is expressed by the equation

$$l_i \Omega = \frac{l_i}{\Delta} \frac{\omega^2 - \omega_0^2}{\omega}$$

where Δ is a positive constant. The series resonator reactances are to be negatives of each other at the band edges of the filter (the reactance at the upper edge, ω_2 is positive), so that $(\omega_2^2 - \omega_0^2)/\omega_2 \Delta = -(\omega_1^2 - \omega_0^2)/\omega_1 \Delta$. Solve for ω_0

$$\omega_0 = \sqrt{\omega_1 \omega_2} \tag{6.3.6}$$

To evaluate Δ, note that setting prototype inductive reactance at $1, -1$ equal to series resonator reactance at ω_2, ω_1, gives

$$(\omega_2^2 - \omega_1 \omega_2)/\omega_2 \Delta = 1$$

so that $\Delta = \omega_2 - \omega_1$. The same results would have been achieved by examining the transformation of a capacitor in the prototype into a branch parallel resonant at ω_0 in the filter. The required frequency transformation is therefore given by the LPR mapping

$$S = \frac{1}{\omega_2 - \omega_1} \frac{s^2 + \omega_0^2}{s} \quad \text{or} \quad \Omega = \frac{1}{\Delta} \frac{\omega^2 - \omega_0^2}{\omega} \tag{6.3.7}$$

Combining the two mappings (6.3.7) and (6.3.1), we obtain

$$Rl_i S = \frac{l_i R}{\Delta} s + \frac{l_i R \omega_0^2}{\Delta} \frac{1}{s} = L_i s + \frac{1}{C_i s}$$

$$\frac{c_j}{R} S = \frac{c_j}{R\Delta} s + \frac{\omega_0^2 c_j}{R\Delta} \frac{1}{s} = C_j s + \frac{1}{L_j s}$$

The values of the transformed inductors and capacitors in the series resonant (subscript "i") and parallel resonant (subscript "j") branches are thus given by

$$L_i = \frac{l_i R}{\Delta} \qquad C_i = \frac{\Delta}{l_i R \omega_0^2}$$
$$C_j = \frac{c_j}{R\Delta} \qquad L_j = \frac{R\Delta}{c_j \omega_0^2} \tag{6.3.8}$$

The passband edges ω_1 and ω_2, and the *center* (not the arithmetic center) of the band, ω_0, are not independent. Their relation is found from eq. (6.3.7) which relates the cutoff points $\Omega = \pm 1$ of the prototype to the bandpass band edges ω_2, ω_1 by $\pm\Delta\omega_k = \omega_k^2 - \omega_0^2$, $k = 1, 2$. Thus

$$\omega_k^2 \mp \Delta\omega_k - \omega_0^2 = 0$$

There are four roots because of the $\mp$ sign. The two positive roots are ω_1, ω_2

$$\begin{aligned} \omega_1 &= -\frac{\Delta}{2} + \sqrt{(\frac{\Delta}{2})^2 + \omega_0^2} \\ \omega_2 &= \frac{\Delta}{2} + \sqrt{(\frac{\Delta}{2})^2 + \omega_0^2} \end{aligned} \tag{6.3.9}$$

The other two roots are associated with a second passband symmetrically located to the left of the origin. Referring to eq. (6.3.6), the "center" frequency, ω_0, is the geometric mean of the upper and lower cutoff points. In fact, the bandpass characteristic possesses what may be termed *geometric symmetry.* This means that any given frequency Ω on the low frequency prototype scale maps into two frequencies on the bandpass scale, $\omega_2 > \omega_0$, $\omega_1 < \omega_0$, and $\omega_1\omega_2 = \omega_0^2$. The transformation of a prototype frequency characteristic with arithmetic symmetry into a distorted response with geometric symmetry may sometimes lead to special difficulties. For example, linear phase for the prototype results in a nonlinear phase characteristic for the bandpass filter. Another point is worth mentioning. Suppose we fix the capacitors in the transformation which takes the prototype capacitors into parallel resonant circuits. That is, choose the transformation to simply add shunt inductance to the capacitance already present. Thus referring to eqs. (6.3.8), if $C_j = c_j/R$, we must have $\Delta = 1$ and the unit bandwidth from d.c. to cutoff in the protoype is conserved, independent of the value of ω_0. If the prototype had cutoff frequencies $\pm\Omega_c$, then the filter capacitance in the parallel resonant arm would be $C_j = c_j\Omega_c/R\Delta$. Maintainig c_j/R unchanged would thus require the passband of the filter to be $\Delta = \Omega_c$ and, again, the bandwidth would be conserved.[1] This procedure also maintains the value of the prototype inductors which transform into series resonators. The result is referred to as the *Principle of Conservation of Bandwidth.*[2]

[1]The invariance property demonstrates the futility of attempting to increase bandwidth by simply cancelling out parasitic reactance with resonant elements.

[2]H. W. Bode, *Network Analysis and Feedback Amplifier Design*, New York, D. Van Nostrand, 1945, pp. 211-214.

(4) Lowpass to bandstop transformation: following a procedure analogous to the bandpass case, series inductors in the prototype are transformed into parallel resonant arms in the bandstop filter (simply "filter") and the shunt capacitors of the prototype into series resonant branches in the filter. Thus,

$$\Omega = -\Delta \frac{\omega}{\omega^2 - \omega_0^2} \tag{6.3.10}$$

where ω_1 and ω_2 are the lower and upper edges of the desired stopband and, as in the bandpass case, we again evaluate $\omega_0^2 = \omega_1\omega_2$, $\Delta = \omega_2 - \omega_1$. We can use eq. (6.3.10) to solve for the positive frequencies $\omega = \omega_1, \omega_2$, corresponding to prototype frequencies $\Omega = 1, -1$. The results are the same as for the bandpass case eq. (6.3.9).

The operation of the filter can be described using eq. (6.3.10), by tracing the filter frequency response with respect to the prototype. As the filter frequency increases from $\omega = 0$ to ω_1, the prototype goes from $\Omega = 0$ to 1. The filter is therefore in a pass mode. As ω increases further from ω_1 to ω_0, Ω increases from 1 to $+\infty$. Thus, corresponding to the prototype, the filter is now in the first portion of its bandstop band. As the filter frequency increases from ω_0 to ω_2, the prototype frequency moves from $\Omega = -\infty$ to $\Omega = -1$. This range therefore corresponds to the remainder of the filter bandstop band. Then as the filter frequency increases from ω_2 to $+\infty$, the prototype frequency moves from -1 to 0, and the filter is once again in a pass mode. As ω continues to increase from $-\infty$ to $-\omega_2$, Ω traverses the prototype pass band from 0 to 1. If we continue further, the filter frequency moves over the negative frequency region of its bandstop band which images the positive frequency response. The transformation of prototype element values to achieve a specified bandstop region for the filter, taking into account frequency and impedance scaling, is (subscript "i", parallel resonant arm, "j" series resonant arm)

$$C_i = \frac{1}{l_i R \Delta} \qquad L_i = \frac{l_i R \Delta}{\omega_0^2}$$

$$L_j = \frac{R}{c_j \Delta} \qquad C_j = \frac{c_j \Delta}{R \omega_0^2}$$

A few further conceptual points should be discussed in connection with frequency transformations. The first is whether, by assuming sufficiently complicated LPR functions, one could design complex multiband filters. The answer is yes, but the design would be impractical, because tuning adjustments are not easily performed except in ladder or other simple structures. A better solution is by cascading independent sections.

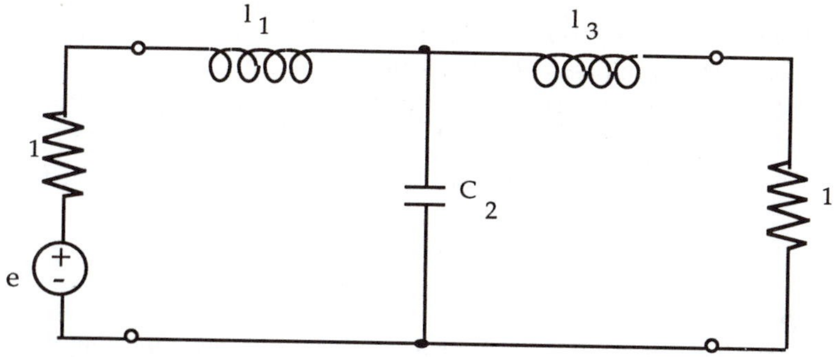

FIGURE 6.3.2
A normalized lowpass filter.

The second is whether frequency transformations involve any limitations with respect to the approximation problem. The answer is no, as long as we are concerned with amplitude approximation; but as mentioned earlier in connection with geometric passband symmetry, the transformations introduce phase distorsions, which require a correction of the transformed filter, not just of the prototype.

Finally, a brief comment on the general design process as carried out in terms of the various frequency transformations. Generally we are given the specifications of the final desired filter, including cutoff frequencies, sharpness of cutoff, and amplitude tolerance in the pass and stop bands. Additionally, for bandpass filters, for example, we must allow for geometric passband symmetry. The specifications are then translated into the equivalent requirements for the prototype, and the rational Transducer Gain approximation is carried out for the lowpass prototype. The lowpass structure is then synthesized. The final filter is determined by appropriate transformation of the prototype elements.

Example 6.3.1

From the lowpass prototype filter of Fig. 6.3.2, with $l_1 = l_3 = 1$, $c_2 = 2$, determine:

(a) a highpass filter with $f_0 = 100\,\text{kHz}$ terminated at both ends on $50\,\Omega$;

(b) a bandpass filter with $R = 50\,\Omega$, $f_1 = 100\,\text{kHz}$, $f_2 = 150\,\text{kHz}$.

Solution

(a) By using the highpass eq. (6.3.5) with $\omega_0 = 2\pi \cdot 100 \cdot 10^3\,\text{rad/s}$, we get

$$C_1 = C_3 = \frac{1}{1 \cdot 50 \cdot 2\pi \cdot 100 \cdot 10^3} 10^6 = 31.8\,\text{nF}$$

$$L_2 = \frac{50}{2 \cdot 2\pi \cdot 100 \cdot 10^3} 10^6 = 39.8\,\mu\text{H}$$

(b) Using bandpass formulae (6.3.8) with $\Delta = 2\pi \cdot 50 \cdot 10^3$, and center frequency $\omega_0^2 = (2\pi)^2 \cdot 1.5 \cdot 10^{10}\,\mathrm{rad}^2/\mathrm{s}^2$, we find

$$L_1 = L_3 = 159.\,\mu\mathrm{H} \qquad C_1 = C_3 = 10.6\,\mathrm{nF}$$

$$C_2 = 127.\,\mathrm{nF} \qquad L_2 = 13.3\,\mu\mathrm{H}$$

□

Example 6.3.2
A lowpass distributed amplifier with series inductance and shunt capacitance equal to the parasitic values imposed by the solid state devices has cutoff frequency Ω_c. The structure is to be redesigned to operate in a bandpass mode with center frequency ω_0. Find the band edge frequencies ω_2, ω_1 for maximum bandwidth.

Solution The parasitic elements should not be increased under the bandpass transformation for this would reduce bandwidth and, obviously, they cannot be decreased. Hence the optimum solution is to retain the parasitic values and conserve bandwidth. Thus $\omega_2 - \omega_1 = \Omega_c$. Substituting $\omega_1 = \omega_0^2/\omega_2$, we have the equation

$$\omega_2^2 - \Omega_c\omega_2 - \omega_0^2 = 0$$

and the positive ω_2 root as well as ω_1 are (also see eqs. (6.3.9))

$$\omega_2 = \frac{1}{2}\left(\Omega_c + \sqrt{\Omega_c^2 + \omega_0^2}\right), \quad \omega_1 = \omega_2 - \Omega_c = \frac{1}{2}\left(-\Omega_c + \sqrt{\Omega_c^2 + \omega_0^2}\right)$$

□

6.4 Specifications for Amplitude Approximation

The concept of an ideal (or quasi-ideal) filter can be rephrased in terms of the two following objectives for transducer gain $T(\omega^2)$:

(a) The filter should have unit transducer gain over the passband(s), and zero gain over the stop band(s);

(b) As the frequency increases, each passband should abruptly change to the adjacent stopband, i.e., the transition regions should have zero bandwidth.

As discussed in Section 6.1, a physically realizable filter cannot satisfy either of the two above requirements. However, guided by the ideal prescription, more realistic specifications can be offered based on the following

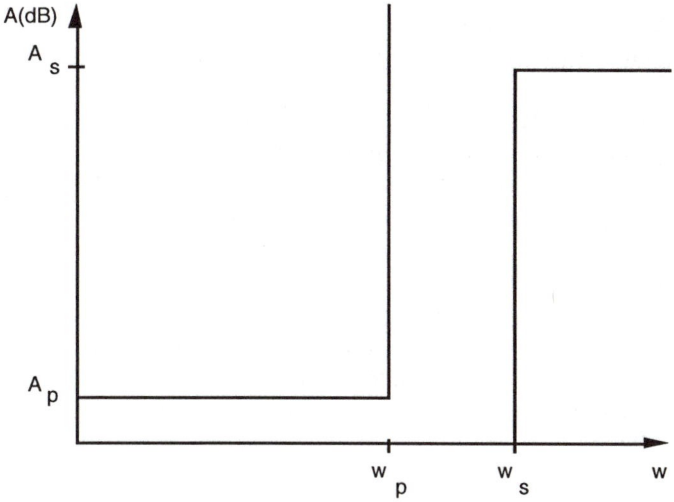

FIGURE 6.4.1
Example of amplitude specifications.

concepts which will be developed for a normalized lowpass prototype and then extended to any filter derived by the transformations described in the previous section.

We consider the *insertion loss*

$$A = -10 \log T(\omega^2) = 10 \log \frac{1}{|S_{21}|^2} \text{ dB}$$

The ideal passband gain is $T = 1$, or a loss of $A = 0$; the ideal stopband gain is $T = 0$, corresponding to a loss of $A = +\infty$. More realistically we take as tolerances

$$\text{(passband)} \;\; A \leq A_p, \quad \text{(stopband)} \;\; A \geq A_s$$

where A_p is a relatively small quantity (for example, of the order of 0.5 dB), and A_s is relatively large (say, for example, 40 dB).

We can also set tolerances for the bandwidth of the transition region. Let ω_p, ω_s correspond to A_p, A_s, respectively, and limit the passband to the interval $0 \leq \omega \leq \omega_p$ and the stopband to the interval $\omega_s \leq \omega < +\infty$, so that the transition region between the two is finite. The situation is depicted in Fig. 6.4.1. The normalization frequency, i.e., corresponding to unit frequency for the prototype, is generally chosen in the transition band, $\omega_p \leq \omega \leq \omega_s$, often at the edge of the passband, ω_p.

When a highpass, bandpass, or bandstop filter is obtained by transformation from a lowpass prototype, the prototype tolerances are maintained for the new frequency bands. Difficulties might arise if the transformation is to produce a multiband filter; in that case, the specifications might differ for the various bands; often then the most restrictive are imposed for all the bands at the price of a possible increase in the total number of filter elements.

The ratio ω_s/ω_p is called the *selectivity*; the ratio A_s/A_p is called the *discrimination*. The selectivity measures the frequency width of the transition region; the discrimination measures the ratio (in dB) of the rejection loss in the stopband to the loss in the passband.

The design procedures to be presented here stem from the Belevitch canonic scattering representation of a lossless two-port as given in eq. (5.4.16). Thus $S_{11} = h/g$, $S_{12} = S_{21} = f/g$ and $S_{22} = -(h_* f)/(g f_*) = \mp h_*/g$ where h, f, g, are polynomials in s. The transducer gain function defines the transmittance relation eq. (6.2.2), which, in turn, can be expressed in terms of the characteristic function ψ and leads to the evaluation of h, f, g. Thus,

$$S_{21*}S_{21} = \frac{f_* f}{f_* f + h_* h} = \frac{1}{1 + \psi_* \psi}, \qquad \psi_* \psi = \frac{h_* h}{f_* f}$$

The most often used approximations generally produce either symmetric, $S_{11} = S_{22}$, or antimetric filters, $S_{11} = -S_{22}$. (See Corollary 5.4.2.) In the case of a symmetric filter the characteristic function ψ is odd since we must have opposite parities for h and f to obtain $S_{11} = S_{22}$. (See definition of S_{22} above.) On the other hand, the parities of h and f must be the same to realize $S_{11} = -S_{22}$. Hence for a symmetric filter, $\psi(j\omega)$ is imaginary. In the antimetric case, $\psi(j\omega)$ is real. If the designs are confined to only symmetric or antimetric filters, it is convenient to express ψ in terms of the normalized function $\hat{\psi}$

$$w = \psi(j\omega) = \epsilon k \hat{\psi}(\omega) = \epsilon k w_1 \tag{6.4.1}$$

Here $\hat{\psi}$ is a real function of ω which is equal to 1 at the normalization frequency, ϵ is a real constant multiplier to set the specified gain level at the normalization frequency, and k is $\pm j$ for symmetric filters, ± 1 for antimetric filters. The sign is dictated by the requirement that $\hat{\psi}$ satisfy eq. (6.4.1) and equal $+1$ at the normalization frequency. The function $w_1 = \hat{\psi}(\omega)$ is rational. It may, of course, reduce to a polynomial, in which case the filter is said to be a *polynomial filter*.

It is often useful to express the function $w_1 = \hat{\psi}(\omega)$ in parametric form

$$\omega = \phi_2(z)$$

$$w_1 = \phi_1(z)$$

where z is a complex variable. Examples of this type of representation will be presented in the following sections. The reader should not be alarmed by the seemingly large number of definitions given above. In an actual example, the different relationships fall out naturally as the steps are carried out.

6.5 Butterworth Approximation

The Butterworth approximation for a lowpass prototype is defined by the following transducer power gain function which obviously satisfies the gain realizability Theorem 6.2.1, since $0 \leq T(\omega^2) \leq 1$,

$$T(\omega^2) = |S_{21}(j\omega)|^2 = \frac{1}{1 + \epsilon^2 \omega^{2n}}$$

where ϵ is the constant introduced in the previous section, eq. (6.4.1), to set the insertion loss at the normalization frequency, usually chosen at $\omega = 1$. The integer n is equal to the degree of the filter, and all n zeros of transmission occur at infinity, so that, as discussed in Section 5.14, the Butterworth filter is realizable as an n element lowpass ladder. Since $T(0) = 1$, the prototype filter is transparent at d.c. so that the load resistance must be equal to that of the generator, that is, they can both be set to 1. It is immediately seen that the characteristic function ψ, eq. (6.2.2), is given by

$$w = \psi(s) = \frac{h(s)}{f(s)} = \epsilon s^n$$

Thus the polynomial f is equal to ± 1 and the polynomial h is equal to $\pm \epsilon s^n$. We set $f = 1$, for since $S_{21} = f/h$, setting $f = -1$ only amounts to a reversal of polarity at the output port. Since f is an even function for all n, and h is even for n even and odd for n odd, the parities of f and h are the same for n even, opposite for n odd. Thus the filter is symmetric for n odd, antimetric for n even, Corollary 5.4.2. The function $\hat{\psi}$, according to eq. (6.4.1), is

$$w_1 = \hat{\psi}(\omega) = \omega^n$$

The insertion loss is given by

$$A = 10 \log(1 + \epsilon^2 \omega^{2n})$$

The curves $T = T(\omega^2)$ are shown in Fig. 6.5.1 for $\epsilon = 1$ and $n = 2, 5, 7$.

The Butterworth function approximation has its first $2n - 1$ derivatives equal to zero, both at the origin and at infinity (where the function itself is also zero). This is the maximum number of derivatives that can be set to zero for a rational function of degree $2n$ and, therefore, the Butterworth gain function is said to be *maximally flat* at zero and infinity. The maximally flat property is easily demonstrated by expanding function T in powers of ω^2 around the origin and in powers of $1/\omega^2$ around infinity. For convenience take $\epsilon = 1$, then

$$T(\omega^2) = 1 - \omega^{2n} + \omega^{4n} - \omega^{6n} + \ldots$$

$$T(\omega^2) = \frac{1}{\omega^{2n}} - \frac{1}{\omega^{4n}} + \frac{1}{\omega^{6n}} - \ldots$$

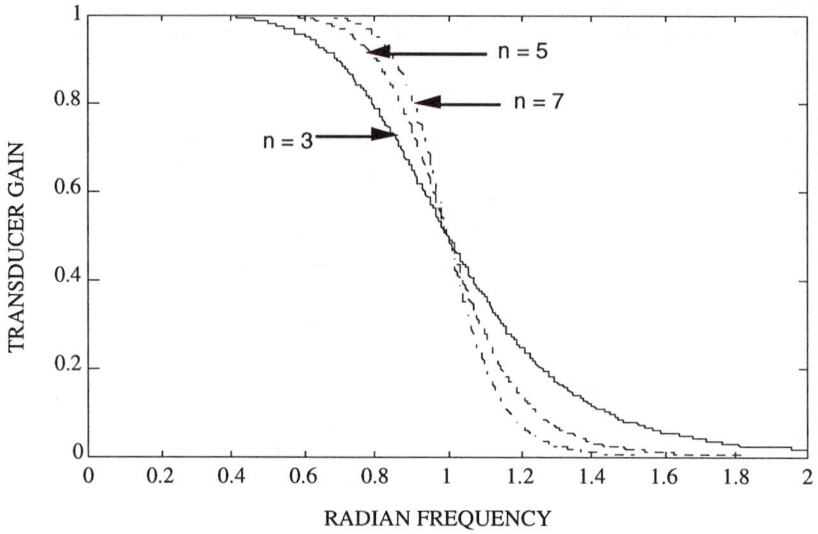

FIGURE 6.5.1
Butterworth's low-pass filter transducer power gain for $\epsilon = 1$ and $n = 3$ (–), 5 (– –), 7 (– ·).

The coefficients of $\omega^{\pm k}$ in the two series are the k-th derivatives at zero and infinity and, since they are zero for $k < 2n$, the maximally flat property follows.

Inspection of Fig. 6.5.1 indicates that the Butterworth response is not highly selective for a reasonable number of filter elements. Thus for $\epsilon = 1$, the insertion loss at $\omega = 1$ is

$$A = -10 \log T(1) = -10 \log \frac{1}{2} = 3.02\,\text{dB}$$

for any n. At, say, $\omega = .9$ in the passband and $n = 7$, the loss only drops to 0.89 dB, whereas at 10% above cutoff, the loss has only increased to 6.8 dB.

To further put the Butterworth response into perspective, consider the design requirements with specified filter tolerances. We require that $A(\omega_p) \leq A_p$ and $A(\omega_s) \geq A_s$. Choose as normalization frequency the edge of the passband so that $\omega_p = 1$. Then for prescribed insertion losses A_p, A_s at the edge of the pass and stop bands

$$\begin{aligned} 10 \log(1 + \epsilon^2) &\leq A_p \\ 10 \log(1 + \epsilon^2 \omega_s^{2n}) &\geq A_s \end{aligned} \tag{6.5.1}$$

We can determine ϵ from the first of eqs. (6.5.1). The second of eqs. (6.5.1) can then be used to calculate n as the lowest integer for which the inequality is satisfied.

We illustrate the procedure by choosing more or less average filter tolerances in the following example.

Example 6.5.1
Design a Butterworth filter with the following specifications:

$$f_p = 95\,\text{kHz} \qquad f_s = 105\,\text{kHz}$$

$$A_p = 0.5\,\text{dB} \qquad A_s = 40\ \text{dB}$$

Solution Normalizing to the frequency f_p at the edge of the passband, take $\omega_p = 1$, so that $\omega_s = f_s/f_p = 1.11$. Using eqs. (6.5.1), we have

$$\epsilon^2 = 10^{0.5/10} - 1 = 0.122$$

$$n = \frac{\log[(10^{40/10} - 1)/\epsilon^2]}{2\log 1.11} = 55$$

□

The large value of n obtained in Example 6.5.1 shows that although the Butterworth filter has a smooth monotone amplitude response, it has only a very modest selectivity and/or discrimination. This is mainly due to the fact that maximum flatness puts strong constraints at the origin and infinity but not in between. Nevertheless, the design of Butterworth filters illustrates in a clear and simple manner all the basic ideas of filtering, for instance, the approximation of the quasi-ideal filter by a sequence of finite filters of increasing degree.

The synthesis of the Butterworth transducer gain function ($\epsilon = 1$) with a lowpass prototype is carried out along lines earlier described in Chapter 5. We wish to compute the input immittance of the filter from $T(\omega^2)$ and then synthesize the prototype by the Darlington procedure, which, in this case, reduces to a Cauer lowpass ladder expansion, since all the transmission zeros are at infinity.

Based on the transducer gain, we have

$$ff_* = 1 \tag{6.5.2}$$

$$gg_* = 1 + (-s^2)^n \tag{6.5.3}$$

We have already chosen $f = 1$ for the factorization of eq. (6.5.2). To factor eq. (6.5.3), we observe that the roots of $gg_* = 0$ are the $2n$-th roots of -1

$$\begin{aligned} s_k &= j\exp j\tfrac{(2k+1)\pi}{2n} = \\ &= -\sin\frac{2k+1}{2n}\pi + j\cos\frac{2k+1}{2n}\pi \end{aligned} \tag{6.5.4}$$

Thus the zeros of $g_* g$ lie on a circle centered at the origin and of unit radius.

Since g must be Hurwitz, its factors are uniquely determined; only the roots whose real part is negative (in the LHP) may be chosen for g. These roots correspond to $k = 0, 1, 2, ...n-1$. Polynomial g is thus constructed as a product of known factors

$$g = \prod_{k=0}^{n-1} (s - s_k)$$

and finally expanded as

$$g = s^n + a_{n-1}s^{n-1} + a_{n-2}s^{n-2} + ... + a_1 s + a_0$$

where $a_0 = 1$ and $a_k = a_{n-k}$. The coefficients a_{n-k} are readily calculated and have been tabulated.[3] The two equations below express the coefficients, both recursively and directly.

$$\frac{a_{k+1}}{a_k} = \frac{\cos(k)\gamma}{\sin(k+1)\gamma}$$

$$a_k = \prod_{i=1}^{k} \frac{\cos(i-1)\gamma}{\sin i\gamma} \tag{6.5.5}$$

where

$$\gamma = \frac{\pi}{2n}$$

Eqs. (6.5.2) and (6.5.3) can be written as

$$S_{21*}S_{21} = \frac{ff_*}{gg_*} = \frac{1}{1 + (-s^2)^n}$$

and, therefore, from the paraunitary condition (see eq. (4.5.7)),

$$S_{11*}S_{11} = 1 - S_{21*}S_{21} = \frac{(-s^2)^n}{1 + (-s^2)^n} \tag{6.5.6}$$

Factorization of eq. (6.5.6) yields

$$S_{11} = \pm \frac{s^n}{g} \tag{6.5.7}$$

If the plus sign is used in eq. (6.5.7), then $S_{11} = 1$ at infinity corresponding to an input open circuit at infinite frequency. That is, the first element of the filter would be a series inductance. If the minus sign is chosen, then

[3] L. Weinberg, *Network Analysis and Synthesis*, New York, McGraw-Hill, 1962, p. 494.

at infinity $S_{11} = -1$ or the filter at its input terminals presents a short circuit at infinite frequency, so that the first element of the filter would be a shunt capacitor. From eq. (6.5.7) and the second of eqs. (5.4.19) the input impedance $z = (1+S_{11})/(1-S_{11})$ is deduced. As pointed out earlier, all n transmission zeros are at infinity so that the impedance is realizable by a lowpass ladder containing only series inductors and shunt capacitors, a total of n elements. The ladder can start with either an inductor or a capacitor depending on the sign chosen for S_{11} and the final termination is a unit resistor.

The element values can be obtained by using a lowpass Cauer continued fraction expansion, Section 5.3. However, the explicit values of the network elements are given by the formula[4]

$$g_r = 2\sin\frac{(2r-1)\pi}{2n} \qquad r = 1, 2, ..., n \tag{6.5.8}$$

where g_r is alternately equal to l_r and c_r, etc. and the first element is l_1 when the sign in eq. (6.5.7) is (+), or is c_1 if the sign is (−).

Example 6.5.2

To illustrate the procedure, consider the Butterworth function for $n = 2$, $\epsilon = 1$. (Despite its simplicity, the example points out all the salient facets of the design process.) Then $T(\omega^2) = 1/(1+\omega^4)$, and

$$S_{12*}S_{12} = \frac{ff_*}{gg_*} = \frac{1}{1+s^4}, \qquad S_{11*}S_{11} = \frac{hh_*}{gg_*} = \frac{s^4}{1+s^4}$$

The two roots of g_*g in the LHP are $-\exp(\pm j\pi/4)$ and, when we multiply the two root factors together, we obtain the Hurwitz polynomial $g = s^2 + \sqrt{2}\,s + 1$. Thus

$$\begin{aligned} S_{21} &= \frac{f}{g} = \frac{1}{s^2+\sqrt{2}\,s+1}, \\ S_{11} &= \frac{h}{g} = \pm\frac{s^2}{s^2+\sqrt{2}\,s+1}, \\ S_{22} &= \frac{-h_*}{g} = -S_{11} \end{aligned}$$

Choose the (+) sign for S_{11}. Then we immediately find z, and it is a trivial matter to carry out the Cauer expansion (note it is not purely reactive, there is a terminating unit resistor)

$$z = \frac{1+S_{11}}{1-S_{11}} = \frac{2s^2+\sqrt{2}\,s+1}{\sqrt{2}\,s+1} = \sqrt{2}\,s + \frac{1}{\sqrt{2}\,s+1}$$

[4] H. J. Orchard, "Formulae for Ladder Filters", *Wireless Engineer*, vol. 30, pp. 3-5, Jan. 1953.

As expected, the (+) sign for S_{11} leads to an input inductance, and $l_1 = \sqrt{2}$. The next element is a shunt capacitor $c_2 = \sqrt{2}$. The final element is a unit resistor in parallel with c_2. The element values can be verified using eq. (6.5.8). If the (−) sign is used, then $c_1 = \sqrt{2}$ in shunt at the input, followed by $l_2 = \sqrt{2}$ and unit resistance in series. As expected, since $S_{22} = -S_{11}$, the filter is antimetric. □

It is easily seen, eq. (6.3.3), that denormalization from the prototype to a lowpass filter with passband edge at f_p and a nonunit value of ϵ can be carried out according to the formulae

$$L_i = \frac{l_i R}{2\pi f_p \sqrt[n]{\epsilon}} \qquad C_i = \frac{c_j}{2\pi f_p R \sqrt[n]{\epsilon}}$$

6.6 Chebyshev Approximation

The Chebyshev approximation for the lowpass prototype transducer gain is defined by the following transducer power gain function

$$T(\omega^2) = |S_{21}(j\omega)|^2 = \frac{1}{1 + \epsilon^2 T_n^2(\omega)} \tag{6.6.1}$$

where $T_n(\omega)$ is the *Chebyshev polynomial* of degree n and ϵ is the weighting constant; once again, n is the degree of the filter.

The n-th degree Chebyshev polynomial is defined by the following equation:

$$T_n(\cos z) = \cos nz$$

where $z = \arccos \omega$ is a complex variable. The function $\cos nz$ is actually a polynomial in $\omega = \cos z$. This is readily seen using the trigonometric multiple angle cosine identities. For example,

$$\begin{aligned}
T_0 &= \cos 0 = &&= 1 \\
T_1 &= \cos z = &&= \omega \\
T_2 &= \cos 2z = 2\cos^2 z - 1 &&= 2\omega^2 - 1 \\
T_3 &= \cos 3z = 4\cos^3 z - 3\cos z &&= 4\omega^3 - 3\omega \\
T_4 &= \cos 4z = 8\cos^4 z - 8\cos^2 z + 1 &&= 8\omega^4 - 8\omega^2 + 1
\end{aligned}$$

Moreover, note that $-\omega = \cos(\pi - z)$, so that $T_n(-\omega) = \cos n(\pi - z) = \pm \cos nz$. The sign is (+) for n even, (−) for n odd. Therefore the Chebyshev polynomials are even or odd corresponding to n an even or odd integer. As a result we can write

$$|\psi(j\omega)|^2 = \epsilon^2 T_n^2(\omega)$$

since we have defined $|\psi(j\omega)|^2$ as an even, nonnegative polynomial. Therefore the Chebyshev transducer gain is realizable; see Theorem 6.2.1.

A useful recursion formula for the Chebyshev polynomials is easily deduced. We have the trigonometric identity $\cos[(n \pm 1)z] = \cos nz \cos z \mp \sin nz \sin z$. Add the two relations defined by the two choices of sign. The sin terms cancel. Identify $T_k = \cos kz$, $\omega = \cos z$, and the result is

$$T_{n+1}(\omega) = 2\omega T_n(\omega) - T_{n-1}(\omega)$$

As a consequence of eq. (6.6.1), when we omit the the amplitude scale factor, ϵ, we have the normalized function $w_1 = \hat{\psi}(\omega)$, as defined in eq. (6.4.1) Thus

$$w_1 = \hat{\psi}(\omega) = T_n(\omega)$$

or in parametric form

$$\begin{aligned} w_1 &= \cos nz \\ \omega &= \cos z \end{aligned} \tag{6.6.2}$$

Eqs. (6.6.2) have a simple geometrical interpretation in the complex z plane. Consider the oriented path consisting first of $\mathrm{C_p}$, the real axis segment $[\pi/2,\ 0]$. $\mathrm{C_p}$ is followed by $\mathrm{C_s}$ consisting of the positive imaginary axis with end points $[0,\ +j\infty)$.

How does ω behave on the path which is the union of $\mathrm{C_p}$ and $\mathrm{C_s}$? Let be $z = x+jy$. The second of eqs. (6.6.2) shows that along the real axis segment, $\mathrm{C_p}$, $\omega = \cos x$ and therefore ω varies from 0 to 1 (though not linearly with x), while along the positive imaginary axis, we have $\omega = \cos jy = \cosh y$ and therefore ω varies from 1 to $+\infty$ along $\mathrm{C_s}$. Take $\omega = 1$ as the normalized edge of the passband. Then $\mathrm{C_p}$ in the z plane maps onto the passband in ω. The imaginary portion of the z axis path, $\mathrm{C_s}$, maps onto the transition and stopband in positive ω.

To complete the mapping we observe that along the path $\mathrm{C_s'}$ in z, consisting of the line parallel to jy from $\pi + j\infty$ to $\pi + j0$, the mapping goes into ω from $-\infty$ to -1. The path $\mathrm{C_p'}$ along x from π to $\pi/2$ corresponds to the ω path from -1 to 0. Thus the union of $\mathrm{C_s'}$ and $\mathrm{C_p'}$ maps into the stop, transition, and pass bands in $-\omega$.

The Theorem summarizes the complete mapping by describing the overall transformation in terms of four separate submapping statements.

THEOREM 6.6.1

(a) *The mapping $\omega = \cos z$ of eq. (6.6.2) takes the semiinfinite rectangular boundary $C \equiv (\pi + j\infty,\ \pi + j0,\ 0 + j0,\ 0 + j\infty)$, onto the ω axis $(-\infty,\ -1,\ 1,\ \infty)$.*

(b) *The semiinfinite strip, R in z, enclosed by C, corresponds to the lower half of the ω plane.*

(c) *The semiinfinite strip R transforms into the right half of the complex $s = j\omega$ plane, and the infinite strip R′ extending from $j\infty$ to $-j\infty$ in z maps onto the entire s plane.*

(d) *We can adjoin R′ with an infinite number of similar strips, all of width π. Each of these strips in z repeats the mapping onto the s plane. The transformation from z to s is termed* simply periodic.

We now turn our attention to the first of eqs. (6.6.2) to see how T_n behaves. On the positive passband path C_p, nx has a net excursion of $n\pi/2$, therefore $w_1(x) = \cos nx$ oscillates between -1 and 1, exhibiting n quarter periods over the positive passband. At the edge of the passband $x = 0$, $\omega = 1$, $T_n(1) = 1$, and $T_n(-1) = \pm 1$, depending on whether n is even or odd. As a further consequence of the parity of $T_n(\omega)$, the Chebyshev polynomials have even or odd symmetry about $\omega = 0$ and, therefore, for the entire passband, $-1 \leq \omega \leq 1$, there are $n/2$ oscillations between the values -1 and 1, with n real roots. Thus *all* the roots of $T_n(\omega)$, which is a polynomial of degree n, are real and all are located between $-1 < \omega < 1$.

The stopband corresponds to the path C_s in the z plane, and here we have

$$\begin{aligned} w_1 &= \cos jny = \cosh ny = \frac{1}{2}(\epsilon^{ny} + \epsilon^{-ny}) \\ \omega &= \cos jy = \cosh y = \frac{1}{2}(\epsilon^{y} + \epsilon^{-y}) \end{aligned} \tag{6.6.3}$$

To calculate the asymptotic behavior of $w_1 = T_n(\omega)$ in the stopband we assume $y \gg 0$ and neglect $\exp(-y)$, and $\exp(-ny)$. Then substitute 2ω into the first of the above equations and obtain

$$w_1 = 2^{n-1}\omega^n \tag{6.6.4}$$

Based on the preceding discussion, we see that in the passband, $-1 \leq \omega \leq 1$, $|\psi(j\omega)|^2 = \epsilon^2 T_n^2(\omega)$ oscillates between 0 and ϵ^2, and has n zeros, all of multiplicity two. Thus over the complete passband, the transducer gain $T(\omega^2)$, eq. (6.6.1), ripples between n peaks of unit gain (at the zeros of $|\psi|^2$), and minima all equal to $1/(1+\epsilon^2)$. The filter is, in fact, termed *an equal ripple filter.* In the stopband $|\psi|^2$ increases monotonically without limit. Correspondingly, the stopband gain falls off monotonically from the passband values and approaches zero, $\omega \to \infty$.

The Chebyshev polynomials have special properties which are important in the theory of approximation and make them optimum for the transducer gain of polynomial filters. Define the monic polynomial $t_n(\omega) = (2^{n-1})^{-1}T_n(\omega)$, with the coefficient of ω^n equal to unity, see eq. (6.6.4).

Then t_n is that monic polynomial among all monic polynomials of degree n, whose maximum deviation from zero in the interval $-1 \leq \omega \leq 1$, is minimum (minimax property). Evidently the minimax deviation is $(2^{n-1})^{-1}$.

The minimax property can be phrased more directly in the context of the characteristic function and transducer gain.

THEOREM 6.6.2

(Capture Property) *Consider the even positive polynomial, of degree $2n$, $W(\omega^2) = |\psi(j\omega|^2 = \epsilon^2 T_n^2(\omega)$ whose maximum tolerance inside the passband, $-1 \leq \omega \leq 1$, is ϵ^2. Then no other positive even polynomial of degree $2n$, which exceeds W anywhere outside the band $-1 \leq \omega \leq 1$, has a tolerance less than ϵ^2 inside the band.*

PROOF The polynomial W is determined by its $n+1$ coefficients and, as discussed above, it has n zeros of multiplicity 2 in $-1 \leq \omega \leq 1$. Let P_{2n} be the competitor polynomial whose tolerance is less than ϵ^2 in the passband. Then, because of the n roots and the equal ripple behavior of W, between zero and ϵ^2, a plot of P_{2n} intersects the W curve $2n$ times in $-1 \leq \omega \leq 1$. P_{2n} is equal to, or interpolates to, W at these crossing points. This is the *capture property* of the Chebyshev polynomials. Only n of the interpolations are independent since the polynomials are even. If P_{2n} exceeds W outside the passband there must be at least one more independent point of interpolation. The number of interpolation points will thus be $n+1$, the number of degrees of freedom which defines W, so that $P_{2n} = W$ and the theorem is proved. ∎

In effect, for a given discrimination, the Chebyshev filter has the highest selectivity of any polynomial filter of the same degree.

The value of $T_n^2(\omega) = \cos^2 nz$ at $\omega = 0$, determines whether or not the Chebyshev filter can be terminated with equal resistances. At d.c. we have $nz = n\pi/2$, therefore if n is even, $T_n^2(0) = 1$. For n odd, $T_n^2(0) = 0$. Therefore, referring to eq. (6.6.1) for n odd $|S_{21}(0)|^2 = 1$; for n even, $|S_{21}(0)|^2 = 1/(1+\epsilon^2)$. At d.c. the filter is transparent, and the input resistance is the normalized load value $r_2 = [1 + S_{11}(0)]/[1 - S_{11}(0)]$. Furthermore, $|S_{11}|^2 = 1 - |S_{21}|^2$, so for n odd, $S_{11}(0) = 0$; for n even $S_{11}^2(0) = \epsilon^2/(1+\epsilon)^2$. We therefore have the normalized load values as

$$(n = 2k+1): \ S_{11}(0) = 0, \ r_2 = 1$$

$$(n = 2k): \ S_{11}(0) = \pm\frac{\epsilon}{\sqrt{1+\epsilon^2}}, \quad r_2 = \frac{\sqrt{1+\epsilon^2} \pm \epsilon}{\sqrt{1+\epsilon^2} \mp \epsilon} \tag{6.6.5}$$

The filter is therefore matched at d.c. if n is odd, mismatched if n is even.

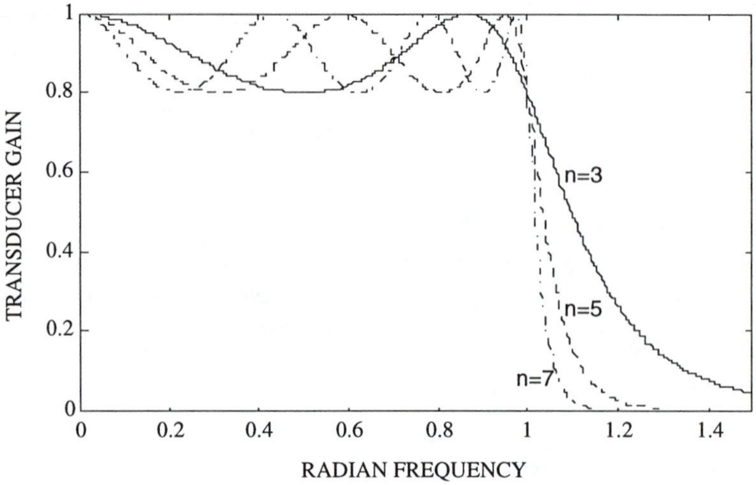

FIGURE 6.6.1
Chebyshev lowpass filter transducer power gain for $\epsilon = 0.5$ and n = 3 (–), 5 (– –), 7 (– ·).

The curves $T = T(\omega^2)$ are shown in Fig. 6.6.1 for $\epsilon = 0.5$ and $n = 2, 5, 7$. Because of the equal ripple property, the amplitude of the oscillations in the passband is constant. However, their frequency increases towards the upper edges of the band.

The specifications require that $A(\omega_p) \leq A_p$ and $A(\omega_s) \geq A_s$. We choose as the normalization frequency the edge of the passband so that $\omega_p = 1$. Then we must satisfy the inequalities

$$\begin{gathered} 10\log(1+\epsilon^2) \leq A_p \\ 10\log[1+\epsilon^2 T_n^2(\omega_s)] \geq A_s \end{gathered} \tag{6.6.6}$$

The first of eqs. (6.6.6) gives the value of ϵ. From the second equation we can calculate n as the lowest integer for which the the second inequality is satisfied. Note that in the stopband $z = jy$, $y = \operatorname{arcosh}\omega$ as in eq. (6.6.3), so that $T_n(\omega) = \cos nz = \cosh ny$. Referring to the transducer gain, we have $1/T(\omega_s) = 1 + \epsilon^2 \cosh^2 ny_s$, or

$$\begin{gathered} ny_s = \operatorname{arcosh}\frac{1}{\epsilon}\sqrt{\frac{1}{T(\omega_s)} - 1} \\ y_s = \operatorname{arcosh}\omega_s \end{gathered} \tag{6.6.7}$$

We illustrate the procedure by the following.

Example 6.6.1
Design a Chebyshev filter with the following specifications:

$$f_p = 95\,\text{kHz} \qquad f_s = 105\,\text{kHz}$$

$$A_p = 0.5\,\text{dB} \qquad A_s = 40\,\text{dB}$$

Solution Normalizing to frequency at the the edge of the passband we obtain $\omega_p = 1$ and $\omega_s = 105/95 = 1.11$. Corresponding to a maximum pass band loss of 0.5 dB, we have $10\log(1+\epsilon^2) = 0.5$ or

$$\epsilon^2 = 10^{.05} - 1 = 0.122$$

Also $10\log(1/T_s) = 40\,\text{dB}$, so $(1/T_s) = 10^4$. Using eq. (6.6.7)

$$ny = \text{arcosh}\,\frac{1}{0.349}\sqrt{10^4 - 1} = 6.35, \quad y = \text{arcosh}\,1.11 = 0.465, \quad n = 14$$

□

The degree of a Butterworth filter having the same amplitude specifications as the foregoing Chebyshev filter was n=55, (Example 6.5.1), a convincing demonstration of the greater efficiency of the equal ripple over the maximally flat design.

The synthesis of the Chebyshev low-pass prototype is carried out along the same lines earlier described for the Butterworth filter. Both follow the design procedures for a two-port, all of whose zeros of transmission are at infinity. The transmittance S_{21} and reflectance S_{11} must be calculated from the transducer gain by spectral factorization. Thus

$$f_* f = 1 \tag{6.6.8}$$

$$g_* g = 1 + \epsilon^2 T_n^2(s/j) = 1 + \epsilon^2 \cos^2 nz \tag{6.6.9}$$

Factorization of eq. (6.6.8) yields $f = \pm 1$; since the choice of minus sign would simply involve a phase reversal, we choose $f = 1$. To determine g we set $g_* g = 0$ in eq. (6.6.9) and compute the roots in the complex s plane corresponding to $nz = (\alpha + j\beta)$

$$\pm j\frac{1}{\epsilon} = \cos nz = \cos(\alpha + j\beta)$$

$$s = j\cos z$$

Thus

$$\pm j\frac{1}{\epsilon} = \cos(\alpha + j\beta) = \cos\alpha\cosh\beta + j\sin\alpha\sinh\beta$$

Equating real and imaginary parts, we obtain

$$\alpha = (k + \tfrac{1}{2})\pi, \quad \sin\alpha = \pm 1$$

$$\beta = \pm\,\text{arsinh}\,\frac{1}{\epsilon}$$

Thus we have

$$z_k = \frac{\alpha + j\beta}{n}$$

and finally all the roots are obtained from $s_k = j \cos z_k$

$$s_k = -\sin\frac{(2k+1)\pi}{2n}\sinh\frac{\beta}{n} + j\cos\frac{(2k+1)\pi}{2n}\cosh\frac{\beta}{n} \tag{6.6.10}$$

where the integer k varies from 0 to $2n-1$. If we write $s_k = \sigma_k + j\omega_k$, then referring to eq. (6.6.10), we get

$$\frac{\sigma_k^2}{\sinh^2\beta/n} + \frac{\omega_k^2}{\cosh^2\beta/n} = \sin^2\alpha/n + \cos^2\alpha/n = 1$$

Thus the zeros of g_*g lie on an ellipse, centered at the origin, whose axes are $a = \sinh\beta/n$ and $b = \cosh\beta/n$. The corresponding roots for the Butterworth filter of degree n are located on a circle at $p_k = -\sin(\alpha/n) + j\cos(\alpha/n)$, eq. (6.5.4). Therefore the Chebyshev roots can be obtained by multiplying the real part of each Butterworth root by a, and the imaginary part by b.

The roots factors of g are uniquely determined since it must be a Hurwitz polynomial. Therefore only those roots of eq. (6.6.10) are selected whose real parts are negative. That is, we choose $k = 0, 1, 2, ...n-1$ and so construct the polynomial g as a product of factors

$$g = \prod_{k=0}^{n-1}(s - s_k)$$

which can be expanded as

$$g = s^n + a_1 s^{n-1} + a_2 s^{n-2} + ... + a_{n-1}s + a_n$$

The coefficients a_{n-k} have been tabulated[5] for various values of ϵ, but it is probably more convenient to calculate them directly by a computer program. From eqs. (6.6.8) and (6.6.9) we find

$$S_{21*}S_{21} = \frac{1}{1 + \epsilon^2 T_n^2(s/j)}$$

and, therefore, applying the paraunitary condition

$$S_{11*}S_{11} = \frac{\epsilon^2 T_n^2(\omega)}{1 + \epsilon^2 T_n^2(\omega)} \tag{6.6.11}$$

[5]L. Weinberg, *Network Analysis and Synthesis*, New York, McGraw-Hill, 1962, p. 516 and ff.

Factorization of eq. (6.6.11) yields

$$S_{11} = \pm \frac{j\epsilon T_n(s/j)}{g} \tag{6.6.12}$$

The choice of the plus or minus sign in eq. (6.6.12) is treated as in the Butterworth filter. At infinity $S_{11} = \pm 1$ and, with the (+) sign, the input impedance is an open circuit, so the first element is a series inductor. With the (−) sign, the first element must be a shunt capacitor. From eq. (6.6.12) and the first of eqs. (5.4.19) the input impedance z is deduced. Since all n transmission zeros are at the infinity, the impedance is realizable by a ladder containing only series inductors and shunt capacitors for a total of n elements. The ladder starts with either an inductor or a capacitor, depending on the sign chosen for S_{11}, and terminates in a resistance whose value is given by eq. (6.6.5).

The values of the network elements are given by the recursion formulae[6]

$$g_1 = \frac{2\sin(\pi/2n)}{\sinh[(1/n)\,\text{arsinh}(1/\epsilon)]}$$

$$g_r g_{r+1} = \frac{4\sin[(2r-1)\pi/2n]\sin[(2r+1)\pi/2n]}{\sinh^2[(1/n)\,\text{arsinh}(1/\epsilon)] + \sin^2(r\pi/n)}$$

where $r = 1, 2, 3, \ldots, n-1$, g_r is alternately equal to l_r, c_r, etc., and g_1 is either a series inductance or a shunt capacitor depending on the sign of S_{11}.

It is easily seen that denormalization from the prototype to a filter with passband edge at f_p can be carried out according to the formulae

$$L_i = \frac{l_i R}{2\pi f_p \sqrt[n]{\epsilon}} \qquad C_i = \frac{c_j}{2\pi f_p R \sqrt[n]{\epsilon}}$$

We finally note that, when $\epsilon = 1$, the minima of transducer gain ripples over the passband and at the band edge are equal to 1/2 (3 dB of attenuation). This is the minimum passband gain (it occurs at the band edge) for the Butterworth filter as well.

The Chebyshev gain function resembles the Butterworth gain in that it rolls off monotonically outside the passband. It also possible to have an equal ripple lowpass transducer gain function which resembles the Butterworth function in the passband rather than in the stopband.

[6] J. O. Scanlan and R. Levy, *Circuit Theory*, vol. 2, Edinburgh, Oliver and Boyd, 1973, p. 511.

Example 6.6.2

Construct a transducer gain function which, like the Butterworth function, has monotonic behavior in the passband, but has an equal ripple response in the stopband.

Solution Consider the function $S(\omega^2) = 1 - T(\omega^2)$, where T is the Chebyshev transducer gain function. The function S is a highpass characteristic, for it oscillates between 0 and ϵ^2 for $-1 \leq \omega \leq 1$, and approaches unity monotonically as $\omega \to \infty$. To transform it to a lowpass characteristic we simply make the mapping $\omega \to 1/\omega$. The result is

$$\hat{T}(\omega^2) = S(1/\omega^2) = \frac{\epsilon^2 T_n(1/\omega^2)}{1 + \epsilon^2 T_n(1/\omega^2)}$$

□

Like the Butterworth gain function, $\hat{T}$ is maximally flat at the origin. and the gain is unity at that point. The equal ripple behavior between 0 and ϵ^2 stretches out in the stopband to infinity. The zeros of transmission are distributed over the stopband; indeed, for n even there is no zero of transmission at infinity. As a result the filter configuration is no longer a lowpass LC ladder. We will encounter a similar situation in the design of elliptic filters which is discussed in the next section.

6.7 Elliptic Approximation

The elliptic approximation realizes an equiripple gain function in both the passband and the stopband. The transducer gain is written

$$T(\omega^2) = |S_{21}|^2 = \frac{1}{1 + \gamma^2 |\hat{\psi}|^2}$$

The function $w_1 = \hat{\psi}(\omega)$ has zeros in the passband (unit gain points), and poles in the stopband (zeros of transmission) located so as to achieve the required equal ripple performance in both pass and stop bands, and is represented in the following parametric form

$$\begin{aligned} w_1 &= \mathrm{Sn}(u, z) \\ \omega &= \mathrm{Sn}(mu, z) \end{aligned} \qquad (6.7.1)$$

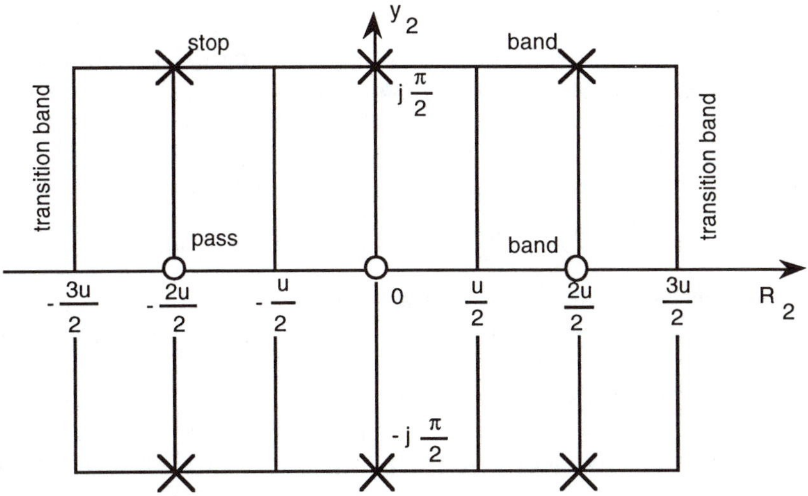

FIGURE 6.7.1
Pole-zero configuration of Sn(u, z).

Here $\mathrm{Sn}(u, z)$ is a *modified elliptic sine*[7] function (i.e., doubly periodic and meromorphic) and m is odd. The odd assumption is not strictly necessary but introduces simplifications in the design, and we will retain it for the remainder of this section, unless otherwise indicated. The choice of m even also introduces the minor inconvenience that the origin in z does not map into the origin of w_1 or ω; in any event, it can be treated analogously to the odd case after replacing z with $z - u/2$ in the first of eqs. (6.7.1).

Before examining the Sn function, it is instructive to review the simpler cosine function used for the Chebyshev transducer gain, $w_1' = \cos z'$, where $z' = nz$, Section 6.6. The pattern of critical points in the z' plane consists of an infinite string of simple zeros at $\pi/2 \pm k\pi/2$, k an odd integer. Thus w_1' has the single period $u = 2\pi$, $(\cos(z' + ku) = \cos z')$ and each cycle is twice the distance between adjacent zeros, the width of a fundamental strip, call it $\mathcal{R}$. The period of $\omega = \cos z = \cos(z'/n)$, is nu, corresponding to a fundamental strip $\mathcal{R}'$, with a width (the passband) of precisely n fundamental $\mathcal{R}$ strips. The mapping along the real axis of $\mathcal{R}'$ is periodic resulting in equal ripple passband behavior. Along the edge of $\mathcal{R}'$, parallel to the imaginary axis, the response increases monotonically producing the Chebyshev stopband behavior.

The pattern of critical points for $\mathrm{Sn}(u, z)$, Fig. 6.7.1, consists of infinite strings of both simple zeros and poles arranged parallel to the real axis. The x separation between adjacent zeros is u, as is the distance between

[7] P. Amstutz, "Elliptic Approximation and Elliptic Filter Design on Small Computers", *IEEE Trans. on Circ. and Sys.*, vol. CAS-25, no. 12, pp. 1001-1011, Dec. 1978.

adjacent poles, and there is a zero at the origin, so that the Sn function is odd and otherwise resembles the sine function along the x axis. Based on the symmetry of the pattern the horizontal period is $2u$, twice the distance between two adjacent zeros or poles. The vertical spacing between each adjacent parallel string of zeros and poles is $\pi/2$, so that the symmetric zero-pole pattern also produces a vertical periodicity of $j\pi$.

$$\mathrm{Sn}(u, z + 2u) = \mathrm{Sn}(u, z), \quad \mathrm{Sn}(u, z + j\pi) = \mathrm{Sn}(u, z)$$

The Sn function is thus *doubly periodic*, and the half period parallel to the real axis is noted as the first entry of the function's argument. Now consider $\omega = \mathrm{Sn}(mu, z)$, with half a fundamental rectangle (the other half is symmetrically located below the real axis), designated as $\mathcal{S}'$. The four corners of the rectangle are $(-mu/2 + j\pi/2), (-mu/2 + j0), (mu/2 + j0), (mu/2 + j\pi/2)$. Let $\mathcal{S}$ denote half a fundamental rectangle for $w_1 = \mathrm{Sn}(u, z)$; then precisely m of the $\mathcal{S}$ rectangles cover $\mathcal{S}'$. Qualitatively as we traverse the periphery of $\mathcal{S}'$, ω varies from $-\infty$ to $+\infty$, and correspondingly $w_1 = \mathrm{Sn}(u, z)$ has m zeros in the passband (including the zero at the origin since m is odd) and m poles in the stopband. In the case of the Chebyshev filter, the mapping includes n zeros on the real axis portion of the semi-infinite rectangle and as a result $T_n(\omega) = \cos n \arccos \omega$ is a polynomial. In the elliptic function case, as we have seen, the function $w_1(\omega) = \mathrm{Sn}[u, z(\omega)]$ includes m zeros and m poles, and therefore $w_1(\omega)$ is a rational function. The symmetry of the zero-pole pattern insures equal ripple behavior for w_1 with $m + 1$ maximum amplitudes each occurring midway between zeros, and $m + 1$ minimum amplitudes each occurring midway between poles. Also note that the rectangle $\mathcal{S}'$ for $\mathrm{Sn}(mu, z)$ maps into the upper half of the ω plane, and each similar rectangle repeats the mapping. An advantage for having chosen m odd is immediately apparent; the mapping has a zero at the origin so that the transducer gain will permit equal resistive terminations. Furthermore, there is a pole at infinity corresponding to a zero of transmission at that point. This simplifies the physical structure of the lowpass filter, since coupled coils *must* be employed when there is no zero of transmission at infinity.

The meromorphic function, $\mathrm{Sn}(u, z)$, is completely defined to within a multiplicative constant by its zeros and poles. The scaling factor can be chosen so that the frequency in the middle of the transition region, i.e., at $z = mu/2 + j\pi/4$, is unity. We can therefore set the scaling so that

$$\mathrm{Sn}(mu, mu/2 + j\pi/4) = 1, \text{ or } \mathrm{Sn}(u, u/2 + j\pi/4) = 1$$

Based on the symmetries of the zero-pole pattern, we can summarize some of the important properties of $w_1 = \mathrm{Sn}(u, z)$ and $\omega = \mathrm{Sn}(mu, z)$:

1. $\mathrm{Sn}(u, 0) = 0$, $\mathrm{Sn}(u, -z) = -\mathrm{Sn}(u, z)$

2. The zeros of w_1 occur at $z_i = lu$; the poles at $z_j = lu + j\pi/2$, $l = 0, \pm 1, \pm 2, \ldots, \pm(m-1)/2$. The zeros and poles are at frequencies

$$\text{(zeros)} \quad \omega_i = \mathrm{Sn}(mu, lu), \qquad \text{(poles)} \quad \omega_j = \mathrm{Sn}(mu, lu + j\pi/2)$$

3. The frequencies at the edges of passband and stopband are

$$\omega_p = \mathrm{Sn}(mu, mu/2), \quad \omega_s = \mathrm{Sn}(mu, mu/2 + j\pi/2)$$

   with corresponding values of $|w_1|$

$$|w_{1p}| = \mathrm{Sn}(u, u/2), \quad |w_{1s}| = \mathrm{Sn}(u, u/2 + j\pi/2)$$

4. The frequency at the center of the transition region is

$$\omega_0 = \mathrm{Sn}(mu, mu/2 + j\pi/4) = 1$$

   and the corresponding magnitude

$$|w_1| = |\,\mathrm{Sn}(u, mu/2 + j\pi/4)| = \mathrm{Sn}(u, u/2 + j\pi/4) = 1$$

Guided by the properties listed above, an expression for $\mathrm{Sn}(u, z)$ can be constructed in terms of elementary functions. We need to find a representation that has the same zero-pole pattern and the same scaling. We base the choice on $\tanh z$ since it is a meromorphic function with a string of alternating simple zeros and poles on $z = jy$. To complete the zero-pole pattern we multiply by $\tanh(lu \pm z)$, $l = 1, 2, \ldots, \infty$, which introduces further zero-pole strings parallel to the y axis and intersecting the real axis at $\pm u, \pm 2u, \ldots$ The result is

$$\mathrm{Sn}(u, z) = \tanh z \prod_{l=1}^{+\infty} \tanh(lu - z)\tanh(lu + z) \tag{6.7.2}$$

The function defined above has the requisite zero-pole pattern. We need only verify the scaling at $z = u/2 + j\pi/4$ to insure the validity of the representation. With $z = u/2 + j\pi/4$ the first two terms of the product eq. (6.7.2) are $\tanh(z)\tanh(u - z)$, which is equal to

$$\tanh(u/2 + j\pi/4)\tanh(u/2 - j\pi/4) = \frac{1 + j\epsilon^{-2a_k}}{1 - j\epsilon^{-2a_k}}\,\frac{1 - j\epsilon^{-2a_k}}{1 + j\epsilon^{-2a_k}} = 1$$

The next two terms are

$$\tanh(u + z)\tanh(2u - z) = \tanh(3u/2 + j\pi/4)\tanh(3u/2 - j\pi/4) = 1$$

Similarly each successive pair is unity, so that the scaling is correct and the infinite product representation of $\mathrm{Sn}(u, z)$ is verified.

We can infer a number of useful properties of the Sn function by inspection of eq. (6.7.2):

1. Since $\tanh(p \pm j\pi/2) = 1/\tanh p$, we have

$$\mathrm{Sn}(u, z + j\pi/2) = \frac{1}{\mathrm{Sn}(u, z)} \tag{6.7.3}$$

2. Since $\tanh x$ is real, so, too, is $\mathrm{Sn}(u, x)$, and using eq. (6.7.3), $\mathrm{Sn}(u, x + j\pi/2)$ is real.

3. In the transition region $z = u/2 + jy$, so that taking pairs of tanh functions in eq. (6.7.3), as we did to verify scaling, gives a sequence of products consisting of a function and its conjugate. Therefore ω and $w_1(\omega)$ as defined by the parametric eqs. (6.7.1) are real in the transition region and by the previous Item, the mapping ω and $w_1(\omega)$ is real over the entire pass, stop, and transition regions.

4. Consider any frequency in the passband, ω_{po}, and the stopband frequency $\omega_{so} = 1/\omega_{po}$. Then

$$\omega_{po} = \mathrm{Sn}(mu, x), \quad \omega_{so} = \mathrm{Sn}(mu, x + j\pi/2)$$

and the associated responses satisfy

$$\hat{\psi}(\omega_{po}) = \frac{1}{\hat{\psi}(\omega_{so})}$$

5. In particular at the edges of the pass- and stopbands

$$\omega_p = \mathrm{Sn}(mu, mu/2) = \frac{1}{\omega_s}, \quad \hat{\psi}(\omega_p) = \mathrm{Sn}(u, u/2) = \frac{1}{\hat{\psi}(\omega_s)}$$

6. Let ω_i (unit passband gain points), ω_j (zeros of transmission) correspond, respectively, to the frequencies at which the zeros and poles occur in the z plane. Then referring to Item 2 of the earlier discussion of Sn properties

$$\omega_i = \mathrm{Sn}(mu, lu), \quad \omega_j = \frac{1}{\omega_i} \quad (l = 0, \pm 1, \pm 2, \ldots \pm \frac{m-1}{2}) \tag{6.7.4}$$

The infinite product of eq. (6.7.2) converges rapidly so it is useful for practical calculations. One approach to preparing a program is to note that the sequence can be terminated when the l_o-th factor is within a preassigned deviation, δ, from unity. The program then terminates at term l_o. If we expand $\mathrm{Sn}(au, bu)$, then

$$l_o = \frac{1}{2ua} \operatorname{arcosh}\left(\cosh \frac{2bu}{\delta}\right)$$

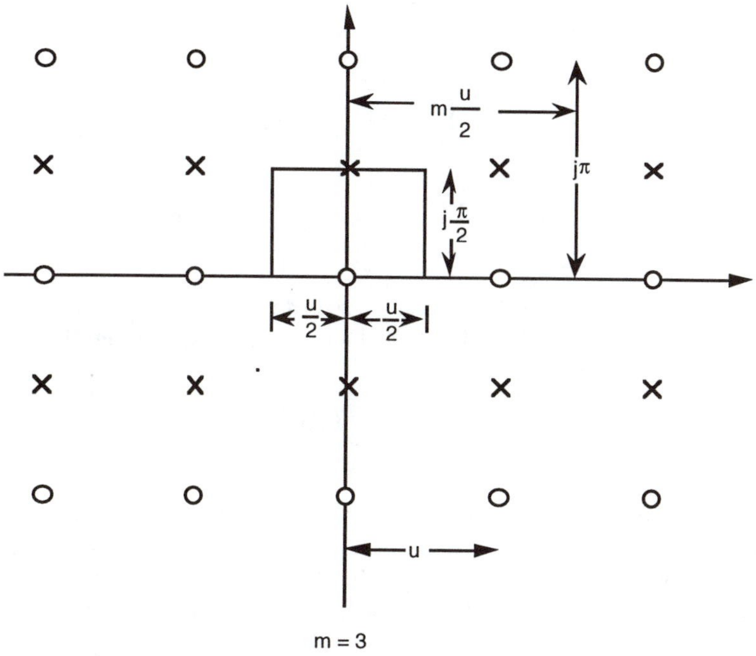

FIGURE 6.7.2
The functions $\omega = Sn(mu, z)$ and $w_1 = Sn(u, z)$ for m=3.

For example, if $u = .95$, $a = 3$, $b = 1$, $\delta = 10^{-6}$, then Sn $= 0.707$ requiring 3 terms. If $a = 1$, $b = .5$, then Sn $= .15$, $l_o = 8$.

The elliptic approximation realizes an equiripple gain function in both the passband and the stopband. It is optimum in the sense that among all rational, realizable functions (even, nonnegative, bounded by unity) of a given degree m, the elliptic gain function $T(\omega^2)$ has simultaneously the minimax deviation from unity in a prescribed passband (band edge ω_p), and the minimax deviation from zero in the stopband (band edge $1/\omega_p$). The result is demonstrated by using the "capture" argument employed for Chebyshev gain functions, Theorem 6.6.2.

All the information for constructing the final gain function with equal ripple passband and stopband behavior is now in place, and typical curves illustrating the parametric representation are shown in Fig. 6.7.2. The rational expression for $\hat{\psi}(\omega)$ for m odd is constructed as below. We also include the result for m even.

$$\hat{\psi}(\omega) = \omega \prod_{i=1}^{(m-1)/2} \frac{\omega_i^2 - \omega^2}{1 - \omega_i^2\omega^2} \qquad m \quad \text{odd}$$
$$\hat{\psi}(\omega) = \prod_{i=1}^{m/2} \frac{\omega_i^2 - \omega^2}{1 - \omega_i^2\omega^2} \qquad m \quad \text{even} \tag{6.7.5}$$

Referring to the odd case, it can be seen that $\hat{\psi}(\omega)$ has zeros at $\omega_0 = 0$, as well as at the ω_i; the poles are at $\omega_j = 1/\omega_i$, including the point at infinity. The rational function

$$\gamma^2 |\hat{\psi}(\omega)|^2 = \frac{h(j\omega)h(-j\omega)}{f(j\omega)f(-j\omega)}$$

and by spectral factorization we then determine $h(s)$, $g(s)$. Similar results hold for the even case, but then there is no zero at the origin or pole at infinity. Note that the polynomial f is always even, while the polynomial h is odd if m is odd, even if m is even. Thus the filters with m odd are symmetric, the filters with m even are antimetric.

The design of elliptic filters is carried out based on the following inequalities

$$A_p \leq 10\log[1 + \gamma^2 \operatorname{Sn}^2(u, u/2)]$$
$$A_s \geq 10\log[1 + \gamma^2 \operatorname{Sn}^2(u, u/2 + j\pi/2)]$$
$$\omega_p = \operatorname{Sn}(mu, mu/2) \tag{6.7.6}$$
$$\omega_s = \operatorname{Sn}(mu, mu/2 + j\pi/2)$$

The steps of the procedure are the following:

1. In the first two of eqs. (6.7.6), we make use of the relation $|\hat{\psi}_p|^2 = \operatorname{Sn}^2(u, u/2)$, and $|\hat{\psi}_s|^2 = 1/|\hat{\psi}|_p^2 = \operatorname{Sn}^2(u, u/2 + j\pi/2)$, and calculate γ
$$\gamma^4 = (10^{A_p/10} - 1)(10^{A_s/10} - 1)$$
2. From the second of eqs. (6.7.6), we calculate $\operatorname{Sn}(u, u/2)$
$$\operatorname{Sn}(u, u/2) = \frac{10^{A_p/10} - 1}{\gamma^2} \tag{6.7.7}$$
3. From eq. (6.7.7) and eq. (6.7.2) we calculate u. A simple solution routine is to program the infinite tanh product with an initial guess for u, and essentially use a cut-and-try procedure, though, of course, a more elegant algorithm can be employed.

4. From the third of eqs. (6.7.6), again using a computer routine, we calculate mu and therefore m, as the minimum integer greater than the ratio mu/m.

Example 6.7.1

Design an elliptic filter with the following specifications:

$$f_p = 90\,\text{KHz} \qquad A_p = 0.5\,\text{dB}$$

$$f_s = 105\,\text{KHz} \qquad A_s = 40\,\text{dB}$$

Solution The results, obtained by executing the steps above, are the following:

$$\gamma = 5.9101 \qquad u = 0.7006 \qquad mu = 4.3814$$

Calculating m as the lowest integer larger than mu/u we obtain $m = 7$. This result should be compared with $n = 55$ for the Butterworth filter, Example 6.5.1, and with the Chebyshev filter, $n = 14$, Example 6.6.1.

The rounding of m either increases mu with u constant, or decreases u with mu constant. The changes are generally modest. If u is held constant the passband edge is increased and the stopband edge is decreased with the same attenuation tolerances. If mu is fixed, the band edges are not modified, but the attenuation at the passband edge is reduced and that at the stopband edge increased. The results for this second case are ($mu = 4.3814$, $u = mu/7 = 0.6259$)

$$A_p = 0.2227\,\text{dB} \qquad A_s = 43.6519\,\text{dB}$$

The transmission zeros, ω_j, in the stop band (poles of $\hat{\psi}$) and the unit gain points in the passband, ω_i, (zeros of $\hat{\psi}$) are calculated from eq. (6.7.4) as

$$\omega_j = 1.8032 \qquad 1.1827 \qquad 1.0620$$

$$\omega_i = 0.5546 \qquad 0.8455 \qquad 0.9416$$

□

Once the rational function $\hat{\psi}$ has been determined from its zeros and poles and from its value (=1) at the cutoff frequency in the center of the transition band, the knowledge of γ completely defines the characteristic function ψ, and one can proceed to the synthesis of the filter according to the Darlington procedure of Chapter 4.

The following simple example, though of low degree, illustrates the various steps in the design of a prototype lowpass elliptic filter.

Example 6.7.2
Let the passband and stopband tolerances be $A_p = 0.5\,\text{dB}$ and $A_s = 20.0\,\text{dB}$. The edge of the passband is $w_p = 0.8$, so that the stopband edge is $w_s = 1/w_p = 1.25$. Determine the S matrix of the filter and the filter elements.

Solution Proceeding exactly as in Example 6.7.1, the rounded values of $m = 3$, $u = 0.990$, $\gamma = 2$ are computed. These can be checked against the specifications. Thus the lowpass band edge is $\omega_p = \text{Sn}(3u, 1.5u) = 0.8134$. The value of $|\hat{\psi}_p|^2 = \text{Sn}^2(u, u/2) = (0.1654)^2$, so that $A_p = 10\log[1 + 4(0.1654)^2] = 0.451\,\text{dB}$, and $A_s = 10\log[1 + \gamma^2/(0.1654)^2] = 21.68\,\text{dB}$. The zeros in the pass band are at $\omega = 0$, and $\omega = \text{Sn}(3u, u) = 0.7284$. The insertion loss in the center of the transition region, $\omega = 1$, is $10\log(1+\gamma^2) = 7.0\,\text{dB}$.

We now have all the data for expressing the transducer gain $T(\omega^2) = 1/(1 + \gamma^2|\hat{\psi}|^2)$, since, referring to eq. (6.7.5)

$$|\psi(j\omega)|^2 = \gamma^2|\hat{\psi}|^2 = 4\frac{\omega^2(0.7284^2 - \omega^2)^2}{(1 - 0.7284^2\omega^2)^2} = \frac{h(j\omega)h(-j\omega)}{f(j\omega)f(-j\omega)}$$

It follows that

$$h(s) = \pm 2s(s^2 + 0.7284^2), \quad f(s) = 0.7284^2 s^2 + 1$$

and

$$gg_* = ff_* + hh_* = -4s^6 - 3.963s^4 - 0.0649s^2 + 1$$

There are formulae[8] for the spectral factorization of gg_*, but a computer root-finding routine is actually more convenient. The three left half plane roots are $s_{1,2} = -0,1898 \pm j0.8598$, and $s_3 = -0.6449$. Thus

$$g = 2(s - s_1)(s - s_2)(s - s_3) = 2s^3 + 2,0489s^2 + 2.0403s + 1$$

The input reflectance of the filter is $S_{11} = h/g$. We will choose the (+) sign for h, which corresponds to $S_{11} = 1$ at $\omega = \infty$, an open circuit, so that the leading element of the filter is a series inductance. The (−) sign starts the filter with a shunt capacitor. We can easily complete the Darlington synthesis presented in Chapter 5. Thus the open circuit impedances of the unterminated filter, eq. (5.5.1) are ("e" and "o" refer to even and odd powers)

$$z_{11} = z_{22} = \frac{g_e + g_o}{g_e - g_o} = \frac{2.0489s^2 + 1}{0.9792s}$$

[8]P. Amstutz, "Elliptic Approximation and Elliptic Filter Design on Small Computers", *IEEE Trans. on Circ. and Sys.*, vol. CAS-25, no. 12, pp. 1001-1011, Dec. 1978.

Now extract from z_{11} an inductor L corresponding to a partial residue at infinity so that the remainder impedance z_1 has a zero at the zero of transmission $j(1/.7284) = j1.3729$.

$$z_{11}(j1.3729) - j1.3729L = 0, \quad L = 1.5506$$

Then the remainder is

$$z_1(s) = z_{11}(s) - sL = \frac{0.7854^2 s^2 + 1}{0.9792s}, \quad L_1 = 0.5419, \quad C_1 = 0.9782$$

The elements L_1, C_1 form a series resonator placed in shunt. Reflectance $S_{22} = -h_*/g = S_{11}$, thus the filter is symmetric as expected since $m = 3$ is odd. The final structure is a mid-series "Tee" consisting of L in series, then a shunt connected branch consisting of L_1 in series with C_1, followed by L in series to complete the "Tee". For higher values of m, a symmetric filter with $S_{11} = S_{22}$ will not be structurally symmetric. □

In general, elliptic filters with m odd can be realized in either mid-series or mid-shunt form. Had the $(-)$ sign been chosen for S_{11} in the Example, the filter would have been a mid-shunt "Pi", with an antiresonant circuit in the series branch. Finally, the reader is referred to the Amstutz paper for a more complete treatment of the entire subject.

6.8 Phase Equalization

The design of the filters so far described has concentrated on realizing a response which approximates the amplitude of a quasi ideal filter according to some specific optimization criterion. Since no constraint has been imposed upon the phase, it is to be expected that this property will deviate in a more or less significant manner from the ideal requirement of phase linearity vs. frequency. Certainly the minimum phase determined by the Hilbert transform of a rather complex amplitude response will assuredly not be linear. The negative derivative of a phase function, $\theta(\omega)$, with respect to ω is the *group delay*, $\tau(\omega)$. Thus

$$\tau = -\frac{d\theta}{d\omega}$$

Corresponding to linear phase, the idealized delay function should be a constant, and we therefore seek a filter transmittance which, in some sense, provides a best approximation to flat delay. The approach we shall follow is to introduce an equalizer section which, without affecting the amplitude response, adds a component to the delay of the existing filter so as to yield an approximately flat composite response.

A basic property of the group delay function for reactance two-ports is given by the following theorem.

THEOREM 6.8.1

The group delay of a reciprocal lossless two-port is always nonnegative.

$$\tau = -\frac{d\theta}{d\omega} \geq 0, \;\; \forall\omega$$

and the delay is an even function.

PROOF Using Belevitch notation, Corollary 5.4.1, the transmittance of the two-port is $S_{21} = f/g$, and $\arg S_{21}(j\omega) = \theta(\omega) = \theta_n - \theta_d$ ("n" and "d" refer to numerator and denominator). Then

$$\tau = -(\frac{d\theta_n}{d\omega} - \frac{d\theta_d}{d\omega})$$

But the numerator f is either even or odd, so its phase on $j\omega$ is constant, hence it does not contribute to the delay. Write the denominator in terms of its even (e) and odd (o) powers, $g(j\omega) = g_e(\omega) + jg_o(\omega)$. Then $\theta_d = \arctan g_o/g_e$ and, letting $X(\omega) = g_e(\omega)/g_o(\omega)$,

$$\tau(\omega) = \frac{d\theta_d}{d\omega} = \frac{1}{1+X^2}\frac{dX}{d\omega}$$

But g is a Hurwitz polynomial, hence according to Theorem 5.2.3, $X(\omega)$ must be a Foster reactance function. It follows, therefore, Theorem 5.2.4, that $dX/d\omega > 0$. Furthermore, since X is odd, its derivative is even, therefore $\tau(\omega) = -\tau(-\omega)$ and the theorem is proved. ∎

The clue to the implementation of delay equalization is given by eq. (6.2.3), which is repeated here for the sake of convenience. We assume the filter and equalizer satisfy reciprocity.

$$S = \begin{pmatrix} S_{11} & S_{21}\eta \\ S_{21}\eta & S_{22}\eta^2 \end{pmatrix}$$

where η is the transfer function of an allpass section inserted between the output port of the filter and the load resistance. As pointed out earlier, the input to the allpass is equal to the load resistance so the original filter remains properly terminated, and the allpass amplitude response is unity $\forall\omega$. The overall transmittance is therefore $S_{21}\eta$ and the gain is $T(-s^2) =$

$S_{21}\eta S_{21*}\eta_* = S_{21}S_{21*}$, the same as that of the filter without the allpass. Accordingly, the amplitude is not changed, but the phase is modified as

$$\arg(S_{21}\eta) = \arg(S_{21}) + \arg(\eta)$$

which, for the sake of future discussions, is rewritten as

$$\theta = \theta_f + \theta_\eta \tag{6.8.1}$$

where θ_f is the phase of the filter, θ_η is the phase of the allpass, and θ is the phase of the cascaded two-port. By taking derivatives with respect ω, eq. (6.8.1) can be rewritten as

$$\tau = \tau_f + \tau_\eta \tag{6.8.2}$$

with an obvious meaning for the symbols.

In practice, the allpass function, η, whose parameters are chosen to provide the corrected delay function, is realized by cascading one or more elementary sections, either of degree one, realizing a transmission zero on the positive real axis (a *Type C section*), or of degree two, realizing a pair of complex conjugate transmission zeros in the right half-plane (a *Type D section*). See Sections 5.11, 5.12. In the next two sections, a procedure is given for the design of delay equalizers.

6.9 Allpass C-Section Phase Equalizers

The phase response of reciprocal allpass C-sections is suitable for the delay equalization of lowpass filters. The allpass has degree one and, therefore, produces a single transmission zero on the positive real axis at $s_0 = \sigma_p$. When terminated in unit load resistances it is matched at the input and output ports. Furthermore, its allpass property is determined by the fact that the transmittance produces phase shift, but its amplitude is unity $\forall\omega$. Consequently, the scattering matrix has the form

$$S = \begin{pmatrix} 0 & \dfrac{s-\sigma_p}{s+\sigma_p} \\ \dfrac{s-\sigma_p}{s+\sigma_p} & 0 \end{pmatrix} \tag{6.9.1}$$

Then $\theta_\eta = \arg\eta(j\omega) = \arg(j\omega - \sigma_p)/(j\omega + \sigma_p)$ is the phase function of the allpass, and

$$\theta_\eta = \pi - 2\arctan\frac{\omega}{\sigma_p}$$

$$\tau_\eta = -\frac{d\theta_\eta}{d\omega} = \frac{2\sigma_p}{\sigma_p^2 + \omega^2}$$

consequently, the total group delay of eq. (6.8.2) is

$$\tau = \tau_f + \frac{2\sigma_p}{\sigma_p^2 + \omega^2}$$

Since lowpass filters show a more or less constant passband amplitude response, and a sharp rolloff beyond cutoff, the Hilbert transform (Section 3.7) qualitatively indicates a group delay (corresponding to minimum phase) which increases with frequency in the passband to reach a peak in the neighborhood of the cutoff frequency, followed by a steady decrease of delay in the stopband. For an allpass C-section, the group delay is monotone decreasing. It is therefore not surprising that a judicious choice of σ_p, the single free parameter in τ_η, can produce the required compensation, yielding a passband group delay for the overall structure whose deviation from flatness is substantially improved. While the choice can be optimized numerically, say, by a least squares routine, usually a cut-and-try procedure yields satisfactory results.

The Darlington procedure, Section 5.11, can be used to synthesize the allpass C-section. First, we find the Belevitch representation of S, Section 5.4. In this case, we simply introduce a common numerator and denominator factor so that the resulting numerator ff_* of $S_{21}S_{21*} = ff_*/gg_*$ is a perfect square and $f = \sqrt{ff_*}$ is even. The result is

$$S_{21} = \frac{f}{g} = \frac{\sigma_p^2 - s^2}{(\sigma_p + s)^2}, \qquad z_{11} = \frac{g_e + h_e}{g_o + h_o} = \frac{s^2 + \sigma_p^2}{2\sigma_p s}$$

where we have applied eq. (5.5.1) to find the open circuit input impedances $z_{11} = z_{22} = g_e/g_o$, of the allpass, noting that $h = 0$. Inductance L is extracted so that the remainder impedance has the zero of transmission, $z_2(\sigma) = z_{11}(\sigma) - L\sigma_p = 0$. Thus

$$L = \frac{z_{11}(\sigma)}{\sigma} = 1, \qquad z_2 = z_{11} - sL = \frac{-s}{2\sigma_p} + \frac{\sigma_p}{2s}$$

The two-port is symmetric and consists of a Tee of three inductors, the two series elements, both $L = 1/\sigma_p$, and shunt element $M = -1/2\sigma_p$, connected to ground through a capacitor $C_1 = 2/\sigma_p$. The three inductors are then replaced by a pair of perfectly coupled coils with self inductances $L_{11} = L_{22} = L + M = 1/2\sigma_p$, and mutual inductance $L_{21} = M = -1/2\sigma_p$, thus forming the type C section, Fig. 5.10.1 (h).[9]

[9]The symmetric type C section is equivalent to a symmetric lattice with no coupled coils. The two series arms of the lattice are inductances each equal to $1/\sigma_p$, and the two cross arms are capacitors each equal to $1/\sigma_p$. A major disadvantage is that the lattice is a four terminal structure without a common ground terminal for the two ports.

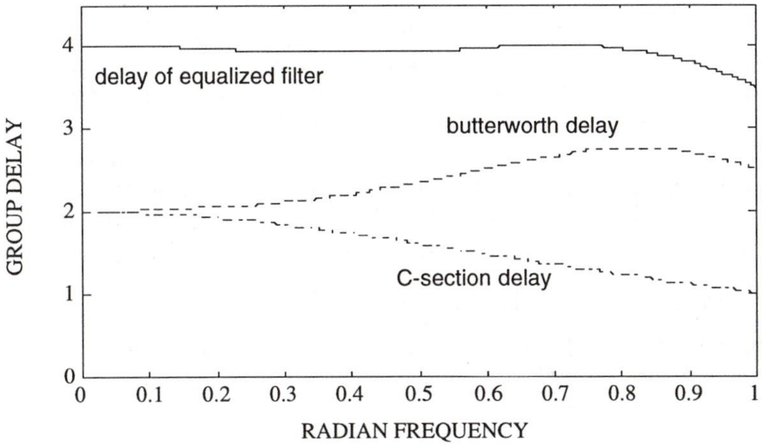

FIGURE 6.9.1
(a) Group delay of a low-pass third order Butterworth filter (−−); (b) group delay of a C-section (− ·); (c) group delay of the cascaded two-port(−).

The following example is a simple illustration of how the zero of transmission of the allpass, σ_p, may be chosen to provide equalization for flat delay.

Example 6.9.1
Design a single section Type C equalizer to compensate the group delay of a third order normalized Butterworth filter.

Solution The transmittance of a third order lowpass prototype Butterworth filter is (Section 6.5)

$$S_{21} = \frac{1}{s^3 + 2s^2 + 2s + 1}$$

Its phase on the $j\omega$ axis is given by

$$\theta_f = -\arctan\frac{2\omega - \omega^3}{1 - 2\omega^2}$$

The group delay of the filter is

$$\tau_f = -\frac{d\theta_f}{d\omega} = \frac{2 + \omega^2 + 2\omega^4}{1 + \omega^6}$$

We seek a value of σ_p, so that the function

$$\tau = \frac{2 + \omega^2 + 2\omega^4}{1 + \omega^6} + \frac{2\sigma_p}{\sigma_p^2 + \omega^2}$$

is as flat as possible in the passband. If we choose $\sigma_p = 1$, we find that for $0 \le \omega \le 0.8$, τ_f varies from a maximum value of 2.74 to a minimum value of 2 (37%), while in the same frequency interval the compensated delay, τ, varies from a maximum value of 4 to a minimum value of 3.93; the variation is reduced from 37% to 1.8%. The group delays of the uncompensated filter, of the C-section, and of the compensated filter vs. normalized angular frequency are plotted in Fig. 6.9.1. The graph of equalized delay for the normalized third order Butterworth filter shows a substantially constant group delay over a passband $0 \le \omega \le 0.8$; the insertion loss at the upper edge of this band is 2.3 dB.

The final filter consists of a three element lowpass ladder Tee section, cascaded with a type C section. □

6.10 Allpass D-Section Phase Equalizers

The reciprocal allpass D-section is useful for the phase equalization of bandpass filters. The section has degree two and produces a pair of complex conjugate transmission zeros in the open right half-plane at $s_p = \sigma_p \pm j\omega_p$. Its scattering matrix has the form

$$S = \begin{pmatrix} 0 & \dfrac{s^2 - 2\sigma_p s + |s_p|^2}{s^2 + 2\sigma_p s + |s_p|^2} \\ \dfrac{s^2 - 2\sigma_p s + |s_p|^2}{s^2 + 2\sigma_p s + |s_p|^2} & 0 \end{pmatrix}$$

The type D allpass can be obtained as a bandpass transformation on the type C allpass, for note that the mapping $s \to (s^2 + |s_p|^2)/2s$ applied to the type C eq. (6.9.1) produces the above equation. We therefore can expect to use it for the equalization of bandpass filters. However, since the transformation produces nonlinear frequency scaling, it would be a mistake to transform the equalized lowpass system and expect the bandpass version to retain the correct delay equalization. Instead, we must directly calculate the parameters of the type D allpass, based on the delay response of the bandpass filter.

The type D transmittance is $\eta = (s^2 - 2\sigma_p s + |s_p|^2)/(s^2 + 2\sigma_p s + |s_p|^2)$ and

$$\theta_\eta = -2 \arctan \frac{2\sigma_p \omega}{|s_p|^2 - \omega^2}$$

$$\tau_\eta = -\frac{d\theta_\eta}{d\omega} = \frac{4\sigma_p(|s_p|^2 + \omega^2)}{(|s_p|^2 - \omega^2)^2 + 4\sigma_p^2\omega^2}$$

Thus eq. (6.8.2) becomes

$$\tau = \tau_f + \frac{4\sigma_p(|s_p|^2 + \omega^2)}{(|s_p|^2 - \omega^2)^2 + 4\sigma_p^2\omega^2} \tag{6.10.1}$$

For bandpass filters, the group delay increases with frequency in the lower stopband to reach a peak in the neighborhood of the lower cutoff frequency, then decreases until near the center of the passband where it has a minimum; from that point on, it increases again until close to the higher cutoff frequency it has a second peak. Beyond the high frequency cutoff, the delay decreases steadily; the delay curve resembles two hills with a valley in between. For an allpass D-section the group delay has a peak at about the center of the passband. It is therefore not surprising that a judicious choice of the two free parameters σ_p and ω_p can produce compensation of the nonlinear phase characteristic of the filter. For the overall structure, we can achieve an approximately constant group delay in the passband, although the results, with the use of a single allpass D-section, are not as good as for the lowpass case. Further improvement can be obtained by cascading additional sections, with peaks slightly shifted above and below the center of the passband. As in case of lowpass filters, cut-and-try procedures can yield satisfactory results provided only one or two equalizers are used. One must also be aware of the fact that allpass sections introduce considerable structural complexity, and this factor must be taken into account in making final design decisions.

The concepts discussed above are illustrated by the following.

Example 6.10.1

Design a single section allpass D-equalizer to compensate for the group delay of a Butterworth filter obtained by bandpass transformation of a third order lowpass prototype. The normalized bandpass cutoff frequencies are $\omega_1 = 3$ and $\omega_2 = 4$.

Solution The transmittance of the normalized bandpass Butterworth filter is

$$S_{21} = \frac{1}{S^3 + 2S^2 + 2S + 1}$$

where

$$S = \frac{1}{\omega_2 - \omega_1}\left(s + \frac{\omega_0^2}{s}\right)$$

The phase, $\arg S_{21}(j\omega)$, after writing $S = j\Omega$, is given by

$$\theta_f = -\arctan\frac{2\Omega - \Omega^3}{1 - 2\Omega^2}$$

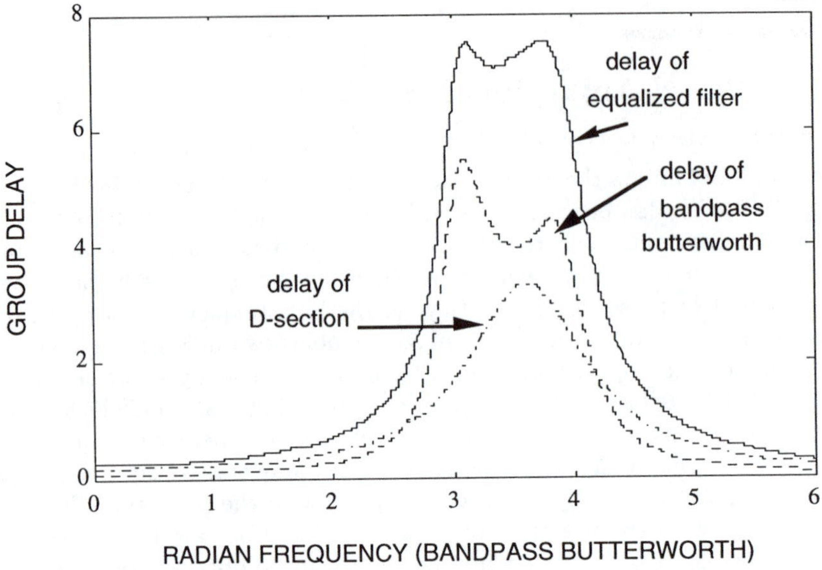

FIGURE 6.10.1
a) Group delay of a band-pass third order Butterworth filter (– –); b) group delay of a D-section (– ·); c) group delay of the cascaded two-port (–).

The group delay of the filter is

$$\tau_f = -\frac{d\theta_f}{d\omega} = \frac{2+\Omega^2+2\Omega^4}{1+\Omega^6}\frac{1}{\omega_2-\omega_1}\left[1+\left(\frac{\omega_0}{\omega}\right)^2\right]$$

Referring to eq. (6.10.1), we seek values of σ_p and ω_p, so that the function

$$\tau = -\frac{d\theta}{d\omega} = \frac{2+\Omega^2+2\Omega^4}{1+\Omega^6}\frac{1}{\omega_2-\omega_1}\left[1+\left(\frac{\omega_0}{\omega}\right)^2\right] + \frac{4\sigma_p(|s_p|^2+\omega^2)}{(|s_p|^2-\omega^2)^2+4\sigma_p^2\omega^2}$$

is as flat as possible in the passband. If we choose $\sigma_p = 0.6$ and $\omega_p = 3.61$, we find that for $3 \leq \omega \leq 4$, the filter delay, τ_f, varies from a minimum value of 3.50 to a maximum value of 5.48, while in the same frequency interval the equalized delay, τ, varies from a minimum value of 5.88 to a maximum value of 7.51; the variation is reduced from 57% to 28%. Additional equalizers would be required for further improvement in the delay response. The group delays of the uncompensated filter, of the D-section, and of the compensated filter vs. normalized angular frequency are shown in Fig. 6.10.1. □

6.11 Bessel Approximation

In this section we consider the formulation of a polynomial, lowpass, prototype filter function in which the emphasis is on linear phase rather than amplitude selectivity. The idealized linear phase transmittance is

$$S_{21} = \epsilon^{-sT} \tag{6.11.1}$$

which is an allpass function exhibiting a constant group delay T, and corresponds to a delay line which retards the signal without amplitude distortion.

Since the ideal transmittance S_{21} is not rational, no finite two-port can realize it. (It is however realizable by a transmission line.) We seek a realizable rational function in the form of $S_{21} = K/D$, where D is a Hurwitz polynomial which approximates $\exp sT$ over the passband, and K is chosen so that $|S_{21}(j\omega)| \leq 1$. Rewrite eq. (6.11.1) as

$$S_{21} = \frac{1}{\epsilon^{sT}} = \frac{1}{\cosh sT + \sinh sT}$$

or

$$S_{21} = \frac{1}{m+n} \tag{6.11.2}$$

where $m = \cosh sT$ and $n = \sinh sT$ are the even and odd parts of the entire function ϵ^{sT}. We have

$$\tanh sT = \frac{n}{m}$$

The hyperbolic tangent admits the well known infinite continuous fraction expansion

$$\tanh sT = \cfrac{1}{\cfrac{1}{sT} + \cfrac{1}{\cfrac{3}{sT} + \cfrac{1}{\cfrac{5}{sT} + \ldots}}}$$

As our approximation, let the expansion be truncated at some finite step, say, at the term $(2N-1)/sT$; then the resulting finite continuous fraction contains exactly N quotients and is an odd rational function of degree N. Since the coefficients are all positive (they are in fact the odd positive integers), the truncated expansion is a reactance function in Cauer form, see eq. (5.2.5), which is equal to the ratio of two polynomials n and m, with n odd and m even.[10] Thus applying the Hurwitz Theorem 5.2.3, $m+n$

[10]That the ratio is as described and not the reverse is easily seen from the fact that the polynomials must converge, when $N \to +\infty$, to the entire functions $\sinh sT$ (odd) and $\cosh sT$ (even).

is a Hurwitz polynomial and a transmittance of the form (6.11.2) (where m and n are now the approximating polynomials) is realizable by an LTI finite, passive, lossless, reciprocal two-port, provided K is chosen so that $|S_{21}(j\omega)| \leq 1$.

The polynomials $D_N(sT) = m + n$ obtained according to the above rules are called *Bessel polynomials*, and their maximum amplitudes on $j\omega$ occur at the origin. Since the procedure for obtaining them directly from the continuous fraction expansion is rather cumbersome, they are preferably calculated by the following iterative formula

$$D_N(sT) = (2N-1)D_{N-1}(sT) + s^2T^2 D_{N-2}(sT), \quad N > 2 \qquad (6.11.3)$$

To use the recursion equation we note that for $N = 1$, inspection of the tanh expansion gives $n = sT$, $m = 1$, and for $N = 2$, $n = 3sT$, $m = 3 + s^2T^2$. Accordingly,

$$D_1 = 1 + sT, \quad D_2 = 3 + 3sT + s^2T^2$$

Filters whose transmittance is obtained by the above procedure have excellent phase linearity in the passband $0 \leq \omega \leq 1$ and a gently decaying amplitude characteristic, so they are lowpass in character. It must be kept in mind that highpass or bandpass frequency transformations are not acceptable, because they introduce nonlinear phase distortion.

Example 6.11.1

Derive the transmittance of the third order Bessel filter.

Solution Using eq. (6.11.3), with the listed expressions for D_1 and D_2, we find

$$D_3 = 5D_2 + s^2T^2D_1 = s^3T^3 + 6s^2T^2 + 15sT + 15$$

The transmittance $S_{21} = K/D_3(s)$ approximates linear phase. The denominator is Hurwitz, so that if $|S_{21}(j\omega)| \leq 1$, it is BR. In turn, this requires that $K \leq D_3(0)$, since the Bessel polynomial has its maximum at the origin. Thus we have

$$S_{21} = \frac{15}{s^3 + 6s^2 + 15s + 15}$$

where we have chosen $T = 1$ and placed the constant 15 in the numerator so as to have $S_{21}(0) = 1$. The transmittance has all its zeros of transmission at infinity with a d.c. value of unity, so it is realizable as a lowpass LC ladder with unit resistive terminations. The amplitude and phase characteristics of the filter are presented in Fig. 6.11.1. The attenuation is down 0.45 dB at $\omega = 1$,[11] so that the transmittance provides a reasonable LC ladder approximation to a delay line, $0 \leq \omega \leq 1$. □

[11]Nature, acting through the Hilbert transform, makes substantial amplitude selectivity and frequency discrimination incompatible with a constant delay minimum phase transfer

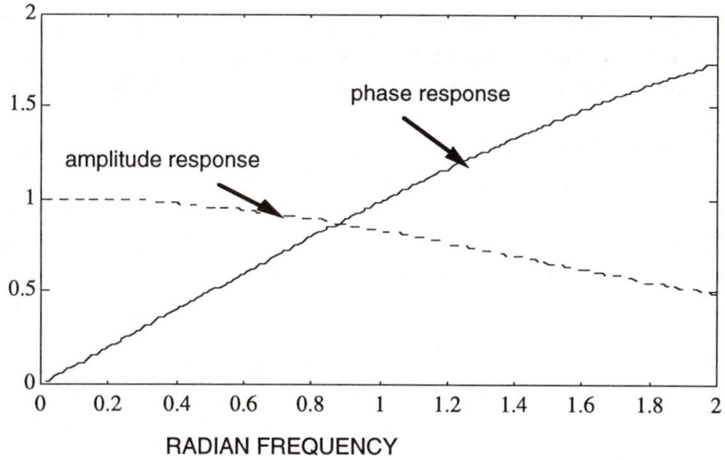

FIGURE 6.11.1
Third order low-pass Bessel filter: (a) amplitude characteristics (− −); (b) phase characteristics (−).

Example 6.11.2

Find the transmittance of a finite allpass two-port whose phase response is approximately linear.

Solution Write the idealized allpass linear phase transmittance as $S_{21} = \exp(-2sT) = \exp(-sT)/\exp(sT)$. Then it is evident that the approximating allpass rational transmittance can be written as

$$S_{21} = \frac{D_N(-sT)}{D_N(sT)}$$

The denominator polynomial is Hurwitz and the numerator is the same polynomial with negative signs for the odd power coefficients, so that the amplitude response is unity $\forall\omega$. The delay is simply twice that of $1/D_N(sT)$. For instance, using the previous example for $N = 3$,

$$S_{21} = \frac{-s^3 + 6s^2 - 15s + 15}{s^3 + 6s^2 + 15s + 15}$$

Of course, the response is no longer realizable by an LC ladder. □

function. For some possible compromises see H. J. Carlin and J. L. C. Wu, "Amplitude Selectivity Versus Constant Delay in Minimum-Phase Lossless Filters", *IEEE Trans. on Circ. and Sys.*, vol. CAS-23, no. 7, pp. 447-455, July 1976.

6.12 Synthesis of Single-Terminated Filters

Single-terminated filters appear in a variety of forms. The core of the system is always a lossless two-port terminated at only one end in a resistor, the other end in a short or open circuit. The source may be associated with either port, i.e., a current source across the resistor or across an open circuit, or a voltage in series with the resistor or in series with a short circuited port. Invariably we are concerned with applying filtering to a response-excitation ratio between the ports, one of the variables being a source quantity, the other an output voltage or current. Thus an open circuit transfer impedance, or a short circuit transfer admittance, or an open-circuit voltage ratio, or finally a short-circuit current ratio, of the augmented two-port, which includes the load resistor, is used to define a gain function, $T(\omega^2)$; this function is to be designed to produce amplitude selectivity and frequency discrimination. We then proceed from T to determine the reactance two-port. This last step is generally carried out by determining an input immittance to the two-port and applying the Darlington synthesis.

Although there are a number of different cases that can occur depending on the source as well as the type of loading at the output port, the procedure is similar for all cases. We will examine a couple of different possibilities, and this should clarify the approach for all the various configurations. Suppose for our first illustration that the input to the lossless two port is a pure current source, i_1, and the output is the voltage v_2, across the resistor r_2 loading port 2, Fig. 6.12.1. The lossless two-port is to filter a prescribed band of frequencies appearing at the input, so they are not transmitted to the output. The filter gain function is defined as $T(\omega^2) = |v_2/i_1|^2 = |z_{21}|^2$, where z_{21} is the open circuit transfer impedance of the two-port consisting of the lossless filter augmented by the resistor r_2 across the output.

Our next step is to find $r(\omega^2)$, the input resistance to the r_2 loaded two-port, from $|z_{21}|^2$, and then determine the input impedance function. Thus consider the input power to port 1 which is given by

$$P_1(\omega) = r(\omega^2)|i_1(j\omega)|^2 \tag{6.12.1}$$

The power delivered to the load is

$$P_2(\omega) = \frac{|v_2(j\omega)|^2}{r_2} \tag{6.12.2}$$

The output voltage v_2 is related to the input current i_1 by the open circuit transfer impedance (transimpedance) z_{21} of the r_2 augmented two-port

$$v_2 = z_{21}\, i_1 \tag{6.12.3}$$

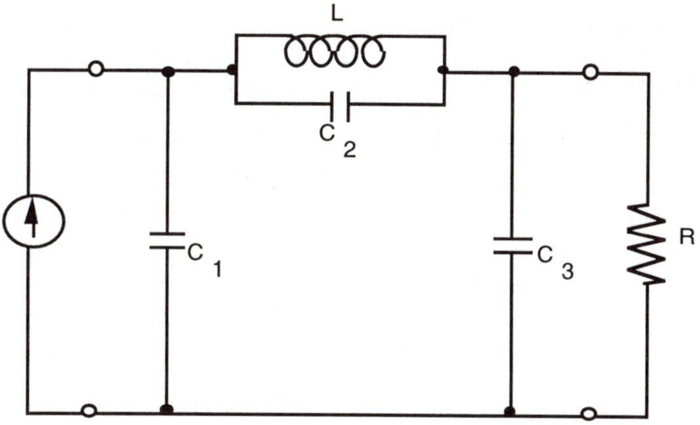

FIGURE 6.12.1
Single-ended filter loaded at port 1 by a current source and at port 2 by a resistor.

Since the two-port is lossless, the input and output powers are equal, $P_1 = P_2$. Hence from eqs. (6.12.1), (6.12.2), and (6.12.3)

$$r(\omega^2)|i_1|^2 = \frac{|v_2|^2}{r_2} = \frac{|z_{21}|^2|i_1|^2}{r_2}$$

or

$$r(\omega^2) = \frac{|z_{21}(j\omega)|^2}{r_2}$$

Therefore the *gain* function is

$$T(\omega^2) = \left|\frac{v_2}{i_1}\right|^2 = |z_{21}(j\omega)|^2 = r_2\, r(\omega^2) \tag{6.12.4}$$

Eq. (6.12.4) can be extended into the complex $s = j\omega$ plane since $r(-s^2) = \frac{1}{2}(z + z_*)$ as

$$T(-s^2) = z_{21}z_{21*} = r_2\,\frac{1}{2}(z + z_*) \tag{6.12.5}$$

Before proceeding to the determination of z from its real part, r, we consider one other illustrative configuration. This time, suppose a current source i_1 is connected across a resistor r_1 at port 1, and the output quantity is the current i_2 flowing into short circuited port 2. The gain function is defined as $T(\omega^2) = |i_2/i_1|^2$. Use the Norton equivalence to change the current source to a voltage source $v_1 = i_1 r_1$ in series with r_1, and consider the short circuit transfer admittance of the lossless two-port augmented by resistor r_1 in series with port 1.

$$y_{21} = i_2/v_1 = i_2/r_1 i_1, \quad \text{and } T = |y_{21}|^2 r_1^2$$

We wish the input immittance to the augmented structure at port 2, since port 1 is terminated in a resistor. By reciprocity, $y_{12} = y_{21}$. Proceed as in the derivation of eq. (6.12.5). Thus a voltage v_2 at port 2 delivers a current i_1 flowing in r_1, $i_1 = y_{12}v_2$. Then by conservation of power $|v_2|^2 g = |i_1|^2 r_1 = |y_{12}v_2|^2 r_1$, so that we obtain $|y_{12}|^2 = |y_{21}|^2 = g/r_1$, where $g(\omega^2)$ is the input conductance at port 2 of the lossless two port terminated in r_1. The gain is therefore $T(\omega^2) = g(\omega^2)r_1$. This time the gain is defined by a conductance function g, measured at the port opposite the fixed resistor. This turns out to be the case for all the different arrangements of singly terminated filters, so that realizability of $T(\omega^2)$ is governed by the following theorem for resistance (it applies equally well to conductance).

THEOREM 6.12.1
Let the rational PR impedance z have as resistance function $r = \Re\, z(j\omega)$. Then $r \geq 0$, $\forall\omega$, and is an even rational function $r(\omega^2)$. Furthermore, $r(\omega^2)$ has no poles on the ω axis.

PROOF The nonnegative property of r follows directly from the PR character of z. Next, since $r = [z(j\omega) + z(-j\omega)]/2$, r is evidently even. Now suppose r has a pole on ω. Since $r(\omega^2)$ is nonnegative, the pole in r must be of even multiplicity, otherwise, the resistance would change sign. But $r(-s^2) = (z+z_*)/2$; therefore, since the pole factor is $(s^2+\omega_0^2)^{2n}$, $n \geq 1$, which is real on $j(\omega)$, it must be present in $z(s)$ as well. But according to Theorem 4.6.5, boundary poles in a PR function must be simple, hence *all* $j\omega$ poles, (a pole at infinity is also forbidden), are excluded from r. ∎

The synthesis of a resistance function which satisfies the preceding theorem can be carried out using the Gewertz procedure.[12]

1. Let the unknown impedance be $z = n/d$ where n and d are assumed to be polynomials of the same degree.

2. The even part of z, i.e., the given resistance $r(-s^2)$, can be written as

$$r(-s^2) = \frac{1}{2}(z + z_*) = \frac{1}{2}\left(\frac{n}{d} + \frac{n_*}{d_*}\right) = \frac{1}{2}\frac{nd_* + n_*d}{dd_*} = \frac{p(-s^2)}{q(-s^2)}$$

3. From the prescribed $T(-s^2) = rr_2$, construct the ratio $r = p/q$.

[12] L. Weinberg, *Network Analysis and Synthesis*, New York, Mc Graw-Hill, 1962, p. 288. There is also an alternate procedure based on partial fractions in H. W. Bode, *Network Analysis and Feedback Amplifier Design*, New York, D. Van Nostrand, 1952, p. 204.

4. To determine d, factor the polynomial $q(-s^2) = dd_*$ and assign to d the LHP root factors. The factorization is unique since d must be a Hurwitz polynomial.

5. To find the numerator polynomial n, use Item 2

$$\frac{1}{2}(nd_* + n_*d) = p$$

Since p and d are known, we can equate coefficients of like powers on the left and right sides of the above equation, so that with p of degree $2m$, we obtain m linear equations for the coefficients of s^k in polynomial n.

6. We have determined the impedance $z = n/d$, and it can be realized by the Darlington synthesis as a cascade of lossless two-port sections terminated on the resistance r_2.

We illustrate the above procedure by the following.

Example 6.12.1
Synthesize a normalized Butterworth filter of degree 3 fed by a current source at port 1 and loaded by a resistor, $r_2 = 1$ at port 2.

Solution We have

$$T(\omega^2) = |z_{21}|^2 = \frac{1}{1+\omega^6} \tag{6.12.6}$$

Factorization of the denominator yields $d = s^3 + 2s^2 + 2s + 1$. Hence we assume for the input impedance z the form

$$z = \frac{a_0 s^3 + a_1 s^2 + a_2 s + a_3}{s^3 + 2s^2 + 2s + 1}$$

We then find

$$\frac{1}{2}(z + z_*) = \frac{-a_0 s^6 + (-2a_0 + 2a_1 - a_2)s^4 + (a_1 - 2a_2 + 2a_3)s^2 + a_3}{1 - s^6} \tag{6.12.7}$$

By equating coefficients of like powers in eqs. (6.12.6) and (6.12.7), according to eq. (6.12.5), we obtain the following coefficient equalities

$$\begin{aligned} a_0 \qquad\qquad\qquad &= 0 \\ 2a_1 \; -a_2 \qquad &= 0 \\ a_1 \; -2a_2 \qquad &= -2 \\ a_3 &= 1 \end{aligned}$$

so that

$$a_0 = 0, \quad a_1 = 2/3, \quad a_2 = 4/3, \quad a_3 = 1/3$$

Thus the impedance z is

$$z = \frac{2s^2 + 4s + 3}{3s^3 + 6s^2 + 6s + 3}$$

and can be expanded into a continued fraction by iterated extraction of poles at infinity, starting with the pole at infinity in $1/z$ of residue $3/2$.

$$z = \cfrac{1}{\frac{3}{2}s + \cfrac{1}{\frac{4}{3}s + \cfrac{1}{\frac{1}{2}s + 1}}}$$

The complete system is made up of an independent current source followed by a lossless Pi section consisting of shunt capacitor $C_1 = 3/2$, series inductor $L_2 = 4/3$, and shunt capacitor $C_3 = 1/2$. The Pi is terminated by a unit resistor. □

7

Transmission Lines

7.1 The TEM Line

This chapter focuses on circuits containing transmission lines which have as their fundamental propagating mode the *Transverse Electromagnetic* (TEM) wave; i.e., the electric (***E***) and magnetic (***H***) field vectors are normal to each other and to the direction of propagation, z. A TEM transmission line is a waveguide which generally consists of two metallic conductors without resistive loss extending in a longitudinally uniform geometry, e.g., parallel wire line, strip line, coaxial line. Multiple conductor (more than two) lines also exhibit TEM modes.

The fundamental mode on a TEM line propagates down to d.c. and is called a *transverse wave* since both ***E*** and ***H*** field vectors lie in a transverse plane perpendicular to the guiding conductors. A classic application of Maxwell's Equations[1] shows that two scalars, respectively, a voltage measured in the transverse plane between the guiding conductors and a current flowing in opposite directions along the conductors parallel to the z-axis, can be derived from the field vectors. It is readily verified that the system is LTI, so that if exponential excitation proportional to $\epsilon^{j\omega t}$ is assumed, then with the system in the steady state, the voltage and current along the line also have exponential time dependence. If, as is usual in LTI system analysis, we suppress the time dependence and exhibit only the phasor response (which depends on distance, z, along the line measured from the source, as well as angular frequency ω), the voltage and current phasors, $v(z, j\omega)$, $i(z, j\omega)$, are given by

$$v = A\epsilon^{-j\beta z} + B\epsilon^{j\beta z} \tag{7.1.1}$$

$$i = \frac{1}{Z_0}\left(A\epsilon^{-j\beta z} - B\epsilon^{j\beta z}\right) \tag{7.1.2}$$

[1]S. Ramo, J. R. Whinnery, T. Van Duzer, *Fields and Waves in Communication Electronics*, New York, Wiley, 1965.

Here the z independent boundary value constants $A(j\omega)$, $B(j\omega)$ depend on the termination at the end of the line, $j\beta \equiv j\beta(\omega)$ is the *propagation function* (β is the *phase function*), and Z_0 is the *characteristic impedance* of the TEM line.

It is instructive to deduce eqs. (7.1.1), (7.1.2) using a model which exhibits the circuit properties of the TEM line. We associate with the two guiding lossless conductors a series inductance and a shunt capacitance per unit length, respectively, L and C. As we move away from the source in the positive z direction, the incremental phasor voltage drop (current in positive z direction, $-dv = -v(z+dz)+v(z)$) between the conductors is $-dv = j\omega Lidz = Zi\,dz$, and the incremental current phasor drop is $-di = -i(z+dz)+i(z) = j\omega Cvdz = Yv\,dz$, where Z and Y are, respectively, the series impedance and shunt admittance per unit length of line. We have the pair of linear, space invariant (so to speak, LSI instead of LTI) differential equations

$$-\frac{dv}{dz} = Zi, \quad -\frac{di}{dz} = Yv \tag{7.1.3}$$

Combining to get two equations, each in an individual variable,

$$\frac{d^2v}{dz^2} = ZYv, \quad \frac{d^2i}{dz^2} = ZYi$$

The last two LSI equations have exponential eigenfunction solutions of the form $\epsilon^{\gamma z}$. Substitute this into either one to get the eigenvalue γ

$$\gamma^2\epsilon^{\gamma z} = ZY\epsilon^{\gamma z}$$

so that the two eigenvalue solutions are

$$\gamma = \pm(\alpha + j\beta) = \pm\sqrt{ZY} \tag{7.1.4}$$

The eigenvalue γ is the *propagation function*, α is the *attenuation function*, and β is the *phase function*, (see eqs. (7.1.1), (7.1.2)). In our case

$$\alpha = 0, \quad \beta = \Im\sqrt{ZY} = \omega\sqrt{LC}$$

The linear combination of the two eigenfunction solutions, corresponding to $\pm\beta$, gives the general solution for (say) $v(z, j\omega)$, and simply retrieves eq. (7.1.1). The current variable $i(z, j\omega)$ is, from the first of eqs. (7.1.3),

$$i = -\frac{1}{Z}\frac{dv}{dz} = \frac{j\beta}{Z}A\epsilon^{-j\beta z} - \frac{j\beta}{Z}B\epsilon^{j\beta z} = i_i + i_r$$

Comparing this with eq. (7.1.2), we have the characteristic impedance

$$Z_0 = \frac{Z}{j\beta} = \frac{j\omega L}{j\omega\sqrt{LC}} = \sqrt{\frac{L}{C}}$$

which is a real positive constant whose value depends on the physical dimensions of the TEM line.

The wave nature of our solution is clarified if we reinsert the exponential time dependence, $\epsilon^{j\omega t}$, into eq. (7.1.1). For any fixed ω

$$v(z,t) = A \exp j(\omega t - \beta z) + B \exp j(\omega t + \beta z) = v_i + v_r \tag{7.1.5}$$

Suppose for a moment that the line has a small resistive loss, $\Re Z = r > 0$, so that referring to eq. (7.1.4), the propagation constant has a positive real part instead of being purely imaginary. According to eq. (7.1.5), it is clear that the first term, v_i, would then attenuate as we move away from the source in the positive z direction. The second term, v_r, increases in amplitude. We can therefore associate the first term with an *incident wave* moving down the line in the forward direction, i.e., away from the source, whereas the second term is a backward or *reflected wave* since it increases as the line termination, where the reflection occurs, is approached.

Now concentrate on the incident wave as we return to the lossless case. Noting that $(\omega t - \beta z) = \beta\,(\omega t/\beta\, - z)$

$$v_i(z,t) = A \exp j\beta \left(\frac{\omega}{\beta} t - z\right) \equiv A \exp j\beta \left[\frac{\omega}{\beta}\left(t + z_1 \frac{\beta}{\omega}\right) - (z + z_1)\right]$$

or

$$v_i(z,t) = v_i \left(z + z_1, t + z_1 \frac{\beta}{\omega}\right) \tag{7.1.6}$$

This forward wave, v_i, remains constant in amplitude ($|A(j\omega)|$) but its phase changes as z increases. Referring to eq. (7.1.6), the phase at a point z and time t takes on the same value at a distance z_1 down the line from z after a time delay $t_1 = z_1\beta/\omega$. This means that the phase front has traveled the distance z_1 at the *phase velocity*

$$v_p = \frac{z_1}{t_1} = \frac{\omega}{\beta} = \frac{1}{\sqrt{LC}} = \frac{1}{t_o} \tag{7.1.7}$$

where $t_o = \sqrt{LC}$ is the delay per unit length. For example, in air dielectric coaxial TEM line, t_o equals the reciprocal of the velocity of light, that is approximately 0.333 ns/m.

The TEM line is *nondispersive.* That is, the velocity of wave propagation is a constant independent of frequency, a consequence of the fact that the phase constant is proportional to frequency. Other modes, e.g., *transverse-electric* (TE) and *transverse magnetic* (TM) modes in rectangular wave guide, are *dispersive* because the velocity of propagation is a function of angular frequency.

Only the forward wave exists on a line which extends to infinity from the source. The impedance measured at any point on the infinite line is given by $Z_\infty(z) = v_i(z)/i_i(z) = Z_0$, the characteristic impedance. It is a constant independent of z, the distance measured away from the source. This immediately indicates that to only propagate a forward travelling wave on a TEM line of *finite* length, one terminates the line in the characteristic impedance Z_0. Under this boundary condition propagation on the finite guide emulates the infinite line. Terminating in an impedance other than Z_0 generates a reflected wave.

It is convenient to transfer the origin from the source end of the line to the termination at the end of the line $z = d$, with distance variable $x = d - z$ measured in the positive sense from load towards source. Then eqs. (7.1.1) and (7.1.2) become

$$v(x) = V_i \epsilon^{j\beta x} + V_r \epsilon^{-j\beta x}$$

$$i(x) = \frac{1}{Z_0}[V_i \epsilon^{j\beta x} - V_r \epsilon^{-j\beta x}]$$

The first term of these equations is the incident traveling wave, whose phase advances as we approach the source. The second term is the reflected traveling wave. The superposition of incident and reflected waves forms a *standing wave* which has maxima and minima phasor amplitudes along the line. The reflection factor on the line is defined as the ratio of the reflected to incident traveling waves,

$$\frac{V_r e^{-j\beta x}}{V_i e^{j\beta x}} = \rho e^{-2j\beta x} \qquad \rho = \frac{V_r}{V_i}$$

so that the reflection factor on a lossless line has a constant amplitude $|\rho|$, but varying phase. The boundary condition at the load is $v(0)/i(0) = Z_L$ or $(V_i + V_r)/(V_i - V_r) = Z_L/Z_0$, and dividing numerator and denominator by V_i

$$\frac{1+\rho}{1-\rho} = \frac{Z_L}{Z_0}, \text{ or } \rho = \frac{Z_L - Z_0}{Z_L + Z_0}$$

and ρ is just the reflection factor of the load impedance normalized to Z_0.

Let $\arg V_i = \phi_i$, $\arg V_r = \phi_r$, and $\theta = \beta x$. Then, when the phases of the travelling waves are equal, $\phi_i + \theta = \phi_r - \theta$, i.e., $\theta = (1/2)(\phi_i - \phi_r)$, the amplitude of $v(x)$ is the sum of the amplitudes, $|V_i| + |V_r|$, a maximum. When the phases of the traveling waves differ by π, the amplitudes subtract and we have a minimum, $|V_i| - |V_r|$. The ratio of maximum to minimum voltage is the *voltage standing wave ratio*, or VSWR. Thus

$$\text{VSWR} = \frac{|V_i| + |V_r|}{|V_i| - |V_r|} = \frac{1 + |\rho|}{1 - |\rho|} \geq 1 \tag{7.1.8}$$

the inequality stemming from $|\rho| \leq 1$, as the reflection factor of a passive load.

7.2 The Unit Element (UE); Richards' Transformation

The presence of incident and reflected waves on a finite, terminated, lossless TEM line suggests that it would be advantageous to analyze the system using scattering variables. Choose the normalizing number equal to the characteristic impedance Z_0. Compute the incident and reflected scattering variable phasors, $a(z)$, $b(z)$ from the voltage and current along the line (Section 4.3).

$$a(z) = \frac{1}{2}\left[\frac{v(z)}{\sqrt{Z_0}} + i(z)\sqrt{Z_0}\right] \qquad b(z) = \frac{1}{2}\left[\frac{v(z)}{\sqrt{Z_0}} - i(z)\sqrt{Z_0}\right]$$

Substituting eqs. (7.1.1), (7.1.2)

$$a(z) = \frac{A}{\sqrt{Z_0}}\epsilon^{-j\beta z} \equiv a(0)\epsilon^{-j\beta z} \qquad b(z) = \frac{B}{\sqrt{Z_0}}\epsilon^{j\beta z} \equiv b(0)\epsilon^{j\beta z} \tag{7.2.1}$$

Introducing the variable $x = d - z$ as in the previous Section, eqs. (7.2.1) become

$$a(x) = a(0)\epsilon^{j\beta x} \qquad b(x) = b(0)\epsilon^{-j\beta x} \tag{7.2.2}$$

where, of course, $a(0)$ and $b(0)$ now refer to the boundary values of incident and reflected scattering phasors measured at the point where the load terminates the line, $x = 0$.

The reflection factor $S(x)$ on the line looking towards the load is the ratio of reflected to incident wave variables

$$S(x) = \frac{b(x)}{a(x)} = \frac{b(0)}{a(0)}\epsilon^{-j2\beta x} = S(0)\epsilon^{-j2\beta x} \tag{7.2.3}$$

In eq. (7.2.3), $S(0)$ is the reflection factor of the terminating load. For example, if the termination is an open circuit, $S(0) = 1$ and $S(x) = \epsilon^{-j2\beta x}$. For a short circuit termination $S(0) = -1$ and $S(x) = -\epsilon^{-j2\beta x}$.

The impedance on the line $Z(x)$ can be readily determined from eq. (7.2.3) using the relations connecting reflection factor and impedance.

$$S(x) = \frac{Z(x) - Z_0}{Z(x) + Z_0} \qquad Z(x) = Z_0\frac{1 + S(x)}{1 - S(x)} \tag{7.2.4}$$

Replace $S(0)$ in eq. (7.2.3) using the first of eqs. (7.2.4) and then substitute into the second. We employ the identity

$$\epsilon^{-j2\theta} = \frac{1 - j\tan\theta}{1 + j\tan\theta} \tag{7.2.5}$$

which is based on Euler's formula. This is applied to obtain the final result relating the load impedance $Z(0) \equiv Z_L$ to the impedance elsewhere on the line, $Z(x)$,

$$Z(x) = Z_0 \frac{Z_L + jZ_0 \tan \beta x}{Z_0 + jZ_L \tan \beta x} \tag{7.2.6}$$

As an immediate check, note that if the termination is $Z_L = Z_0$, then $Z(x) = Z_0$, $\forall x$. The purely reactive impedances on a line terminated in a short circuit $Z_L = 0$, or an open circuit $Y_L = 1/Z_L = 0$ are readily found using eq. (7.2.6)

$$(Z_L = 0) : Z(x) = jZ_0 \tan \beta x \qquad (Z_L = \infty) : Z(x) = -jZ_0 \cot \beta x \tag{7.2.7}$$

Another useful property is that a one quarter wavelength line, ($\beta x_0 = 2\pi x_0/\lambda_g = \pi/2$) acts as an impedance inverter. Here λ_g is the guide wavelength. Referring to eq. (7.2.6), $\tan \beta x_0 = \infty$, and

$$Z(x_0) = \frac{Z_0^2}{Z_L}, \qquad \frac{x_0}{\lambda_g} = \frac{1}{4} \tag{7.2.8}$$

The *unit element* (UE), so-called because it is employed as a building block in the design of distributed circuits, is a two-port defined as a lossless TEM transmission line of prescribed finite length d. The characteristic impedance of the line, Z_0, is the circuit parameter which defines the element (d is fixed). The scattering matrix, $\boldsymbol{S} = (S_{ij})$, of the UE normalized to Z_0 is easily found. Terminate port 2 in Z_0 so that $b(0) = 0$, then $S(0) = 0$ in eq. (7.2.3). Based on this equation, the reflection factor at the input of the terminated UE is $S_{11} = S(d) = S(0)\exp(-j2\beta d) = 0$. Similarly, terminating port 1 in Z_0, we find $S_{22} = 0$. To find S_{21} note that when the port 2 termination is Z_0, then $a_2 = b(0) = 0$, and the two-port scattering equations, $(b_1,\ b_2)' = \boldsymbol{S}(a_1,\ a_2)'$, yield $b_2 = S_{21}a_1$. The variable, b_2, *reflected from port 2* of the UE is equal to the wave variable *incident* on the load, $b_2 = a(0)$. The incident wave on port 1 of the UE is given by eq. (7.2.2) as $a_1 = a(d) = a(0) \exp j\beta d$. Thus

$$S_{21} = \frac{b_2}{a_1} = \epsilon^{-j\beta d} = S_{12}$$

since the UE is reciprocal. Now make the substitution from eq. (7.1.7) $j\beta d = j\omega t_o d = s\tau$. The last term continues $j\omega$ into the complex frequency s-plane, and uses the *delay length* of the line defined as $\tau = t_o d$.

The scattering matrix of the UE can therefore be written

$$\boldsymbol{S} = \begin{pmatrix} S_{11} & S_{12} \\ S_{21} & S_{22} \end{pmatrix} = \begin{pmatrix} 0 & \epsilon^{-s\tau} \\ \epsilon^{-s\tau} & 0 \end{pmatrix}_{s=j\omega} = \begin{pmatrix} 0 & \epsilon^{-j\omega\tau} \\ \epsilon^{-j\omega\tau} & 0 \end{pmatrix} \tag{7.2.9}$$

If eq. (7.2.9) is compared with eq. (7.2.5) an almost trivially simple idea emerges which has extremely fruitful consequences for distributed parameter circuit design. Just make the frequency transformation (*Richards' Transformation*[2]) from ω to a new variable Ω

$$j\Omega = j\tan\frac{\omega\tau}{2} = \tanh\frac{j\omega\tau}{2}, \quad \text{or } \lambda = \tanh\frac{s\tau}{2} \tag{7.2.10}$$

where the second expression is a transformation of the complex s plane into a complex λ plane. Then using eqs. (7.2.5) and (7.2.10)

$$\epsilon^{-s\tau} = \frac{1-\lambda}{1+\lambda} \tag{7.2.11}$$

so that eq. (7.2.9) becomes

$$\boldsymbol{S} = \begin{pmatrix} 0 & \dfrac{1-\lambda}{1+\lambda} \\ \dfrac{1-\lambda}{1+\lambda} & 0 \end{pmatrix} \tag{7.2.12}$$

and the S matrix of the UE now involves *rational* functions of a (transformed) frequency variable rather than exponential functions. More generally, if a TEM line has a delay length which is commensurate to that of the UE, i.e., equal to $m\tau$ where m is any integer, then since

$$\epsilon^{-s(m\tau)} = \left(\frac{1-\lambda}{1+\lambda}\right)^m$$

the S (or immittance) matrix of a commensurate line depends only on rational functions of λ. It should also be evident that the scattering or immittance matrix of an n-port circuit formed by the arbitrary interconnection of resistors and commensurate UE elements is rational in λ. Some of the interconnected elements may, of course, be *stubs*, i.e., commensurate UE's terminated in short or open circuits.

Let $s = \sigma + j\omega$, $\lambda = \Sigma + j\Omega$, then Richards' transformation becomes

$$\lambda = \Sigma + j\Omega = \tanh\frac{s\tau}{2} = \frac{1-\epsilon^{-\sigma\tau}\epsilon^{-j\omega\tau}}{1+\epsilon^{-\sigma\tau}\epsilon^{-j\omega\tau}}$$

Examination of the equation shows that whenever $\sigma > 0$ then $\Sigma > 0$, and when s is purely imaginary, $\lambda = j\tan\frac{\omega\tau}{2}$,is purely imaginary as well. We therefore have the following Lemma.

[2]P. I. Richards, "Resistor-transmission-line circuits", *Proc. IRE*, vol. 36, pp. 217-220, Feb. 1948.

LEMMA 7.2.1
The transformation $\lambda = \tanh \frac{s\tau}{2}$ *maps the closed right half of the s plane into the closed right half of the* λ *plane.*[3]

Now recall (Chapter 4) that all the properties of BR and PR functions and matrices (whether rational or not) stem from their behavior in the closed right half plane. The following theorem, fundamental for the application of Richards' transformation, may therefore be stated.

THEOREM 7.2.1
Let $\mathcal{N}$ *be an n-port consisting of the interconnection of commensurate UE's, and resistors with BR scattering matrix,* $\boldsymbol{S}_1(s)$. *Then under Richards' transformation,* $\lambda = \tanh \frac{s\tau}{2}$, $\mathcal{N}$ *has a scattering matrix* $\boldsymbol{S}_2(\lambda)$ *which is BR and rational. A similar result applies to the PR immittance matrices of* $\mathcal{N}$ *under the same transformation.*

The use of the half-argument $s\tau/2$ is required so that the transmittances, S_{ij} become rational in λ. Frequently the emphasis is on the reflectances, S_{jj}, which, in most practical applications, remain rational even when the transformation is on $s\tau$. In the latter case, because the degree in λ of the scattering and immittance functions is reduced by half, it is often convenient to use the *modified Richards' transformation*[4]

$$\lambda = \tanh s\tau \tag{7.2.13}$$

Referring to eq. (7.2.7) we see that under the above transformation short and open circuited stubs of delay length τ have input impedances

$$Z_{\text{short}} = Z_0\lambda, \qquad Z_{\text{open}} = \frac{Z_0}{\lambda}$$

The stubs correspond to inductance and capacitance in the transformed domain.

As a consequence of the above theorem, Richards' transformation permits a useful class of distributed circuits to be treated by the methods of lumped network theory. The next two examples illustrate some techniques for the analysis of UE circuits. Further analysis and synthesis methods will be discussed in subsequent sections.

[3]Note that the inverse transformation $s = \frac{2}{\tau} \operatorname{artanh} \lambda$, being multivalued, maps the RHP in λ into horizontal RHP strips in s.

[4]Usually both half and full argument forms are referred to as Richards' transformation and are distinguished by their context.

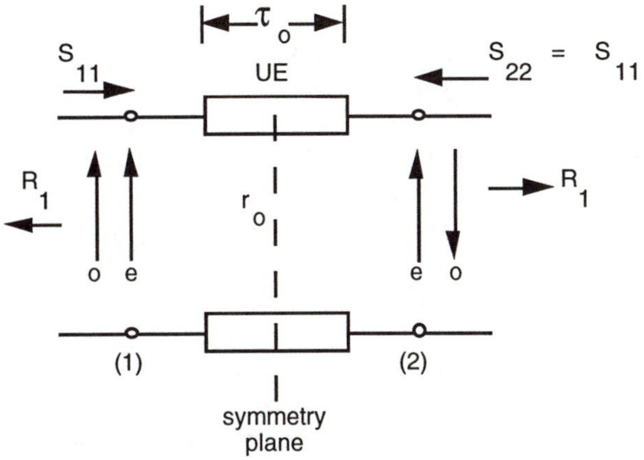

FIGURE 7.2.1
Unit Element mismatched to terminations.

Given an impedance Z, it is convenient, as in the following example and subsequent applications, to express its reflectance $S^{(j)}$, normalized to Z_j, in terms of its reflectance $S^{(i)}$, normalized to Z_i. Thus

$$S^{(i)} = \frac{Z - Z_i}{Z + Z_i} \qquad \text{and} \qquad Z = Z_i \frac{1 + S^{(i)}}{1 - S^{(i)}}$$

Or

$$S^{(j)} = \frac{Z - Z_j}{Z + Z_j} = \frac{Z_i \dfrac{1 + S^{(i)}}{1 - S^{(i)}} - Z_j}{Z_i \dfrac{1 + S^{(i)}}{1 - S^{(i)}} + Z_j} = \frac{(Z_i - Z_j) + S^{(i)}(Z_i + Z_j)}{(Z_i + Z_j) + S^{(i)}(Z_i - Z_j)}$$

Let

$$k_{ij} = \frac{Z_i - Z_j}{Z_i + Z_j} = -k_{ji} \tag{7.2.14}$$

Substituting eq. (7.2.14), we obtain the desired result

$$S^{(j)} = \frac{k_{ij} + S^{(i)}}{1 + S^{(i)} k_{ij}} \tag{7.2.15}$$

Example 7.2.1
Find the scattering matrix of the Unit Element shown in Fig. 7.2.1, i.e. of the UE of characteristic impedance r_0 and delay length τ_0, but connected to a system of impedance R_1 at the two ports.

Solution The simplest approach to this problem is to exploit the symmetry properties of the circuit. The UE is symmetric about a plane halfway down the line. If symmetric (even) excitation (e) is applied to the ports, the current at the symmetry plane is zero; in effect, an open circuit appears at $\tau_0/2$. If antimetric (odd) excitation (o) is applied, then the voltage is zero at the symmetry plane and a short circuit appears at $\tau/2$. The scattering matrix of the R_1 terminated UE will be found by superimposing the even and odd solutions for the structure.

Suppose an incident wave a_1 is applied to port 1. This corresponds to even excitation $a_1/2$, $a_1/2$, plus odd excitation $a_1/2$, $-a_1/2$ applied to the ports. We compute the reflected wave responses $b_1 = b_{1e} + b_{1o}$, $b_2 = b_{2e} + b_{2o}$, so as to determine the reflectance $S_{11} = S_{22} = b_1/a_1$, and transmittance $S_{21} = S_{12} = b_2/a_1$. For the odd mode, the input reflectance (short at mid-plane) at ports 1 and 2 is $S_o = -\exp(-s\tau)$, as in eq. (7.2.3) with $S(0) = -1$, $2j\beta x = s\tau$. Under even excitation (open at mid-plane), $S_e = \exp(-s\tau)$. These reflectances are normalized to r_0 of the UE. If we normalize to R_1 using eq. (7.2.15),

$$S_{11e} = \frac{k_{01} + e^{-s\tau}}{1 + k_{01}e^{-s\tau}} \qquad S_{11o} = \frac{k_{01} - e^{-s\tau}}{1 - k_{01}e^{-s\tau}}$$

For even excitation both incident signals are $a_1/2$, whereas for odd excitation the inputs are $a_1/2$, $-a_1/2$. Then $\boldsymbol{b} = \boldsymbol{S}\boldsymbol{a}$ are the two-port scattering relations to be applied at port 1.

$$b_{1e} = S_{11}\frac{a}{2} + S_{21}\frac{a}{2} = S_{11e}\frac{a}{2}$$

$$b_{1o} = S_{11}\frac{a}{2} - S_{21}\frac{a}{2} = S_{11o}\frac{a}{2}$$

Therefore,

$$S_{11e} = S_{11} + S_{21}, \qquad S_{11o} = S_{11} - S_{21}$$

Adding and subtracting these relations, the result is

$$S_{11} = \frac{1}{2}(S_{11e} + S_{11o}), \quad S_{21} = \frac{1}{2}(S_{11e} - S_{11o}) \tag{7.2.16}$$

These equations characterize any structurally symmetric two-port and are a consequence of *Bartlett's bisection theorem* originally published in 1930 for impedances (also generalized as Babinet's principle in electromagnetics). If S_{11e}, and S_{11o} are substituted in eq. (7.2.16), the final results are

$$S_{21} = \frac{(1 - k_{01}^2)\epsilon^{-s\tau}}{1 - k_{01}^2\epsilon^{-2s\tau}} \tag{7.2.17}$$

and

$$S_{11} = S_{22} = \frac{k_{01}(1 - \epsilon^{-2s\tau})}{1 - k_{01}^2\epsilon^{-2s\tau}} \tag{7.2.18}$$

The above scattering parameters can be transformed into the λ domain using either eq. (7.2.10) (half angle transformation) or eq. (7.2.13), the modified transformation. Under $\lambda = \tanh(s\tau/2)$, or $\epsilon^{-s\tau} = (1-\lambda)/(1+\lambda)$, eqs. (7.2.17), (7.2.18) become

$$S_{11}(\lambda) = \frac{k_{01}}{1-k_{01}^2} \frac{4\lambda}{\lambda^2 + 2\lambda \dfrac{1+k_{01}^2}{1-k_{01}^2} + 1}, \qquad \epsilon^{-s\tau} = \frac{1-\lambda}{1+\lambda} \tag{7.2.19}$$

and

$$S_{21}(\lambda) = \frac{1-\lambda^2}{\lambda^2 + 2\lambda \dfrac{1+k_{01}^2}{1-k_{01}^2} + 1}, \qquad \epsilon^{-s\tau} = \frac{1-\lambda}{1+\lambda} \tag{7.2.20}$$

Under the modified transformation (full angle)

$$S_{11}(\lambda) = \frac{k_{01}}{1-k_{01}^2} \frac{2\lambda}{1 + \lambda \dfrac{1+k_{01}^2}{1-k_{01}^2}}, \qquad \epsilon^{-2s\tau} = \frac{1-\lambda}{1+\lambda} \tag{7.2.21}$$

and

$$S_{21}(\lambda) = \frac{\sqrt{1-\lambda^2}}{1 + \lambda \dfrac{1+k_{01}^2}{1-k_{01}^2}}, \qquad \epsilon^{-2s\tau} = \frac{1-\lambda}{1+\lambda} \tag{7.2.22}$$

□

For the half angle transformation, eqs. (7.2.19), (7.2.20), all scattering functions are rational. In the case of the modified transformation equations (7.2.21), (7.2.22), the transmittance $S_{21}(\lambda)$ is irrational, but the reflectance is still a rational function. However, for the full angle transformation the two scattering parameters are substantially simpler than for the half angle case, i.e., the degrees are reduced by a factor of two. In practice, especially for the design of cascaded UE systems, the modified transformation is usually used. As we shall see, it is possible, despite transmittance irrationality, to carry out a synthesis using the rational reflectance function.

The analysis of a circuit comprising UE's and resistance is illustrated in the next example. In this and subsequent discussions, it will sometimes be useful to employ the impedance and admittance matrices of a UE. We have the unit normalized scattering matrix $\boldsymbol{S}$ of the UE in the λ domain, eq. (7.2.11). The $\boldsymbol{Z}$ matrix for a UE of characteristic impedance R_0 is therefore $\boldsymbol{Z} = R_0(\boldsymbol{I}+\boldsymbol{S})(\boldsymbol{I}-\boldsymbol{S})^{-1}$, where $\boldsymbol{I}$ is the 2×2 identity matrix. Then with $\lambda = \tanh\theta/2$, $\theta = j\omega\tau$

$$\boldsymbol{Z} = R_0 \begin{pmatrix} 1 & \dfrac{1-\lambda}{1+\lambda} \\ \dfrac{1-\lambda}{1+\lambda} & 1 \end{pmatrix} \begin{pmatrix} 1 & -\dfrac{1-\lambda}{1+\lambda} \\ -\dfrac{1-\lambda}{1+\lambda} & 1 \end{pmatrix}^{-1} = R_0 \begin{pmatrix} \dfrac{1+\lambda^2}{2\lambda} & \dfrac{1-\lambda^2}{2\lambda} \\ \dfrac{1-\lambda^2}{2\lambda} & \dfrac{1+\lambda^2}{2\lambda} \end{pmatrix}$$

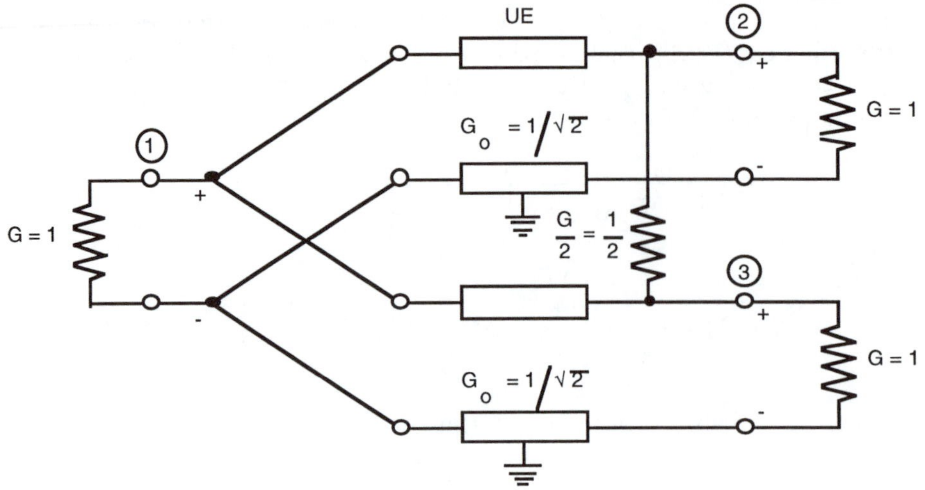

FIGURE 7.2.2
Equivalent circuit for Wilkinson power divider.

and with $G_0 = 1/R_0$, the admittance matrix is

$$\boldsymbol{Y} = \boldsymbol{Z}^{-1} = G_0 \begin{pmatrix} \dfrac{1+\lambda^2}{2\lambda} & -\dfrac{1-\lambda^2}{2\lambda} \\ -\dfrac{1-\lambda^2}{2\lambda} & \dfrac{1+\lambda^2}{2\lambda} \end{pmatrix} = \begin{pmatrix} y_1 \; y_2 \\ y_2 \; y_1 \end{pmatrix} \tag{7.2.23}$$

The expressions in terms of the full angle $\lambda = \tanh\theta$ are as follows:

$$Z_{11} = Z_{22} = R_0 \frac{1}{\lambda} \qquad Z_{21} = Z_{12} = R_0 \frac{\sqrt{1-\lambda^2}}{\lambda}$$

$$Y_{11} = Y_{22} = G_0 \frac{1}{\lambda} \qquad Y_{21} = Y_{12} = -G_0 \frac{\sqrt{1-\lambda^2}}{\lambda}$$

Example 7.2.2
The Wilkinson power divider,[5] with an equivalent circuit as shown in Fig. 7.2.2, is a relatively simple transmission line structure which functions as a three-port matched power splitter at a frequency where the electrical length of the UE is one-quarter wavelength. Demonstrate this result, and find the frequency response to illustrate the wideband properties of the device.

[5]E. J. Wilkinson, "An N-Way Hybrid Power Divider", *IRE Trans. on Microwave Th. and Tech.*, vol. 8, pp. 116-118, Jan. 1960.

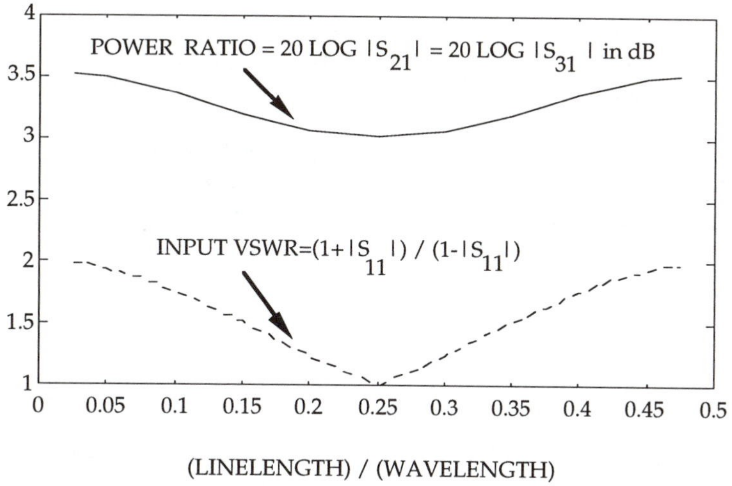

FIGURE 7.2.3
Frequency response of Wilkinson power divider.

Solution Referring to the equivalent circuit, the power divider has a network graph with three independent nodes. The datum node is one side of the UE's, which might be the ground plane of microstrip or the outer casing of coaxial line. The UE's introduce coupling between the node to datum branches, defined by y_1 and y_2 of the admittance matrix of a UE, eq. (7.2.23). Using the methods of nodal analysis, Section 2.9, the 3×3 nodal admittance matrix can be written by inspection. Thus

$$\boldsymbol{Y} = \begin{pmatrix} 2y_1 & y_2 & y_2 \\ y_2 & y_1 + \dfrac{G}{2} & -\dfrac{G}{2} \\ y_2 & -\dfrac{G}{2} & y_1 + \dfrac{G}{2} \end{pmatrix}$$

Assume $G = 1$ for the load and source conductances. Then the scattering matrix of the system is $\boldsymbol{S} = (\boldsymbol{I}_3 - \boldsymbol{Y})(\boldsymbol{I}_3 + \boldsymbol{Y})^{-1}$. The most direct way of determining $\boldsymbol{S}$ is to implement the calculation with a simple computer routine, say, using the commercial program MATLAB. The divider requires that the normalized characteristic admittance of the UE be $G_0 = 1/\sqrt{2}$. When the electrical length of each UE is one-quarter wavelength, $\theta = \pi/4$, and $\lambda = \tanh j\theta = j \tan \pi/4 = j$, so that $y_1 = 0$, $y_2 = j/\sqrt{2}$, and $\boldsymbol{S}$ is computed as

$$\boldsymbol{S} = \frac{1}{\sqrt{2}} \begin{pmatrix} 0 & -j & -j \\ -j & 0 & 0 \\ -j & 0 & 0 \end{pmatrix}$$

so that at the quarter wavelength point, precisely one-half the available power at port 1 is delivered to each of ports 2 and 3 ($|S_{21}|^2 = |S_{31}|^2 = 0.5$). Also the input to port 1 is matched, i.e., the voltage standing wave ratio, eq. (7.1.8), in the input line, VSWR $= (1 + |S_{11}|)/(1 - |S11|) = 1$. The frequency response is readily determined by calculating $\boldsymbol{Y}(\lambda)$ and then computing the elements of $\boldsymbol{S}(\lambda)$. The results are shown on Fig. 7.2.3, where fraction of available power in dB delivered to ports 2 and 3, as well as input mismatch expressed as VSWR, are plotted against electrical length of the UE's. □

7.3 Richards' Theorem: UE Reactance Functions

Richards' Theorem provides the basis for the synthesis of distributed structures made up of cascaded UE sections of differing characteristic impedances. We consider a reflectance $S(\lambda)$ which is rational and BR under the full angle Richards Transformation, $\lambda = \tanh s\tau$ or $\epsilon^{2s\tau} = (1+\lambda)/(1-\lambda)$. S is termed a *lambda function*. The problem at hand is to determine the possibility of extracting a UE from $S(\lambda)$ so as to leave a remainder which is also a BR λ-function.

Let a prescribed BR λ-function be $S_1^{(0)}(\lambda)$, where the superscript corresponds to positive real normalization impedance r_0, and the subscript indicates the terminal plane location. Now extract a UE of characteristic impedance r_1, and determine the resulting reflectance measured at the output of the UE, i.e., at terminal plane 2, and normalized to r_1. Thus designating the reflectance at port 1 with normalization r_0 as $S_1^{(0)}$, we transform this function to the output of the UE using the relation eq. (7.2.3), $S_2^{(1)} = S_1^{(1)}\epsilon^{2s\tau}$. Next invoke the change in normalization formula, eq. (7.2.15), which expresses $S_1^{(1)}$ in terms of normalization number r_0 rather than r_1. The result is

$$S_2^{(1)} = S_1^{(1)}\epsilon^{2s\tau} = \frac{k_{01} + S_1^{(0)}}{1 + k_{01}S_1^{(0)}}\,\frac{1+\lambda}{1-\lambda} \tag{7.3.1}$$

The amplitude of the remainder reflectance $S_2^{(1)}$ on the $\lambda = j\Omega$ boundary is unchanged, since $|(1+j\Omega)/(1-j\Omega)| = 1$. Furthermore, note that $S_1^{(1)}$ is BR for it is simply the prescribed $S_1^{(0)}$ with a different real positive normalization number. Evidently were it not for the simple pole introduced in the RHP at $\lambda = 1$, the remainder, $S_2^{(1)}$, would be a BR λ-function. We now show that the RHP pole factor can always be canceled if the characteristic impedance of the extracted UE is suitably chosen.

Let k_{01} be chosen so that a numerator zero is introduced at $\lambda = 1$ by setting $k_{01} = -S_1^{(0)}(1)$. Then

$$k_{01} = \frac{r_0 - r_1}{r_0 + r_1} = -S_1^{(0)}(1), \quad \text{or} \quad r_1 = r_0 \frac{1 + S_1^{(0)}(1)}{1 - S_1^{(0)}(1)} = z_1(1) > 0 \qquad (7.3.2)$$

In eq. (7.3.2) we have solved for the nonnormalized input impedance z_1 (at $\lambda = 1$) in terms of the input reflection factor, $S_1^{(0)}$. Since the input impedance is PR, its value, $z_1(1)$ as shown, is a positive real number, equal to the characteristic impedance of a UE $r_1 = z_1(1)$ which when extracted leaves a remainder which is a BR reflectance. This result is known as Richards' Theorem.[6]

THEOREM 7.3.1

Given a rational BR λ function reflectance, $S_1^{(0)}$, and corresponding PR impedance, z_1. A UE, whose characteristic impedance $r_1 = z_1(1) > 0$, can always be extracted leaving a remainder reflectance $S_2^{(1)}$, which is also a BR rational λ function. Equivalently the remainder expressed as an impedance, $z_2(\lambda)$, is a rational PR function.

$$S_2^{(1)} = \frac{S_1^{(0)}(\lambda) - S_1^{(0)}(1)}{1 - S_1^{(0)}(1)S_1^{(0)}(\lambda)} \, \frac{1+\lambda}{1-\lambda}, \quad z_2(\lambda) = z_1(1)\frac{z_1(\lambda) - \lambda z_1(1)}{z_1(1) - \lambda z_1(\lambda)} \qquad (7.3.3)$$

The first equation cited in the theorem is obtained by replacing k_{01} with $-S_1^{(0)}(1)$ in eq. (7.3.1). The equation for z_2 is found by calculating the impedance corresponding to reflectance $S_2^{(1)}$.

But there is still a fly in the ointment. Under repeated extractions, the process does not converge,since in eq. (7.3.3), the degree of the remainder reflectance is unchanged by line extraction, for it was lowered by 1 due to cancellation of numerator factor $1 - \lambda$, but promptly raised again by 1 because of the numerator factor $1 + \lambda$. The degree would be lowered if a root factor at $\lambda = -1$ could be introduced into the denominator of $S_2^{(1)}$, but there are no other free parameters available. We can, however, make the process converge for an important special class of functions, namely, those corresponding to lossless one-ports. For the lossless case, the reflectance is LBR, hence $S_1^{(0)} S_{1*}^{(0)} = S_1^{(0)}(\lambda) S_1^{(0)}(-\lambda) = 1, \forall\lambda$, so that in the denominator $1 - S_1^{(0)}(1)S_1^{(0)}(-1) = 0$, precisely the condition needed to cancel the numerator root at $\lambda = -1$. We have seen that the extraction always leads to a BR remainder; therefore, starting with an LBR reflectance from

[6] P. I. Richards, "Resistor-transmission-line circuits", *Proc. IRE*, vol. 36, pp. 217-220, Feb. 1948.

which a lossless UE is extracted, the result must be an LBR remainder, which moreover is reduced in degree, so the iterated process terminates in an open or short circuit.

THEOREM 7.3.2
Given a rational LBR reflectance $S_1^{(0)}(\lambda)$. Repeated UE extractions according to Richards' Theorem gives a realization consisting of a cascade of a finite number of UE's of different characteristic impedances terminated in an open or short circuit when the remainder reflectance is ± 1.

Example *7.3.1*
Consider a one-port consisting of a unit characteristic impedance UE with output open circuited (termed an open circuited *stub*), in series with a short circuited UE of characteristic impedance 2. Find a cascaded UE equivalent (also termed a *stepped transmission line*) for the one port.

Solution The open circuited stub has input impedance $\coth s\tau = 1/\lambda$, and the shorted stub has impedance 2λ, so the series connection results in the one-port LPR impedance $z_1 = (1+2\lambda^2)/\lambda$. This is a reactance function and the series connection of the stubs is a Foster realization. (Section 5.2). We are to find the equivalent cascade of UE's. We can extract the UE's employing either the reflectance or impedance form of eq. (7.3.3). Use impedance, then the first UE has characteristic impedance $r_1 = z_1(1) = 3$, and the remainder is

$$z_2 = 3\frac{\dfrac{1+2\lambda^2}{\lambda} - 3\lambda}{3-(1+2\lambda^2)} = 3\frac{1-\lambda^2}{2\lambda(1-\lambda^2)}$$

The expected common factors $(1+\lambda)(1-\lambda)$ are present. After cancellation, the input impedance to the next UE is $z_2 = 3/(2\lambda)$. The second extraction step can be omitted, since it is clear that the second line in the cascade is simply a short circuited UE of characteristic impedance 3/2. The final result is a stepped transmission line of two UE's, with characteristic impedances $r_1 = 3$ and $r_2 = 3/2$, and a short circuit at the output port. □

7.4 Doubly Terminated UE Cascade

An important group of distributed parameter components such as filters, wideband transformers, and couplers, can be realized using stepped transmission line two-ports connected at input and output to TEM transmission

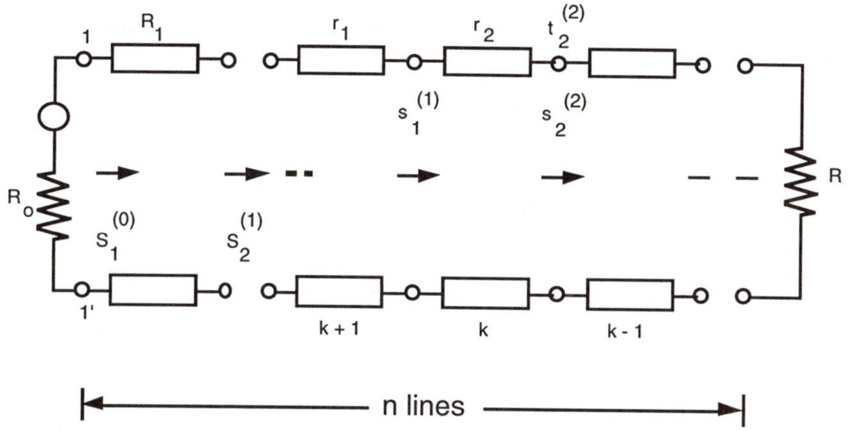

FIGURE 7.4.1
Stepped transmission line.

lines, corresponding to resistive terminations. The following theorem is basic for the design of these structures.

THEOREM 7.4.1
Let $S_{21}(\lambda)$, $\lambda = \tanh s\tau = \Sigma + j\Omega$, be the transmittance of a doubly terminated two-port. The necessary and sufficient conditions that S_{21} be realizable as a cascade of n UE's (stepped transmission line) is that

$$\text{(a)} \quad S_{21}S_{21*} = \frac{(1-\lambda^2)^n}{P_n(-\lambda^2)}, \qquad \text{(b)} \quad 0 \le |S_{21}(j\Omega)|^2 \le 1 \qquad (7.4.1)$$

where $P_n(\Omega^2)$ is an even polynomial of degree $2n$ with real coefficients.

PROOF Necessity: We are given a stepped line with n UE's and wish to show that the transmittance satisfies eq. (7.4.1 (a)). We proceed by induction. Refer to Fig. 7.4.1, and let the reflectance measured at the output of the k-th UE (counted back from the output) be $s_2^{(2)}(\lambda)$, normalized to the characteristic impedance r_2 of line k. Since the UE's form a lossless two-port, the transmittance $t_2^{(2)}$ of the $k-1$ line chain, normalized to r_2 at the input, R at the output of the chain, satisfies $t_2^{(2)}t_{2*}^{(2)} = 1 - s_2^{(2)}s_{2*}^{(2)}$, and the induction hypothesis is that $t_2^{(2)}t_{2*}^{(2)}$ has the form of eq. (7.4.1 (a)) with $n = k-1$. It will be noted that this assumption implies that $t_2^{(2)}t_{2*}^{(2)}$ and $s_2^{(2)}s_{2*}^{(2)}$ are rational functions of the same degree, and that the numerator and denominator polynomials of these functions are of equal degree. We must next show that at the output of line $k+1$ of characteristic impedance r_1, the transmittance $t_1^{(1)}$ of the k line chain normalized to r_1 and R, satisfies eq. (7.4.1 a) with $n = k$.

Referring to eq. (7.2.12), the reflectance at the output of line k can be stepped back to the input of line k. Thus $s_1^{(2)} = s_2^{(2)}(1-\lambda)/(1+\lambda)$. We now apply the change in normalization, eq. (7.2.15), to obtain $s_1^{(1)}$.

$$s_1^{(1)} = \frac{k_{21} + s_1^{(2)}}{1 + k_{21}s_1^{(2)}} = \frac{(1+\lambda)k_{21} + s_2^{(2)}(1-\lambda)}{(1+\lambda) + (1-\lambda)k_{21}s_2^{(2)}} = \frac{(1+\lambda)k_{21} + s_2^{(2)}(1-\lambda)}{D(\lambda)}$$

The equality of numerator and denominator degrees has been retained but the degrees are increased by one. Since $t_1^{(1)}t_{1*}^{(1)} = 1 - s_1^{(1)}s_{1*}^{(1)}$, the transmittance product has its numerator and denominator degrees increased by two, and

$$t_1^{(1)}t_{1*}^{(1)} = \frac{(1-\lambda^2)(1-k_{21}^2)(1-s_2^{(2)}s_{2*}^{(2)})}{DD_*} = \frac{(1-\lambda^2)(1-k_{21}^2)}{DD_*}t_2^{(2)}t_{2*}^{(2)}$$

The induction hypothesis has $(1-\lambda^2)^{k-1}$ as the numerator of $t_2^{(2)}t_{2*}^{(2)}$. Referring to the last equation, moving one UE towards the source (k UE's) raises the factor to $(1-\lambda^2)^k$, and accounts for the total degree of the k line numerator. We have already shown that the denominator degree has been raised by two. To complete the induction proof, we need only establish a numerator of $1-\lambda^2$ for the transmittance product of the first UE at the output end of the cascade, and show that numerator and denominator degrees of this initial UE are equal. This is immediately verified by eq. (7.2.22) which gives the transmittance of a single UE with arbitrary normalization. Thus eq. (7.4.1 (a)) is established for an n line cascade. The inequality is valid as well since the system is passive.

Sufficiency: The transmittance satisfies eq. (7.4.1 (a)); we must demonstrate a stepped line synthesis assuming source impedance R_0 and load impedance R. Proceed according to the Darlington approach, and using the Belevitch notation (Section 5.4), set $S_{11}S_{11*} = 1 - S_{21}S_{21*} = hh_*/gg_*$, a rational function. Then since by hypothesis, $0 \le |h(j\Omega)|^2/|g(j\Omega)|^2 \le 1$, we can use spectral factorization to obtain $S_{11} = h/g$, where h, and g are polynomials in λ, g is Hurwitz, and S_{11} is BR.

Refer to Fig. 7.4.1, but note a slightly modified notation for line extraction. We now remove a line of characteristic impedance R_1 at the input (terminal plane $1, 1'$) using Richards' Theorem 7.3.1. For convenience, $S_{11} \equiv S_1^{(0)}$. Then the remainder reflectance function normalized to R_1 is

$$S_2^{(1)} = \frac{S_1^{(0)}(\lambda) - S_1^{(0)}(1)}{1 - S_1^{(0)}(\lambda)S_1^{(0)}(1)}\,\frac{1+\lambda}{1-\lambda}$$

The reflectance is BR since Richards' Theorem guarantees cancellation of the $1-\lambda$ factor when the proper R_1 UE is removed. However, for convergence under repeated line removals, it is necessary that the denominator have canceling root factors at $\lambda = -1$ as well. We consider this possibility

$$1 - S_1^{(0)}(\lambda)S_1^{(0)}(1)|_{\lambda=-1} = 1 - S_{11}(\lambda)S_{11}(-\lambda)|_{\lambda=\pm 1} = S_{21}(\lambda)S_{21}(-\lambda)|_{\lambda=\pm 1}$$

But, referring to eq. (7.4.1 (a))

$$S_{21}(\lambda)S_{21}(-\lambda)|_{\lambda=\pm 1} = \left.\frac{(1-\lambda^2)^n}{P_n(-\lambda^2)}\right|_{\lambda=\pm 1} = 0$$

Therefore the $1+\lambda$ cancellation occurs, the reflectance degree is reduced by 1, and we can repeat the extraction n times corresponding to the factor $(1-\lambda^2)^n$ in $S_{21}S_{21*}$, ending with the terminating resistor when we reach degree zero. ∎

Example 7.4.1

Let $\lambda = \tanh s\tau = \Sigma + j\Omega$, and consider the transducer gain function for unit resistor terminations

$$T(\Omega^2) = \frac{(1+\Omega^2)^2}{1+7\Omega^2+\Omega^4}$$

Find a UE realization of the gain.

Solution It is evident that the prescribed transducer gain satisfies Theorem 7.4.1, so that we can obtain a stepped line realization. We let $\lambda = j\Omega$ and write $S_{21}S_{21*} = (1-\lambda^2)/(1-7\lambda^2+\lambda^4)$ so that

$$S_{11}S_{11*} = 1 - S_{21}S_{21*} = \frac{-5\lambda^2}{1-7\lambda^2+\lambda^4}$$

Reflectance $S_{11} = h/g$ and $h = \pm\sqrt{5}\lambda$. In this elementary example, the Hurwitz polynomial g can be found by a simple device that avoids factorization since a quadratic has no RHP roots if its coefficients are positive. Thus

$$(1+a\lambda+\lambda^2)(1-a\lambda+\lambda^2) = 1-7\lambda^2+\lambda^4$$

Equating coefficients of λ^2, $a = 3$. Therefore if we choose the (+) sign for h

$$S_{11} = \frac{\sqrt{5}\lambda}{1+3\lambda+\lambda^2}$$

We might also note that we take $S_{21} = f/g = (1-\lambda^2)/h$. Here $n = 2$ in $S_{21}S_{21*}$; were n odd, S_{21} would have an irrational numerator, although the reflectance would remain rational. The normalized generator and load terminations are both unity since $S_{21}(0) = 1$. From S_{11}, with $r_0 = 1$, we find the input impedance

$$z_1 = \frac{1+S_{11}}{1-S_{11}} = \frac{1+(3+\sqrt{5})\lambda+\lambda^2}{1+(3-\sqrt{5})\lambda+\lambda^2}$$

Using Richards Theorem 7.3.1, the characteristic impedance of the first line $r_1 = z_1(1) = (\sqrt{5}+1)/(\sqrt{5}-1)$, and the input to the second line (omitting much of the algebra) is

$$z_2 = r_1 \frac{z_1 - \lambda r_1}{r_1 - \lambda z_1} = r_1 \frac{(1-\lambda^2)(1+r_1\lambda)}{(1-\lambda^2)(r_1+\lambda)} = r_1 \frac{1+r_1\lambda}{r_1+\lambda}$$

Again apply Richards' Theorem, $r_2 = z_2(1) = r_1$ and the remainder is

$$z_3 = r_1 \frac{r_1 \dfrac{1+r_1\lambda}{r_1+\lambda} - \lambda}{1 - \lambda \dfrac{1+r_1\lambda}{r_1+\lambda}} = r_1 \frac{1-\lambda^2}{r_1(1-\lambda^2)} = 1$$

as expected. □

The final realization consists of two UE's in cascade, both of the same characteristic impedance $r_1 = r_2 = (\sqrt{5}+1)/(\sqrt{5}-1)$, with unit load termination. The system is symmetric as expected since $S_{22} = -h_*/g = S_{11}$. A check on the realization is to note that at $\Omega = \infty$, the UE's are quarter wavelength lines. Using the inversion property, eq. (7.2.8), the unit load is transformed to r_1^2 by the output line, and then inverted again to unity by the input UE, so that $S_{11}(\infty) = 0$, as required by the prescribed reflectance. Finally, it should be noted that once the stepped line property of $S_{21}S_{21*}$ has been established, a numerically superior computational procedure to that illustrated above is to compute the parameters of the *unterminated* structure, as in the Darlington synthesis, and apply the lossless version of Richards Theorem 7.3.2.[7]

7.5 Stepped Line Gain Approximations

An n line cascaded UE structure has a transducer gain based on the form for $S_{21}S_{21*}$ given in eq. (7.4.1).

$$T(\Omega^2) = \frac{(1+\Omega^2)^n}{P_n(\Omega^2)}, \quad \Omega = \tan\theta, \quad \theta = \omega\tau$$

Several points ought to be noted in connection with the gain equation. Along the real frequency axis, Richards' variable has a complete excursion

[7] J. Komiak and H. J. Carlin, "Improved Accuracy for Commensurate-Line Synthesis", *IEEE Trans. on Microwave Th. and Tech.*, pp 212-215, April 1976.

of $-\infty < \Omega < +\infty$, corresponding to $-\pi/2 < \theta < \pi/2$. Since the system is periodic in θ, the gain response will repeat in real frequency for every interval of $\theta = k\pi$, but the approximation problem is concerned with performance in the Ω domain, that is, as the electrical length of the UE goes from 0 to one quarter wavelength. The restrictive form of the gain function precludes any real frequency zeros of transmission since the only zero of the numerator is at $\Omega = \pm j$. In particular, since the numerator and denominator of T are even polynomials of the same degree, there is no zero of transmission at $\Omega = \infty$, which limits the resources for finding approximations for filters of large discrimination and selectivity. Nevertheless, stepped lines can be designed to provide useful components, and we consider below a procedure for solving the approximation problem.

By a change in variable, the transducer gain can be put into the form of $T = 1/D$, where D is a polynomial, thereby simplifying the mechanics of the approximation problem. Note the identity, $1 + \Omega^2 = 1 + \tan^2\theta = 1/\cos^2\theta$, and

$$P_n(\Omega^2) = \sum_{k=0}^{n} a_{2k}\Omega^{2k} = \sum_{k=0}^{n} a_{2k}\tan^{2k}\theta = \sum_{k=0}^{n} a_{2k}\frac{\sin^{2k}\theta}{\cos^{2k}\theta}$$

Then

$$\begin{aligned} T(\Omega^2) &= \left[\cos^{2n}\theta\sum_{k=0}^{n} a_{2k}\frac{\sin^{2k}\theta}{\cos^{2k}\theta}\right]^{-1} \\ &= (a_0\cos^{2n}\theta + a_2\cos^{2n-2}\theta\sin^2\theta + \ldots + a_{2n}\sin^{2n}\theta)^{-1} \end{aligned}$$

Now make the replacement $\sin^{2k}(\theta) = (1 - \cos^2\theta)^k$, and introduce the new variable $x = \alpha_1\cos\theta$, where α_1 is a constant which will be used for setting bandwidth. Note that for $0 \le \Omega < +\infty$, or $0 \le \theta < \pi/2$, we have $\alpha_1 \ge x > 0$, and the gain function expressed in terms of x satisfies

$$0 < T(x^2) = \left[\sum_{k=0}^{n} A_{2k}x^{2k}\right]^{-1} \le 1, \quad 0 < |x| \le \alpha_1, \quad x = \alpha_1\cos\theta$$

where the A_{2k} are constants to be determined by the approximation procedure.

Substituting $\cos^{2k}\theta = (1 - \sin^2\theta)^k$, with $y = \alpha_2\sin\theta$, a similar polynomial form results

$$0 < T(y^2) = \frac{1}{D(y^2)} \le 1, \quad 0 < |y| \le \alpha_2, \quad x = \alpha_2\sin\theta$$

where $D(y^2) = B_0 + B_2y^2 + \ldots + B_{2n}y^{2n}$. The variable y is suitable for lowpass problems, since $\theta = \omega\tau = 0$ transforms into $y = 0$. On the other hand, a prescribed band centered about $x = 0$ corresponds to θ

centered about $\pi/2$, so that x can be considered a "bandpass" variable. In subsequent discussions we will use the single symbol x for either case; the context will make clear whether the lowpass or bandpass variable is being employed.

The Butterworth and Chebyshev approximations, discussed in Sections 6.5 and 6.6, can be immediately applied to stepped line transducer gain problems.

$$T(x^2) = \frac{1}{1 + x^{2n}} \qquad \text{Butterworth} \tag{7.5.1}$$

$$T(x^2) = \frac{1}{1 + \epsilon^2 T_n^2(x)} \qquad \text{Chebyshev} \tag{7.5.2}$$

The maximally flat and equal ripple properties of the two transducer gain forms are retained although the frequency scaling is distorted by the change in variable. Furthermore, the realizability restrictions for a cascade of UE's are satisfied. That is, when the x variable is transformed back to Ω, the transmittance representation satisfies Theorem 7.4.1. The λ domain is restored (and vice versa) by the following substitutions. In the case of the lowpass variable $x = \alpha \sin\theta$

$$x^2 = \alpha^2 \sin^2\theta = \alpha^2 \frac{1}{\csc^2\theta} = \frac{\alpha^2 \tan^2\theta}{1 + \tan^2\theta} = \frac{\alpha^2\Omega^2}{1 + \Omega^2} \tag{7.5.3}$$

$$\lambda^2 = -\Omega^2 = \frac{x^2}{x^2 - \alpha^2} \tag{7.5.4}$$

In the case of the bandpass variable

$$x^2 = \alpha^2 \cos^2\theta = \alpha^2 \frac{1}{\sec^2\theta} = \frac{\alpha^2}{1 + \Omega^2} \tag{7.5.5}$$

$$\lambda^2 = \frac{x^2 - \alpha^2}{x^2} \tag{7.5.6}$$

We can obtain $T(\Omega^2) = |S_{21}(j\Omega)|^2 \to S_{21}S_{21*}$ by using the above equations. Then, in principle, spectral factorization can be employed to find the reflectance S_{11}, and the stepped line is synthesized by Richards' Theorem as discussed in the preceding Section. Generally it is more convenient to carry out the spectral factorization in the x domain and subsequently find the appropriate LHP roots in λ. In the process one must not forget that a polynomial is defined by its roots *and* a multiplicative constant; omitting the latter is a common cause of error.

To illustrate these ideas consider the Butterworth gain function with $x = \alpha\cos\theta$. Let $u = x^2$, with u_k designating the n roots of $1 + u^n = 0$, $u_k = \exp j(2k-1)\pi/n$, $k = 1, 2, \ldots, n$. Then apply eq. (7.5.6) to find the roots $\lambda_k^2 = (x_k^2 - \alpha^2)/x_k^2$. The complete spectrum of roots is $\pm\sqrt{\lambda_k^2}$, and from these we select those λ_k roots which are in the LHP. Recall that

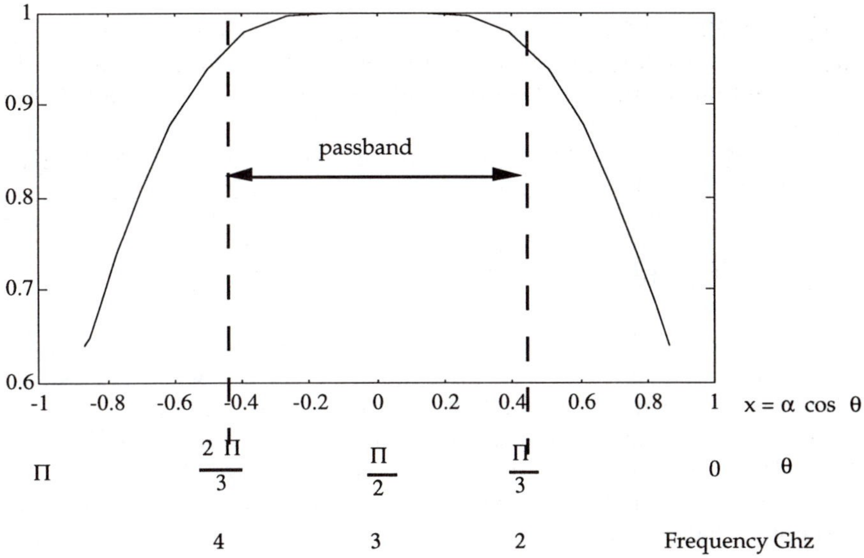

FIGURE 7.5.1
Transducer gain of two section 1:4 Butterworth impedance transformer.

in the Belevitch notation, $S_{21} = f/g$, and $S_{11} = h/g$. The polynomial g contains the LHP roots and, referring to eqs. (7.5.5) and (7.5.1), has a unit multiplier since this is the coefficient of λ^n. The expression for transmittance is therefore

$$S_{21} = \frac{(\sqrt{1-\lambda^2})^n}{(\lambda-\lambda_1)\ldots(\lambda-\lambda_n)} = \frac{f}{g} \qquad \Re\,\lambda_k < 0 \tag{7.5.7}$$

Again referring to eqs. (7.5.5), (7.5.1), and using $S_{11}S_{11*} = 1 - S_{21}S_{21*}$, the input reflectance is

$$S_{11} = \frac{\pm\alpha^n}{g} \tag{7.5.8}$$

In the case of the lowpass variable $x = \alpha \sin\theta$, the multiplier is $\sqrt{1+\alpha^{2n}}$, and the transmittance and reflectance have the forms

$$S_{21} = \frac{(\sqrt{1-\lambda^2})^n}{\sqrt{1+\alpha^{2n}}(\lambda-\lambda_1)\ldots(\lambda-\lambda_n)} = \frac{f}{g} \quad \Re\,\lambda_k < 0 \tag{7.5.9}$$

$$S_{11} = \frac{\pm\alpha^n\lambda^n}{g} \tag{7.5.10}$$

Stepped Line Impedance Transformers

The transducer gain of wideband cascaded UE impedance transformers is best expressed in terms of the bandpass variable $x = \alpha \cos\theta$, for at d.c. the lines have zero electrical length and the input reflectance must be exactly that due to the mismatched load and generator impedances. The impedances can be matched about a center frequency chosen at the point where the electrical length of the UE's are one quarter wavelength, i.e., $\theta = \pi/2$. For example, suppose the load and generator impedances to be matched are R_1, R_2. Then for a one line matching transformer choose the characteristic impedance of the UE as $r_1 = \sqrt{R_1 R_2}$, for then, by the inversion property of a quarter wave line, eq. (7.2.8), the input impedance is $r_1^2/R_2 = R_1$. Let us consider the broadbanding of a quarter wave transformer by using a stepped transmission line whose transducer gain function is the Butterworth characteristic of eq. (7.5.1). Using the bandpass variable, the response must be symmetric about $\theta = \pi/2$, where the system is matched and the reflectance is set to zero. Thus $x = 0$ at the center of the passband.

Suppose that the gain is $T(x_L)$ at the edge of the passband, θ_L, where $\theta_L < \pi/2$. Then by symmetry about $\pi/2$, the same loss occurs at $\theta_H = \pi - \theta_L$, see Fig. 7.5.1. This means that $x_H = \alpha\cos(\pi - \theta_L) = -x_L$, so the gain is a lowpass characteristic, symmetric in x. To find line length d, we note that due to the arithmetic symmetry in $\theta = \omega\tau$, the midband frequency occurs at the arithmetic mean between θ_L, and θ_H and, at this midband point, the UE's are one quarter wavelength.

The two other free parameters are α and n. At d.c., $x = \alpha$, and the reflectance at the input is $S_0 = (R_2 - R_1)/(R_2 + R_1)$, so that $T(\alpha) = 1 - |S_0|^2 = 4R_1R_2/(R_1 + R_2)^2$ and, given the source and load impedances, this equation must be satisfied exactly. At the passband edge $x_L = \alpha\cos\theta_L$, the transducer gain is prescribed as T_L; this specification may be modified to adjust for the integer value of n. The two following equations therefore define the values of n and α.

$$T(\alpha) = \frac{4R_1R_2}{(R_1+R_2)^2} = \frac{1}{1+\alpha^{2n}} \qquad T(x_L) = \frac{1}{1+\alpha^{2n}\cos^{2n}\theta_L} \tag{7.5.11}$$

Then

$$n = \frac{1}{2}\,\frac{\log\left[\dfrac{1}{T(x_L)} - 1\right] - \log\left[\dfrac{1}{T(\alpha)} - 1\right]}{\log(\cos\theta_L)} \tag{7.5.12}$$

The value of n is determined as the lowest integer equal to or greater than the solution of eq. (7.5.12). The value of α is then found from the first of eqs. (7.5.11). As a result the second equation is not precisely satisfied; however, the resultant tolerance at x_L is improved slightly. With all parameters evaluated, the scattering functions $S_{21}(\lambda)$, $S_{11}(\lambda)$ can be determined by eqs. (7.5.7) and (7.5.9), (7.5.8) and (7.5.10). The stepped line transformer is then synthesized by applying Richards' Theorem.

The following example illustrates the procedure.

Example 7.5.1
Determine a stepped line transformer when normalized source and load impedances are $R_1 = 1$, $R_2 = 4$, respectively. The VSWR is to be less than 1.5 over the band 2 Ghz to 4 Ghz.

Solution The quarter wave point occurs at the arithmetic center of the band which is equal to 3 Ghz, and corresponds to a free space wavelength of 10 cm (a useful figure to remember). Therefore the length of each UE is $d = 10/4 = 2.5$ cm. The value of $\theta = \pi/2$ at midband, so that at the lower edge of the passband (2 Ghz) $\theta_L = (2/3)(\pi/2) = \pi/3$. At 2 Ghz, therefore, $x_L = \alpha \cos\theta_L = \alpha/2$. For a reflectance magnitude of S, the VSWR $= (1+S)/(1-S) = 1.5$, so that $S = 0.2$ at 2 Ghz. Therefore $T(x_L) = 1 - S^2 = 0.960$. At d.c. the gain is set by the mismatch of the terminations and using the first of eqs. (7.5.11), $T(\alpha) = 4(1\cdot 4)/(1+4)^2 = 0.64$. Substituting these values into eq. (7.5.12) gives $n = 1.88$. Choose $n = 2$. Then by the first of eqs. (7.5.11), $\alpha = 0.866$. Then the second equation of eqs. (7.5.11) gives $T(x_L) = 0.9343$, or a VSWR $= 1.45$, a slight improvement over the specified 1.5.

The transmittance function can now be found. Based on the Butterworth gain function, eq. (7.5.1), the roots $x_k^2 = \pm j$, and using eq. (7.5.6), we obtain the four roots $\lambda_k \to \pm 1.0607 \pm j0.3535$. The two LHP roots are $\lambda_{1,2} = -1.0607 \pm j0.3535$. Substituting into eq. (7.5.8), the input reflectance is

$$S_{11} = \frac{h}{g} = \frac{0.75}{\lambda^2 + 2.1213\lambda + 1.25}$$

The easiest way to synthesize the stepped line is to calculate the input impedance

$$z_1 = \frac{g_e + h_e}{g_o - h_o} = \frac{2+\lambda^2}{2.1213\lambda}$$

Applying Richards' Theorem, the two lines have characteristic impedances $r_1 = \sqrt{2}$, $r_2 = 2\sqrt{2}$.

As a check, the inversion property of a quarter wave line can be applied to verify that the input impedance is $2/(8/4) = 1$. One further point; the back end reflectance is $S_{22} = -h_*/g = -S_{11}$, so the two-port should be antimetric. An antimetric stepped line has the following property:[8] if the line characteristic impedances starting from the input are $r_1, r_2, \ldots, r_{n-1}, r_n$, then

$$r_n = \frac{R_1 R_2}{r_1}, \quad r_{n-1} = \frac{R_1 R_2}{r_2} \quad \ldots$$

[8] H. J. Carlin, "Distributed Circuit Design With Transmission Line Elements", *Proc. IEEE*, vol. 59, no. 7, pp. 1059-1079, July 1971.

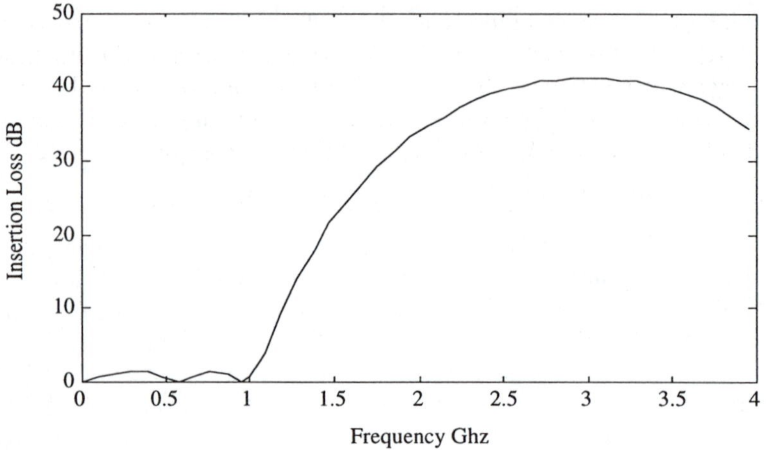

FIGURE 7.5.2
Insertion Loss of equal ripple filter with 5 UE's. (Passband scale exaggerated.)

The values of r_1 and r_2 satisfy this relation, and the response of the transformer is shown on Fig. 7.5.1. □

Equal Ripple Stepped Line Filters

The lowpass variable $x = \alpha \sin\theta$ is appropriate for the transfer function of a stepped line filter. We choose the Chebyshev transducer gain expression, $T(x)$, eq. (7.5.2). The parameters of line length L, bandwidth constant α, ripple factor ϵ, and number of sections n, are set as follows:

1. Since the gain function is even in x, the fundamental band runs from $0 \le \theta < \pi/2$. Thereafter, the response repeats over succeeding $\pi/2$ intervals. At the edge of the fundamental band $L/\lambda_g = 1/4$, and this determines the length of the UE's.

2. Let $\theta_L < \pi/2$ correspond to the prescribed frequency at the edge of the passband. As a matter of convenience, α is defined so that the normalized frequency variable x is unity at this point. Thus $x = \alpha \sin\theta_L = 1$, or $\alpha = 1/\sin\theta_L > 1$.

3. Let T_p be the prescribed minimum value of equal ripple gain within the passband, $0 \le x \le 1$. Exactly as for lumped filters, $T_p = 1/(1+\epsilon^2)$, and this determines ϵ.

4. To determine n, let T_s be the prescribed gain at the edge of the fundamental band where $x = \alpha$. We can then approximate the Chebyshev polynomial by its asymptotic behavior, which is just the

highest degree term present in $T_n(x)$, eq. (6.6.4). Since $\alpha > 1$, $T(\alpha) \approx 1/(1+\epsilon^2 2^{2(n-1)}\alpha^{2n})$. We now calculate n as the smallest integer which satisfies $T(\alpha) \leq T_s$. The value can, of course, be checked using the actual Chebyshev polynomial. Furthermore, it should be kept in mind that for equal terminations n must odd, eq. (6.6.5).

Once the Chebyshev gain function has been constructed in terms of the lowpass variable $x = \alpha \sin\theta$, the analytic reflectance $S_{11}(\lambda)$ can be determined and the stepped line synthesized by using Richards' Theorem. The reflectance is calculated from the gain function by using spectral factorization as discussed above for the Butterworth transformer. Generally factorization is carried out in the x domain using the explicit root formulas for the Chebyshev case; the λ_k roots may then be computed from the x_k roots. Consider the denominator root factors of the gain $T(x)$.

$$G(x) = 1 + \epsilon^2 T_n^2(x) = 2^{2(n-1)}\epsilon^2(x^2 - x_1^2)(x^2 - x_2^2)\ldots(x^2 - x_n^2)$$

since the factor $2^{2(n-1)}$ is the coefficient of x^{2n} in $T_n^2(x)$, and $\pm x_k$ are the roots of $G(x)$ from eq. (6.6.10). Then using eq. (6.6.10), the root factors in Ω^2 are found from

$$G(x) = 2^{2(n-1)}\epsilon^2 \prod_{k=1}^{n}\left(\frac{\alpha^2\Omega^2}{1+\Omega^2} - x_k^2\right) =$$

$$= \frac{2^{2(n-1)}\epsilon^2}{(1+\Omega^2)^n}\prod_{k=1}^{n}(\alpha^2 - x_k^2)\left(\Omega^2 - \frac{x_k^2}{\alpha^2 - x_k^2}\right)$$

The transmittance is $S_{21}(\lambda) = h/g$, so that substituting $\lambda^2 = -\Omega^2$ in the above equation

$$S_{21} = \frac{(\sqrt{1-\lambda^2})^n}{F\prod_{k=1}^{n}(\lambda - \lambda_k)} \qquad F = \epsilon 2^{n-1}\sqrt{\prod_{k=1}^{n}(x_k^2 - \alpha^2)} \tag{7.5.13}$$

The roots λ_k of the Hurwitz polynomial $g(\lambda)$ are selected from the $2n$ quantities $\pm\sqrt{x_k^2/(x_k^2 - \alpha^2)}$, where $k = 1, 2, \ldots n$, and we use the n values with negative real parts. Note that there are no real roots for the x_k^2, which satisfy $G(x) = 0$, only complex conjugate pairs. Therefore $\prod_{k=1}^{n}(x_k^2 - \alpha^2)$, as a squared amplitude, is real and positive, as is F.

The reflectance is calculated from

$$1 - T(x) = \frac{\epsilon^2 T_n^2(x)}{1 + \epsilon^2 T_n^2(x)}$$

In order to determine numerator h, in $S_{11} = h/g$ we first compute the roots X_k of $T_n(x)$ and transform these into the roots in the λ plane, very much as in the preceding discussion. The equation $T_n(X_k) = \cos n \arccos X_k = 0$ becomes $\cos nX_k = 0$, so that the n roots are

$$T_n(X_k) = 0,\ X_k = \cos[(\pi/2n) + (k\pi/n)],\ k = 0, 1, \ldots, n-1 \tag{7.5.14}$$

The roots are all real and since $T_n(x)$ is either even or odd, we have $T_n(x) = \prod(x - X_k)(x + X_k) = \prod(x^2 - X_k^2)$, with one root at $X_k = 0$ if n is odd. Proceeding exactly as in the denominator factorization we obtain for h, when n is even (the factor $(1 - \lambda^2)^{n/2}$ is canceled by the denominator).

$$h = \pm F_e \prod_{k=1}^{n/2} \left(\lambda^2 + \frac{X_k^2}{\alpha^2 - X_k^2} \right)$$

where $F_e = \epsilon 2^{n-1} \prod_{k=1}^{n/2} (\alpha^2 - X_K^2)$. For n odd

$$h = \pm F_o \lambda \prod_{k=1}^{(n-1)/2} \left(\lambda^2 + \frac{X_k^2}{\alpha^2 - X_k^2} \right) \tag{7.5.15}$$

where $F_o = \prod_{k=1}^{(n-1)/2} (\alpha^2 - X_K^2)$.

The roots are real and $X_k^2 < 1$, whereas $\alpha^2 > 1$, so that $F_{e,o} > 0$, and the finite roots of h are all purely imaginary, i.e. $\pm jX_k/\sqrt{\alpha^2 - X_k^2}$, corresponding to the real frequency unit gain points in the passband.

Example 7.5.2

Determine a stepped line equal ripple filter with the following specifications: (1) Edge of fundamental band at 3 Ghz. (2) Edge of passband at 1 Ghz. 3) Maximum loss in passband: $-10 \log T_p = 0.42$ dB. (4) Minimum loss at edge of fundamental band: $-10 \log T_s = 40$ dB.[9]

Solution Referring to items 1-4 listed above:

1. Quarter wave point ($\theta = \pi/2$) is at 3 Ghz (10 cm); or L=2.5 cm.
2. Edge of passband is at 1 Ghz ($\theta = \pi/6$), or $\alpha = 1/\sin \pi/6 = 2$.
3. $T_p = 1/1.1015$; $\epsilon^2 = 0.1$ gives 0.41 dB maximum passband loss.
4. $1 + 0.1 \cdot 2^{2(n-1)} 2^{2n} = 1/T_s \geq 10^4$. Then $n = 5$. The actual loss at this point is $10 \log[1 + 0.1\, T_5^2(2)] = 41$ dB.

[9] I first gave this Example in a 1971 paper (loc. cit.). It must possess special charms, for it was later copied identically, solution and all, in a much later text by another author, though without attribution and without the present explanatory discussion which has not appeared elsewhere. H. J. C.

We can now apply eqs. (7.5.13), (7.5.14), (7.5.15), to the Chebyshev polynomial $T_5(x) = 16x^5 - 20x^3 + 5x$, and determine $S_{11} = h/g$:

$$S_{11} = \frac{114.5\lambda^5 + 44.27\lambda^3 + 3.1\lambda}{114.5\lambda^5 + 83.21\lambda^4 + 74.48\lambda + 28.89\lambda^2 + 8.53\lambda + 1}$$

Although the coefficients of λ^5 are shown equal, they actually differ in the fourth decimal place to account for the fact that the reflectance is slightly less than unity at the edge of the fundamental band ($\lambda = j\infty$). This introduces numerical difficulties if the open circuit $z_{11} = (g_e + h_e)/(g_o - h_o)$ is used, due to the subtraction in the denominator. Instead it is preferable to use the short circuit input admittance $y_{11} = (g_e - h_e)/(g_o + h_o)$, eq. (5.5.3). When Richards' Theorem is applied to y_{11}, the characteristic admittances $g_k = 1/r_k$ are computed. Performing the calculation we can determine the stepped line characteristic impedances normalized to unity

$$r_1 = 3.18,\ r_2 = 0.443,\ r_3 = 4.38,\ r_4 = 0.443,\ r_5 = 3.18$$

Since h is odd and f is even, we have $S_{22} = S_{11}$; therefore, the filter is symmetric as indicated by the r_k. The response is shown on Fig. 7.5.2. □

7.6 Transfer Functions for Stepped Lines and Stubs

It was pointed out in the preceding Section that a major limitation for a cascaded UE transfer function is that it cannot possess zeros of transmission on the real frequency axis, the origin and the point at infinity included. If short or open circuited UE stubs are incorporated into the structure, then real frequency zeros may be realized. A UE terminated in an open circuit has zero impedance at its input at the quarter wave point ($\lambda = j\infty$); placed in shunt with the line, it produces a zero of transmission at infinity. A short circuited UE in series with the line also produces a zero at infinity. The dual arrangement (e.g., shorted UE across the line) introduces a zero of transmission at the origin. A zero can also be introduced at a finite frequency by employing a double stub, i.e., a cascade of two UE's terminated in an open or short circuit. For instance, refer to Example 7.3.1, where it was shown that a double stub terminated in an open circuit has an input impedance of the form $z = (\lambda^2 + a^2)/b\lambda$, so that when placed across the line produces a zero of transmission at $\lambda = \pm ja$.

A straightforward method to realize lowpass frequency selective filters with UE's and stubs is to employ a useful circuit equivalence, the Kuroda Identity.[10]

[10]First introduced in Japan, where much of the early work on UE design was carried

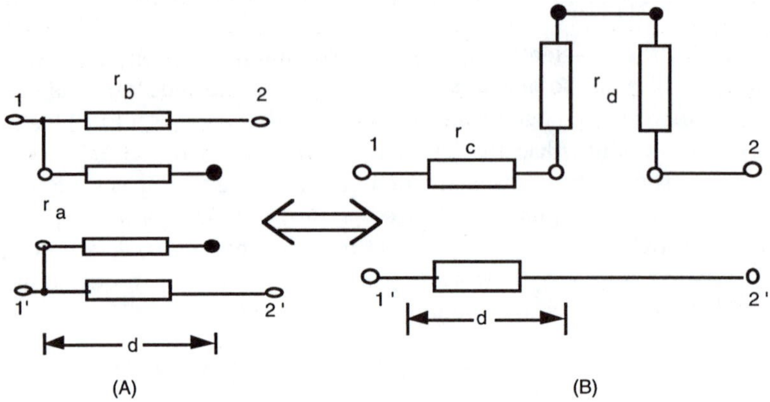

FIGURE 7.6.1
Kuroda's lowpass line-stub identity. $r_c = (r_a r_b)/(r_a + r_b)$, $r_d = r_b^2/(r_a + r_b)$, and $r_a = r_c[1 + (r_c/r_d)]$, $r_b = r_c + r_d$.

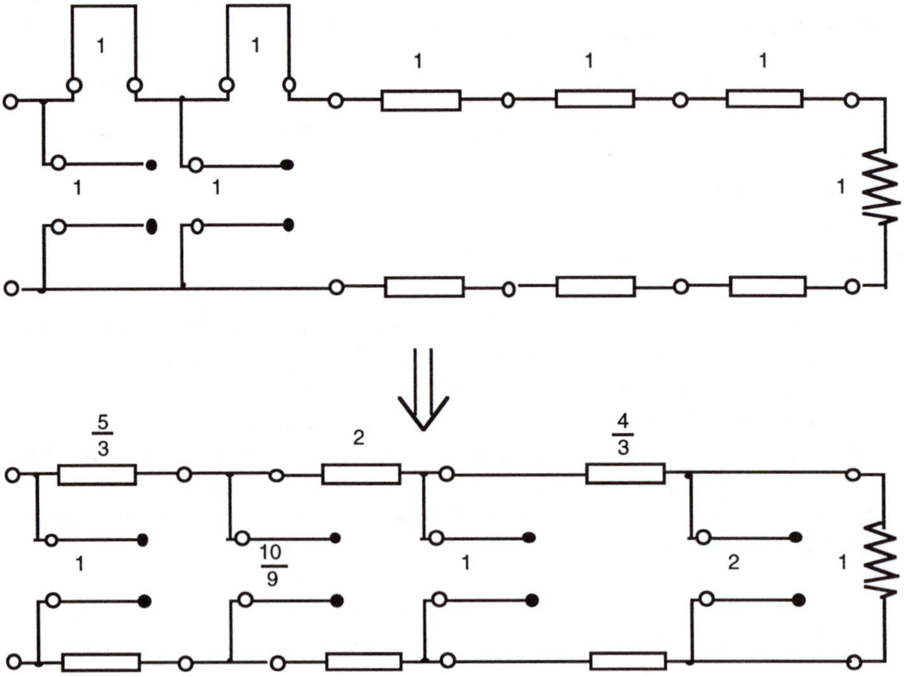

FIGURE 7.6.2
Illustration of Kuroda identity.

Referring to Fig. 7.6.1, we note that both structures have a single zero of transmission at infinity, so that the transmittances have the same numerator, i.e., in $S_{21} = f/g$, $f = \sqrt{1-\lambda^2}$. The denominators are of degree 2. Equivalence therefore depends on the equality of the input reflectances, that is the equivalence of the polynomials h, and g. In turn, this means we only have to look at the open circuit input impedances z_{11} and z_{22}. The calculation of these functions involves the series and parallel combinations of UE elements, and the input impedance, z_1 of a UE of characteristic impedance r_0, terminated in z_2. Referring to eq. (7.2.6), with $\lambda = \jmath \tan \beta d$

$$z_1 = r_0 \frac{z_2 + r_0\lambda}{r_0 + z_2\lambda}$$

Then for Fig. 7.6.2 A

$$z_{11A} = \frac{1}{\lambda}\frac{r_a r_b}{r_a + r_b}, \qquad z_{22A} = \frac{r_b}{r_a + r_b}\left(\frac{r_a}{\lambda} + r_b\lambda\right)$$

and for Fig. 7.6.2 B

$$z_{11B} = \frac{r_c}{\lambda}, \qquad z_{22B} = \frac{r_c}{\lambda} + r_d\lambda$$

Equating the appropriate impedances

$$r_c = \frac{r_a r_b}{r_a + r_b}, \qquad r_d = \frac{r_b^2}{r_a + r_b} \tag{7.6.1}$$

Noting that $r_a/r_b = r_c/r_d$, it is easy to invert eqs. (7.6.1) to obtain

$$r_a = r_c\left(1 + \frac{r_c}{r_d}\right), \qquad r_b = r_c + r_d$$

These four relations define the Kuroda identity. For a zero of transmission at the origin, there is no similar identity relating the positions of a series open circuited stub and a shunt short circuited stub.

The Kuroda identity may be used for lowpass filter design by first constructing an all-stub structure that realizes a specified transducer gain. The stubs are then shifted along a UE chain, using the identity to permit spatial separation between adjacent stubs. For example, with $\Omega = \tan \omega\tau$, consider the lowpass Chebyshev transducer gain

$$T(\Omega^2) = \frac{1}{1 + \epsilon^2 T_n^2(\Omega)}$$

out. K. Kuroda, "A method to derive distributed constant filters from lumped constant filters", *Joint Conv. Elec. Inst. Japan*, Kansai, Oct. 1952, ch. 9-10.

Of course, this function is not in a form realizable by cascaded UE's. Instead of using Richards' Theorem, we proceed exactly as though we were dealing with a lumped system. The complex variable $\lambda = j\Omega$ is introduced and, in the usual fashion of Darlington synthesis, Section 6.6, we employ spectral factorization to find the reflectance $S_{11}(\lambda) = h/g$. This is realized by a doubly terminated LC ladder, whose inductive series impedances have frequency dependence proportional to $\lambda = j\tan\omega\tau$, whereas the shunt capacitive impedances have frequency response proportional to $1/\lambda = -j\cot\omega\tau$. The structure in the true frequency domain is therefore a ladder of shorted UE stubs in the series arms, and open circuited UE stubs in the shunt arms. Unfortunately they are all connected together at the same terminal plane and the circuit is not practically feasible. The problem of how to spatially separate the stubs is solved by using Kuroda's identity. Suppose the load impedance normalized to r_0 is unit, and the stub impedances are normalized to r_0 as well. Place in front of the load a cascade of UE's all of characteristic impedance unity. The load terminating the stub filter still remains unit. Using Kuroda's identity as many times as necessary, move the stubs down the chain so that each is located at a different terminal plane separated from its neighbors by a UE on either side. The resultant stub-stepped-line structure realizes the identical transducer gain of the all stub filter. Fig. 7.6.2 illustrates the process, carried out so that all the stubs are open circuited UE's across the line, a convenient arrangement for microstrip. The choice of unity for all the initial stub characteristic impedances was only for illustrative purposes.

Maximally Flat Gain Functions for Stubs and UE's

The process just described for using the Kuroda identity to realize a lowpass filters with a multiple zero of transmission at infinity employs the UE's as spatial separators. They function as dummy elements without affecting the filter selectivity. We now consider the construction of a lowpass gain function in which stubs and UE's all contribute to the response.

First consider the construction of a lowpass ($x = \alpha\sin\theta$) Butterworth gain function so that it contains zeros of transmission at $\Omega = \infty$, i.e. at $x = \alpha$. Modify eq. (7.5.1) by replacing $x^n \to x^n\Omega^q$ so that $T(x^2) = (1 + x^{2n}\Omega^{2q})^{-1}$, or since $x^2 = (\alpha\Omega)^2/(1+\Omega^2)$, the transducer gain is

$$T(\Omega^2) = \frac{1}{1 + \dfrac{\alpha^{2n}\Omega^{2(q+n)}}{(1+\Omega^2)^n}} = \frac{(1+\Omega^2)^n}{P_{n+q}(\Omega^2)} \tag{7.6.2}$$

where $P_{n+q}(\Omega^2)$ is an even polynomial of degree $2(n+q)$.

The gain function has the requisite zero of transmission at infinity of order q, and is maximally flat there to order $2q$. Furthermore, in the neighborhood of $\Omega = 0$, it behaves like $(1 + \alpha^{2n}\Omega^{2(q+n)})^{-1}$ so it is maximally flat at the origin to order $2(q + n)$.

For the bandpass case, $x = \alpha \cos\theta$ and the gain function is modified by $x^n \to x^n/\Omega^q$. Therefore, since $x^2 = \alpha^2/(1 + \Omega^2)$,

$$T(\Omega^2) = \frac{1}{1 + \dfrac{\alpha^{2n}}{(1+\Omega^2)^n\Omega^{2q}}} = \frac{\Omega^{2q}(1+\Omega^2)^n}{P_{n+q}(\Omega^2)} \tag{7.6.3}$$

In this case there is a zero of transmission of order q at the origin, and the gain is maximally flat at the center of the passband.

We realize these gain functions in the usual fashion. The input reflectance is computed using spectral factorization. But now the synthesis includes the extracting of UE's according to Richards' Theorem, and also the removal of poles at the origin or infinity. In the case of the lowpass function, impedance poles at infinity correspond to series connected short circuited stubs. Admittance poles correspond to open circuited shunt stubs. By alternating a stub removal with a UE removal we realize spatial separation between stubs, and both the stubs and the UE's contribute to the selectivity. In the band pass case poles are removed at the origin. These would correspond to series connected open circuited stubs or short circuited shunt stubs. The gain functions shown in eqs. (7.6.2), (7.6.3) are realized with q stubs and n UE's, and realizability as stub-UE chains is assured if the functions have the forms shown and are bounded by zero and unity.

Equal Ripple Gain Function for Stubs and UE's

The second forms shown in eqs. (7.6.2), (7.6.3) can be considered as the general expression for a stub-UE realization. We now address the problem of constructing a gain function in either of these forms so that the passband response is equal ripple in character, rather than maximally flat.[11]

Consider first the lowpass case where $x = \alpha \sin\theta = \cos\phi$, $\theta = \omega\tau$. We will show that the following transsducer gain function has equal ripple passband behavior. Further on making the substitution, eq. (7.5.4), $x^2 = \alpha^2\Omega^2/(1 + \omega^2)$, $\omega = \tan\theta$, the gain function T will be transformed into the second form shown in eq. (7.6.2), hence realizable by UE's and stubs.

$$T = \frac{1}{1 + \epsilon^2 \cos^2(n\phi + q\zeta)} \qquad \cos\zeta = x\sqrt{\frac{\alpha^2 - 1}{\alpha^2 - x^2}} \tag{7.6.4}$$

[11]Some of the following material is based on: H. J. Carlin and W. Kohler, "Direct synthesis of band-pass transmission line structures." *IEEE Trans. on Microwave Th. and Tech.*, vol. MTT-13, no. 3, pp. 283-297, May 1965.

We first show the equal ripple character of the gain[12] over the passband $-1 \leq x \leq 1$. Correspondingly $\phi = \arccos x$, $n\phi$ has a net monotonic excursion from $n\pi$ to zero. Were ζ not present, this would account for the equal ripple response of the Chebyshev gain function with n unit gain points. In the present case, $\cos\zeta$ also varies monotonically from -1 to 1 on a nonlinear scale in x. Thus $q\zeta$ introduces an additional monotone excursion of $q\pi$ in the argument of the cosine of eq. (7.6.4). Consequently the gain function still has equal ripple passband behavior, but now with $n+q$ unit gain points.

In order to express eq. (7.6.4) in terms of Ω, use the trigonometric identity $\cos(n\phi + q\zeta) = \cos n\phi \cos q\zeta - \sin n\phi \sin n\zeta$. Then with $T_n(x)$ the n-th order Chebyshev polynomial

$$\cos n\phi = \cos n \arccos x = T_n(x) = T_n\left(\frac{\alpha\Omega}{\sqrt{1+\Omega^2}}\right)$$

Referring to eq. (7.6.4), and using $x = \alpha \sin\theta$, $\cos\zeta = \beta \tan\theta = \beta\Omega$, where $\beta = \sqrt{\alpha^2 - 1}$

$$\cos n\zeta = \cos n \arccos \beta\Omega = T_q(\beta\Omega)$$

We also have

$$\sin n\phi \sin n\zeta = \sqrt{1-x^2}\,\sqrt{1-\zeta^2}\, U_{n-1}\left(\frac{\alpha\Omega}{\sqrt{1+\Omega^2}}\right) U_{q-1}(\beta\Omega)$$

where U_{j-1} are Chebyshev polynomials of the second kind, even or odd as $j-1$ is even or odd. We have

$$\sqrt{1-x^2}\,\sqrt{1-\zeta^2} = \frac{1-(\alpha^2-1)\Omega^2}{\sqrt{1+\Omega)^2}}$$

If we set $f = \cos n\phi \cos q\zeta - \sin n\phi \sin n\zeta = f_1 - f_2$, then

$$f_1 = T_n\left(\frac{\alpha\Omega}{\sqrt{1+\Omega^2}}\right) T_q(\beta\Omega) = \frac{M_{n+q}(\Omega)}{(\sqrt{1+\Omega^2})^n}$$

and

$$f_2 = \frac{1-(\alpha^2-1)\Omega^2}{\sqrt{1+\Omega^2}} U_{n-1}\left(\frac{\alpha\Omega}{\sqrt{1+\Omega^2}}\right) U_{q-1}(\beta\Omega) = \frac{N_{n+q}(\Omega)}{(\sqrt{1+\Omega^2})^n}$$

where M_{n+q}, N_{n+q} are even or odd polynomials of degree $n+q$. Finally, therefore,

$$T(\Omega^2) = \frac{1}{1+\epsilon^2 f^2} = \frac{(1+\Omega^2)^n}{P_{n+q}(\Omega^2)}$$

[12] C. B. Sharpe, "A general Tchebycheff rational function", *Proc. IRE*, vol. 42, no. 2, pp. 452-457, Feb. 1954.

which is the desired form.

Example 7.6.1

Find the parameters of a line-stub Chebyshev lowpass filter with an equal number of UE's and stubs. The edge of the fundamental band is at 3 Ghz, and the passband loss tolerance is 0.2 dB. The edge of the passband is at 1 Ghz, with an insertion loss of at least 50 dB at 2 Ghz.

Solution The minima of the passband gain ripples are $T_p = 1+\epsilon^2$. Take $\epsilon = 0.2$; then the passband loss tolerance is $10\log 1.04 = 0.17$ dB. At the edge of the fundamental band (wavelength 10 cm) where the zeros of transmission occur, $\theta = \pi/2$ so that the length of each UE is $d = 10/4 = 2.5$ cm. At the edge of the passband (1 Ghz), $\theta = \pi/6$; then $\alpha = 1/\sin(\pi/6) = 2$. At 2 Ghz we have $\theta = \pi/3$ and $x = \alpha\sin(\pi/3) = \sqrt{3}$. In order to determine n, we use the asymptotic expressions for the Chebyshev polynomials T_k and U_k. These are the highest power terms in the polynomial expansions. Thus for $x > 1$

$$T_k(x) \approx 2^{k-1}x^k$$

and

$$U_k(x) \approx 2^{k-1}x^{k-1}$$

Applying these expressions for the case that $n = q$

$$f^2 = \cos^2(n\phi + n\zeta) \approx \frac{1}{4}\left[(2x)^4\frac{\alpha^2-1}{\alpha^2-x^2}\right]^n, \quad x > 1$$

and if we take integer $n = 3$, then at $x = \sqrt{3}$, the loss is $10\log(1+f^2) = 59$ dB. □

The Chebyshev UE-stub bandpass transducer gain function with $x = \cos\phi = \alpha\cos\theta$ is derived in the same way as the lowpass function, only now $\cos\zeta = \sqrt{\alpha^2-1}/\Omega$. As a result the bandpass gain function has the form

$$T(\Omega^2) = \frac{1}{1+\epsilon^2\cos^2(n\phi+q\zeta)} = \frac{\Omega^{2q}(1+\Omega^2)^n}{P_{n+q}(\Omega^2)} \qquad x = \alpha\cos\theta \qquad (7.6.5)$$

Example 7.6.2

Determine the form of the bandpass equal ripple gain function as function of x for a filter of 2 UE's and one stub. Take $\alpha = 2$.

Solution Since $n = 2$, $q = 1$, we have $f = \cos(2\phi+\zeta) = \cos 2\phi\cos\zeta -$

$\sin 2\phi \sin\zeta$. Then

$$\cos 2\phi = 2\cos^2\phi - 1 = 2x^2 - 1 \qquad \cos\zeta = x\sqrt{\frac{\alpha^2 - 1}{\alpha^2 - x^2}}$$

$$\sin 2\phi = 2\cos\phi\sin\phi = 2x\sqrt{1 - x^2} \qquad \sin\zeta = \alpha\sqrt{\frac{1 - x^2}{\alpha^2 - x^2}}$$

Substituting $\alpha = 2$

$$f^2 = \frac{2}{4 - x^2}[(28 + 16\sqrt{3})x^6 - (44 + 24\sqrt{3})x^4 + (19 + 8\sqrt{3}x^2]$$

When $x^2 = \alpha^2/(1 + \Omega^2)$ is substituted into the gain function $(1 + \epsilon^2 f^2)^{-1}$, the denominator is an even polynomial, $P_3(\Omega^2)$, and the numerator is $(4 - x^2)(1 + \Omega^2)^3 = 4\Omega^2(1 + \Omega^2)^2$. The final result is therefore in the form of eq. (7.6.5). □

A bandpass gain function has zeros of transmission at zero frequency. For example, a single shorted stub placed in shunt with the line produces a simple zero of transmission. Additional shunting short circuited stubs, no matter where they appear along the UE chain, cannot raise the order of the zero since at d.c. they function as a single stub. The same is true for open circuited stubs in series. For higher order multiplicity zeros, realization must proceed by the extraction of shunt short circuited stubs alternated with series open circuited stubs. In the synthesis process, Richards' Theorem is used to separate stubs by UE's.

The realization of prescribed terminations needs some comment. In the lowpass case, the terminations can be readily prescribed since the UE-stub cascade is transparent at d.c. In the bandpass case, the filter is opaque at d.c., so that the loads are indeterminate at this point.[13] One approach to load control is to synthesize the bandpass function and then modify the resultant structure to secure the prescribed load resistances. The bandpass Kuroda identity shown in Fig. 7.6.3 (derived in the same manner used for Fig. 7.6.1) is useful for adjusting the load resistance. To illustrate the idea, consider the transducer gain of the preceding example whose response can be visualized with the following data. Taking $\epsilon^2 = 0.02$ gives a passband tolerance of $\approx 0.1\,\text{dB}$. Since $\alpha = 2$, the edges of the passband ($x = \pm 1$) correspond to $\theta_{h,l} = \pi/2 \pm \pi/6$. The response is down $\approx 3\,\text{dB}$ at $1.07\theta_h$, and at the symmetrical point below $\pi/2$. If the transducer gain is synthesized by first extracting an impedance pole at the origin, the result is as shown in Fig 7.6.4 A, and the normalized termination is 18.9. All or a portion

[13]This problem does not arise for bandpass filters determined from lowpass prototypes by frequency transformation.

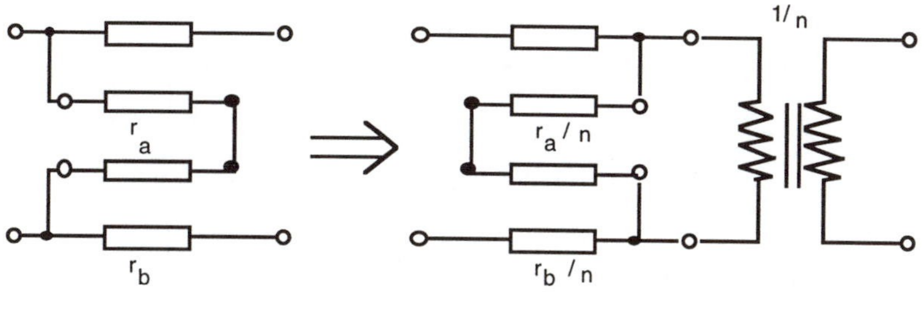

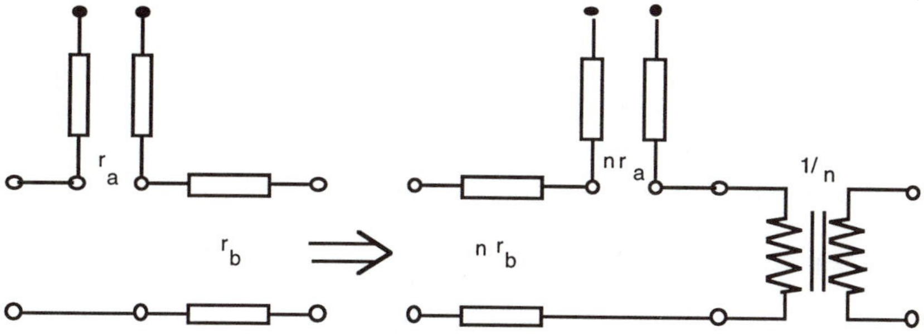

FIGURE 7.6.3
Kuroda bandpass identity. $n = (r_a + r_b)/r_a$.

of the stub can be shifted down the chain to reduce the load resistance by applying Fig. 7.6.3 A. In this example, if the entire stub is transferred to the next UE junction as in Fig. 7.6.4 B the load resistance becomes unity, equal to the source resistance. In any event, it has probably not escaped the reader's notice that with the stub at the input the structure functions as a remarkably efficient impedance transformer.

7.7 Coupled UE Structures

A structure consisting of a number of parallel conductors with a ground return can be described as a set of TEM transmission lines coupled by distributed series inductance and distributed shunt capacitance. Examples would be multiconductor parallel strips above a ground plane or parallel wires enclosed by a coaxial sheath. A system of this type can be analyzed by using the matrix analog of the differential equations for a single TEM line. Thus, we can rewrite eq. (7.1.3) for the voltage, $\boldsymbol{v}$, and current, $\boldsymbol{i}$,

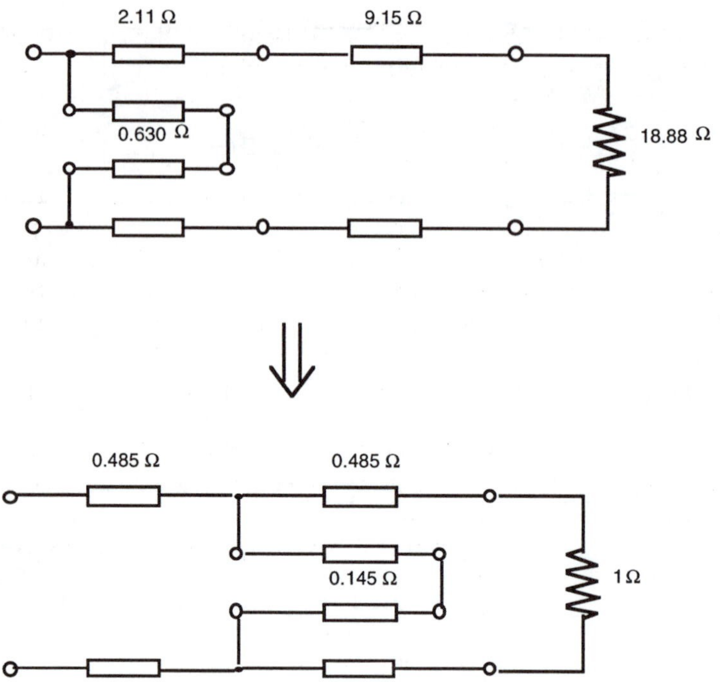

FIGURE 7.6.4
Load adjustment by Kuroda bandpass identity.

vectors along the line as

$$-\frac{d\boldsymbol{v}}{dz} = \boldsymbol{Z}\boldsymbol{i}, \quad -\frac{d\boldsymbol{i}}{dz} = \boldsymbol{Y}\boldsymbol{v}$$

where the $\boldsymbol{Z} = \boldsymbol{L}s$, $\boldsymbol{Y} = \boldsymbol{C}s$, and $\boldsymbol{L}$, $\boldsymbol{C}$ are, respectively, the distributed coupling inductance and capacitance matrices per unit length, with $\boldsymbol{LC} = \mu\epsilon\boldsymbol{I}_n$, $1/\sqrt{\mu\epsilon}$ is the velocity of propagation of light, and $\boldsymbol{I}_n$ is the $n \times n$ identity matrix. We will not be concerned with further analysis of the general equations,[14] since we will only consider here cascades of four-ports, each consisting of a pair of coupled UE's. We will further particularize the discussion to four-port chains which realize broadband *directional couplers*, since, as we shall see, these chains can be reduced to the case of stepped uncoupled transmission lines.

Directional couplers are important components in distributed systems. They are used to sample the incident and/or reflected waves on a transmission line, have a variety of applications, and come in an amazing number

[14]The general material is covered in several standard texts; also a short straightforward exposition is given by G. I. Zysman and A. Matsumoto "Properties of Microwave C-Sections", *IEEE Trans. on Circ. Th.*, vol. CT-12, no. 1, pp. 74-82, March 1965.

of different physical embodiments.[15] Directional couplers are members of a class of lossless four-ports, including the hybrid coil, which were introduced as *biconjugate* or *maximum output* networks by Campbell and Foster in a classic paper.[16] The fundamental relation of maximum output networks to microwave circuits was first given by Montgomery and Dicke.[17]

Consider first the group of reciprocal lossless n-ports which are simultaneously matched at all ports when the structure is terminated in prescribed resistive loads. Such structures are Campbell and Foster's maximum output networks.[18] The $\boldsymbol{S}$ matrix therefore has all its diagonal elements $S_{jj} = 0$. Matched n-ports have many unusual properties which stem from the unitarity of the scattering matrix at real frequencies, $\boldsymbol{S}_*\boldsymbol{S} = \boldsymbol{I}_n$.[19]

For example, a matched lossless reciprocal three-port is not physically realizable; a matched lossless nonreciprocal three-port is realizable, but only as a circulator for which the $\boldsymbol{S}$ matrix has a single unit amplitude off-diagonal element in every row and column; a frequency independent matched lossless reciprocal n-port is only possible if n is even. The matched, lossless reciprocal four-port defines a directional coupler. The structure of its $\boldsymbol{S}$ matrix is shown below

$$\boldsymbol{S} = \begin{pmatrix} 0 & S_{12} & 0 & S_{14} \\ S_{12} & 0 & S_{23} & 0 \\ 0 & S_{23} & 0 & S_{34} \\ S_{14} & 0 & S_{34} & 0 \end{pmatrix} \tag{7.7.1}$$

We have imposed reciprocity ($S_{ij} = S_{ji}$) on the matrix, though the general structure is valid even in the nonreciprocal case. The pattern of matrix entries shows that all ports are matched, that ports 1 and 3 are decoupled from each other, and that ports 2 and 4 are decoupled from each other; in fact, as a consequence of unitarity the decoupled property must follow if all four ports are prescribed as matched. If we were given only that ports 1 and 2 were matched and decoupled, then it would also follow, again from unitarity, that ports 3 and 4 were matched and decoupled. Of course, under a changed port numbering, the pattern of zeros entries would be different, e.g., ports 1 and 4 decoupled and 2 and 3 decoupled. What is important is that there are always two distinct port pairs such that the two members of

[15] R. Levy, "Directional couplers", in *Advances in Microwaves*, New York and London, Academic Press, pp. 115-207, 1966.

[16] G. A. Campbell and R. M. Foster, "Maximum Output Networks for Telephone Substation and Repeater Circuits", *Trans. AIEE*, vol. 39, 1920.

[17] C. G. Montgomery, R. H. Dicke, and E. M. Purcell, *Principles of Microwave Circuits*, New York, McGraw-Hill Book Co., 1948.

[18] V. Belevitch, "2n-Terminal Networks for Conference Telephony", *Elec. Communication*, vol. 27, p. 231, Sept. 1950.

[19] H. J. Carlin and A. B. Giordano, *Network Theory: An Introduction to Reciprocal and Nonreciprocal Circuits*, Englewood Cliffs, NJ, Prentice Hall, 1964.

a pair are decoupled from each other. Note also in eq. (7.7.1) that applying unitarity for the first two rows $|S_{12}|^2 + |S_{14}|^2 = 1$, and $|S_{12}|^2 + |S_{23}|^2 = 1$, so that $|S_{14}|^2 = |S_{23}|^2$. Similarly using rows 2 and 3, $|S_{12}|^2 = |S_{34}|^2$. We can summarize these properties by displaying the prototype matrix, $\boldsymbol{C}$, of a directional coupler, whose components are the transducer gains between the various ports:

$$\boldsymbol{C} = \begin{pmatrix} 0 & T_1 & 0 & T_2 \\ T_1 & 0 & T_2 & 0 \\ 0 & T_2 & 0 & T_1 \\ T_2 & 0 & T_1 & 0 \end{pmatrix} \tag{7.7.2}$$

Thus only two transducer gain functions define the operation of the ideal directional coupler.

Stepped Line Directional Couplers

As a basis for realizing a directional coupler, consider a pair of coupled TEM lines made up of two parallel conductors and a ground return. A fixed length of the structure constitutes a pair of coupled UE's, and its four ports consist of the two conductor-to-ground terminal pairs at input and output.[20] Analysis of the propagation properties of this waveguide is much simplified if a symmetrical arrangement of the conductors is assumed. For example, refer to Fig. 7.7.1 (a), (b), which shows a symmetric arrangement of two conductor strips with respect to ground that has extensive microwave application. The diagram schematically indicates the electric field distribution for two TEM modes which ideally are not coupled to each other. For the even mode, the two strips behave as though they were a single conductor, with the ground return the second conductor. For the odd mode, the TEM guide is the two conductor line formed by the two strips. Each of these separate TEM modes has its own characteristic impedance, but we assume the propagation constants of both modes to be the same. Superposition of the two uncoupled mode responses gives the net result.

Suppose at the input of the microstrip (ports 1 and 2), port 1 is fed from an R_1 source with an incident wave, a. The other three ports are terminated in resistors, R_1 (or transmission lines of characteristic impedance R_1). Both modes will be excited at ports 1, 2 and their separate responses superimposed to obtain a solution. The even excitation at ports 1, 2 is $a/2$, $a/2$. The odd excitation is $a/2$, $-a/2$. Adding these quantities gives the prescribed excitation of a at port 1, 0 at port 2. The situation is shown in Fig. 7.7.1 (c). The coupled system is then treated by superposition after analyzing the even and odd mode UE's in (d), (e) of the figure.

[20]E. M. T. Jones and J. T. Bolljahn, "Coupled strip transmission lines and directional couplers", *IRE Trans. on Microwave Th. and Tech.*, vol. MTT-4, pp. 75-81, 1956.

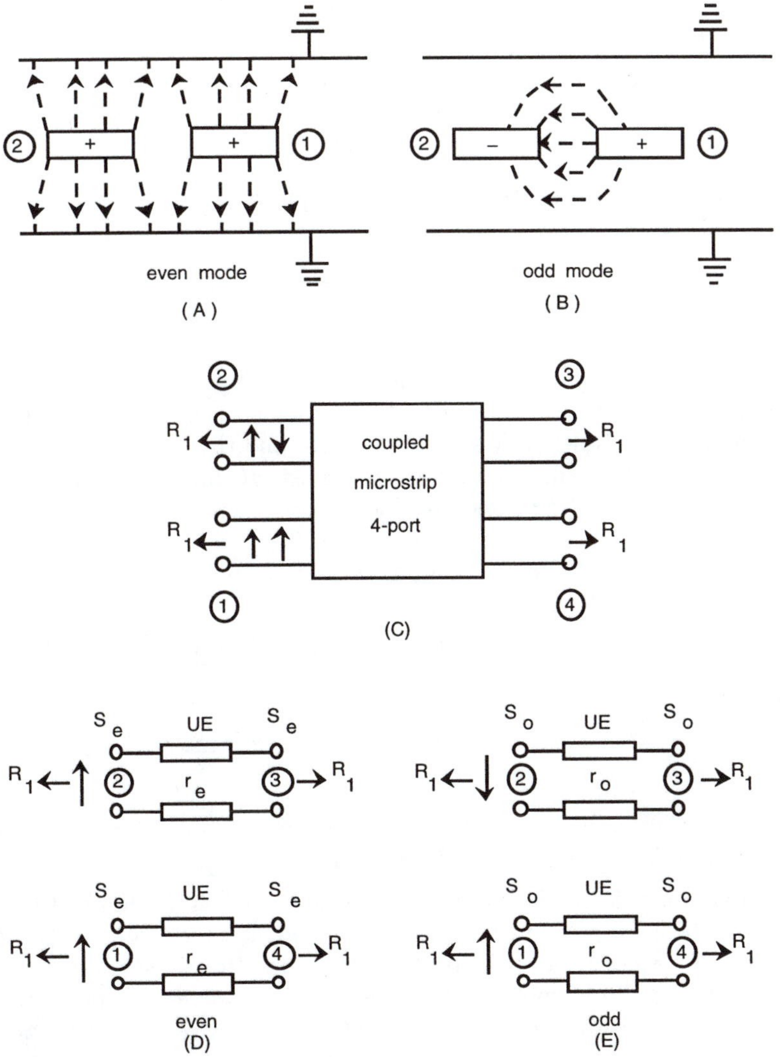

FIGURE 7.7.1
(A),(B): Cross section of coupled stripline with even and odd mode field representation. (C): 4-Port excitation. (D),(E): Even-odd mode decomposition.

We have already derived the scattering matrix of a UE terminated in equal loads, which differ from the UE characteristic impedance, Example 7.2.1. Thus for the even mode UE, the characteristic impedance is r_e, and the reflectance $S_{11e} = S_{22e} \equiv S_e$ is given by eq. (7.2.21) ($\lambda = \tanh \omega\tau$)

$$S_e = \frac{k_{e1}}{1-k_{e1}^2} \frac{2\lambda}{1+\lambda\dfrac{1+k_{e1}^2}{1-k_{e1}^2}} \qquad k_{e1} = \frac{r_e - R_1}{r_e + R_1} \tag{7.7.3}$$

The transmittance for the even mode guide $S_{21e} = S_{12e} \equiv H_e$ is given by eq. (7.2.22);

$$H_e = \frac{\sqrt{1-\lambda^2}}{1+\lambda\dfrac{1+k_{e1}^2}{1-k_{e1}^2}}$$

The odd mode UE has characteristic impedance r_o, and the equations for reflectance S_o and transmittance H_o are the same as above except subscript "o" replaces "e". Now apply superposition to find the reflected wave, b_1, at port 1, and the S_{11} reflectance of the four-port. If the circuit is to function as a directional coupler we require that $S_{11} = 0$. That is

$$S_{11} = \frac{b_1}{a} = \frac{1}{2}(S_e + S_o) = 0$$

Referring to eq. (7.7.3), it is clear that the above equation will hold if $k_{e1} = -k_{o1}$. Thus

$$k_{e1} = \frac{r_e - R_1}{r_e + R_1} = -k_{o1} = -\frac{r_o - R_1}{r_o + R_1}$$

which is satisfied when

$$r_e r_o = R_1^2 \tag{7.7.4}$$

The same analysis can be used when an incident wave is applied to each of the other three ports in turn, and it is evident that when eq. (7.7.4) is satisfied, the reflectances at ports 2, 3, 4 will be zero as well, and the four port must be a directional coupler. The transmittances from port 1 to the other ports are readily computed. Thus $b_2 = (a/2)S_e + (-a/2)S_o$, so that

$$S_{21} = \frac{b_2}{a} = \frac{1}{2}(S_e - S_o) = \frac{k_{e1}}{1-k_{e1}^2} \frac{2\lambda}{1+\lambda\dfrac{1+k_{e1}^2}{1-k_{e1}^2}} \tag{7.7.5}$$

since $k_{o1}^2 = k_{e1}^2$. Similarly,

$$S_{31} = \frac{1}{2}(H_e - H_o) = 0$$

For port 4, $b_4 = (a/2)H_e + (a/2)H_o$.

$$S_{41} = \frac{1}{2}(H_e + H_o) = \frac{\sqrt{1-\lambda^2}}{1 + \lambda\dfrac{1+k_{e1}^2}{1-k_{e1}^2}} \tag{7.7.6}$$

Given the three preceding relations and that the ports are matched, it is evident from the symmetry of the structure that the entries of its scattering matrix fall precisely into the pattern of the directional coupler matrix of eq. (7.7.1). Moreover the derivation shows that the isolation between pairs of ports as well as the matched property are independent of frequency (this is really a consequence of the assumption of decoupled even and odd modes). However, the coupling expressions *are* frequency dependent, but as indicated in eq. (7.7.2) there are only two independent transmittance amplitudes, which in this case, again due to structural symmetry and reciprocity, reduce to only two complex transmittance functions, $S_{21} = S_{43}$, and $S_{41} = S_{32}$.

Since the ports are matched for R_1 terminations, when the four-ports are cascaded (output ports 3, 4 connected to input ports 1, 2), the resultant four-port with R_1 terminations remains a directional coupler provided that for each j-th section the even and odd characteristic impedances satisfy $r_{ej}r_{oj} = R_1^2$. In operation the input of the chain is at port 1, and most of the input power is delivered to port 4. The transmittance between ports 1 and 2 accounts for only a small fraction of the input power, and it is this coupled signal to port 2 which is a sample of the incident wave at port 1. The transmittances $S_{21j} = S_{12j}$ and $S_{41j} = S_{14j}$ for each section of the chain are given by eqs. (7.7.5), (7.7.6), and can be thought of as belonging to a symmetric UE two-port whose reflectance is S_{21j}, whose transmittance is S_{41j}, and whose characteristic impedance r_{oj} (or r_{ej}) is mismatched to R_1. The behavior of the transmittances for the overall cascade of four-ports is therefore defined by a stepped line of cascaded UE's, with characteristic impedances r_{oj} and R_1 terminations. The problem of designing a wideband directional coupler whose coupling functions remain approximately constant over the band is therefore reduced to the design of a stepped transmission line.

The design of the stepped line has some special features which require slight revisions of the functions used earlier in this chapter for cascaded UE problems. We first note that the transducer gain of the coupler, T_2, (i.e., fraction of available power transmitted from port 1 to 4, the main channel) should approximate a constant slightly less than unity over the passband. Over this same band, it is required that the fraction of available power delivered to port 2 (the *coupling function*) be a small and approximately constant quantity $T_1 = 1 - T_2$. At zero frequency (and for any $\theta = \omega\tau = n\pi$), the coupling to port 2 must reduce to zero with all available power transmitted to port 4.

The following response functions for the UE stepped line satisfy these requirements. Consider a coupling function $S \equiv T_1$ prescribed to yield a gain to port 2 of S_c at the edges of the passband, where $x = \alpha \cos\theta_0$. The associated gain to port 4 is $T_a = 1 - S_c$; the prescribed coupling loss is therefore $C = -10\log S_c$, and $A = -10\log T_a$. The maximally flat gain function to port 4 is $T \equiv T_2$,

$$T(x) = \frac{1}{1 + \alpha^{2n} - x^{2n}} \tag{7.7.7}$$

T is realizable since $x^2 \leq \alpha^2$. The associated coupling function is $S(x) = 1 - T(x)$; S is a squared reflectance amplitude, T a squared transmittance amplitude. We note the following properties of T. At d.c., $x = \alpha$ and $T(\alpha) = 1$, $S(\alpha) = 1 - T(\alpha) = 0$. At the passband edge, $x = x_p = \alpha\cos\theta_o$,

$$T(1) = T_a = \frac{1}{1 + x_p^{2n}}$$

At midband (the quarter wave point), $\theta = \pi/2$, $x = 0$, and

$$T(0) = \frac{1}{1 + \alpha^{2n}}$$

The coupling function S also exhibits maximally flat behavior with a coupling of $10\log[1 - T(0)]$ at midband.

An equal ripple gain function is similarly constructed using the Chebyshev polynomials, $T_n(x)$. In this case we set the passband edges at $x = \pm 1$, with $\alpha = 1/\cos\theta_0 > 1$.

$$T(x) = \frac{1}{1 + \epsilon^2[T_n^2(\alpha) - T_n^2(x)]}$$

T is realizable since $x^2 \leq \alpha^2$, and $T_n^2(x)$ is monotone increasing outside the passband ($x^2 > 1$), with $T_n^2(x) \leq T_n^2(\alpha)$. Since $x = \alpha$, when $\theta = 0$, there is no coupling to port 2 at d.c.

The synthesis of the stepped line to give the r_{ej} is straightforward using Richards' theorem. The odd impedances are given by $r_{oj} = R_1^2/r_{ej}$.

Example 7.7.1

Determine the even and odd characteristic impedances for a Butterworth directional coupler. The coupling is to approximate 10 dB over a 2:1 passband, $\pi/3 \leq \theta \leq 2\pi/3$. At $\theta_0 = \pi/3$, the coupling loss should not exceed 10.5 dB.

Solution First try $n = 2$. At $x = 0$, $S(0) = 0.1$ corresponding to a midband coupling loss of $C = 10$ dB. Then $T(0) = 0.9$, and referring to the gain function eq. (7.7.7), $\alpha^2 = 1/3$. At the passband edge, $x_p^2 = (\alpha \cos\theta_0)^2 = 1/12$, $T(\alpha/2) = 0.9057$, thus $-10\log[1 - T(\alpha/2)] = 10.25$ dB. Thus $n = 2$, $\alpha^2 = 1/3$ is a satisfactory choice. As a comparison, note that for $n = 1$, the band edge coupling is 11.14 dB, and for $n = 3$ it is 10.06 dB. Summarizing the response for $n = 2$, the coupling at $\theta = \pi/2, \pi/3, \pi/4, \pi/6$, is 10 dB, 10.25 dB, 11.1 dB, and 13.3 dB, respectively.

The coupling function is $S(x) = 1 - T(x) = (\alpha^{2n} - x^{2n})/(1 + \alpha^{2n} - x^{2n})$. After spectral factorization, the reflectance of the stepped line in the λ plane, ($x^2 = \alpha^2/(1 - \lambda^2)$) which defines the port 1 to port 2 coupling, is given as[21]

$$S_{21} = \frac{1}{\sqrt{10}} \frac{\lambda(\lambda + \sqrt{2})}{\lambda^2 + 1.974\lambda + 0.9487} = h/g$$

We can apply Richards' theorem directly to S_{21}, or compute $z = (g_e + h_e)/(g_o - h_o)$ and then apply Richards's theorem. The result is $r_{e1} = 1.48$, $r_{e2} = 1.07$. Using $r_e r_o = R_1^2 = 1$, $r_{o1} = 0.676$, $r_{o2} = 0.936$. □

Allpass Type C Section

We have already shown that a short or open circuited stub consisting of two cascaded UE's can be used to introduce a zero of transmission on the imaginary λ axis. By employing coupled TEM lines, it is also possible to introduce zeros of transmission on the $\Re\lambda = \Sigma$ axis and in the interior of the right half λ plane. In fact N. Saito, has shown that the complete analog of the Darlington lumped element synthesis with Type C, Brune, and Type D sections can be carried out by realizing the different section types with coupled transmission lines.[22] As a result, coupled line filters can be realized with equal ripple passband and stopband response.[23]

Coupled lines can also be used for interdigital filters (an array of coupled lines with the conductor terminations alternately open and short circuited),[24] and it may be noted that these structures generally have equivalent circuits consisting of cascaded UE's and stubs, so that their synthesis may be treated by the methods of Section 7.6.

[21]The notation can cause some confusion. Recall that S_{21}, as discussed earlier, is a transmittance of a four-port coupler, but is also identified as the *reflectance* of a stepped transmission line lossless two port. As such, it can be expressed in Belevitch notation as the ratio of polynomials, $S_{21} = h/g$.

[22]N. Saito, "Coupled-Line Filters", *Theory and Design of Microwave Filters and Circuits*, New York, Academic Press, 1970, Ch. 7.

[23]M. C. Horton and R. J. Wenzel, "General Theory and Design of Optimum Quarter-Wave TEM Filters", *IEEE Trans. on Microwave Th. and Tech.*, vol. MTT-13, no. 3, pp. 316-327, May 1965.

[24]G. L. Matthaei, "Interdigital band-pass filters", *IEEE Trans. on Microwave Th. and Tech.*, vol. MTT-10, pp. 479-491, Nov. 1962.

Another application of coupled lines is in the design of linear phase TEM filters and flat delay equalizers.[25] Coupled line sections, which are the counterparts of lumped C and D allpass sections, can be employed to realize transmission line equalizers. As a simple illustration of how coupled lines are used in these various applications, we derive a type C allpass section using the symmetric microstrip four-port discussed in the preceding analysis of coupled stepped line directional couplers.

Refer to Fig. 7.7.1 and assume that the ends of the two conductor strips are connected together at output ports 3 and 4. As a result, the odd mode propagating guide is terminated in a short circuit. The strips float above the ground plane, consequently the effect is to terminate the even mode in an open circuit. The resultant structure is now a lossless two-port, with input at 1, output at 2, and terminations R_1. As in our earlier analysis, port 1 is excited with incident signal a_1; then the even mode excitation for ports 1 and 2 is $a_1/2$, $a_1/2$; for the odd mode, $a_1/2$, $-a_1/2$. We compute the reflected or output wave at port 2 using superposition. The input impedance at port 2 in the even mode (open) is z_e/λ, for the odd mode (short) $z_o\lambda$, so the port 2 even and odd mode reflectances are

$$S_e = \frac{r_e - R_1\lambda}{r_e + R_1\lambda} \qquad S_o = \frac{r_o\lambda - R_1}{r_o\lambda + R_1}$$

The total reflected output signal at port 2 is $b_2 = b_e + b_o$ so that with all impedances normalized to R_1,

$$b_2 = \frac{a_1}{2}\left(\frac{r_e - \lambda}{r_e + \lambda} - \frac{r_o\lambda - 1}{r_o\lambda + 1}\right) = a_1 \frac{r_e - r_o\lambda^2}{(r_e + \lambda)(r_o\lambda + 1)}$$

The transmittance is $S_{21} = b_2/a_1$, and if we use the above equation inserting our earlier assumption that $r_e\, r_o = 1$,

$$S_{21} = \frac{r_o(r_e^2 - \lambda^2)}{(r_e + \lambda)(r_o\lambda + 1)} = \frac{r_o(r_e - \lambda)}{r_o\lambda + 1} = \frac{r_e - \lambda}{r_e + \lambda}$$

an allpass type C section with a zero at $\lambda = r_e$.

The phase shift of the allpass is $\phi(\Omega) = -2\arctan(r_e/\Omega)$ and since $\Omega = \tan\theta$, the delay is given by

$$\tau = -\frac{d\phi}{d\omega} = -\frac{d\phi}{d\Omega}\frac{d\Omega}{d\omega} = 2\tau_0 \sec^2\theta \frac{r_e}{r_e^2 + \Omega^2} = \frac{2\tau_0 r_e}{1 + (r_e^2 - 1)\cos^2\theta}$$

where $\theta = \omega\tau_o$, and τ_0 is the delay of the coupled UE microstrip.

[25] W. Steenart, "A Contribution to the Synthesis of Distributed Allpass Networks", *Proc. of Symposium on Generalized Networks*, New York, PIB Press, 1966, pp. 173-192.

8

Broadband Matching I: Analytic Theory

8.1 The Broadbanding Problem

The filtering problem, as discussed in Chapter 6, addresses the design of a lossless reciprocal two-port, which realizes a prescribed attenuation and/or phase characteristic when inserted between given *resistive* generator and load terminations. The transfer function and circuit elements of the filter are entirely at the disposal of the designer, subject only to the constraints of losslessness.

On the other hand, an important class of problems requires that either the load or the generator or both be not purely resistive, and a lossless two-port must be designed to operate between these prescribed frequency dependent terminations. Another case, in which a portion of the circuit is fixed and frequency dependent, occurs when a two-port (often an amplifier) is prescribed and lossless equalizers are to be placed between it and resistive input and output loads, so as to optimize system performance.

Each of the problems can be viewed in terms of the realization of a resistively terminated system consisting of the cascade of several structures, some of which are prescribed. The fixed portions of the system impose constraints on the design. For example, when a frequency dependent one-port load is given, the load impedance can be replaced, according to Theorem 5.7.1, by a lossless reciprocal two-port terminated in a unit resistor (its *Darlington equivalent*). The problem is then to design, say a filter, to be placed in cascade with the fixed Darlington two-port. In the amplifier equalization problem, the fixed part is the central block of a cascade structure consisting of three sections, two of which must be designed subject to the constraints of the amplifier.

In most cases, we wish to realize filtering characteristics like those discussed in Chapter 6. This means approximately matching, over a given passband, a resistive generator to a resistive load in the presence of the fixed sections. These are generally called *broadband matching* problems,[1] or more or less synonymously *gain-bandwidth problems*, since achievable gain levels are limited by the parameters of the fixed sections. When only one of the terminations is frequency dependent, we speak of the *single matching problem*; when both are frequency dependent, we have the *double matching* problem; the *two-port matching* problem is when a prescribed two-port must be matched to a resistive generator and load.

There are two general approaches to the solution of broadband matching problems. The classic procedure is through analytic *Gain-Bandwidth Theory* which extends to frequency variable terminations the techniques of *Insertion Loss Theory* which were described in earlier chapters, and optimizes a prescribed transducer gain function in the presence of the fixed portions of the system. The second type of solution is accomplished by numerical optimization. Here the objective is generally chosen as the achievement of maximum level of minimum gain within the passband, and the procedure is generally referred to as the *Real Frequency Technique* (RFT),[2] since the results are obtained by algorithmic means directly applied in the real frequency domain with only nominal application of analytic function theory.

In this chapter we first present Gain-Bandwidth Theory for the case of a resistive generator and a frequency dependent load (*single matching problem*). That theory was initially developed by Fano[3] using the concept of the Darlington equivalent and later reformulated and made rigorous by Youla[4] using complex normalization. We shall present a slight modification of Youla's approach. The problem of analytic double matching theory (frequency dependent generator and load impedances) will then be considered, and the solution will be shown to follow from the methods of single matching theory.

Instructional accounts of gain-bandwidth theory are available primarily treating the case of single matching and the Fano-Youla theory, [5] as well as some special applications.[6] The analytic solution of the double-matching

[1]Narrow band matching is usually accomplished by matching the generator to the load with a lossless section making the two impedances *conjugate complexes* of each other at a single frequency. By the theory of functions, conjugate matching is impossible over any finite frequency band.

[2]H. J. Carlin, "A New Approach to Gain-Bandwidth Problems", *IEEE Trans. on Circ. and Sys.*, vol. CAS-23, no. 4, pp. 170-175, Apr. 1977.

[3]R. M. Fano, "Theoretical Limitations on the Broad-Band Matching of Arbitrary Impedances", *J. Franklin Inst.*, vol. 249, no. 1, pp. 57-83, Jan. 1950, and no. 2, pp. 139-154, Feb. 1950.

[4]D. C. Youla, "A New Theory of Broadband Matching", *IEEE Trans. on Circ. Th.*, vol. CT-11, no. 1, pp. 30-50, March 1964.

[5]Wai-Kai Chen, *Theory and Design of Broadband Matching Networks*, Oxford, Pergamon Press, 1976, 432 p.

[6]C. Dehollain, *Adaptation d'impédance à large bande*, Lausanne, Presses Polytechniques et Universitaires Romandes, 1996, v+260 p.

problem has also been published.[7]

A different approach to both theory and numerical techniques has been developed by J. W. Helton[8], who treats broadband matching as an existence and approximation problem in H^∞ (the vector space of functions which are analytic and bounded in $\Re s > 0$). According to this method, realizability is ascertained by testing a set of matrices for positive definiteness. However, no actual realization examples have been published so far, so that it is not yet possible to evaluate the practical feasibility of the method.

8.2 The Chain Matrix of a Lossless Two-Port

As a preliminary to complex normalization, we describe the *chain matrix* $\boldsymbol{K}$ of a lossless two-port. This matrix relates the input and the output port quantities through the equation

$$\begin{pmatrix} v_1 \\ i_1 \end{pmatrix} = \begin{pmatrix} A & B \\ C & D \end{pmatrix} \begin{pmatrix} v_2 \\ -i_2 \end{pmatrix} = \boldsymbol{K} \begin{pmatrix} v_2 \\ -i_2 \end{pmatrix} \tag{8.2.1}$$

Matrix $\boldsymbol{K}$ can be deduced, like any other port description, from the general representation, eq. (4.2.1), provided the (2×2)-minor whose columns correspond to v_1 and i_1 is nonzero. This condition is evidently fulfilled whenever there is transmission from the input to the output and this is certainly true in the context of matching equalizers. The output current is denoted as $-i_2$, as a matter of convenience, since the usual port description has i_2 polarized into port 2, but our concern is with a succeeding section of the cascaded chain whose input is, say, v_3, i_3, where $v_3 = v_2$, $i_3 = -i_2$. Incidentally, this designation results in the overall matrix description, $\boldsymbol{K}$, of a group of cascaded two-ports with individual $\boldsymbol{K}_j$ matrices as $\boldsymbol{K} = \boldsymbol{K}_1\boldsymbol{K}_2\boldsymbol{K}_3 \ldots$.

To derive the properties of the chain matrix, we assume that the two-port possesses an impedance matrix $\boldsymbol{Z}$ (see Section 4.2, in particular eq. (4.2.3) and the following discussion). This assumption simplifies the calculation, but the results are entirely general and not a consequence of the existence

[7] H. J. Carlin, S. B. Yarman, "The Double Matching Problem: Analytic and Real Frequency Solutions", *IEEE Trans. on Circ. and Sys.*, vol. CAS-30, no. 1, pp. 15-28, Jan. 1983.

[8] J. W. Helton, "Broadbanding: Gain Equalization Directly from Data", *IEEE Trans. on Circ. and Sys.*, vol. CAS-28, no. 12, pp. 1125-1137, Dec. 1981.

of $\boldsymbol{Z}$. The two-port equations, written on an impedance basis, are

$$\begin{pmatrix} v_1 \\ v_2 \end{pmatrix} = \begin{pmatrix} z_{11} & z_{12} \\ z_{21} & z_{22} \end{pmatrix} \begin{pmatrix} i_1 \\ i_2 \end{pmatrix} \tag{8.2.2}$$

By solving eqs. (8.2.2) for v_1 and i_1 and identifying the coefficients with those of eqs. (8.2.1), we find

$$\begin{aligned} A &= \frac{z_{11}}{z_{21}} & B &= \frac{\det \boldsymbol{Z}}{z_{21}} \\ C &= \frac{1}{z_{21}} & D &= \frac{z_{22}}{z_{21}} \end{aligned} \tag{8.2.3}$$

Reciprocity requires that $z_{12} = z_{21}$. (See Corollary 2.14.4.) Losslessness requires that $\boldsymbol{Z}$ be paraskew. (See Corollary 4.6.5.) Thus $z_{11} = -z_{11*}$, $z_{22} = -z_{22*}$, and $z_{12} = -z_{12*}$ (the last condition presumes reciprocity). Thus eqs. (8.2.3) yield

$$\det \boldsymbol{K} = 1 \tag{8.2.4}$$

and a necessary condition for a lossless two-port

$$\begin{aligned} A &= A_* & B &= -B_* \\ C &= -C_* & D &= D_* \end{aligned} \tag{8.2.5}$$

Eqs. (8.2.4) and (8.2.5) will be used to derive a useful version of the scattering matrix which is normalized to complex rather than real quantities. Otherwise stated, complex normalization is with respect to terminations which are PR impedances rather than resistors.

8.3 Complex Normalization

In Section 4.4 the scattering matrix of an n-port was normalized to a set of n positive real numbers representing n purely resistive terminations. The procedure can be extended to complex numbers with positive real parts and even to dissipative PR functions, which is equivalent to assuming that the n-port terminations are passive immittances taken either at a fixed frequency or over the infinite frequency band.

The theory of complex normalization is particularly applicable to broadband matching, because, roughly speaking, our objective can be viewed as the construction of an equalizer whose reflectance when normalized to a given load impedance approximates zero over the passband. This statement

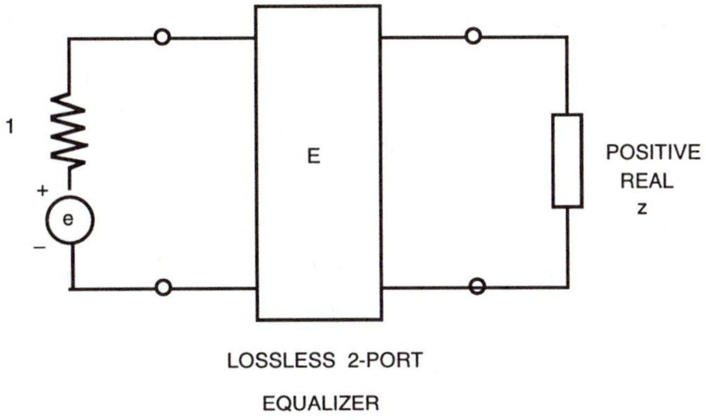

FIGURE 8.3.1
Matching of a resistive source to a frequency-dependent load.

is not as straightforward as it sounds, because of the problems of physical realizability associated with a complex normalized reflectance function. We shall see that an appropriate complex normalization method is directly related to Darlington's theorem that a PR impedance can always be represented as a lossless two-port terminated in a resistor. For our purposes it is particularly relevant to consider the complex normalization of the reflectance of a passive one-port. Extension to the n-port case is not difficult, but need not be considered here.

We consider the circuit of Fig. 8.3.1 where the two-port $\mathcal{E}$ is inserted between a resistive generator and a frequency dependent load, whose (nonnormalized) impedance z is represented by its Darlington equivalent $\mathcal{D}$. Let S_2 be the unit normalized reflectance at the output port of the two-port formed by the cascade of $\mathcal{E}$ and $\mathcal{D}$. The corresponding (nonnormalized) impedance into this port is Z_2. We wish to express S_2 in terms of Z, the (nonnormalized) impedance looking into port 2 of $\mathcal{E}$, when it is terminated at the input (source) port with a unit resistor.[9] To this aim, we represent $\mathcal{D}$ by its chain matrix $\boldsymbol{K}$, eq. (8.2.1). Then $Z_2 = v_2/i_2$ can be expressed in terms of $Z = v_1/(-i_1)$ by the bilinear transformation

$$Z_2 = \frac{DZ + B}{CZ + A}$$

so that we have

$$S_2 = \frac{\dfrac{DZ + B}{CZ + A} - 1}{\dfrac{DZ + B}{CZ + A} + 1}$$

[9]The choice of unit termination rather than a resistance, R, is merely a matter of convenience; it does not influence the definition of complex normalized reflectance.

By simple algebra, the above equation can be rewritten as

$$S_2 = \frac{D-C}{D+C} \frac{Z - \dfrac{A-B}{D-C}}{Z + \dfrac{A+B}{D+C}}$$

Now, since $\mathcal{D}$ is lossless, we have $A = A_*$, $B = -B_*$, $C = -C_*$, and $D = D_*$ by eq. (8.2.5). Thus the last equation becomes

$$S_2 = \frac{D_* + C_*}{D+C} \frac{Z - \dfrac{A_* + B_*}{D_* + C_*}}{Z + \dfrac{A+B}{D+C}} \tag{8.3.1}$$

We note that the input impedance, z, of the unit terminated two-port $\mathcal{D}$, is obtained from eq. (8.2.1) as the ratio $z = v_1/i_1$ with $v_2/(-i_2) = 1$,

$$z = \frac{A+B}{C+D} = \frac{M_1 + N_1}{M_2 + N_2}$$

The impedance z is rational, as are A, B, C, D (i.e. not polynomials). In the above equation, however, M_1, M_2, N_1, N_2 *are* polynomials, with M_j even, N_j odd. To obtain the explicit form of the polynomials M_j, N_j we examine r, the even part of z. Thus we have

$$r = \frac{1}{2}(z + z_*) = \frac{1}{(D+C)(D_* + C_*)} = \frac{M_1 M_2 - N_1 N_2}{M_2^2 - N_2^2} \tag{8.3.2}$$

As is usual in the Darlington procedure, if necessary we introduce common factors in the numerator and denominator of eq. (8.3.2) so that all RHP roots (these are necessarily paired with LHP images since $r > 0$) are of even multiplicity, and the numerator becomes a perfect square. We can perform spectral factorization on the numerator polynomial since it is even and positive. Thus

$$D + C = \frac{M_2 + N_2}{n_*} \equiv \frac{d}{n_*} \qquad D_* + C_* = \frac{M_2 - N_2}{n} \equiv \frac{d_*}{n} \tag{8.3.3}$$

where n is any factorization of $M_1 M_2 - N_1 N_2$ such that all of its RHP and LHP zeros (if any) are paired with image zeros placed in n_*, and d is a Hurwitz polynomial. At this point eq. (8.3.1) can be rewritten as

$$S = S_2 = \frac{n_*}{n} \frac{d_*}{d} \frac{Z - z_*}{Z + z} = \frac{w}{w_*} \frac{Z - z_*}{Z + z} \tag{8.3.4}$$

where

$$w = \frac{n_*}{d} \tag{8.3.5}$$

and

$$\frac{w}{w_*} = \frac{n_*}{n}\frac{d_*}{d} = \frac{n_*}{n}b$$

b is an allpass or Blaschke product, eq. (5.6.3), formed on the RHP zero factors of d_*. Note further that it follows from eqs. (8.3.2), (8.3.3), and (8.3.5) that

$$r = ww_*$$

The function S has all the properties of a BR reflectance, i.e.:

(a) it is analytic in the RHP and devoid of poles on the imaginary axis;

(b) $1 - S^*(j\omega)S(j\omega) \geq 0 \quad \forall$ real ω;

(c) $S^*(s) = S(s^*)$.

This follows immediately from our method of constructing S as equal to S_2, a unit normalized BR reflectance, and we therefore consider S as the BR reflectance of the one-port of impedance Z normalized to the passive impedance (or the PR function) z.

The incident and reflected waves, may be defined as follows[10]

$$2w\,a = v + zi$$
$$2w_*\,b = v - z_*i$$

since dividing the two equations and referring to eq. (8.3.4) yields $S = b/a$, the reflectance of Z complex normalized to the impedance it faces, z.

As an alternate approach we can start with the above equations for complex normalized incident and reflected waves and then obtain the second expression of eq. (8.3.4) as the *definition* of the reflectance S normalized to a dissipative PR function. The equality of S and S_2 would then be expressed by the following theorem[11,12]

THEOREM 8.3.1

Let S be the complex normalized reflectance of a PR impedance Z normalized to a prescribed dissipative PR impedance z.

$$S = \frac{w}{w_*}\frac{Z - z_*}{Z + z}$$

[10] v and i represent nonnormalized voltage and current.

[11] H. J. Carlin, S. B. Yarman, "The Double Matching Problem: Analytic and Real Frequency Solutions", *IEEE Trans. on Circ. and Sys.*, vol. CAS-30, no. 1, pp. 15-28, Jan. 1983.

[12] R. Pauli, "Darlington's Theorem and Complex Normalization", *Int. J. Circ. Th. Appl.*, vol. 17, no. 4, pp. 429-446, Oct. 1989.

Let S_2 be the unit normalized reflectance seen looking in at the output port of the Darlington equivalent ($\mathcal{D}$) of z, when $\mathcal{D}$ is terminated at its input by Z. Then $S = S_2$.

Some interesting consequences of the method of complex normalization used here are given below:

1. We can compute the different port voltages and currents from the expressions for complex normalized a and b

$$v = \frac{z_*}{w_*}a + \frac{z}{w}b$$

$$i = \frac{a}{w_*} + \frac{b}{w}$$

   The power flow across any terminal pair of the lossless cascade of $\mathcal{D}$ and $\mathcal{E}$ may be computed as $P = (1/2)(vi^* + v^*i)$, and using the above equations, we can show that $P = a^*a - b^*b$, just as with real normalization.[13] Indeed, real normalization can be viewed as an example of complex normalization for the special case that one of the two impedances is a resistor.

2. Since the system is lossless, P is invariant as we progress along the chain. It follows that the paraconjugate expression $a_*a - b_*b$ is also unchanging along the chain (of course a and b are appropriately normalized depending on which two impedances are facing each other).

3. When the system is (conjugate) matched, $Z = z^*$ and $b = 0$ at every junction of the cascade. Then $P_A = a_*a$ is the available source power and P_A as well as a is invariant along the chain. When the system is mismatched ($b \neq 0$), a remains invariant (a_*a is still the available power) along the cascade. The presence of b reduces the net power flow from the available level, and b is also invariant along the cascade.

4. The polynomials n and d depend only on the load impedance, a basic observation for setting up the broadband matching constraints. The choices available for RHP or LHP factors of n are just a consequence of the nonuniqueness of spectral factorization when computing the Darlington representation of the load (see the Spectral Factorization Lemma in Chapter 5, Section 5.7). The most convenient choice is $n = n_* = \sqrt{M_1M_2 - N_1N_2}$ for, as discussed earlier, we can use superfluous

[13]This idea, which led to the concept of complex normalization, is due to H. W. Bode, *Network Analysis and Feedback Amplifier Design*, New York, D. Van Nostrand Co., 1945, p. 364.

factors to make nn_* a perfect square. With this choice $n/n_* = 1$, and eq. (8.3.4) reduces to

$$S = b\,\frac{Z - z_*}{Z + z} \tag{8.3.6}$$

where the allpass Blaschke product $b = d_*/d$, is formed on the RHP pole factors in z_*. This is the function that Youla uses in his procedure, and the one we shall also employ.

8.4 The Gain-Bandwidth Restrictions

We first describe analytic gain bandwidth theory for the case of single matching. The system is schematically shown in Fig. 8.3.1. The passive load is represented by its Darlington equivalent, the lossless two-port $\mathcal{D}$ terminated on a unit resistor, while the two-port $\mathcal{E}$ represents the matching section to be designed.

At the outset, as in the case of doubly terminated filters (Chapter 6), a transducer gain function, $T(\omega^2)$, is prescribed, except that now it contains some unknown parameters, in particular gain level, which are to be optimized subject to physical realizability and the constraints of the load. Since the overall two-port comprising the cascade of $\mathcal{E}$ and $\mathcal{D}$ is lossless, we can obtain from T the reflectance expression $S_2 = S$. It must be emphasized that these are alternate representations of the same function; S is the complex normalized form in terms of Z and z, and S_2, the unit normalized version measured at port 2 of $\mathcal{D}$. In the following discussion when we speak of S, we are invoking properties of the reflectance which are relevant to the complex normalized form. We proceed as follows:

(a) Calculate $|S(j\omega)|^2 = 1 - T(\omega^2)$;

(b) Construct the analytic extension of $|S(j\omega)|^2$ as S_*S;

(c) From S_*S determine S by spectral factorization; it contains the unknown parameters which are to be optimized.

Note that the last step gives $S = S_2$ within a Blaschke product (i.e., an allpass). The Darlington procedure of synthesis, carried out from the output of the overall two-port and moving towards the source, must take into account the fact that the entire section $\mathcal{D}$ is fixed. We will show that this imposes constraints which force power series coefficients of reflectance S to satisfy certain equalities and (possibly) inequalities at the transmission zeros of the load, represented by the two-port $\mathcal{D}$.

Recall the definition of transmission zero, eq. (5.6.2), as closed RHP (i.e.,including imaginary axis) zeros of F.

$$F = 2b\frac{z + z_*}{(z+1)^2} = \frac{4rb}{(z+1)^2} \tag{8.4.1}$$

Our first step is to catalog the various kinds of transmission zeros.[14]

DEFINITION 8.4.1 *Let s_0 denote any transmission zero of z. Then s_0 belongs to one of the following mutually exclusive classes:*

(a) Class A: $\Re s_0 > 0$, *the transmission zero lies in the open RHP.*

(b) Class B: $\Re s_0 = 0$ *and* $z(j\omega_0) \neq \infty$, *the transmission zero lies on the imaginary axis at a point where the load impedance z is finite (possibly zero).*

(c) Class C: $\Re s_0 = 0$ *and* $z(j\omega_0) = \infty$, *the transmission zero lies on the imaginary axis at a point where the load impedance has a pole.*

The basic Gain-Bandwidth Theorem given below relates properties of the system reflectance to the load zeros of transmission, and as discussed in the preceding Section, we choose $n = n_*$, so that the representation of S has the form given by eq. (8.3.6), i.e.,$S = b(Z - z_*)/(Z + z)$. This equation leads to the replacement of Z by $Z + z - z$.

$$b - S = \frac{2rb}{Z + z} \tag{8.4.2}$$

We now prove the following theorem.

THEOREM 8.4.1
A function $T = T(\omega^2)$, such that $0 \leq T(\omega^2) \leq 1$, $\forall\omega$, can be realized as the transducer power gain of finite lossless equalizer $\mathcal{E}$, inserted between a resistive generator (whose resistance is unit normalized) and a frequency dependent load (whose impedance is a dissipative PR rational function z), if and only if, at each transmission zero of the load, s_0, of order k, the coefficients of the Taylor expansion of $b - S$ about s_0 satisfy the following constraints:

Class A: $\Re s_0 > 0$

$$b_h = S_h \quad (h = 0, 1, 2, ..., k-1) \tag{8.4.3}$$

[14]Youla uses the closed RHP zeros of r/z for his definition of transmission zeros. This differs from our definition, which is in conformity with that employed in Darlington synthesis problems.

Class B: $\Re s_0 = 0,\ z(s_0) \neq \infty$

$$b_h = S_h \qquad (h = 0, 1, 2, ..., 2k-2) \tag{8.4.4}$$

$$\frac{b_{2k-1} - S_{2k-1}}{(2rb)_{2k}} \geq 0 \tag{8.4.5}$$

Class C: $\Re s_0 = 0,\ z(s_0) = \infty$

$$b_h = S_h \qquad (h = 0, 1, 2, ..., 2k-2) \tag{8.4.6}$$

$$\frac{b_{2k-1} - S_{2k-1}}{(2rb)_{2k-2}} \leq \frac{1}{c_{-1}} > 0 \tag{8.4.7}$$

In eq. (8.4.7), c_{-1} is the residue of z at the $s_0 = j\omega_0$ pole, otherwise, the subscript identifies a Taylor coefficient.

Before proceeding with the actual proof, we first present a few pertinent preliminaries. We are concerned with the load zeros of transmission since at these points the load strongly constrains system performance. Thus in the statement of the theorem, since b is only a function of z, all constraints on the overall system function S depend on the load alone. If we compare the function F in the definition of transmission zero, eq. (8.4.1), with $b - S$ in eq. (8.4.2), it is evident that a zero in F will also appear in $b - S$, though it may be with lower multiplicity because of possible cancellations. That is, for a transmission zero s_0 of order k in the RHP, we can write the Taylor series for F and $b - S$ as

$$F = A_k(s - s_0)^k + A_{k+1}(s - s_0)^{k+1} + \ldots$$
$$b - S = B_k(s - s_0)^k + B_{k+1}(s - s_0)^{k+1} + \ldots$$

A zero on the $j\omega$ boundary is always of even order, hence it appears with multiplicity $2k$ though counted as a transmission zero of order k. Thus

$$F = A_{2k}(s - j\omega_0)^{2k} + A_{2k+1}(s - j\omega_0)^{2k+1} + \ldots$$
$$b - S = B_{2k}(s - j\omega_0)^{2k} + B_{2k+1}(s - j\omega_0)^{2k+1} + \ldots$$

In the series for $b - S$, the first nonvanishing power may be $k-1$ (or $2k-1$), if a cancellation occurs. We use the following Lemma.

LEMMA 8.4.1
A zero of transmission cannot appear in S nor in the Blaschke product

$$b = \frac{d_*}{d}, \quad \text{where } 2r = z + z_* = \frac{n}{d} + \frac{n_*}{d_*}$$

The zero appears in the transmittance so it cannot be present in the reflectance. Next suppose b had the zero, then it would appear in d_*, and referring to eq. (8.4.1) would cancel from F, for since z is PR, $z+1$ has no zeros or poles in $\Re s \geq 0$ to provide alternate cancellations. In other words, a zero in b cannot be a transmission zero. As a consequence of the lemma, $b-S$ can only go to zero at a transmission zero as a result of Taylor coefficient cancellations.

We return to the main body of the proof.

PROOF Necessity. The load and equalizer are given and we wish to prove the constraints of the theorem. In view of the lemma, we examine the behavior of b and S at a load transmission zero by expanding both sides of eq. (8.4.2) in Taylor series about each transmission zero. Refer to F in eq. (8.4.1) and note that if s_0 belongs to Class A and is of order k, it can only appear as a zero of the same order in r, since neither b nor $(z+1)^2$ can reduce the order of the zero. The same reasoning applies to the expression $b-S = 2rb/(Z+z)$. Therefore the RHP zeros of r and $b-S$ coincide and have the same order. Thus the first $k-1$ coefficients in the Taylor expansion of $b-S$ about s_0 vanish. Otherwise stated, the first $k-1$ Taylor coefficients in the b and S expansions about s_0 must be equal.

If s_0 belongs to Class B, then, referring to F and our preliminary discussion, a transmission zero of order k appears as a multiplicity $2k$ zero in r. This means that if $Z+z$ is not zero at s_0, there is no zero factor cancellation, and $b-S$ has the s_0 zero to at least order $2k$. In this case the inequality eq. (8.4.5) becomes an equality. In the event that the equalizer is such that $Z(s_0)+z(s_0)=0$, termed a *degeneracy*,[15] then, since a $j\omega$ pole must be simple with a positive residue, the denominator zero is simple and near the zero $Z+z \approx a_1(s-s_0)$, with $a_1>0$. It is clear that the degeneracy reduces the multiplicity of the zero in $b-S$ to $2k-1$. Moreover near s_0, after cancelling the zero due to $Z+z$, we can write (using subscripts to denote the appropriate Taylor coefficient)

$$(b-S)_{s\to s_0} = \left(\frac{2rb}{Z+z}\right)_{s\to s_0} \approx \frac{(2rb)_{2k}(s-s_0)^{2k-1}}{a_1}$$

Equating Taylor coefficients

$$\frac{b_{2k-1}-S_{2k-1}}{(2rb)_{2k}} = \frac{1}{a_1} \geq 0$$

If s_0 belongs to Class C, the transmission zero of order k results from a double pole in the denominator of F (since z has a simple pole), combined

[15]An example would be a load whose input is a shunt inductance facing another shunt inductance which is part of the equalizer. Thus $Z+z=0$ at $s=0$.

with a zero of order $2k-2$ in r. This means that $b - S = 2br/(Z+z)$ has a zero of order $2k-1$, since the denominator pole raises the order of the zero in r by 1. If $Z(s_0) \neq \infty$ then near s_0, $Z+z \approx c_{-1}/(s-s_0)$ (residue $c_{-1} > 0$), and equating Taylor coefficients, we have the constraint of eq. (8.4.7) with the equal sign. A degeneracy for the class C case occurs when Z has a pole at s_0, and for $s \to s_0$ this results in $Z + z \approx (C_{-1} + c_{-1})/(s - s_0)$, both residues positive. Then

$$\frac{b_{2k-1} - S_{2k-1}}{(2rb)_{2k-2}} = \frac{1}{C_{-1} + c_{-1}} \leq \frac{1}{c_{-1}}$$

which is eq. (8.4.7). The result includes the case of a pole at ∞.

Sufficiency. To prove sufficiency, we need only show that the output impedance Z of the unit terminated equalizer is PR after the load has been removed in the Darlington synthesis (see Section 5.7) of S_2. Then Z is also realized as a reactive two-port terminated on a unit resistor, and this two-port is just the equalizer $\mathcal{E}$.

From eq. (8.4.2), we obtain

$$Z = \frac{2br}{b - S} - z \tag{8.4.8}$$

The real part $\Re Z(j\omega) = R(\omega)$ is given by

$$R(\omega) = \frac{r(1 - |S(j\omega)|^2)}{|b(j\omega) - S(j\omega)|^2}$$

This can be seen using eq. (8.4.8) to calculate $R(\omega) = 1/2(Z(j\omega) + Z^*(j\omega))$ taking into account the fact that $bb^* = 1$ on the imaginary axis. Evidently $R(\omega) \geq 0$, $\forall\omega$. Consider now

$$Y_1 = \frac{1}{Z_1} = \frac{b - S}{2br} = \frac{1}{Z + z}$$

We prove that this function is PR. We have $\Re Y_1 = \Re(Z+z)/|Z+z|^2$. Since $R(\omega) \geq 0$, and z is PR, it follows that $\Re Y_1(j\omega) \geq 0$. We must now show that Y_1 is analytic in $\Re\, s > 0$, with only simple boundary poles and positive residues at these poles. Both b and S are analytic in $\Re s \geq 0$. Thus closed RHP poles of Y_1 occur only at the zeros of $2br$. If there is a zero of $2br$ in the RHP, it must coincide with a Class A transmission zero. But in that case, Y_1 is regular at this zero, since for class A, $b - S$ and $2rb$ have zeros of the same order, which cancel in Y_1. Thus Y_1 is analytic everywhere in $\Re\, s > 0$. If $2br$ has a zero on $j\omega$ it may occur where z is finite or zero (Class B), or where z has a pole (Class C); this exhausts all the possibilities. At a transmission zero of Class B, Y_1 has a simple pole with positive residue as given by eq. (8.4.5). At a transmission zero of Class C, Y_1 has a simple zero with positive coefficient given by $1/c_1$ in eq. (8.4.7). Thus Y_1 has a nonnegative real part on the $j\omega$-axis, is analytic in the RHP, and its poles on the $j\omega$-axis are simple with positive residues. Therefore Theorem 4.6.5 and Theorem 4.6.3 insure that Y_1 and $Z_1 = 1/Y_1$ are PR. Now eq. (8.4.8) yields $Z = Z_1 - z$. It has already been shown that $\Re Z(j\omega) \geq 0$; moreover, Z is analytic in the RHP since both the PR Z_1 and z are analytic there. The Class C transmission zeros coincide with the $j\omega$ poles of z. We must check that at these poles Z has positive residues despite the presence of $-z$. We have seen that Y_1 has a zero and therefore Z_1 a pole at a class C zero. Thus the residue of Z at the pole is $\text{Res}(Z_1) - \text{Res}(z)$, or

$$\text{Res}(Z) = \frac{(2rb)_{2k-2}}{b_{2k-1} - S_{2k-1}} - c_{-1}$$

which by eq. (8.4.7) is nonnegative. We have therefore shown that Z is PR. ∎

We have earlier pointed out that if the highest order coefficient constraint for Class B and C transmission zeros applies as a strict inequality, the matching section is said to be degenerate. Physically degeneracy means that a transmission zero of the load is only partially extracted in its Darlington equivalent and that the remainder is left in the equalizer.

Additionally it should be noted that in order to determine a reflectance $S = S_2$ which satisfies the gain-bandwidth (GBW) constraints of the Theorem, we may write S in the form

$$S = \eta \hat{S}$$

where η is an analytic allpass and $\hat{S}$ is a minimum phase function. If $\hat{S}$ contains sufficient parameters to satisfy the GBW restrictions, η can be chosen equal to ± 1; otherwise, its parameters can be used to supplement those of $\hat{S}$ for satisfying the restrictions.

Finally, in choosing a transducer gain function whose parameters are to be determined in accordance with the GBW restrictions, it is obviously necessary that the gain contain the zeros of transmission of the load with at least the same multiplicity.

We illustrate the application of the GBW Theorem with the following example which applies to a load with a transmission zero at infinity; accordingly $T(\omega^2)$ has been chosen to have a multiple zero of transmission at infinity.

Example 8.4.1

Given a parallel RC load to be matched with a lossless equalizer to a source of internal resistance R, determine the parameters of an optimum Butterworth gain function which describes the overall equalizer-load system.

$$T(\omega^2) = \frac{K}{1 + \epsilon^2 \omega^{2n}} \tag{8.4.9}$$

where $0 < K \leq 1$.

Solution The Darlington equivalent of the load is simply the shunt capacitor C. The load impedance is $z = R/(1 + RCs)$, and there is a simple Class B transmission zero ($k = 1$) at infinity, since $z(\infty) = 0$.

We have

$$z = \frac{R}{1 + RCs}$$

$$r = \frac{R}{1 - R^2C^2s^2}$$

$$b = \frac{1 - RCs}{1 + RCs}$$

and therefore

$$2rb = \frac{2R}{(1 + RCs)^2}$$

Moreover, we choose

$$\eta = \prod_{i=1}^{m} \frac{\lambda_i - s}{\lambda_i^* + s} \tag{8.4.10}$$

where the λ_i's are real or complex conjugate zeros (as yet unknown) of the allpass η with $\Re\lambda > 0$.

From eq. (8.4.9), we obtain by analytic continuation

$$T(-s^2) = \frac{K}{1 + (-1)^n \epsilon^2 s^{2n}}$$

and therefore

$$S_* S = 1 - T(-s^2) = (1 - K) \frac{1 + (-1)^n \dfrac{\epsilon^2}{1-K} s^{2n}}{1 + (-1)^{2n} \epsilon^2 s^{2n}}$$

We perform spectral factorization and obtain

$$\hat{S} = -\sqrt{1-K} \cdot$$

$$\cdot \frac{1 + a_1 \left(\dfrac{\epsilon}{\sqrt{1-K}} \right)^{\frac{1}{n}} s + \ldots + a_{n-1} \left(\dfrac{\epsilon}{\sqrt{1-K}} \right)^{\frac{n-1}{n}} s^{n-1} + \dfrac{\epsilon}{\sqrt{1-K}} s^n}{1 + a_1 \epsilon^{\frac{1}{n}} s + \ldots + a_{n-1} \epsilon^{\frac{n-1}{n}} s^{n-1} + \epsilon s^n}$$

In the equation above, both numerator and denominator are Butterworth polynomials (all roots chosen in the LHP) in the variables $[\epsilon/(1-K)]^{1/n} s]$

and $\epsilon^{1/n}s$, respectively. Thus both are Hurwitz polynomials and the reflectance, $\hat{S}$, is minimum phase.

We choose η as in eq. (8.4.10) and proceeding according to the GBW Theorem for a class B zero, we expand $S = \eta\hat{S}$, b, and $2rb$ about infinity

$$S = -1 + \left(a_{n-1} \frac{1 - (\sqrt{1-K})^{\frac{1}{n}}}{\epsilon^{\frac{1}{n}}} + 2\sum_{i=1}^{m} \lambda_i \right) \frac{1}{s} + o\left(\frac{1}{s^2}\right)$$

$$b = -1 + \frac{2}{RCs} + o\left(\frac{1}{s^2}\right)$$

$$2rb = \frac{2}{RC^2 s^2} + o\left(\frac{1}{s^4}\right)$$

The gain-bandwidth restrictions (8.4.4) and (8.4.5) give

$$b_0 = S_0$$

$$\frac{b_1 - S_1}{(2rb)_2} \geq 0$$

The first equation is identically satisfied ($-1 = -1$), and the second yields

$$\frac{2}{RC} \geq a_1 \frac{1 - (\sqrt{1-K})^{\frac{1}{n}}}{\epsilon^{\frac{1}{n}}} + 2\sum_{i=1}^{m} \lambda_i$$

The coefficient a_{n-1} is obtained from the first of eqs. (6.5.5) as

$$a_1 = \frac{1}{\sin \dfrac{\pi}{2n}}$$

and therefore the above inequality can be rewritten as

$$(1-K)^{\frac{1}{2n}} \geq 1 - 2\epsilon^{\frac{1}{n}} \sin(\pi/2n) \left(\frac{1}{RC} - \sum_{i=1}^{n} \lambda_i \right) \tag{8.4.11}$$

We distinguish two cases.

1. $2\epsilon^{\frac{1}{n}} \sin(\pi/2n)/RC \geq 1$. By properly choosing the λ_i's ($\lambda_i = 0$, $\forall i$, is a simple and perfectly acceptable choice), we can satisfy inequality (8.4.11) with $K = 1$.

2. $2\epsilon^{\frac{1}{n}} \sin(\pi/2n)/RC < 1$. The solution with all $\lambda_i = 0$ gives $K = K_0 < 1$. On the other hand, with some $\lambda_i \neq 0$ the summation of all the allpass roots must be positive since they are all in the RHP and are real or in conjugate pairs. The resulting solution would therefore be $K < K_0$, hence we set $\eta = 1$.

The optimum gain level therefore satisfies

$$K \leq 1 - \left[1 - 2\epsilon^{\frac{1}{n}} \frac{\sin(\pi/2n)}{RC}\right]^{2n} \tag{8.4.12}$$

Therefore, for a given bandwidth (here assumed normalized to unity), the optimum gain K is inversely proportional to the RC product. In terms of a *gain-bandwidth restriction*, this means that the larger the parasitic capacitance, the smaller is the achievable gain level.

If in eq. (8.4.12) we take the limit for $n \to \infty$, by taking into account the well known definition of the exponential

$$\epsilon^{am} = \lim_{m \to \infty} \left(1 + \frac{a}{m}\right)^m$$

we obtain

$$K \leq 1 - \exp\left(-\frac{2\pi}{RC}\right) \tag{8.4.13}$$

Eq. (8.4.13) expresses the ultimate upper bound for the d.c. gain K when the equalizer contains infinitely many elements and its transfer function amplitude is constant in the passband, zero in the stopband. This important result, originally due to Bode but by a different procedure,[16] shows that the upper bound of K is fixed by the value of the parasitic capacitance C. If in eq. (8.4.13) cutoff is at ω_0 instead of unity, C must be replaced by $C\omega_0$. Consequently an increase of the capacitance C with no reduction in the d.c. gain K is possible only at the expense of a reduction in bandwidth, ω_0. □

Example 8.4.2
Equalize the load shown in Fig. 8.4.1 (a) to a unit resistance generator for the Butterworth gain

$$T = \frac{K}{1 + \omega^2}$$

Solution The load impedance is

$$z = \frac{4 + s}{1 + s}$$

[16] H. Bode, *Network Analysis and Feedback Amplifier Design*, New York, Van Nostrand, 1945, p. 367.

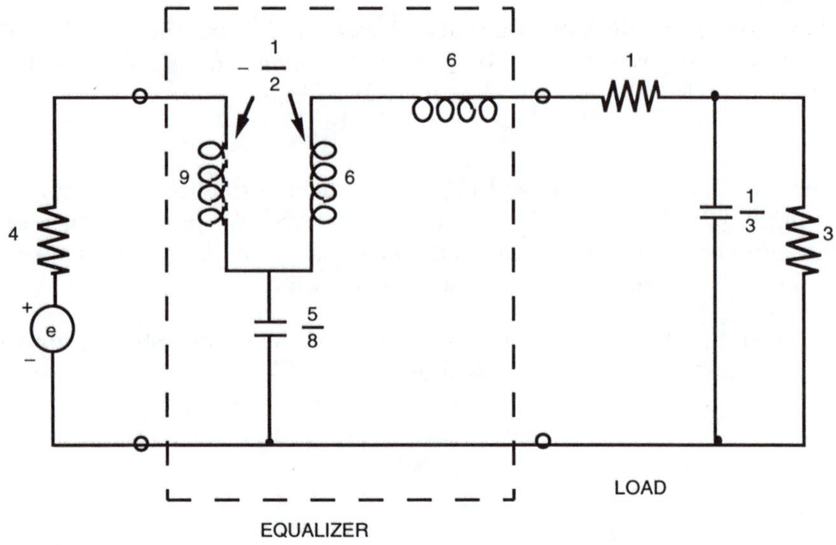

FIGURE 8.4.1
Butterworth equalization of a load with a RHP transmission zero.

Its transmission zeros, as given by eq. (5.6.5), are the RHP roots of the equation

$$4 - s^2 = 0$$

and therefore $s_0 = 2$, a Class A zero. We look for a solution with $K = 1$. Thus we have

$$SS_* = 1 - \frac{1}{1 - s^2} = \frac{-s^2}{1 - s^2}$$

The minimum phase factorization of the above equation yields

$$S = \pm\frac{s}{s + 1}$$

The allpass b is

$$b = \frac{1 - s}{1 + s}$$

Since the transmission zero at $s_0 = 2$ is simple, the Taylor coefficients, according to eq. (8.4.3), must satisfy $b_0 = S_0$ which gives $\pm 2/3 = -1/3$, an evidently impossible condition. Thus we introduce a suitable allpass η into the reflectance S. Write

$$S = \frac{s}{s + 1}\,\frac{\mu - s}{\mu + s}$$

Then the restriction $b_0 = S_0$ becomes

$$-\frac{1}{3} = \frac{2}{3}\,\frac{\mu - 2}{\mu + 2}$$

which gives $\mu = 2/3$. One might also choose the $(-)$ sign for S and obtain $\mu = 6$; however, we retain the first choice. Applying eq. (8.4.8) we find the output impedance of the unit terminated equalizer

$$Z = \frac{2br}{b - S} - z = \frac{6s^2 + 15s + 4}{s + 1}$$

Impedance Z can be realized by Darlington's procedure. First, we extract the pole at infinity

$$Z_1 = Z - 6s = \frac{9s + 4}{s + 1}$$

By using eq. (5.6.5) again, we note that Z_1 has a transmission zero at $s = 2/3$ (the RHP root of the equation $4 - 9s^2 = 0$). Thus we extract a Darlington's C-section according to the procedure described in Section 5.11. This section is composed of two perfectly coupled coils, with self-inductances $L_1 = 6$ and $L_2 = 9$, and mutual inductance $M = 3.6$, in series with a capacitance $C = 5/8$, the combination connected across the line.

The C-section is required because of the requirement for an allpass, η, introduced into S. It is easy to verify that, even if a d.c. gain $K < 1$ were acceptable, the only value that would satisfy gain-bandwidth restrictions without an allpass would be $K = 0$! Thus the equalizer design based on Analytic Theory necessarily leads to the use of coupled coils, an ordinarily undesirable constraint. We shall see in Section 9.1 that the same problem can be solved numerically with a superior gain response and a simpler equalizer. □

8.5 The Gain-Bandwidth Restrictions in Integral Form

The gain-bandwidth restrictions can be presented in *integral form* involving only $j\omega$ boundary behavior of b and S. Furthermore, the integral relations lead directly to expressions for attainable *amplitude* response. This approach avoids the use of limit processes such as employed at the end of the previous section, where the maximum gain level was determined as the limit of a Butterworth transfer function of infinite degree.

In order to obtain the restrictions on amplitude, we note that $\ln S = \ln|S| + j\phi$, and therefore we seek the GBW constraints in terms of the logarithms of the pertinent functions. Start with eq. (8.4.2) $b - S = 2br/(Z+z)$, or $S/b = 1 - 2r/(Z + z)$, and take the logarithm of both sides.

$$\ln\left(\frac{S}{b}\right) = \ln\left(1 - \frac{2r}{Z + z}\right)$$

At a transmission zero $2r/(Z+z) \to 0$, so that we can expand the above expression in a logarithmic series.

$$\ln\left(\frac{S}{b}\right) = \ln S - \ln b = -\frac{2r}{Z+z} + \frac{1}{2}\left(\frac{2r}{Z+z}\right)^2 - \ldots$$

Thus the zeros of $\ln S - \ln b$ coincide with those of $2r/(Z+z)$, which in turn are the transmission zeros. (The absence of b in the numerator is irrelevant since b does not contain transmission zeros.) We conclude that the gain-bandwidth restrictions on $\ln S$ in terms of $\ln b$ are the same as those of S vs. b given in the Gain-Bandwidth Theorem 8.4.1. We therefore restate the Theorem in terms of logarithms.

As a matter of notation, denote β_i, Σ_i, ρ_i as the coefficients of the Taylor expansions, respectively of $\ln b$, $\ln S$, r around a transmission zero. Thus we have the following Theorem.

THEOREM 8.5.1

Let k be the order of a transmission zero, s_0. Then the coefficients of the Taylor expansion of $\ln b - \ln S$ about s_0 satisfy the following constraints:

Class A: $\Re s_0 > 0$

$$\beta_h = \Sigma_h \qquad (h = 0, 1, 2, \ldots, k-1) \tag{8.5.1}$$

Class B: $\Re s_0 = 0,\ z(s_0) \neq \infty$

$$\beta_h = \Sigma_h \qquad (h = 0, 1, 2, \ldots, 2k-2) \tag{8.5.2}$$

$$\frac{\beta_{2k-1} - \Sigma_{2k-1}}{\rho_{2k}} \geq 0 \tag{8.5.3}$$

Class C: $\Re s_0 = 0,\ z(s_0) = \infty$

$$\beta_h = \Sigma_h \qquad (h = 0, 1, 2, \ldots, 2k-2) \tag{8.5.4}$$

$$\frac{\beta_{2k-1} - \Sigma_{2k-1}}{\rho_{2k-2}} \leq \frac{1}{c_{-1}} > 0 \tag{8.5.5}$$

In eq. (8.5.5), c_{-1} is the residue of z at the pole, s_0.

The restrictions above are to be put into integral form in order to obtain expressions directly in terms of $\ln|S|$. We use the representation derived from the Hilbert transform (3.7.3) that we rewrite here as

$$\ln S(s) = \ln \eta(s) + \frac{2s}{\pi}\int_0^{+\infty} \frac{\ln|S(j\Omega)|}{s^2+\Omega^2}\, d\Omega \tag{8.5.6}$$

In this equation, $S = \eta\hat{S}$, where η is an analytic allpass and $\hat{S}$ is minimum phase, but in the integrand we have used $|S(j\omega)| = |\hat{S}(j\omega)|$. For the sake of convenience, we set

$$\frac{2s}{\pi(s^2+\Omega^2)} = \Psi(s,\Omega)$$

so that eq. (8.5.6) becomes

$$\ln S = \ln\eta + \int_0^{+\infty} \Psi(s,\Omega)\ln|S(j\Omega)|d\Omega \tag{8.5.7}$$

We expand both sides of (8.5.7) in a Taylor series about a zero of transmission, $s = s_0$, and by equating coefficients formally obtain

$$\Sigma_h = \eta_h + \frac{1}{h!}\int_0^{+\infty} \Psi^{(h)}(s_0,\Omega)\ln|S(j\Omega)|d\Omega \tag{8.5.8}$$

where η_h is the h-th Taylor coefficient in the expansion of $\ln\eta(s_0)$, and $(1/h!)\Psi^{(h)}(s_0,\Omega)$ is the coefficient in the expansion of ψ at s_0 (superscript "(h)" indicates the h-th derivative).

We now discuss the validity of eq. (8.5.8) with regard to the convergence of the integral on the right hand side. First note that the denominator of $\Psi^{(h)}(s_0,\Omega)$ is $(s_0^2+\Omega^2)^{h+1}$, so that for a Class B or Class C zero of transmission with $s_0 = j\omega_0$, $\psi^{(h)}$ has a pole of order $h+1$ on $j\Omega$. Also note that the β_k coefficients for Class B and C zeros require the expansion of $\ln b(j\omega) = \ln|b(j\omega)| + j\arg b = j\arg b$, since $|b(j\omega)| = 1$. Now consider the logarithmic allpass sum, when $s = j\omega$, for $\ln b = j\theta(\omega)$. Then, since $d\theta/ds = (d\theta/d\omega)1/j$,

$$\frac{d^h\ln b}{ds^h} = j\frac{d^h\theta(\omega)}{d\omega^h}\frac{1}{j^h}$$

at the transmission zeros on the imaginary axis, β_h is real for h odd, and is imaginary for h even. In particular, at $\Omega = 0$ and $\Omega = \infty$, all nonzero coefficients must be real; therefore, for this case, all even coefficients must vanish.[17]

For a Class A transmission zero of order k, s_0 is in the RHP, so that $\psi(s_0,\Omega)$ has no poles on the boundary. The integral is, in fact, the classic formula for the Taylor coefficients of a function analytic in the closed RHP. The coefficients $\Sigma_h = \beta_h$, for h from 0 to $k-1$.

If s_0 is a Class B transmission zero, then by Theorem 8.5.1, the Taylor coefficients of $\ln|S(j\Omega)|$ must equal those of $\ln|b(j\Omega)|$ up to order $2k-2$

[17]Explicit formulas for the various Taylor expansions are given in the literature, for example, see D. C. Youla, "A New Theory of Broadband Matching", *IEEE Trans. on Circ. Th.*, vol. CT-11, no. 1, pp. 30-50, March 1964.

(this is implied in the equality of coefficients for $\ln b(s)$ and $\ln S(s)$). But since $\ln|b(j\Omega)| = 0$, *all* its coefficients are zero, hence those of $\ln|S(j\Omega)|$ are zero up to $2k-2$, and so $\ln|S(j\Omega)|$ has a zero of order $2k-1$, which cancels the pole, order $h+1$, in $\psi^{(h)}$ up to and including $h = 2k-2$. For this range of h, the integral converges and the Taylor coefficients, Σ_h, are therefore well defined. For a zero of transmission at $s_0 = 0$ or $s_0 = \infty$, the expansion of $\ln S(j\Omega)$ in Ω or $1/\Omega$, respectively, has only even order coefficients; thus in this case, *all the coefficients* up to $2k-1$ are zero (i.e., a zero of order $2k$) and the integral converges for $h = 2k-1$. Following the same line of argument we arrive at similar conclusions for a zero of Class C.

Thus we have the following Theorem for the *gain-bandwidth restrictions in integral form.*

THEOREM 8.5.2

Let k be the order of a transmission zero, s_0. Then the coefficients of the Taylor expansion of $\ln b - \ln S$ about s_0 satisfy the following constraints:

Class A: $\Re s_0 > 0$

$$\beta_h = \eta_h + \int_0^{+\infty} \Psi^{(h)}(s_0, \Omega) \ln|S(j\Omega)| d\Omega \qquad (h = 0, 1, 2, ..., k-1) \tag{8.5.9}$$

Class B: $\Re s_0 = 0,\ z(s_0) \neq \infty$

$$\beta_h = \eta_h + \int_0^{+\infty} \Psi^{(h)}(s_0, \Omega) \ln|S(j\Omega)| d\Omega \qquad (h = 0, 1, 2, ..., 2k-2) \tag{8.5.10}$$

$$\frac{\beta_{2k-1} - \Sigma_{2k-1}}{\rho_{2k}} \geq 0 \tag{8.5.11}$$

Class C: $\Re s_0 = 0,\ z(s_0) = \infty$

$$\beta_h = \eta_h + \int_0^{+\infty} \Psi^{(h)}(s_0, \Omega) \ln|S(j\Omega)| d\Omega \qquad (h = 0, 1, 2, ..., 2k-2) \tag{8.5.12}$$

$$\frac{\beta_{2k-1} - \Sigma_{2k-1}}{\rho_{2k-2}} \leq \frac{1}{c_{-1}} \geq 0 \tag{8.5.13}$$

When $s_0 = 0$ or $s_0 = \infty$, Σ_{2k-1} can always be evaluated by its integral representation.

In the following example we apply the integral form of the GBW restrictions to the limiting case of absolutely flat gain level in the passband.

Example 8.5.1
Given a load consisting of a resistance R in parallel with a capacitance C, find the gain-bandwidth restrictions on an ideal low-pass equalizer placed between the load and a source of internal resistance R, so that the system has a flat gain of K in the normalized passband $[0, 1]$, and zero gain in the stopband.

Solution As in Section 7.3, the load has a Class B first order transmission zero at infinity. S is a piecewise constant function which will be assumed minimum phase, since we have seen that the introduction of an allpass into S reduces the gain level. Since $\eta = 1$, we have $S = -\sqrt{1-K}$ in the passband, and -1 in stopband.[18] The expansion of ψ at ∞ is

$$\frac{2}{\pi}\frac{s}{s^2+\Omega^2} = \frac{2}{\pi}\left(\frac{1}{s} - \frac{\Omega^2}{s^3} + \frac{\Omega^4}{s^5} - \dots\right) \tag{8.5.14}$$

Since $\ln|S| = 0$ for $\Omega > 1$, we have

$$\Sigma_0 = j\pi$$

$$\Sigma_1 = \frac{2}{\pi}\int_0^{+\infty} \ln|S(j\Omega)|d\Omega = \frac{1}{\pi}\int_0^1 \ln|S(j\Omega)|^2 d\Omega$$

$$= \frac{1}{\pi}\ln(1-K)$$

Also we have the expansions around infinity for $\ln b$ and r

$$\ln b = \ln\frac{1-sRC}{1+sRC} = j\pi - \frac{2}{RCs} + O(1/s^2)$$

$$r = \frac{R}{1-R^2C^2s^2} = -\frac{1}{RC^2s^2} + O(1/s^4)$$

By equating the various coefficients according to eqs. (8.5.10), and (8.5.11) we find

$$j\pi = j\pi$$

$$\frac{-2/RC - (1/\pi)\ln(1-K)}{-1/RC^2} \geq 0$$

[18] The requirement that the gain be zero in the stopband, together with the condition that the equalizer be lossless, implies $S^2 = 1$ or $S = \pm 1$. The minus sign is dictated by the fact that $S = -1$ at ∞ since the capacitor is a short circuit at infinite frequency.

and finally

$$\frac{1}{\pi}\ln(1-K) \geq -\frac{2}{RC}$$

or

$$K \leq 1 - \epsilon^{-2\pi/RC}$$

The above result for maximum flat gain level with an RC load coincides with that obtained in the previous section for the limiting case of a Butterworth equalizer of infinite degree. But the reader will appreciate how much simpler it is to apply the GBW constraints in integral form directly on the amplitude response, rather than going through a lengthy rational approximation followed by a limiting process. □

8.6 Example: Double Zero of Transmission

When a load imposes more than one constraint, the GBW restrictions lead to simultaneous nonlinear equations. These are usually difficult or impossible to solve explicitly, especially if the load is of more than moderate complexity, since there are a variety of acceptable parameter choices, and it is not evident which will lead to an optimum solution. As a simple illustration we consider the following example.

Example 8.6.1

Consider an LCR load consisting of a series inductance L, followed by shunt C in parallel with R. The source resistance is R. Determine the optimum lowpass solution for flat gain, K, in the passband $(0, \omega_c)$, zero gain in the stopband.

Solution There is a double zero of transmission ($k = 2$) at $s_0 = \infty$. The load impedance has a pole at ∞, so that s_0 is a Class C zero. The load impedance is

$$z = sL + \frac{(1/C)}{s + 1/RC}$$

Then the allpass, formed on the poles of z_*, is

$$b = \frac{s-\gamma}{s+\gamma}, \quad \gamma = \frac{1}{RC}$$

To apply the integral restrictions, we determine the appropriate power series at $1/s = 0$.

$$\ln b = \frac{-2\gamma}{s} - \frac{2}{3}\frac{\gamma^3}{s^3} + \ldots$$

$$r = R - \frac{1}{RC^2 s^2} + \dots$$

We take $S = \eta\hat{S}$, where $\ln\eta = \sum \ln(s - \mu_j)/(s + \mu_j)$, $\Re\mu_j > 0$, and $\hat{S}$ is minimum phase. Then

$$\ln\eta = -\frac{2}{s}\sum_j \mu_j - \frac{2}{3s^3}\sum_j \mu_j^3 + \dots$$

Since the μ_j are real or occur in conjugate pairs, the η_h are real. Furthermore $\sum_j \mu_j > 0$. The power series for $\psi(1/s)$ is given by eq. (8.5.14) and, clearly, $\Sigma_0 = \beta_0 = 0$. Since the even series coefficients are zero, the non-trivial constraints correspond to $h = 1$, and $h = 2k - 1 = 3$, eqs. (8.5.12) and (8.5.13), respectively. With $h = 1, 3$

$$\Sigma_1 = \frac{2}{\pi}\int_0^\infty \ln|S|d\omega - 2\sum \mu_j = \beta_1 = -2\gamma$$

$$\frac{\beta_3 - \Sigma_3}{\rho_2} = (\frac{1}{-2/RC^3})\left(\frac{-2\gamma^3}{3} + \frac{2}{\pi}\int_0^\infty \Omega^2 \ln|S|d\Omega + \frac{2}{3}\sum_j \mu_j^3\right)$$
$$\leq \frac{1}{c_1} = \frac{1}{L}$$

For flat passband gain K, we let $A = \ln|1/S|^2 = \ln 1/(1-K)$ be a constant from 0 to ω_c, and zero beyond. Then integrating and collecting terms

$$A\omega_c = \frac{2\pi}{RC} - 2\pi\sum_j \mu_j \tag{8.6.1}$$

$$\frac{A\omega_c^3}{3} \leq 2\pi\left(\frac{1}{LRC^3} - \frac{1}{3(RC)^3} + \frac{1}{3}\sum_j \mu_j^3\right) \tag{8.6.2}$$

If we can satisfy the two constraints without RHP zeros in S, that is all $\mu_j = 0$, $\eta = 1$, then restriction eq. (8.6.1) would indicate that the optimum response is the same as for an RC load, i.e., $K = 1 - \exp[-2\pi/(\omega_c RC)]$, as in Section 8.5. This would mean that the inductance L is small enough so that the second constraint can be satisfied using a degeneracy, that is placing additional series inductance into the equalizer. So to satisfy the restrictions without RHP zeros requires

$$\frac{\omega_c^2}{RC} \leq \frac{3}{LRC^2} - \frac{1}{(RC)^3} \tag{8.6.3}$$

We can then increase L until the above equation is satisfied with the equal sign.

If eq. (8.6.3) is not satisfied we must introduce RHP zeros. It suffices to choose a single real RHP zero, $\mu_1 = \sigma > 0$. This increases the RHS of eq. (8.6.2). At the same time, the RHS of eq. (8.6.1) is decreased so that the achievable GBW product for flat gain is reduced. We now seek a solution for the two constraint equations using the equal sign when the RHP zero is present. It is simplest to substitute $A\omega_c$ from the first equation into the second. We obtain

$$\sigma^3 + \omega_c^2\sigma - \left[\frac{\omega_c^2}{RC} - \frac{3}{LRC^2} + \frac{1}{(RC)^3}\right] = 0$$

Since now there is no degeneracy, we are postulating that the inequality of eq. (8.6.3) no longer holds. Therefore the bracketed term of the above equation will be positive, and a positive solution for σ is guaranteed.

Once σ is found we can then determine A and K. Evidently frequency scaling no longer holds since the cutoff frequency ω_c enters non-linearly. Physically this is because the load should be viewed as a sealed black box. Nothing inside it can be changed. Frequency scaling for the system cannot be carried out since the contents of the box cannot be altered.

To complete the example,consider the following numerical illustration (taken from Fano's original paper)

$$L = 2.3, \quad C = 1.2, \quad R = 1, \quad \omega_c = 1$$

If we check eq. (8.6.3), the LHS is 0.833 and the RHS is 0.327; the inequality is not satisfied, so a degeneracy will not yield a solution. With a RHP zero introduced, the cubic equation for σ is

$$\sigma^3 + \sigma - 0.506 = 0$$

The positive real root is $\sigma = 0.427$. From eq. (8.6.1) we have $A = 2.55$. Then $1/(1 - K) = \exp(2.55)$, and the flat gain level is $K = 0.923$. □

The case of finite Butterworth and Chebyshev equalizers for the type of LCR load considered here has been presented in detail in the literature.[19]

8.7 Double Matching

In the last three sections Gain-Bandwidth Theory has been applied to single matching problems. Here the extension to double matching is pre-

[19] W. K. Chen, "Synthesis of Optimum Butterworth and Chebyshev Broadband Impedance Matching Networks", *IEEE Trans. on Circ. and Sys.*, vol. CAS-24, pp. 152-169, Apr. 1977.

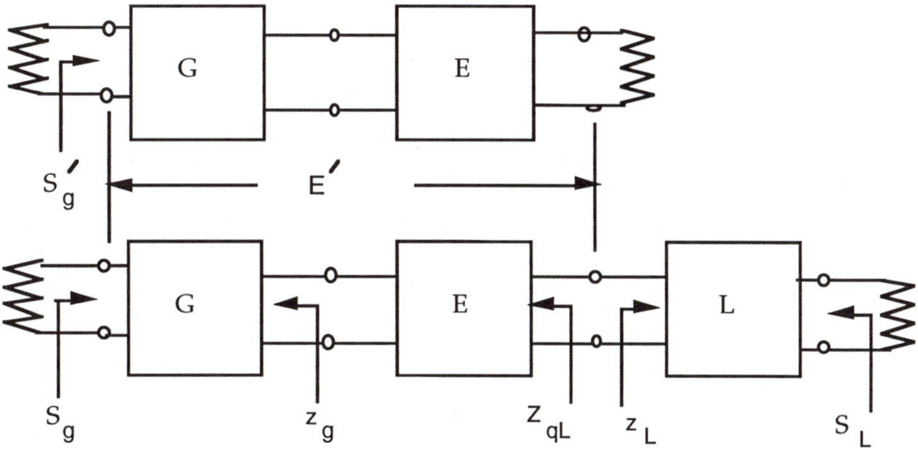

FIGURE 8.7.1
Double matching system.

sented. That is, both source and load impedances are fixed and frequency dependent. A lossless two-port matching equalizer, to be placed between source and load, is to be designed. We can show that the gain-bandwidth restrictions can be phrased in terms of single matching restrictions individually applied at source and load. We will assume that the system is not *doubly degenerate*. By this we mean that if generator and load have a common zero of transmission at s_0, say, of order n_g, n_l, respectively, then with the equalizer in place, the overall zero of transmission is at least of order $n_g + n_l$; the equalizer may increase the order, but not decrease it. An example of double degeneracy would be an RC load and generator, and a shunt capacitor as equalizer. Then at infinity $n_g + n_l = 2$ and, with equalizer present, the order of the zero for the system is reduced to 1, since the three shunt capacitors can be combined into one. The doubly degenerate case is far removed from practicality, and because of the complications it introduces we exclude it. There is thereby no significant loss of generality. Further discussion can be found in the literature.[20]

The following theorem presents the gain bandwidth restrictions for double matching. The system consists of prescribed impedances z_g and z_l, separated by a lossless equalizer $\mathcal{E}$. The reactive two ports of the Darlington equivalents for generator and load impedances are designated $\mathcal{G}$, $\mathcal{L}$. The discussion below can be followed with the aid of Fig. 8.7.1.

[20] D. C. Youla, H. J. Carlin, and B. S. Yarman, "Double Broadband Matching and the Problem of Reciprocal Reactance $2n$-Port Cascade Decomposition", *Int. J. Circ. Th. Appl.*, vol. 12, no. 3, p. 269-281, July 1984.

THEOREM 8.7.1

Let a rational transducer gain function $T(\omega^2)$ be prescribed where $0 \leq T \leq 1$. Let PR dissipative impedances, z_g and z_l, for source and load be given, and the zeros of transmission of these impedances be contained in T. Then, assuming no double degeneracies, the necessary and sufficient condition that the gain T be physically realizable by the system consisting of the source z_g, the load z_l, and a lossless equalizer E placed between source and load is that the single matching gain bandwidth restrictions (Theorem 8.4.1) be simultaneously satisfied at the source and load ports of the equalizer.

PROOF Necessity: We are given the physical system of z_g, E, and z_l, realizing the system gain function, T. Then by the broadband matching Theorem 8.4.1, the gain bandwidth restrictions must be satisfied at the ports of the equalizer.

Sufficiency: The hypothesis for sufficiency is that the gain bandwidth restrictions are simultaneously satisfied at the ports of the equalizer. We must show that source and load impedances separated by a realizable lossless equalizer can realize the prescribed gain function. Replace z_g by the Darlington reactance two-port $\mathcal{G}$ resistively terminated at the source, and z_l by the resistively terminated Darlington two-port $\mathcal{L}$. At the (unit) terminating load resistance the unit normalized input reflectance to the system of three cascaded two-ports is S_l. The gain bandwidth restrictions of the load port of $\mathcal{E}$ are based on the z_l transmission zeros and these are the zeros of the function

$$b_l - S_l = \frac{2b_l r_l}{Z_l + z_l}$$

where b_l is the allpass formed on the denominator zeros of z_{l*}, Z_l is the input impedance to $\mathcal{E}$ (z_g terminated) measured at the load end, and $2r_l = z_g + z_{g*}$. The gain bandwidth restrictions are that k Taylor coefficients of S_l are equal to those of b_l (the final restriction may be an inequality for a simple degeneracy) at a simple or multiple zero of transmission of z_l, and these Taylor coefficients depend *only on the load.* Since the restrictions are satisfied, if we start at the resistor terminating $\mathcal{L}$, the reflectance S_l computed from T can be synthesized by the cascade of the prescribed $\mathcal{L}$ followed by a realizable lossless two-port, call it $\mathcal{E}'$. That is the two-port, $\mathcal{E}'$, terminated in z_l, and in a unit resistance at the source end realizes the gain function T.

The question remains whether $\mathcal{E}'$ is realizable by the z_g Darlington two port, $\mathcal{G}$, followed by a realizable lossless equalizer two-port, $\mathcal{E}$. When the load z_l was in place, the hypothesis stated that S_g satisfied a set of restrictions similar to those stated above for S_l. When the load, z_l, is removed and replaced by unit resistor, the input unit normalized reflectance at the source resistor changes to, say, S_g', *but the z_g gain-bandwidth restrictions* *remain satisfied* for S'_g, since these only depend on z_g. Thus we are assured that S'_g may be realized by $\mathcal{G}$ followed by a realizable lossless two-port, the required equalizer $\mathcal{E}$ terminated in unit resistor. The overall system gain for $\mathcal{G}$, $\mathcal{E}$, and $\mathcal{L}$, is T. ∎

To apply the Theorem for simultaneous satisfaction of the GBW restrictions at load and generator, the overall gain, T, is specified with unknown parameters. Then by spectral factorization the unit normalized reflectances S_g, and S_l are determined. The two reflectances are related as in the Belevitch representation of a lossless two port when the transmittance is f/g.

$$S_g = \frac{h}{g}, \quad S_l = -\frac{h_* f}{g f_*}$$

The Taylor coefficient restrictions are imposed on S_g and S_l, and the resulting equations are solved simultaneously to obtain the undetermined parameters. The process is illustrated by the following example.

Example 8.7.1

Given a parallel RC load to be matched with a lossless equalizer to a parallel RC source, determine the parameters of a Butterworth gain function which describes the overall system.

Solution We proceed as in Section 8.4. The gain level constant K appearing in the gain function is to be maximized subject to the GBW restrictions.

$$T = \frac{K}{1 + \epsilon^2 \omega^{2n}}$$

Note that the larger is ϵ, the smaller is the bandwidth, say, defined by ω at the gain level $T = K/2$. To allow the option af an extra free parameter in satisfying the GBW restrictions, we include an allpass transmission zero common to all elements of the scattering matrix. (Note that if the allpass appears in one reflectance of the $\boldsymbol{S}$ matrix, realizability requirers that it appear in all elements.) By spectral factorization of T, the expressions for the reflectances at the two ports of the $\mathcal{G}$, $\mathcal{E}$, $\mathcal{L}$ system as given in eq. (4.2.1) are

$$S_l = -\frac{\lambda - s}{\lambda + s}\sqrt{1-K} \cdot$$
$$\cdot \frac{1 + a_1 \left(\frac{\epsilon}{\sqrt{1-K}}\right)^{\frac{1}{n}} s + \ldots + a_{n-1}\left(\frac{\epsilon}{\sqrt{1-K}}\right)^{\frac{n-1}{n}} s^{n-1} + \frac{\epsilon}{\sqrt{1-K}} s^n}{1 + a_1 \epsilon^{\frac{1}{n}} s + \ldots + a_{n-1} \epsilon^{\frac{n-1}{n}} s^{n-1} + \epsilon s^n}$$

for the equalizer reflectance at the load port, and at the generator

$$S_g = -\frac{\lambda - s}{\lambda + s}\sqrt{1-K} \cdot$$

$$\cdot \frac{1 + a_1 \left(\frac{\epsilon}{\sqrt{1-K}}\right)^{\frac{1}{n}} (-s) + \cdots + a_{n-1} \left(\frac{\epsilon}{\sqrt{1-K}}\right)^{\frac{n-1}{n}} (-s)^{n-1} + \frac{\epsilon}{\sqrt{1-K}}(-s)^n}{1 + a_1 \epsilon^{1/n} s + \cdots + a_{n-1}\epsilon^{(n-1)/n} s^{n-1} + \epsilon s^n}$$

As in Section 8.4 we expand S_l, b_l, and $2r_l b_l$ about infinity; and again for for S_g, b_g, and $2r_g b_g$. We obtain

$$S_l = -1 + \left(a_{n-1}\frac{1-(\sqrt{1-K})^{\frac{1}{n}}}{\epsilon^{\frac{1}{n}}} + 2\lambda\right)\frac{1}{s} + o\left(\frac{1}{s^2}\right)$$

$$b_l = -1 + \frac{2}{R_l C_l s} + o\left(\frac{1}{s^2}\right)$$

$$2r_l b_l = \frac{2}{R_l C_l^2 s^2} + o\left(\frac{1}{s^4}\right)$$

and

$$S_g = -1 + \left(a_{n-1}\frac{1+(\sqrt{1-K})^{\frac{1}{n}}}{\epsilon^{\frac{1}{n}}} + 2\lambda\right)\frac{1}{s} + o\left(\frac{1}{s^2}\right)$$

$$b_g = -1 + \frac{2}{R_g C_g s} + o\left(\frac{1}{s^2}\right)$$

$$2r_g b_g = \frac{2}{R_g C_g^2 s^2} + o\left(\frac{1}{s^4}\right)$$

The gain-bandwidth restrictions (8.4.4) and (8.4.5) give

$$b_{l0} = S_{l0}$$

$$\frac{b_{l1} - S_{l1}}{(2r_l b_l)_2} \geq 0$$

and

$$b_{g0} = S_{g0}$$

$$\frac{b_{g1} - S_{g1}}{(2r_g b_g)_2} \geq 0$$

The first equations are identically satisfied ($-1 = -1$), and the second yield

$$\frac{2}{R_l C_l} \geq a_{n-1}\frac{1-(\sqrt{1-K})^{\frac{1}{n}}}{\epsilon^{\frac{1}{n}}} + 2\lambda \tag{8.7.1}$$

and

$$\frac{2}{R_g C_g} \geq a_{n-1}\frac{1+(\sqrt{1-K})^{\frac{1}{n}}}{\epsilon^{\frac{1}{n}}} + 2\lambda \tag{8.7.2}$$

where we have taken $1/R_gC_g > 1/R_lC_l$. The coefficient a_{n-1} is obtained from the first of eqs. (6.5.5) as

$$a_1 = \frac{1}{\sin \dfrac{\pi}{2n}}$$

and therefore the above inequalities can be rewritten as

$$(1-K)^{\frac{1}{2n}} \geq 1 - 2\epsilon^{\frac{1}{n}} \sin(\pi/2n)\left(\frac{1}{R_lC_l} - \lambda\right)$$

and

$$(1-K)^{\frac{1}{2n}} \geq -1 + 2\epsilon^{\frac{1}{n}} \sin(\pi/2n)\left(\frac{1}{R_gC_g} - \lambda\right)$$

We now seek a solution of the two above expressions using equal signs rather than the inequalities. Let $\alpha_n = \epsilon^{1/n}\sin(\pi/2n)$, $\sigma_1 = 1/R_1C_1$, and $\sigma_2 = 1/R_2C_2$. Then, by subtracting the equations, we find

$$\lambda = \frac{1}{2}\left(\sigma_1 + \sigma_2 - \frac{1}{\alpha_n}\right) > 0, \quad (1-K)^{\frac{1}{2n}} = |\sigma_1 - \sigma_2|\,\alpha_n \tag{8.7.3}$$

where we have used the absolute value since the solution is indifferent to whether σ_1 or σ_2 is associated with load or generator. We can also obtain a solution if we set $\lambda = 0$, i.e., no additional allpass employed. In that case eq. (8.7.3) no longer applies and given n we must choose ϵ so that $\alpha_n = 1/(\sigma_1 + \sigma_2)$, or

$$\epsilon^{\frac{1}{n}} = \frac{1}{(\sigma_1 + \sigma_2)\sin(\pi/2n)}, \quad K = 1 \tag{8.7.4}$$

□

Some unusual consequences of double matching can be illustrated by the present example. Consider the original constraints, eqs. (8.7.1), (8.7.2). If the first is subracted from the second, then

$$\sigma_2 - \sigma_1 \geq \frac{(1-K)^{1/2n}}{\alpha_n}$$

The flat gain limit requires $n \to \infty$. In that case with $\sigma_2 > \sigma_1$, the numerator approaches unity, the denominator approaches zero, and the limit does not exist. If $\sigma_2 = \sigma_1$, then $K = 1$, but refer to eq. (8.7.4) to see that again the limit does not exist. Similar results hold whatever sequence of finite equalizers is chosen to approximate the flat limit as the number of equalizer elements becomes infinite. *Flat gain over any finite passband cannot be obtained in double matching problems.*[21]

[21]H. J. Carlin, P. P. Civalleri, "On Flat Gain with Frequency-Dependent Terminations", *IEEE Trans. Circ. Sys.*, vol. CAS-32, no. 8, pp. 827-839, Aug. 1985.

Nor can we come arbitrarily close to flat gain, excluding only the final limit. If the bandwidth is fixed, say, at close to unity, $\epsilon \approx 1$, then the last equation (or eq. (8.7.4)) indicates a maximum value of n which can satisfy the inequality. This can be interpreted as meaning that the gain resulting from the matching process is strictly limited to a significant deviation from flatness as measured in this case by maximum allowable n. If the gain function were the Chebyshev response, the comparable result would be a minimum achieveable gain ripple which becomes larger with larger time constant terminations. Single matching has no such restrictions. As indicated in the examples of Sections 8.4 and 8.6 for a resistive generator and a complex load, a realizable lossless equalizer can always be found so that the system gain approaches absolute flatness to within an arbitrarily small tolerance over a prescribed passband.

In the next chapter, numerical techniques for broadband double-matching are considered. Although the flat gain restriction cannot be violated even using a Real Frequency Technique, optimization programs can give results which come close to allowable gain flatness. This is because the RFT is carried out without an initially prescribed form of analytic transfer function.

9

Broadband Matching II: Real Frequency Technique

9.1 Introduction

Filtering and broadbanding problems were discussed in Chapters 6 and 8 from the point of view of an analytically prescribed transfer function. Gain vs. frequency specifications were translated at the outset into a gain vs. frequency rational function from which, through the machinery of spectral factorization, a rational PR impedance Z was calculated and finally realized according to Darlington's Theorem as a lossless two-port terminated on a resistor. In the case of filters doubly terminated by resistors, there are no special constraints on the transfer function other than those imposed by losslessness, and the analytically defined gain functions, e.g., Butterworth, Chebyshev, elliptic, lead to general solutions which are practical and readily implemented. Furthermore the properties of the transfer function directly imply optimal properties of the system vis-à-vis minimum passband loss and maximum selectivity.

When the load(s) are not pure resistors, we have a constrained optimization problem, and the choice of a transfer function from the types used in doubly terminated filter design no longer assures optimum gain and selectivity. In effect the degrees of freedom implied in the approximation, and yielding such properties as maximal flatness or equal ripple behavior are wasted in that they are not addressed to optimizing minimum passband loss or selectivity. Also, once the gain functional form is chosen, the design process becomes rigid; the gain-bandwidth restrictions must be satisfied exactly by the transmittance, and this usually introduces unwanted features in the design. Suppose, for example, that the approximation process has produced a gain function with a transmission zero in the RHP not included in the load; in such a case, the equalizer will inevitably contain a C-section, which requires perfectly coupled coils, a structure that designers would prefer to avoid.

Another difficulty stems from the fact that the class of problems available for solution by analytic means is relatively limited, since the difficulties of implementing the constraints mount very rapidly when the load becomes even moderately complex. In fact, no complete analytic solutions have been published for loads more complicated than the LCR case treated in Section 8.6.

The Real Frequency Technique (RFT), on the other hand, is a numerical method, and generates the rational PR function Z, which is the input impedance to the equalizer, directly from the approximation process. No rational transfer function is prescribed. The optimization can directly impose objectives, such as the minimization of maximum loss in the passband. Furthermore, since no transfer function is specified in advance, degrees of freedom available in the approximation process can be spent to satisfy practical topologic constraints imposed *a priori* on the equalizer structure, such as realization in ladder form without coupled coils. Moreover, bandpass problems can be handled directly without resorting to the limitations of frequency transformation techniques.

The salient features of the RFT are:

(a) The rational PR input impedance function Z of the equalizer is generated directly by an approximation procedure carried out at real frequencies, with no prescribed transfer function or function-theoretic steps involved.

(b) The load need not be specified as a circuit model or, analytically, it can be represented by numerical data.

(c) The approximation method can vary in its details depending on the problem at hand and on the required precision. That is, the RFT implies the design of a matching equalizer which achieves the numerical optimization of the gain of a transmission system partially constrained by fixed components, but it does not prescribe any one specific optimization routine.

An elementary illustration may serve to highlight the comparison between the analytic and numerical approaches. In Example 8.4.2, the second degree Butterworth gain $K/(1+\omega^2)$ was optimized for an RCR load by the analytic method, resulting in $K = 1$ and an equalizer containing a pair of coupled coils. The equalizer and load are shown in Fig. 9.1.1 (a). The response is presented in Fig. 9.1.1 (c) (curve A). We now employ the RFT approach with an optimization objective of minimum loss in the passband ($0 \leq \omega \leq 1$). Let $G(\omega^2)$, be the unknown conductance function seen looking into the back end of the equalizer when it is terminated at the input in the resistive load ($r = 4$) of the analytic design. Assume that the equalizer is an LC ladder (no all-pass present), so that G is the inverse of an even polynomial,

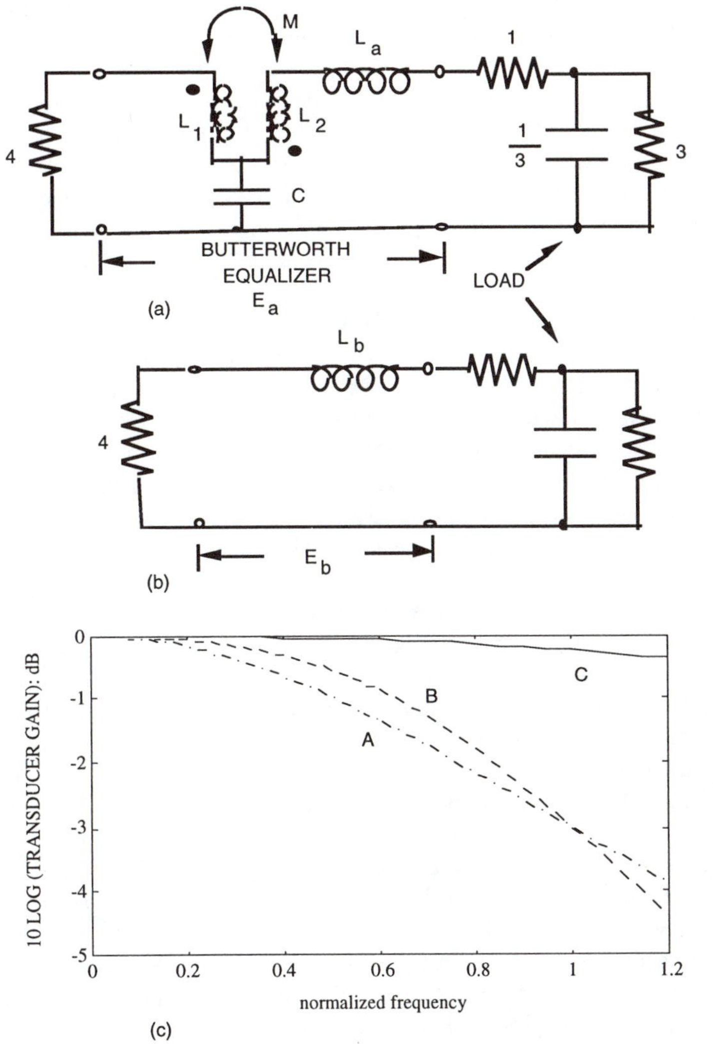

FIGURE 9.1.1
Analytic and RFT equalization of RCR load. (a) Analytic equalizer; (b) RFT equalizer; (c) curve A - analytic; curve B - RFT, 3 dB at band edge; curve C - RFT, min-max loss in passband.

and choose it of degree 2. The details are not important at this point (they will be presented in later sections); the final result is an equalizer which simply consists of a series inductance, $L = 1.47$. The resultant system gain is shown as curve C of Fig. 9.1.1 (c), with maximum passband loss of 0.22 dB. If we seek discrimination comparable with the analytic result, the RFT equalizer can be chosen to give 3.0 dB of loss at $\omega = 1$. The result is an equalizer inductor of $L = 7.64$, with a gain shown as curve B. Even in this case the passband gain exceeds that of the analytic result, and the response falls off more rapidly in the stopband so that the discrimination is superior. Most important, the RFT equalizer is a ladder structure with no coupled coils. One might suggest that the cards are stacked against the analytic method in this example because of the choice of transfer function, but that is exactly the point; when the analytic procedure is used there is no known method for prescribing the best form of transfer function consistent with a given load. In effect, the analytic procedure squanders its resources (circuit elements) to achieve special gain function properties (in this case, maximum flatness leading to an all-pass section), whereas the RFT directly optimizes max-min passband gain without superfluous elements.

One simple and appealing version of RFT is the "Line-Segment" approximation technique, used mainly for single matching problems. The transducer gain is expressed in terms of the load and the equalizer impedances facing each other, z and Z. The resistance $R = \Re Z$ is approximated at real frequencies by a piecewise linear function; the slopes of the straight line segments are free parameters that are adjusted by an optimization technique so as to obtain a gain as flat as possible in the passband, and with the maximum achievable minimum value. In the process of approximation, the reactance, $X = \Im Z$, is easily calculated from the line segment representation of resistance R with the Hilbert transform technique described in Section 3.8; both R and X are linear combinations of a set of functions specific to each, but with the *same coefficients.* Once the line-segment approximation of R is determined, it is approximated in turn by a rational function, $\hat{R}$, from which a minimum reactance function, $\hat{Z}$, is derived, and then finally realized as a lossless two-port terminated on the generator impedance according to Darlington's theory. The Line-Segment Technique is particularly convenient as an interactive approach, in which the rational optimization process is undertaken only at the end; it is recommended for the simplicity of the Hilbert transform procedure in constructing the imaginary from the real part. But one can employ other modes of approximation. For example, one may optimize a rational $\hat{R}$ function with undetermined coefficients from an intelligent initial guess, and use the Gewertz method to construct $\hat{Z}$ from its real part.

In double matching problems one can represent the equalizer by its unit normalized scattering matrix and express the overall system gain in terms of the elements of the $\boldsymbol{S}$ matrix, i.e., by applying the Belevitch form of eq. (4.2.1), one seeks a determination of the coefficients of the polynomial denominator g. This procedure, which can also be applied in single matching problems, is generally more accurate than the line segment approach but requires polynomial factorization and does not lend itself easily to interactive design. One could also proceed by setting the resistance of the unit terminated equalizer as the initialized function to be optimized.

Other types of GBW problems, such as matching an amplifier to both generator and load impedances through a pair of equalizers at input and output, can be solved in the spirit of RFT; the output reflectances of the two equalizers are determined by the approximation process, taking into account the prescribed two-port. The equalizers are finally realized by the Darlington synthesis procedure.

We discuss these problems and various RFT methods of solution in the following sections.

9.2 Single Matching

We illustrate the line-segment approximation technique[1] by applying it to the solution of a single-ended broadband matching problem. The symbols for the various quantities are the same as those defined in Section 8.3 and illustrated in Fig. 8.3.1. In particular, the load impedance is z and the unknown equalizer impedance is Z.

We have seen in Section 8.3, eq. (8.3.4), that when S is the complex normalized output reflectance of the equalizer (normalized to z), and S_2 the unit normalized output reflectance of the overall two-port consisting of the equalizer cascaded with the Darlington equivalent two-port of the load, then $S = S_2$. Since the overall two-port is lossless, we have

$$T(\omega^2) = |S_{21}(j\omega)|^2 = 1 - |S_2(j\omega)|^2 = 1 - |S(j\omega)|^2 \qquad (9.2.1)$$

Combining eqs. (9.2.1) and (8.3.6), we obtain the following expression for system gain:

$$T(\omega^2) = \frac{4Rr}{|Z+z|^2} \qquad (9.2.2)$$

where

$$R = R(\omega) = \Re Z, \qquad r = r(\omega) = \Re z$$

[1]H. J. Carlin, "A New Approach to Gain-Bandwidth Problems", *IEEE Trans. on Circ. and Sys.*, vol. CAS-24, no. 4, pp. 170-175, April 1977.

We assume for R a line segment approximation of the form

$$R(\omega) = R_0 + \sum_{k=1}^{n} D_k a_k(j\omega) \tag{9.2.3}$$

where $D_k = R_k - R_{k-1}$ is the resistance excursion of the k-th line segment and

$$a_k(\omega) = \begin{cases} 1 & \omega_k \le \omega \\ \dfrac{\omega - \omega_{k-1}}{\omega_k - \omega_{k-1}} & \omega_{k-1} \le \omega < \omega_k \\ 0 & \omega < \omega_{k-1} \end{cases} \tag{9.2.4}$$

Each term $D_k a_k$ represents a function which is 0 for $\omega < \omega_{k-1}$, linearly increases from 0 to D_k between ω_{k-1} and ω_k, and has the constant value D_k for $\omega \ge \omega_k$. Note that D_k may be positive, negative, or zero depending on whether the function increases, decreases, or remains constant in the interval $[\omega_{k-1}, \omega_k)$. Any shape of the equalizer output resistance $R = R(\omega)$ can be reproduced to good approximation, provided a sufficient number of line segments is used. It is evident that $R_0 = R(0)$, i.e. R_0 is the d.c. value of the resistance. Also the resistance R must obviously be zero beyond some sufficiently high frequency (the system is not able to transfer power at arbitrarily high frequencies). Assume this final frequency is ω_n, then since the resistance at frequency ω_n is zero,

$$R_0 + \sum_{k=1}^{n} D_k = 0 \tag{9.2.5}$$

which is a dependency reducing the number of unknowns by one. If the given load resistance $r = r(\omega)$ is in the form of numerical (experimental) data, it can be represented in the same manner, obviously with a different set of coefficients, say, r_0 and d_k. In this case there is no need for an approximating circuit model which would be necessary in the analytic approach. Of course, if the load *were defined* by a rational expression or a circuit model, we could use the load resistance values at the break points which in effect is a line segment approximation.

The reactances X and x are deduced from the resistances R and r by using Hilbert transforms according to eqs. (3.8.3) and (3.8.4); X has the form

$$X(\omega) = \sum_{k=1}^{n} D_k b_k(\omega) \tag{9.2.6}$$

and

$$b_k(\omega)] = \frac{1}{\pi(\omega_k - \omega_{k-1})}\{[(\omega + \omega_k)\ln|\omega + \omega_k| + (\omega - \omega_k)\ln|\omega - \omega_k|] - \\ -[(\omega + \omega_{k-1})\ln|\omega + \omega_{k-1}| + (\omega - \omega_{k-1})\ln|\omega - \omega_{k-1}|]\}$$

Note that the unknown D_k's appear linearly in both R and X. When obtained from experimental data for load resistance, x has the same form with the D_k's replaced by d_k's. Again, if the load is specified by a model or by an analytic expression, we merely introduce the reactance values at the break points without the need of the Hilbert transform.

We are thus able to construct the expressions of both Z and z as the sum of their real and imaginary parts at real frequencies. Thus $T(\omega^2)$, as given by eq. (9.2.2), depends on the $n+1$ unknown parameters R_0 and D_k, $k = 1, 2, \ldots, n$, (eq. (9.2.5) reduces the number of independent unknowns to n) that must be determined to minimize, according to some suitable norm, the difference between $T(\omega^2)$ and the prescribed gain curve $T_0(\omega^2)$.

The actual equalizer design process consists of three main steps:

1. Optimization of the gain function $T(\omega^2)$ as an approximation of the ideal gain function $T_0(\omega^2)$. Essentially one must choose a suitable function space and find the D_k's which make the error norm a minimum:
$$||T(\omega^2) - T_0(\omega^2)|| = \min$$
The choice of the function space is not particularly critical in the cases of practical interest and may be suggested by collateral aspects, such as computational simplicity. Choosing the space as L^2 has the advantage that numerical analysis provides efficient algorithms for nonlinear least squares. Other choices are obviously possible, for example, that of the space of measurable essentially bounded functions.[2]

2. Approximation of the line segment representation of $R(\omega)$ by an even rational function, $\hat{R}(\omega)$. This step allows the imposition of topology constraints on the equalizer; for example, if $\hat{R}(\omega)$ is chosen to be the inverse of a polynomial, the network structure will be a lowpass ladder. (See Example 2.12.2.) The least square method can be also applied with advantage to finding the rational resistance approximation.

3. Synthesis of the equalizer: from the real rational resistance $R(\omega)$ the causal impedance Z is derived by Gewertz' method. (See Section 6.12). The resulting impedance is then realized by Darlington's synthesis procedure, yielding a lossless two-port, the equalizer, terminated on the source resistance. If the source resistance is prescribed, that requirement can be built into the approximation procedure. Otherwise, the use of an ideal transformer as the terminating section of the equalizer may be necessary.

[2]J. W. Helton, "Broadbanding: Gain Equalization Directly from Data", *IEEE Trans. on Circ. and Sys.*, vol. CAS-28, no. 12, pp. 1125-1137, Dec. 1981.

The optimization of the gain function may be carried out by a variety of appropriate algorithms, including nonlinear programming. However, at least for an initial insight, an interactive approach may be quite satisfactory. At first the d.c. gain level is set and the value of R_0 is calculated from eq. (9.2.2) as

$$R_0 = \left(\frac{2}{T(0)} - 1 + \sqrt{\left(\frac{2}{T(0)} - 1 \right)^2 - 1} \right) r_0$$

The initial guess for the resistance values at the break points, consequently the associated D_k, is chosen to sustain the d.c. gain by assuming reactance cancellation. The actual reactances corresponding to the resulting line segments are then calculated and the D_k readjusted with the reactance taken into account and the minimum gain over the band noted. The d.c. gain is then increased and the procedure repeated until the minimum gain can no longer be increased.

The procedure is quite effective despite its simplicity, since:

1. The gain at any frequency is relatively weakly influenced by the line segments remote from the frequency in question.

2. When the total reactance $X + x$ is less than about $0.3\,(R + r)$, its effect on gain T is small, since then the total impedance $Z + z$ is approximately equal to $R + r$.

3. The gain level at any point within the interval $[\omega_{k-1}, \omega_k]$ can be adjusted by simply varying D_k.

Example 9.2.1

Equalize for maximum gain level over the passband, the LCR load shown in Fig. 9.2.1 (a) with $R = 1$, $L = 2.3$, and $C = 1.2$.

Solution This example was first treated analytically by Fano.[3] The solution by RFT is carried out by first determining an optimum piecewise linear approximation of the equalizer resistance R, based on some predetermined breakpoint frequencies. We choose a five segment approximation based on the following angular frequencies: $\omega_0 = 0$, $\omega_1 = 0.25$, $\omega_2 = 0.50$, $\omega_3 = 0.75$, $\omega_4 = 1.00$, and $\omega_5 = 1.25$, corresponding to the following initial guess resistance values: $R_0 = 2.2$, $R_1 = 1.9$, $R_3 = 1.5$, $R_3 = 1.2$, $R_4 = 0.6$, and $R_5 = 0$. To avoid an optimized solution with negative resistance values,

[3] R. M. Fano, "Theoretical Limitations on the Broad-Band Matching of Arbitrary Impedances", *J. Franklin Inst.*, vol. 249, no. 1, pp. 57-83, Jan. 1950, and no. 2, pp. 139-154, Feb. 1950.

we have weighted the error function (absolute difference between the ideal flat gain and the computed gain) by a factor of ten whenever a negative value appears.

The piecewise linear resistance is approximated in turn by an even rational function, $\hat{R}$, that we assume to be the inverse of an even polynomial (all zeros of transmission at infinity), so that the resultant equalizer will be an LC lowpass ladder without coupled coils. A polynomial of degree six (third order zero, $n = 3$), corresponding to a three element equalizer, turns out to be a reasonable choice.

The results may be summarized as follows:

1. The approximated rational resistance for the RFT ($R(0) = 2.2$) is given by

$$\hat{R} = \frac{2.2}{1 + 2.49\omega^2 - 4.44\omega^4 + 4.31\omega^6}$$

The equalizer impedance $\hat{Z}$ is then calculated from $\hat{R}$ using the Gewertz procedure. Thus

$$\hat{Z} = \frac{2.2 + 3.31s + 1.61s^2}{1 + 3.321s + 4.27s^2 + 2.076s^3}$$

2. If analytic gain-bandwidth theory is applied to a Chebyshev equalizer with degree $n = 4$, the resulting system transmittance contains an allpass zero and the resultant passband gain has a minimum value of $T_{\text{min}} = 0.831$ with a ripple of $10 \log T_{\text{max}}/T_{\text{min}} = 0.27\,\text{dB}$. If $n = 3$, the minimum passband gain is $T_{\text{min}} = 0.772$ with a ripple of 0.5 dB. As a yardstick for comparison, we have already shown in Section 8.6 that the value of absolutely flat gain level over unit band $0 \leq \omega \leq 1$ for the load under consideration is $T_0 = 0.924$.

3. The resultant system gain for the RFT equalizer with $n = 3$ has $T_{\text{min}} = 0.849$ with a ripple of 0.15 dB. These figures are significantly better than those of the analytically calculated Chebyshev function for $n = 3$. They even surpass the Chebyshev results for $n = 4$. Moreover the Chebyshev equalizer requires coupled coils; the RFT equalizer does not.

4. The gain vs. angular frequency curve for the RFT derived system is shown in Fig. 9.2.1 (b). The equalizer elements, calculated from $\hat{Z}$ by pole extraction at infinity, are $r_g = 2.200$, $c_1 = 0.3517$, $l_2 = 2.909$, $c_3 = 0.9223$.[4]

[4]The numerical results shown for this example are from: H. J. Carlin and P. Amstutz, "On Optimum Broad-Band Matching", *IEEE Trans. on Circ. and Sys.*, vol. CAS-28, no. 5, pp. 401-405, May 1981.

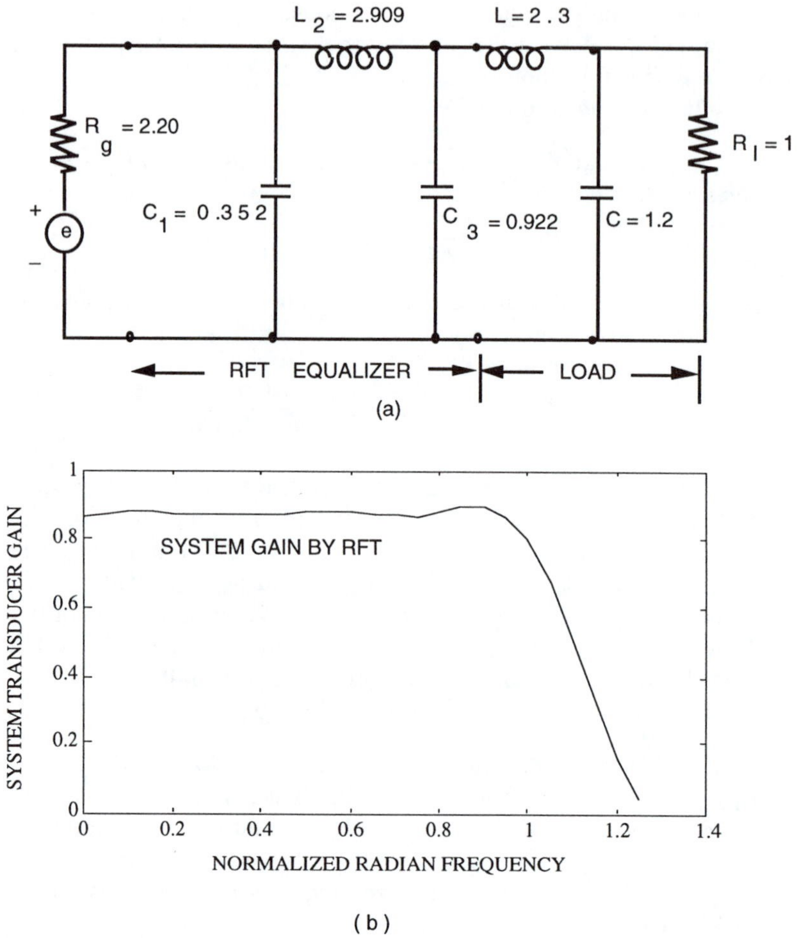

FIGURE 9.2.1
(a) Equalization of an LCR load; (b) gain vs. frequency curve.

5. If, in the RFT, we choose $R_0 < 1$, the response is markedly inferior to the case for $R(0) > 1$, $n = 3$. The calculated results for this case are given in the literature for $n = 3$ and $n = 4$.[5]

□

In this example the line segment approximation technique yields a gain characteristic which is an improvement over the response of the analytic method. Furthermore, the RFT allows the specification *a priori* of equalizer topology, typically a ladder structure, avoiding the need for coupled coils.

9.3 Transmission Line Equalizers

Numerical methods can be readily applied to distributed parameter matching problems. Indeed, some of the initial papers on the application of the RFT were to microwave amplifiers using equalizers consisting of transmission line elements, that is to say, Unit Elements (UE's) and stubs.[6] The general procedure, when a UE-stub equalizer is to be employed, is to express the input of the equalizer by a functional form which represents the appropriate type of transmission line configuration as discussed in Chapter 7. The gain is then optimized subject to the constraints imposed by the transmission line topology. The method is very similar to that used when the topology of the equalizer is a ladder structure. In that case, as discussed in Section 9.2, the input form used for the equalizer is a resistance function which is the inverse of a polynomial.

We consider the case in which the equalizer is to be a lowpass configuration of stepped lines and stubs. The transducer gain using the Richards variable, $\lambda = j\Omega = j\tan\omega\tau$ has the form, eq. (7.6.2),

$$T(\Omega^2) = \frac{(1+\Omega^2)^n}{P_{n+q}(\Omega^2)}$$

where P is an even order polynomial of degree $2(n+q)$, n is the number of UE's, and q the number of stubs (shorted stubs in series, or open circuited stubs in shunt).

[5] H. J. Carlin, "A New Approach to Gain-Bandwidth Problems", *IEEE Trans. on Circ. and Sys.*, vol. CAS-23, no. 4, pp. 170-175, Apr. 1977.

[6] H. J. Carlin and J. J. Komiak, "A New Method of Broad-band Equalization Applied to Microwave Amplifiers." *IEEE Trans. on Microwave Th. and Tech.*, vol. MTT-27, no. 2, pp. 93-99, Feb. 1979.
Also see C. Dehollain, *Adaptation d'impédance à large bande*, Lausanne, Presses Polytechniques et Universitaires Romandes, 1996.

As an alternate to line segment approximation, RFT optimization can be carried out directly in terms of the rational input conductance G (or resistance R).[7] In this problem all boundary zeros of transmission are at infinity. Then the form of the expression for G is set by the fact that it contains all the zeros of transmission, except that if the admittance has a pole at infinity, the multiplicity of the zero in G at infinity is two less than the order of the corresponding zero in T. (A zero of transmission at infinity of order n, has multiplicity $2n$ in T.) It is therefore clear that the conductance has the same functional form as the transducer gain

$$G = \frac{(1+\Omega^2)^n}{D_{n+q'}(\Omega^2)} \tag{9.3.1}$$

and $q' = q$, the number of stubs, unless the admittance has a pole at infinity, in which case $q' = q - 1$. A similar result holds for the resistance function.

A procedure for an RFT optimization program in terms of the rational conductance function is then:

1. Initialize the even positive denominator polynomial, D. A convenient way to do this is to start with real coefficients a_k and form the polynomial $A(\Omega) = a_0 + a_1\Omega + \ldots + a_m\Omega^m$. Then express $D_{n+q'}$ as

$$D(\Omega^2) = \frac{1}{2}[A^2(\Omega) + A^2(-\Omega)] > 0 \quad \forall\Omega \tag{9.3.2}$$

2. Find the equalizer admittance, $Y = n/d$ defined by $G = \Re Y(j\Omega)$, using the Gewertz method, Section 6.12. To do this, determine the spectral factorization of $D(-\lambda^2) = dd_*$, where d is a Hurwitz polynomial. The conductance in terms of $Y = n/d$ is $\mathcal{E}(nd_*/dd_*)$, (where $\mathcal{E}(\cdot)$ means "even part of"). Thus the numerator polynomial is determined by using $nd_* = (1-\lambda^2)^n$ and equating coefficients of like powers.

3. The gain of the system is given by

$$T = \frac{4Gg}{|Y+y|^2}$$

where g and y are the conductance and admittance of the prescribed load.

[7] If the line segment approach is used, it has the advantage of providing useful physical insight for the initial guess in the optimization process. However a further step of rational approximation is required.

4. Starting with the initial guess for A, a least squares optimization routine such as the MATLAB Levinson-Marquardt program, may be used to minimize $|T - T_0|^2$ over the passband, where T_0 is the flat passband gain level to be approximated. In running the program, different choices may be used for T_0 to obtain the most desirable curve shape.

5. With the admittance Y determined, the stepped line-stub equalizer may be synthesized using Richards' Theorem as discussed in Section 7.5.

Example 9.3.1

Consider a microwave termination consisting of a step discontinuity followed by a line of unit characteristic impedance. The discontinuity is represented as an equivalent capacitance $C = 2.5$, normalized so that the angular frequency $\omega = 1$ corresponds to the Richards frequency variable $\Omega = \tan\omega\tau = 1$ (in other words, assume $\tau = \pi/4$). Note that if the true cutoff frequency is, say, $f_c = 4\,\text{GHz}$, then the true capacitance is $2.5/2\pi 4 \cdot 10^9 \approx 100\,\text{pF}$. Determine a single stepped line-single stub lowpass equalizer for the band $0 \le \omega \le 1$.

Solution In this problem, the load is lumped, i.e., $y = 1 + j\omega C$ and the equalizer is a distributed structure. Using eq. (9.3.1) with $n = 1$ and $q = 1$[8]

$$G = \frac{1 + \Omega^2}{d_0 + d_1\Omega^2 + d_2\Omega^4}$$

The optimization program gives $A = 0.2182\Omega^2 + 0.6478\Omega + 0.333$, and the corresponding denominator for G, eq. (9.3.2) is $D = 0.0476\Omega^4 + 0.5650\Omega^2 + 0.111$. After spectral factorization to determine the Hurwitz denominator, the Gewertz method follows, and the equalizer admittance looking towards the source, is

$$Y = \frac{1.7149\lambda + 1}{0.2182\lambda^2 + 0.7104\lambda + 0.333}$$

If Richards' theorem is applied, a UE of characteristic impedance $Z_1 = 0.465$ may be extracted. The remainder is an open circuited stub of characteristic impedance $Z_2 = 1.23$, in parallel with a terminating resistance of $R_G = 1/3$. A stepped line transformer may be used if the source termination is to be unity. The gain response of the system is shown in Fig. 9.3.1. The minimum gain over the band $0 \le \omega \le 1$ is $T_0 = 0.75$. □

[8] We choose G (rather than R) to define the equalizer according to eq. (9.3.1). This avoids the degeneracy of load capacitance in parallel with a capacitive equalizer stub.

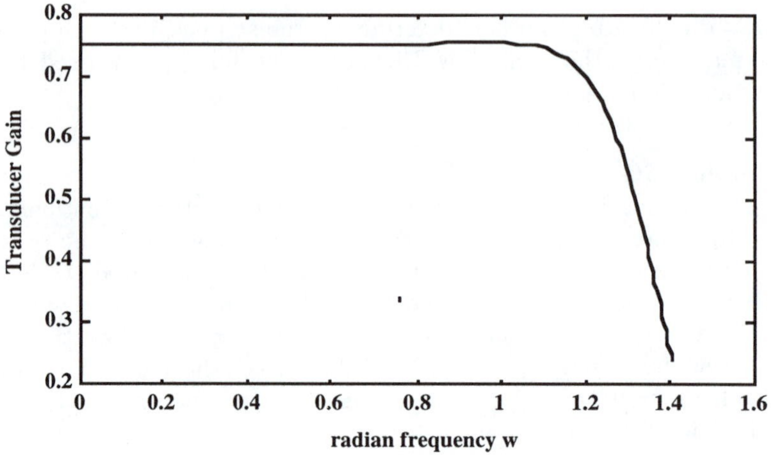

(a)

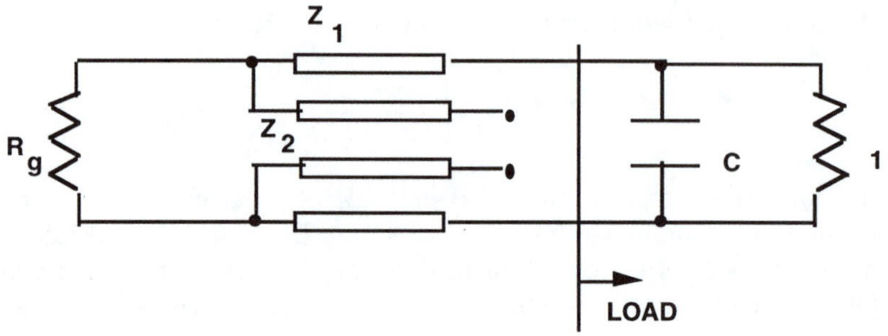

(b)

FIGURE 9.3.1
Line-stub equalization of RC load. (a) Gain response. (b) $C = 2.5$, $Z_1 = 0.465$, $Z_2 = 1.23$, $R_G = 1/3$.

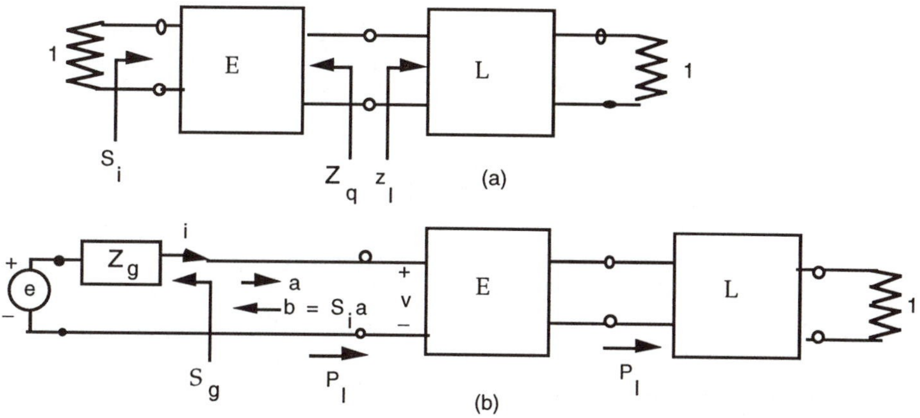

FIGURE 9.4.1
Structure for double matching. (a) Equalizer-load configuration. (b) Complete system.

9.4 Double Matching

We consider the problem of matching a load of impedance z_l to a generator of impedance z_g (both impedances being passive and non-Foster).[9]

The circuit is shown in Fig. 9.4.1 (b). The structures $\mathcal{G}$ and $\mathcal{L}$ are the lossless Darlington two-ports for the prescribed generator and load impedances. $\mathcal{E}$ is the lossless equalizer two-port. The rational impedance of the equalizer measured at the output port of $\mathcal{E}$ when it is unit terminated at the input is Z_e. In the RFT procedure that follows we will initialize $R_e = \Re Z_e > 0$ as a ratio of even polynomials in ω.

Denote the unit normalized scattering matrices of $\mathcal{G}$, $\mathcal{E}$, and $\mathcal{L}$ as $\boldsymbol{G}$, $\boldsymbol{E}$, and $\boldsymbol{L}$, respectively. Let e_1 be the unit normalized input reflectance of $\mathcal{E}$ with the load z_l in place (z_l is the input to $\mathcal{L}$ when it is terminated at its output on a unit resistor). The wave quantities at the output of $\mathcal{G}$ satisfy the equation

$$a_{g2} = e_1 b_{g2}$$

The second scattering equation of two-port $\mathcal{G}$ yields

$$b_{g2} = g_{21}a_{g1} + g_{22}a_{g2} = g_{21}a_{g1} + g_{22}e_1 b_{g2}$$

or

$$b_{g2} = \frac{g_{21}}{1 - g_{22}\, e_1}\, a_{g1} = a_{e1} \tag{9.4.1}$$

[9]The design technique described in this section is based on: H. J. Carlin, S. B. Yarman, "The Double Matching Problem: Analytic and Real Frequency Solutions", *IEEE Trans. on Circ. and Sys.*, vol. CAS-30, no. 1, pp. 15-28, Jan. 1983.

The power delivered to the equalizer is $P_e = (1 - |e_1|^2)\,|a_{e1}|^2$. For our lossless system, the available generator power is $P_A = |a_{g1}|^2$, and the power to the load impedance is $P_l = P_e$. Combining these relations with eq. (9.4.1), the transducer gain of the system is

$$T = \frac{P_l}{P_A} = \frac{P_e}{P_A} = \frac{|g_{21}|^2(1 - |e_1|^2)}{|1 - g_{22}e_1|^2} = \frac{(1 - |g_{22}|^2)\,(1 - |e_1|^2)}{|1 - g_{22}e_1|^2} \tag{9.4.2}$$

where the last expression reflects the losslessness of $\mathcal{G}$.

Presuming we have found Z_e from the initialized R_e, our next step is to calculate the resulting e_1 to insert into eq. (9.4.2). We have already seen, eq. (8.3.6), in our discussion of complex normalization, that the complex normalized Youla reflectance between a generator and load, (which involves a Blaschke product b, that regularizes the complex normalized reflectance), is the same as the unit normalized reflectance measured at the resistively terminated input port. Referring to Fig. 9.4.1 (a), and taking $Z_e = N/D$, the equation we seek is therefore

$$e_1 = b\,\frac{z_l - Z_{e*}}{z_l + Z_e}, \qquad b = \frac{D_*}{D} \tag{9.4.3}$$

We may now employ the gain eq. (9.4.2) to set up an algorithm for RFT double matching. For this purpose we initialize the resistance function of the equalizer $R_e = \Re\, Z_e$

$$R_e = \frac{A_0 + A_1\omega^2 + \ldots + A_k\omega^{2k}}{1 + B_1\omega^2 + \ldots + B_n\omega^{2n}}$$

If all transmission zeros are at infinity or at d.c., then the above equation reduces to

$$R_e = \frac{A_0}{1 + B_1\omega^2 + \ldots + B_n\omega^{2n}}$$

or

$$R_e = \frac{A_0\omega^{2k}}{1 + B_1\omega^2 + \ldots + B_n\omega^{2n}}$$

with $k \le n$. Then we choose one of the following for the form of the initialized resistance function:

$$2k \quad \begin{cases} k = 0 & \text{lowpass} \\ 0 < k < n & \text{bandpass} \\ k = n & \text{highpass} \end{cases}$$

$A_i,\ B_i$	initial guess for the coefficients of the rational function R_l
$z_g(j\omega_i)$	generator impedance values for a discrete set of angular frequencies
$z_l(j\omega_i)$	load impedance values for the same set of angular frequencies

The algorithm for calculating the gain consists of the following steps.

1. Having chosen one of the above forms, make an initial guess for the unknown coefficients. Note that since realizability requires R_e to be even and nonnegative, the initialization of the denominator polynomial can be carried out using the method of Section 9.3, eq. (9.3.2) to insure these properties.

2. Find the roots of the polynomial $DD_* = 1 - B_1 s^2 + \ldots + (-1)^n s^{2n}$.

3. Construct the Hurwitz polynomial D whose roots are those found in the preceding step which have negative real part.

4. Use the Gewertz method, Section 6.12, to reconstruct Z_e from R_e.

5. Compute the loaded equalizer input reflectance e_1 from Z_e by using eq. (9.4.3).

6. Compute the transducer power gain T using eq. (9.4.2).

The algorithm can be coupled with an optimization routine whose objective function is constant gain within the passband, zero outside.

Example 9.4.1
Equalize for maximum flat gain the RL load and the RLC generator shown in Fig. 9.4.2 (a) with $r_l = 1$, $l_l = 1.2$, $r_g = 1$, $l_g = 0.87$, $c_g = 3$.

Solution With the equalizer determined by the above procedure, the system consists of generator impedance, equalizer, and load impedance, as shown in Fig. 9.4.2 (a).

The element values are: $C_1 = 3.753$, $L_2 = 0.893$, $C_3 = 3.403$, $n = 0.6480$. The maximum and the minimum gain are $T_{\max} = 0.922$ and $T_{\min} = 0.794$ with a ripple of 0.649 dB.

A Chebyshev equalizer[10] for the same terminations can be designed by assuming the gain function of degree six

$$T = \frac{K}{1 + \epsilon^2 T_6^2(\omega)}$$

and applying the analytic procedure of Section 8.7. As a result, a degenerate equalizer (not shown) is obtained consisting of a ladder whose elements are $C_1 = 4.014$, $L_2 = 0.898$, $C_3 = 3.861$, $L_4 = 0.350$, $n = 0.504$. Degeneracy is produced by the inductance L_4, which is series connected through the ideal transformer with the inductance l_l of the load. The maximum and the minimum gain are $T_{\max} = 1$ and $T_{\min} = 0.649$ with a ripple of 1.877 dB.

[10] Carlin and Yarman. loc. cit.

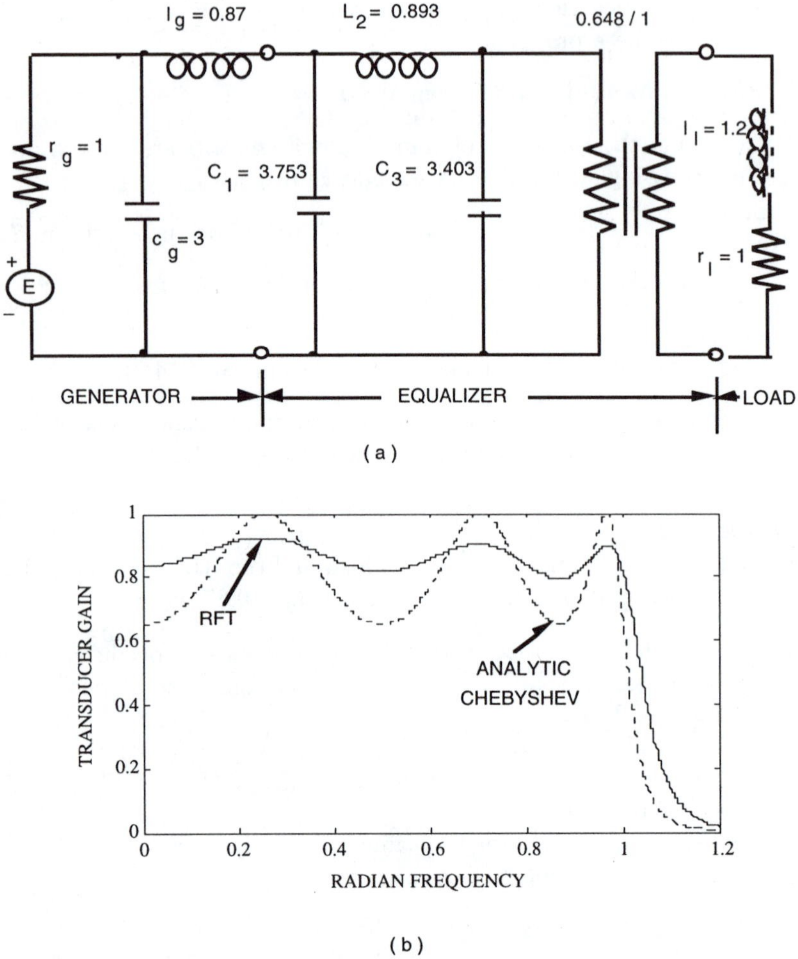

FIGURE 9.4.2
(a) Equalization of an RL load and an RLC generator. (b) Gain vs. frequency curve for the RFT (–) and the Chebyshev equalizer (– –).

Gain vs. frequency curves for both the RFT and Chebyshev equalizers are shown in Fig. 9.4.2 (b). □

The detailed procedure for optimization we have presented is just one possibility, but by no means the only mode of operation. For example, it is possible to proceed in terms of the unit normalized reflectance of the equalizer rather than the resistance function. (See Section 9.5).

In any case, numerical techniques, used either very simply in an interactive manner or in connection with optimization procedures, yield superior results compared to the analytic solution. Furthermore, the RFT can be carried out even when the complexities of the problem preclude the analytic procedure. Nevertheless, the Analytic Theory retains its importance since it gives fundamental insight into the gain-bandwidth performance limitations of linear systems.

9.5 Double Matching of Active Devices

A broadband, multistage microwave amplifier is built up of, say, n cascaded identical devices (for example, field effect transistors, i.e., FET's). Between each pair of devices, and between the first and last devices and their respective final terminations, lossless *equalizer* two-ports are introduced to provide compensation for the device parasitics, and obtain a gain vs. frequency response as flat and high valued as possible in a passband with good selectivity characteristics in the stopband.

As seen from the ports of any equalizer, the problem is that of double matching; in fact, across the input port of the equalizer is a Thévenin equivalent impedance of the generator and, at the output port, a load impedance. The problem is to equalize the generator to the load for high level flat gain taking into account the amplification properties of the system connected to the equalizer. As long as a specific equalizer is considered, we can refer to the structure described in Fig. 9.4.1, where $\mathcal{G}$ and $\mathcal{L}$ now represent the front and the back portions of the chain, while $\mathcal{E}$ represents the equalizer. Note that the entire structure can be considered as terminated in unit resistors. If the generator and load impedances are passive, then $\mathcal{G}$ and $\mathcal{L}$ are lossless Darlington two-ports as discussed in Sections 8.7 and 9.4. In our present discussion, $\mathcal{G}$ and/or $\mathcal{L}$ are generally active rather than lossless structures.

We present a method which is somewhat different than that described in Section 9.4. The gain optimization involves initialization of a reflectance function rather than an impedance and is particularly suited for the type of problem at hand which may include active devices, though it can serve perfectly well as an RFT procedure alternate to that described in Section 9.4 for a passive system.[11, 12, 13]

We first revise the gain formula of eq. (9.4.2) to make it more convenient for use in a procedure which calls for initialization of the equalizer reflectance. Denote the unit normalized scattering matrices of $\mathcal{G}$, $\mathcal{E}$, and $\mathcal{L}$ as $\boldsymbol{G}$, $\boldsymbol{E}$, and $\boldsymbol{L}$, respectively, and once again refer to Fig. 9.4.1.

We have shown in eq. (9.4.1) that the incident and reflected wave quantities with respect to $\mathcal{E}$ are related by $b_{g2} = \dfrac{g_{21}}{1 - g_{22}\, e_1} a_{g1}$. Proceeding in exactly the same manner, the scattering matrix equations give

$$b_{e2} = \frac{e_{21}}{1 - e_{22}\, l_{11}} a_{e1}$$

Substituting $a_{e1} = b_{g2}$, the previous equation yields

$$b_{e2} = \frac{g_{21}}{1 - g_{22}\, e_1} \frac{e_{21}}{1 - e_{22}\, l_{11}} a_{g1}$$

The output power to the $\mathcal{L}$ resistive termination is

$$P_L = |l_{21}|^2 |b_{e2}|^2$$

The generator available power is

$$P_A = |a_{g1}|^2$$

so that, finally, the overall gain $T = P_L/P_A$ is given by

$$T = \frac{|g_{21}|^2}{|1 - g_{22}\, e_1|^2} \frac{|e_{21}|^2}{|1 - e_{22}\, l_{11}|^2} |l_{21}|^2 \tag{9.5.1}$$

where the input reflectance to the terminated equalizer, in terms of the scattering parameters $\boldsymbol{E} = \boldsymbol{E}' = (e_{ij})$, is

$$e_1 = e_{11} + \frac{e_{21}^2\, l_{11}}{1 - e_{22}\, l_{11}}$$

[11] The material presented in this section is based on: B. S. Yarman and H. J. Carlin, "A Simplified "Real Frequency" Technique Applied to Broad-Band Multistage Microwave Amplifiers", *IEEE Trans. on Microwave Th. and Tech.*, vol. 30, no. 12, pp. 2216-2222, Dec. 1982.

[12] Another RFT approach for amplifier equalization, which uses resistive line segment approximations is described in the paper by W. Jung and J. Wu "Stable Broad-Band Microwave Amplifier", *IEEE Trans. on Microwave Th. and Tech.*, vol. 38, no. 8, pp. 1079-1085, Aug. 1990.

[13] H. J. Carlin and J. J. Komiak, "A New Method of Broad-Band Equalization Applied to Microwave Amplifiers", *IEEE Trans. on Microwave Th. and Tech.*, vol. 27, no. 2, pp. 93-99, Feb. 1979, uses complex rather than unit normalization.

Note that if the load is a passive impedance we can use $|l_{21}|^2 = 1 - |l_1 1|^2$ in eq. (9.5.1); similarly, if the generator impedance is passive,

We will apply the gain formula (9.5.1) to the optimization of the $\mathcal{G}$, $\mathcal{E}$, $\mathcal{L}$ system by initializing the equalizer scattering parameter e_{11}. To do this, refer to the Belevitch representation of a lossless 2-port scattering matrix, eq. (4.2.1), and write

$$e_{11} = \frac{h}{g} = \frac{h_0 + h_1 s + \ldots + h_n s^n}{g_0 + g_1 s + \ldots + g_n s^n}$$

$$e_{22} = \pm\frac{h_*}{g} = -(-1)^k \frac{h_0 - h_1 s \ldots + (-1)^n h_n s^n}{g_0 + g_1 s + \ldots + g_n s^n}$$

$$e_{21} = \frac{f}{g} = \pm\frac{s^k}{g_0 + g_1 s + \ldots + g_n s^n}$$

with $k < n$, so as to define a lossless two-port with k transmission zeros at the origin and $n - k$ at infinity.

We construct a subroutine to calculate the gain $T(\omega^2)$ starting from the row vector $\boldsymbol{h} = (h_0, h_1, \ldots h_n)$, whose elements are the initialized set of polynomial coefficients which are to be finally determined by the optimization process. Consider as a typical example the case where the generator impedance is passive and the load is an active device. Then the inputs are

n	Degree of the polynomial h (numerator of e_{11});
k	Degree of the polynomial f (numerator of e_{21});
$\boldsymbol{h}$	Coefficients of the polynomial h (ascending powers);
$g_{22}(j\omega)$	Generator unit normalized reflectance;
$l_{11}(j\omega)$	Load unit normalized reflectance;
$\lvert l_{21}(j\omega)\rvert$	Load (active element) unit normalized.*transmittanceamplitude.*

The algorithm is the following.

1. Generate the polynomial $g_* g = h_* h + s^{2k}$.
2. Find the roots of $g_* g$.
3. Form the Hurwitz polynomial g from the LHP roots of $g_* g$.
4. Construct e_{11}, e_{21}, and e_{22} according to eq. (4.2.1), as summarized above.
5. Compute the gain T using eq. (9.5.1).

6. Repeat the procedure for a suitable set of frequencies.

7. Optimize $||T - T_0||$ over the band.

Note that the realizability of $\mathcal{E}$ is guaranteed by the algorithm, since h/g is always realizable for any initialization of h. If load and generator impedances are passive, then $\mathcal{G}$ and $\mathcal{L}$ are lossless and the power to the load termination is $P_e = P_L$. In that case taking $|g_{21}|^2 = 1 - |g_{22}|^2$, $|l_{21}|^2 = 1 - |l_{11}|^2$, the above algorithm can be used as an alternate to that of Section 9.4.

The ideal gain characteristic, T_0, may be built into the routine as, say, $T_0 = K$ in the passband, $T_0 = 0$ in the stopband, or may be supplied as an input. The constant K should be chosen by cut-and-try. Too low a value would not fully exploit the gain potential of the device; on the other hand, too high a value would exaggerate the passband ripple.

As a result of the optimization process, polynomials h and g are found. Thus the input impedance $Z_e = (g + h)/(g - h)$ of the equalizer terminated on a unit resistance is known and $\mathcal{E}$ is synthesized by Darlington's procedure.

We illustrate the application of the above procedure to the design of a single-stage amplifier. We first consider a structure including the generator ($\mathcal{G}$ lossless) and the device ($\mathcal{L}$) with the equalizer ($\mathcal{E}$) in between. The device is terminated on the load resistor at the output. The equalizer is designed by the above procedure to optimize the gain of the resulting two-port. A further improvement can be obtained by inserting a second equalizer between the device and the load. The optimization would then be repeated by identifying the new $\mathcal{G}$ (nonpassive) as the generator cascaded with the first equalizer and the device. The new $\mathcal{E}$ is the second equalizer, which is simply terminated on the load resistance, i.e., $l_{11} = 0$, $l_{21} = 1$. Somewhat higher gain may be achievable by optimizing both equalizers simultaneously rather than in separate steps.[14]

Example 9.5.1

Design a single-stage FET amplifier[15] with

Generator	$g_{22} = 0$ $(50\,\Omega)$
Load	$l_{11} = 0$ $(50\,\Omega)$
Passband	$11.7\,\text{GHz} \le f \le 12.2\,\text{GHz}$ (X-Band)
Device	Mitsubishi FET, MGF-2124 package

Solution The device scattering parameters over the band are reported in Table I of the paper quoted in footnote 11. Both the front and the back end

[14] W. Jung and J. Wu, *loc. cit.*

[15] The following example is taken from: H. J. Carlin and B. S. Yarman, *loc. cit.*

equalizers are realized as a series inductor cascaded with a shunt capacitor. For the first equalizer, one obtains

$$e_{11}^{(1)} = \frac{0.404s + 1.0065s^2}{1 + 1.475s + 0.0065s^2}$$

leading to $L_1 = 1.23\,\text{nH}$ (in series with source resistor) followed by shunt $C_2 = 0.28\,\text{pF}$. For the second

$$e_{11}^{(2)} = \frac{-0.461s + 0.129s^2}{1 + 0.686s + 0.129s^2}$$

yielding $L_3 = 0.146\,\text{nH}$ (in series with the device) and $C_4 = 0.3\,\text{pF}$ (across the load resistor). The gain over the passband is

$$T(\omega) = 2.26 \pm 0.26\,\text{dB}$$

□

The extension to any number, n, of stages is readily accomplished. Let a single stage be defined as above (generator, equalizer, device). The gain of a k-stage chain, T_k, is expressed in terms of that of the $k-1$-chain, T_{k-1}, by eq. (9.5.1) that we rewrite in the following form.

$$T_k = T_{k-1}\frac{|e_{21_k}|^2|l_{21_k}|^2}{|1 - g_{22_k}e_{1_k}|^2|1 - e_{22_k}l_{11_k}|^2}$$

since $|g_{21_k}|^2 = T_{k-1}$. The process starts by optimizing the first stage, then the cascade of the first with the second (with the first fixed), and so on. Finally, the back-end equalizer terminated in the final resistive load is designed. As a further refinement one might optimize all equalizers simultaneously, taking the previous multistage design as the initialization, but it is questionable whether the improvement in response would justify the added complexity of computation.

Appendix A

Analytic Functions

A.1 General Concepts

It is assumed that the reader is familiar with the definition of complex numbers and the arithmetic operations which apply to them, but for the sake of completeness a brief review of these topics initiates Appendix A.[1]

Start with the concept of a complex number as an ordered pair of real numbers, $c = (a, b) = a + jb$, the second form (*algebraic* or *rectangular* form) being the one usually adopted. The *real part* of c is a, and b is the *imaginary part.* The sum and the difference operations are defined by $c \pm c' = (a \pm a', b \pm b') = (a \pm a') + j(b \pm b')$. The rule for multiplication (chosen early by mathematicians because it leads to such interesting consequences) is $cc' = (aa' - bb', ab' + a'b) \equiv (aa' - bb') + j(ab' + a'b)$. This result is precisely achieved if we take $j = \sqrt{-1}$, and the usual rules of real number arithmetic are employed for $(a + jb)(a' + jb')$ (with $j^2 = -1$). Rectangular rather than ordered pair notation is generally used.

It is also useful to employ the *polar form* for a complex number. Thus $c = a + jb = \rho \exp j\theta$, with the *modulus* $\rho = |c| = \sqrt{a^2 + b^2}$, the *argument* or *phase* $\arg c = \theta$ with $\tan\theta = b/a$. The rectangular form for $\exp j\theta$ can be expressed as

$$\epsilon^{j\theta} = \cos\theta + j\sin\theta \qquad \text{(Euler's equation)}$$

which can be directly applied to verify the equivalence of the polar and rectangular forms.

[1]A broad treatment of the subject of this Appendix is to be found in classic works such as: E.T. Whittaker and G.N. Watson, *A Course of Modern Analysis*, Cambridge, At the University Press, Fourth edition reprinted 1978, Part I; E. C. Titchmarsh, *The Theory of Functions*, Oxford, Oxford University Press, 2nd ed., 1968; K. Knopp, *Theory of Functions*, 5 vols., New York, Dover, 1945.

Complex numbers are usefully represented as *vectors* in a plane (the *complex plane*), in which two orthogonal co-ordinate axes are drawn, the *real axis* and the *imaginary axis*; the real part of the number, a, is the component of the representative vector along the real axis, while the imaginary part b is the component along the imaginary axis. Thus the modulus is the Euclidean norm of the vector; the argument is the angle that the latter forms with the real axis. This representation is known as the *Argand diagram.*

The set of all complex numbers in a complex plane defines a *complex variable*, whose label identifies the plane. Thus we shall speak of complex planes z, w, s, and so on, to denote the complex planes associated with the sets of complex numbers z, w, s, etc.

Consider now two complex planes z and w. A rule which associates to each point of some subset of the complex plane z a point of a subset of the complex plane w defines a *function of a complex variable*, i.e., $w = f(z)$. The rule is usually based on the rules of complex arithmetic and the properties of real functions. The subset of the complex plane z in which the function is defined is called its *domain*; the subset of points of the complex plane w which correspond through the function to the domain is called its *range.* Occasionally the domain or the range or both can coincide with the whole of their respective planes.

A function of a complex variable $w = f(z)$ is equivalent to a pair of real functions of two real variables. In fact, if we write $z = x + jy$ and $w = u + jv$, the equation $w = f(z)$ can always be written as

$$\begin{aligned} u &= u(x, y) \\ v &= v(x, y) \end{aligned}$$

For example, the function $w = \cos z = \cos(x+jy)$ results in $u = \cos x \cosh y$ and $v = -\sin x \sinh y$.

If polar form is used, the two real functions appear as modulus and argument, $w = f(z) = r(x, y) \exp j\theta(x, y)$. As an example, consider $w(z) = \sqrt{z} = r \exp j\phi$, where $z = \rho \exp j\theta$; then $r = \sqrt{\rho}$, and $\phi = (\theta + 2k\pi)/2$, where $k = 0, 1$ yields the two opposite values of the square root. Using Euler's equation and the half-angle formulas for $\cos \theta/2$, $\sin \theta/2$, we can express the result in rectangular form

$$w(z) = u(x, y) + jv(x, y) = \pm \left(\sqrt{\frac{\sqrt{x^2 + y^2} + x}{2\sqrt{x^2 + y^2}}} + j\sqrt{\frac{\sqrt{x^2 + y^2} - x}{2\sqrt{x^2 + y^2}}} \right)$$

where the double sign still corresponds to the two opposite determinations of the square root.

In the complex plane, it is convenient to regard the set of all directions as converging to a single point, the *point at infinity.* The z plane can therefore be considered as the projection of a sphere, onto an euclidean plane. One point of the sphere, which can be considered the South Pole, makes contact with the origin of the complex plane. By means of latitude rays emanating from the North Pole of the sphere every point of z can be made to correspond in a one-to-one fashion to a point of the sphere. The North Pole of the sphere (an actual specific point) represents the point at infinity of the z plane. Thus a neighborhood of the point at infinity in z is simply represented as a finite north polar cap.

We consider the meaning of the statement that $f(z)$ tends to the limit λ when z tends to z_0, written as

$$\lim_{z \to z_0} f(z) = \lambda$$

The statement means that, for any given $\epsilon > 0$, a $\delta(\epsilon) > 0$ can be found, such that it is $|f(z) - \lambda| < \epsilon$ for any z satisfying $|z - z_0| < \delta$. If λ is infinite, the statement means that for any $M > 0$ a $\delta(M)$ can be found, such that $|z - z_0| < \delta$ implies $|f(z)| > M$. When we have a *finite* limit λ with z tending to infinity, this means that for any $\epsilon > 0$ an $R(\epsilon) > 0$ can be found, such that it is $|f(z) - \lambda| < \epsilon$ for any z satisfying $|z| > R$. The case of an *infinite* limit when z tends to infinity is carried out in a similar fashion.

The derivative of a function of a complex variable is formally defined by the limit

$$f'(z) = \lim_{z \to z_0} \frac{f(z) - f(z_0)}{z - z_0}$$

which naturally extends the concept of derivative of a function of one real variable. The difficulty here is that such a limit may depend not only on the point z_0 but also on the direction in which the complex increment $z - z_0$ is taken. If we impose the reasonable requirement that the derivative of a function of a complex variable only be a function of the point z_0, where the derivative is determined, we must demand that the limit defining the derivative be independent of the path along which z approaches z_0. It turns out that the necessary and sufficient condition for the derivative to be the same for *every* limit path is that the derivatives taken along just the two paths parallel to the x and y axes be equal. This leads to the pair of equations

$$\frac{\partial u}{\partial x} = \frac{\partial v}{\partial y} \tag{A.1.1}$$

$$\frac{\partial u}{\partial y} = -\frac{\partial v}{\partial x} \tag{A.1.2}$$

which are known as the *Cauchy-Riemann* (CR) conditions.

A function of a complex variable which possesses a derivative in the sense above in a region $\mathcal{R}$ is said to be *analytic* in that region. A function, which is analytic almost everywhere in the complex plane or a subset of it, is said to be *generally analytic* or simply *analytic*. For example, it is easily verified that the function $w(z) = \cos(x + jy)$ is analytic. Thus both sides of eq. (A.1.1) equal $-\sin x \cosh y$ and both sides of eq. (A.1.2) equal $-\cos x \sinh y$, so the CR conditions are satisfied everywhere in the finite z plane.

It follows immediately from the CR conditions that both u and v, the real and the imaginary parts of the function w, satisfy the Laplace equation

$$\frac{\partial^2 \phi}{\partial x^2} + \frac{\partial^2 \phi}{\partial y^2} = 0 \quad \phi = u, \text{ or } \phi = v$$

that is, $u(x,y), v(x,y)$ are *harmonic functions.* The existence of the second derivatives is not in question because, as will be seen later, wherever a function is analytic, it possesses derivatives of all orders.

A simple closed curve (i.e., no multiple or crossover points) γ divides the plane into an *internal* and an *external* region (Jordan Separation Theorem). The external region is taken as the one which contains the point at infinity. The positive direction on γ is chosen so that the internal region is to the left as the curve is traversed in the positive direction. Therefore a direction which is positive with respect to one of the two complementary regions is negative with respect to the other.

A region $\mathcal{R}$ is *connected* if any two of its points can be joined by a regular curve lying in the region. Consider the set of curves lying within the region with their extreme points on its boundary and suppose we cut the region along any of such curves. The region is *simply connected* if no more than one cut is necessary to make it to lose its connection; it is *doubly connected* if no more than two cuts are necessary to make it to lose its connection, a.s.o. A multiply connected region (with connection of order n) is easily visualized as one including n holes.

A.2 Integration of Analytic Functions

The line integral can be defined for functions of a complex variable as follows. Consider, in the complex plane z, a curve γ resulting from the union of a finite number of regular arcs. Choose arbitrarily on such a curve a finite sequence of points, say, $z_1, z_2, \ldots, z_n$. Construct the sum

$$S = \sum_{i=1}^{n-1} f(z_i)(z_{i+1} - z_i)$$

Let be $\delta = \max_i |z_{i+1} - z_i|$. If for δ tending to zero, S tends to a finite limit, then $f(z)$ is said to be *integrable* along the curve γ, and the limit I

of the sum S is written as

$$I = \int_\gamma f(z)dz \tag{A.2.1}$$

It is easily seen (by writing z and dz in complex form) that the integral of eq. (A.2.1) can be expressed by a pair of real curvilinear integrals

$$I = \int_\gamma (udx - vdy) + j\int_\gamma (vdx + udy) \tag{A.2.2}$$

which can serve as an alternative definition.

If γ is a *closed curve* in a simply connected region $\mathcal{R}$, then a basic theorem of analytic functions can be shown to be true. This is the *Cauchy Integral Theorem*, namely, that in a region where $f(z)$ is analytic the closed line integrals are zero.

THEOREM A.2.1

(Cauchy Integral Theorem) *If $f(z)$ is analytic in some simply connected region $\mathcal{R}$, then its integral extended to any closed curve lying within the region is zero.*

$$\oint_\gamma f(z)dz = 0$$

One method of proving the result is to use Green's theorem to transform the line integrals of eq. (A.2.2) to surface integrals over the area bounded by the closed curve γ. The CR conditions are then invoked for the integrands and the result follows.

Based on this result, it is clear that for an open integration path the necessary and sufficient conditions that the line integral of eq. (A.2.1), or equivalently the two curvilinear integrals of eq. (A.2.2), be independent of the integration path between two given points in some region $\mathcal{R}$ is that $f(z)$ be analytic everywhere in the simply connected region $\mathcal{R}$. Thus we have the following theorem.

THEOREM A.2.2

The function $F(z)$, known as a primitive *of $f(z)$, is defined in a simply connected region $\mathcal{R}$ according to the equation below.*

$$F(z) = \int_{z_0}^{z} f(z)dz$$

The path of integration lies entirely in $\mathcal{R}$ and the initial point z_0 is fixed while the end point z is variable. Then the primitive $F(z)$ exists, is an analytic function, and is unique, if and only if $f(z)$ is analytic in $\mathcal{R}$.

Theorems A.2.1 and A.2.2 are clearly equivalent.

Primitives based on different values of z_0 differ from one another by complex constants. The derivative of a primitive of a function is the function itself, $F'(z) = f(z)$.

The derivative and integral operators have, for analytic functions, the same formal properties as for functions of a real variable. Thus derivatives and integrals of analytic functions can be calculated by using the rules of Calculus. For example, the derivative of z^n is nz^{n-1} and the indefinite integral of $\sin z$ is $-\cos z + c$ where c is an arbitrary complex constant.

A.3 The Cauchy Integral Formula

We consider a closed curve γ, resulting from the union of a finite number of regular arcs, within some simply connected region $\mathcal{R}$ where $f(z)$ is analytic. The function $f(z)$, at each point z belonging to the region internal to the curve, can be represented in terms of its values on the boundary by the following integral, the *Cauchy Integral Formula.*

$$f(z) = \frac{1}{2\pi j} \oint_\gamma \frac{f(\zeta)}{\zeta - z} d\zeta \tag{A.3.1}$$

The result above can be easily proved by considering the equation

$$\frac{1}{2\pi j} \oint_\gamma \frac{f(\zeta) - f(z)}{\zeta - z} d\zeta = 0 \tag{A.3.2}$$

The integral in the left side of the equation above is zero because its integrand is analytic everywhere in $\mathcal{R}$; the zero of $\zeta - z$ in the denominator does not produce a singularity because it is cancelled by the zero of $f(\zeta) - f(z)$ in the numerator. Eq. (A.3.2) can be rewritten as

$$f(z)\frac{1}{2\pi j} \oint_\gamma \frac{d\zeta}{\zeta - z} = \frac{1}{2\pi j} \oint_\gamma \frac{f(\zeta)}{\zeta - z} d\zeta$$

Note that neither one of the integrals in both sides of the equation above need to be zero, because both integrands have a singularity at $\zeta = z$. Consider now a small circle c with center in z and radius ϵ. We can connect a point of this circle with a point of γ by a simple regular nonintersecting arc γ_1; thus we create a composite closed path, formed by γ, traversed in the counterclockwise sense, by c traversed in the clockwise sense, and γ_1, traversed once in the one sense, once in the other. The path leaves the internal region at its left and, therefore, is traversed in the positive sense.

In such a region the function $1/(\zeta - z)$ is analytic and, therefore, its integral along the composite path is zero. Since the two contributions of γ_1 in the opposite senses to the integral cancel mutually, we are left with the equation

$$\frac{1}{2\pi j}\oint_{\gamma}\frac{d\zeta}{\zeta - z} = \frac{1}{2\pi j}\oint_{c}\frac{d\zeta}{\zeta - z} \tag{A.3.3}$$

where c is now traversed counterclockwise as γ. The integral in the right side of eq. (A.3.3) is easily calculated as equal to 1, and this proves eq. (A.3.1).

Cauchy's integral formula shows that the very fact that a function of a complex variable satisfies the CR conditions endows it with a rich internal structure; the formula itself shows that the boundary values completely define the function everywhere in the internal region. Moreover, it can be shown that Cauchy's formula can be differentiated any number of times under the integral sign. Thus $f(z)$ is indefinitely differentiable within its domain of analyticity and its n-th derivative can be expressed as

$$f^n(z) = \frac{n!}{2\pi j}\oint_{\gamma}\frac{f(\zeta)}{(\zeta - z)^{n+1}}d\zeta \tag{A.3.4}$$

A.4 Laurent and Taylor Expansions

The Cauchy integral formula allows us to expand a function $f(z)$ in a series of ascending and descending powers of $z - z_0$ within an annular region with center z_0 and radii ρ_1 and ρ_2, $\rho_1 < \rho_2$, provided $f(z)$ is analytic within the annulus.

The expansion has the form

$$f(z) = \sum_{k=0}^{\infty} c_k(z - z_0)^k + \sum_{k=1}^{\infty}\frac{c_{-k}}{(z - z_0)^k} \tag{A.4.1}$$

where the coefficients are given by

$$c_k = \frac{1}{2\pi j}\oint_{\gamma}\frac{f(\zeta)}{(\zeta - z_0)^{k+1}}d\zeta \tag{A.4.2}$$

and by

$$c_{-k} = \frac{1}{2\pi j}\oint_{\gamma} f(\zeta)(\zeta - z_0)^{k-1}d\zeta \tag{A.4.3}$$

respectively.

Eqs. (A.4.2) and (A.4.3) can be put together in the single form (A.4.2), allowing k to assume all integer positive and negative values. An easy justification of these expressions is obtained by substituting eq. (A.4.1) into Cauchy's integral formula (A.3.1) with $z = z_0$, integrating termwise, and observing that $\frac{1}{2\pi j} \oint_\gamma \frac{d\zeta}{(\zeta - z_0)^k}$, where γ is a closed curve encircling z_0, is 1 or 0 according to whether $k = 1$ or $k \neq 1$.

Note that in the definition of the annular region where the Laurent's expansion is valid, radii ρ_1 and ρ_2 are not unique values, because the only restriction is that within the region the function be analytic. We can however consider the union of all such annular regions; its radii r_1 and r_2 are, respectively, the inf and the sup of the sets of the ρ_1's and the ρ_2's. Such a union defines the largest convergence region of the expansion and is therefore said to be the *convergence annulus* of the Laurent expansion of $f(z)$ around z_0. Within the convergence annulus, the series converges absolutely; within any closed set internal to the annulus, it converges uniformly.[2] On each of the circles of radii r_1 and r_2 there is at least one point where $f(z)$ is not analytic. This sets a natural boundary of the annulus which determines the domain of definition of the function. Such points are called *singular points* or *singularities.*

If the radius r_1 is 0, the annulus degenerates into a circle excluding its center. In such a case, the function $f(z)$ is analytic in a neighborhood of point z_0; in fact, in all the interior of the circle of center z_0 and radius r_2, except z_0 itself. Thus z_0 is a *singular* point and the Laurent expansion is said to be *around* or *in the neighborhood of* z_0.

The terms with ascending powers of $z - z_0$ (including the constant one) form the *regular component,* those with descending powers the *characteristic or principal component* of the expansion.

If the characteristic component includes a finite number of terms, say up to $c_{-n}/(z - z_0)^n$, the singular point is a *pole* of *order* n. Otherwise, it is an *essential singularity.*

Poles and essential singularities are *isolated singularities,* but do not cover the full set of critical points that an analytic function can exhibit. The immediately more complicated are the *nonisolated singularities* which can occur as accumulation or limit points of either poles or essential singularities. Also continuous lines or regions entirely covered by singularities may be encountered. A nonisolated singularity does not admit of any series expansion around it.

If there is no characteristic component present, the expansion reduces to a *Taylor series* and the point z_0 is a point of analyticity; in such a case it

[2] A series converges absolutely at a point if the series of the moduli of its terms converges there. A series converges uniformly in a closed set if for any $\epsilon > 0$ there is a k_0, depending on ϵ but not on z, such that for all $k > k_0$, $|S - S_k| < \epsilon$, S being the sum of the series, S_k the sum of its first k terms. In effect, the rate of convergence is the same at all points within the region.

is easily recognized from eqs. (A.3.4) and (A.4.2) that

$$c_k = \frac{f^k(z_0)}{k!}$$

The Laurent expansion can be readily defined around the point at infinity. We use as our guide *The Principle of Behavior at Infinity*, namely, that the behavior assigned to $f(z)$ at $z = \infty$ is that which $\phi(t) = f(1/t)$, $t = 1/z$, exhibits at $t = 0$. Thus consider the transformation $t = 1/z$, which brings the point at infinity in the z plane into the origin of plane t. Now expand $\phi(t) = f(1/t)$ in a Laurent series around $t = 0$.

$$\phi(t) = \sum_{k=0}^{\infty} c_{-k} t^k + \sum_{k=1}^{\infty} \frac{c_k}{t^k} \tag{A.4.4}$$

If in equation (A.4.4) $1/z$ replaces t, the following expansion is obtained

$$f(z) = \sum_{k=0}^{\infty} \frac{c_{-k}}{z^k} + \sum_{k=1}^{\infty} c_k z^k \tag{A.4.5}$$

In the expansion (A.4.5) the terms in $1/z^k$ represent the *regular* component, the terms in z^k, the *characteristic* one. The definitions of *poles* and *essential singularities* apply in the same way as for an expansion around a finite point.

In most cases, series coefficients are calculated directly by using some special device, instead of applying formulas (A.4.2) and (A.4.3) directly. Consider, as an example, the function

$$w(z) = \frac{1}{z-1} + \frac{1}{z-2} \tag{A.4.6}$$

Around $z = 1$, the function is represented as the sum of the characteristic component $1/(z-1)$ and the regular component, which is a Taylor series for $1/(z-2)$ at the point $z = 1$ (powers of $(z-1)$) where this term is analytic; around $z = 2$, the two terms interchange their roles. Hence the function has simple poles at either point and no other singularities. Thus it can be expanded in Taylor series around the origin or around infinity, or in Laurent series within the annulus delimited by the circles with center in the origin and radii equal to 1 and 2, respectively.

The Taylor series around the origin is obtained by expanding each fraction of eq. (A.4.6) in a Maclaurin's series (power series at the origin) in terms of z and grouping terms.

$$w(z) = \sum_{k=0}^{+\infty} \left[-\left(1 + \frac{1}{2^{k+1}}\right) \right] z^k$$

The Taylor series around infinity is obtained by expanding both fractions in Maclaurin series in terms of $t = 1/z$ and grouping terms.

$$w(z) = \sum_{k=1}^{+\infty} \left(1 + 2^{k-1}\right) \frac{1}{z^k}$$

The Laurent series in the annulus bounded by circles of radii 1 and 2 is obtained by expanding the term $1/(z-1)$ in negative powers of z (valid for $|z| > 1$) and the term $1/(z-2)$ in positive powers of z (valid for $|z| < 2$).

$$w(z) = \sum_{k=1}^{\infty} \frac{1}{z^k} - \sum_{k=0}^{\infty} \frac{z^k}{2^{k+1}}$$

Note that in network theory we encounter functions of the form $f(z) = az + g(z)$, where $g(z)$ is analytic at $z = \infty$. Then $f(z)$ has a simple pole at $z = \infty$.

As another example, consider the well known Maclaurin expansion of $\exp z$

$$\epsilon^z = \sum_{k=0}^{+\infty} \frac{z^k}{k!}$$

Since its radius of convergence is infinite, it can be considered either as a Taylor's series around the origin or a Laurent's series around infinity. In the first case all terms belong to the regular component, in the second, the regular component includes just the constant 1 (corresponding to $k = 0$) while all other terms (an infinite number of them) are in the characteristic component. The point at infinity is thus an essential singularity.

A.5 The Theorem of Residues

Consider the Laurent expansion of a function $f(z)$ around some point z_0. The coefficient c_{-1} of $1/(z - z_0)$ is called the *residue* of $f(z)$ at z_0. If the function is analytic at some point, its residue there is zero. The residue at infinity is the coefficient c_{-1} of z in the expansion (in powers of z) of $f(z)$ around infinity (use the "principle of behavior at infinity" setting $t = 1/z$).

Consider some closed curve γ in the complex plane encircling, say, r isolated singular points (poles or essential singularities) of the function $f(z)$. The following statement holds, where $c_{-1}^{(i)}$ is the residue at the i-th singularity.

THEOREM A.5.1

(Theorem of Residues) *Let γ be a closed regular curve in an open set where the function $f(z)$ is analytic with the exception of at most a finite number of points inside the region $\mathcal{R}$ bounded by the curve. The line integral of $f(z)$ over γ, divided by $2\pi j$, is equal to the sum of the residues at the isolated singular points encircled by γ.*

$$\frac{1}{2\pi j}\oint_{\gamma} f(z)dz = \sum_{i=1}^{r} c_{-1}^{(i)}$$

The theorem of residues can be proved by surrounding each singularity with a small circle γ_i and connecting γ_1 and γ_2 by a line s_1, γ_2 and γ_3 by a line s_2, etc. This composite curve is then joined to γ by another line s_0. The total contour formed by γ, the γ_i, and the s_i and s_0, each of the s_0, s_i traversed once in one sense and once in the other, forms a closed curve inside which there are no singularites. Thus the integral of $f(z)$ around the complete contour is zero by the Cauchy's Theorem. Furthermore, since the paths s_i and s_0 are traversed, first in the positive and then in the negative direction, their contributions to the integral cancel. The remaining components, which sum to zero, consist of the integral in the positive sense around γ, and the integrals in the negative sense around the γ_i. The latter are transposed to the right side of the equation and, since each individual γ_i integral may be replaced by $2\pi j c_{-1}^{(i)}$, we obtain the final result of the theorem.

A.6 Zeros, Poles, and Essential Singularities of Analytic Functions

A function $f(z)$ has a *zero* of *order* n at z_0 if and only if it can be represented in a neighborhood of z_0, say, C, as

$$f(z) = (z - z_0)^n g(z)$$

where $g(z)$ is analytic, single valued, and nonzero in C.

It is easily verified that the definition above is equivalent to the statement that, in the Taylor's series of $f(z)$ around z_0, the first nonzero coefficient is that of power $(z - z_0)^n$.

A function $f(z)$ has a *pole* of *order* n in z_0 iff it can be represented in a neighborhood of z_0, say, C, as

$$f(z) = \frac{g(z)}{(z - z_0)^n} \tag{A.6.1}$$

where $g(z)$ is analytic, single valued, and nonzero in C.

It is immediately seen that this definition is equivalent to that based on the degree of the characteristic component (Section A.4).

Thus the following theorem holds true.

THEOREM A.6.1

If $f(z)$ has a zero of order n at z_0, its reciprocal $g(z) = 1/f(z)$ has a pole of order n there; conversely, if $f(z)$ has a pole of order n at z_0, its reciprocal $g(z)$ has a zero of order n there.

As an example, consider the function $f(z) = 1 - \cos z$. By expanding $\cos z$ in Maclaurin series, it is immediately seen that the function has a double zero at the origin, and the function $g(z)$ is easily identified (still in the form of a series). The reciprocal function $h(z) = 1/(1 - \cos z)$ has a double pole at the origin; the new $g(z)$ is the reciprocal of the old one.

The results for the case in which $f(z)$ has an essential singularity at z_0 are somewhat different. It can be seen that $1/f(z)$ has neither a point of analyticity nor a pole at z_0, and must exhibit either an essential or a nonisolated singularity.

For example, the function $w(z) = \exp(1/z)$ has an essential singularity at the origin, and so has its reciprocal $\exp(-1/z)$. On the other hand, $w(z) = \sin(1/z)$ has an essential singularity at the origin, while its reciprocal has an accumulation point of poles there.

It is evident from eq. (A.6.1) that an analytic function $f(z)$ is not bounded around a pole.

The performance in a neighborhood of an essential singularity is considerably more complicated and is described by the following.

THEOREM A.6.2

(Picard's Theorem) *If a single valued function has an essential singularity, then, in the neighborhood of the singularity, it takes on every value, with one possible exception, an infinite number of times.*

Hence the function oscillates wildly around an essential singularity; in particular, it has no limit there. The reader can verify this by studying the performance of the function $f(z) = \exp(1/z)$ in a small circular neighborhood of the essential singularity at the origin. In this case, for example, the function takes on all values except the exceptional value $f(z) = 0$.

A.7 Some Theorems on Analytic Functions

The importance of singularities in characterizing analytic functions is evidenced by the following.

THEOREM A.7.1
(Liouville's Theorem) *A uniform (i.e., single-valued) analytic function having no singularities in the entire complex plane (including the point at infinity) must be a constant.*

PROOF If the function $f(z)$ has no singularities in the finite part of the plane, it can be expanded in a Taylor's series around the origin. Such a series has an infinite convergence radius since, otherwise, the convergence circle would have a singularity on its circumference. Thus the Taylor's expansion around the origin can be reinterpreted as a Laurent's expansion around infinity. Such an expansion has a regular component which is a constant c_0, while the characteristic component given by the terms with ascending powers of z must disappear, since the point at infinity is not singular. Only the constant remains and the theorem is proved. ■

The extremal properties of an analytic function are stated in several theorems, all essentially based on a single one, the *Theorem of Maximum Modulus.*

THEOREM A.7.2
(Theorem of Maximum Modulus) *If a function $f(z)$ is analytic in an open set containing a region $\mathcal{R}$ and its boundary C, its modulus reaches its maximum on the boundary.*

By applying the theorem above to $g(z) = 1/f(z)$, a similar result for the minimum modulus is obtained.

THEOREM A.7.3
(Theorem of Minimum Modulus) *If a function $f(z)$ is analytic in an open set containing a region $\mathcal{R}$ and its boundary C, with no zeros inside $\mathcal{R}$, the modulus attains its minimum on the boundary.*

In the "minimum" theorem, the further requirement that $f(z)$ be nonzero within $\mathcal{R}$ is dictated by the necessity that $g(z) = 1/f(z)$ also be analytic in $\mathcal{R}$. (Recall that a zero of $f(z)$ is a pole of $g(z)$ of the same order.)

By applying the theorem of the maximum modulus to $\exp(\pm f(z))$ and to $\exp(\pm jf(z))$, we obtain similar theorems for maxima and minima of the real and imaginary parts of $f(z)$.

THEOREM A.7.4
(Theorem of the Maximum and Minimum of the Real and Imaginary Parts) *If a function $f(z)$ is analytic in an open set containing a region $\mathcal{R}$ and its boundary C, the real and imaginary parts of $f(z)$ reach their maxima and minima on the boundary.*

A.8 Classification of Analytic Functions

Analytic functions can be conveniently classified according to the nature of their singularities. That the singularities deeply influence the properties of a function is made evident by the fact that a uniform function with no singularities, according to Liouville's Theorem, reduces to a constant.

Functions can be divided into *algebraic* and *transcendental* depending on whether their definition involves only algebraic operations or not. An example of an algebraic function is $\sqrt{z}$, of a transcendental function, $\ln z$.

Functions can be also classified according to whether they are *single valued* (uniform) or *multivalued.* Multivalued functions include the logarithm and functions like z^α, where α is not a real integer. To verify this, set $z = z\exp(j2n\pi)$ where n is any integer and use Euler's equation.

A function with no singularities in the finite part of the plane is an *entire function.* Examples of entire functions are provided by the exponentials and their linear combinations, in particular, the circular and hyperbolic sines and cosines. If the function has an essential singularity at infinity, it is a *transcendental entire function*; if it has a pole of order n at infinity, it is a *polynomial of degree n.* A function like $\sin z$ is a transcendental entire function.

A function that is the ratio of two entire functions is said to be *meromorphic.* Its only singularities in the finite part of the plane are poles; the point at infinity can be either a point of accumulation of poles or an essential singularity, or even be a pole or a point of analyticity; in the latter case, the number of poles is necessarily finite. The function $\tan z$ has a pole accumulation point at infinity; the function $(\sin z)/z^3$ has an (isolated) essential singularity at infinity. If both entire functions are polynomials, the meromorphic function is *rational.* In the rational case, the difference between the degrees of the numerator and the denominator is the order of the pole or the zero at infinity depending on whether the difference is positive or negative; if it is zero, the function has a finite nonzero value at infinity. The *degree* of a rational function is defined as the greater of the degrees of the numerator and denominator polynomials.

The classification above is clearly not exhaustive; other types of functions, like those that have accumulation points of singularities in the finite part of the plane, have no specific names.

A.9 Multivalued Functions

Multivalued functions are such that to each value of z several (possibly infinitely many) values of $w = f(z)$ are associated. The simplest examples of such functions are $\sqrt{z}$ and $\ln z$; for a given z, the first yields two values of w, the second infinitely many. This is shown by the following equations

$$\begin{aligned}\sqrt{z} &= \sqrt{|z|}\exp j\left(\frac{\theta + 2k\pi}{2}\right) \\ \ln z &= \ln|z| + j\theta + j2k\pi\end{aligned} \tag{A.9.1}$$

where θ is the argument of z. In the first equation, when the integer k assumes the values 0, 1, we get two distinct branches of the square root; all the other integer values bring us back to one or the other of these two branches. In the second equation, on the contrary, a different branch of the logarithm corresponds to each value of the integer k.

The mapping between planes z and w for multivalued functions is made single-valued by properly modifying the structure of complex plane z. We can conceive of the z plane as a (finite or infinite) set of superposed layers, on each of which a point corresponds to a single value of the function. The individual layers can be isolated in trivial cases like that of a function having the values $+1$ and -1 everywhere. In all interesting cases, each layer can be reached from any other along a path which encircles a singularity specific to multivalued functions (the *branch point*) a sufficient number of times. Branch points are *algebraic* if they connect a finite number of layers (or *branches*); otherwise, they are *transcendental.* In the examples above there are two branch points, one in the origin, the other at infinity; such branch points are algebraic for $\sqrt{z}$, transcendental for $\ln z$. The surface consisting of the various branches connected together at branch points is called a *Riemann surface.*

On a Riemann surface, a branch is separated from the others by connecting each branch point with the point at infinity by a curve (*branch-cut*). For example, for the function $w = \sqrt{z}$ we can regard values of z such that $0 \le \arg z < 2\pi$ as on one branch of a Riemann surface distinct from values

$2\pi \leq \arg z < 4\pi$ located on the second branch. The two-sheeted surface is joined together at $z = 0$ (a branch point) and the branches are separated by the branch cut consisting of the positive real axis. As we circle the origin, we move from one branch or back to the other every time the branch cut is crossed. The cut may, of course, be defined as any regular curve connecting zero and infinity, but the branch points (in this case zero and infinity) are fixed. In general if, as we follow a curve which circles the point at infinity and returns to the starting point without encircling any branch points in the finite part of the plane, the final and the initial value of the function coincide, then the point at infinity is not a branch point and the set of branch cuts can be redefined so as to omit the point at infinity. By using the concept of a Riemann surface, many of the properties of single valued functions can be extended to multivalued functions.

Some functions may be either uniform or exhibit algebraic or transcendental multivaluedness depending on values of parameters they contain. An example is the function $w(z) = z^\alpha$. This function is uniform if α is integer, while if α is rational (i.e., a ratio of integers) it is defined on a Riemann surface having q leaves, where q is the minimum denominator of α; thus the origin and the point at infinity are algebraic branch points. On the other hand, if α is not rational or, in general, if it is complex, the Riemann surface has infinitely many branches, so that the origin and the point at infinity are transcendental branch points.

Integration of single-valued functions in regions where they have singular points may generate multivalued functions. For example, we can define the logarithm by the following equation:

$$\ln z = \int_1^z \frac{d\zeta}{\zeta} \tag{A.9.2}$$

To have a single-valued result, we must perform the integration in a simply connected region where the integrand is analytic. Thus in eq. (A.9.2) we must exclude the origin and any contour surrounding it, by connecting it with the point at infinity by a cut. Suppose we choose a cut along the negative real axis. Then all points of the plane can be reached along integration paths which do not cross the cut, hence do not encircle the origin, and the value of the logarithm is independent of the particular path chosen. In particular, we can stipulate that such a branch corresponds to $k = 0$ in eq. (A.9.1), or to the fact that logarithms of real positive numbers are real. Whenever we reach a point z by a path traversing the cut k times, the imaginary part of the logarithm is increased by $2k\pi$.

A.10 The Logarithmic Derivative

The *logarithmic derivative* of $f(z)$ is $d\ln f(z)/dz = f'(z)/f(z)$. Its integral along a closed curve in the complex plane has a very important meaning, which is formalized in the following.

THEOREM A.10.1

(Theorem of the Argument) *Let γ be a closed regular curve in an open set where the function $f(z)$ is analytic with the exception of at most a finite number of poles inside the region $\mathcal{R}$ bounded by the curve; moreover, let $f(z)$ have no zeros on γ. Then the integral around the closed curve γ of the logarithmic derivative of $f(z)$ is given by*

$$\frac{1}{2\pi j}\oint_{\gamma}\frac{f'(z)}{f(z)}dz = N_0 - N_\infty \tag{A.10.1}$$

where N_0 and N_∞ are the number of zeros and poles of $f(z)$ inside γ, counted according to their respective multiplicities.

For a rational function, having a finite number of poles in the entire complex plane, the result holds for both the internal and the external regions of γ with opposite signs. Therefore a rational function has as many poles as zeros in the entire complex plane (including the point at infinity). In particular, a polynomial of degree n has a pole of order n at infinity; hence it must have n zeroes in the finite part of the complex plane. This is the *Fundamental Theorem of Algebra.*

As an example, the function of eq. (A.4.6) is $f(z) = (2z-3)/(z-1)(z-2)$. It has a simple zero at $z = 3/2$ and two poles in the finite part of the plane and hence must have a simple zero at infinity. This is easily verified since it behaves as $2/z$ as $z \to \infty$.

The Theorem of the Argument can be put into an alternate form by writing the integrand in eq. (A.10.1) as $d\ln f(z) = d\ln|f(z)| + jd\arg f(z)$. When integration is carried out along a closed curve, the real part gives a null contribution. On the other hand, the integral of the imaginary part is $\arg f(z) = 2\pi N$ where N is the difference between the number of counterclockwise and clockwise circulations about the origin in $f(z)$, as z traverses the curve γ once in the positive sense, i.e., N is the net number of rotations about the origin in the $f(z)$ plane (positive if counterclockwise and vice versa). Thus we conclude that under the same conditions applying to the Theorem of the Argument:

THEOREM A.10.2

The net number of rotations N of $f(z)$ around the origin when z makes one counterclockwise traversal of the closed curve γ is equal to the difference between the number of zeros and the number of poles in the internal region bounded by the curve.

This theorem leads directly to the well known *Nyquist criterion for stability.*[3]

A.11 Functions with a Finite Number of Singularities

Consider a uniform analytic function $w(z)$ with a finite number of isolated singularities. Let such singular points (necessarily poles and/or essential singularities) be $z_1, z_2, ..., z_n$. Let the function $G(1/(z-z_k))$ be the characteristic component of $f(z)$ around singularity z_k. The function $f_1(z) = f(z) - G(1/(z-z_1))$ has all the singularities of $f(z)$ except z_1; the function $f_2(z) = f_1(z) - G(1/(z-z_2))$ has all singularities of $f(z)$ except z_1 and z_2; and so on. If the procedure is continued until all singularities are extracted, we are left with a function, $G(z)$, having, at most, a singularity at infinity, i.e., an entire function (Section A.8). Hence the function $f(z)$ can be represented as

$$f(z) = G(z) + \sum_{k=1}^{n} G\left(\frac{1}{z-z_k}\right) \tag{A.11.1}$$

that is, as the sum of its characteristics.

If, in particular, $f(z)$ is rational (no singularities except for a finite number of poles), $G(z)$ reduces to a polynomial whose degree is the difference between the degrees of the numerator and the denominator of $f(z)$. When the difference is negative, $G(z)$ is not present. Each of the characteristics $G(1/(z-z_k))$ contains a finite number of terms, m_k, equal to the highest degree of $1/(z-z_k)$, hence to the order of the pole at z_k.

In the rational case, with a total of n poles in the interior of the z plane, eq. (A.11.1) reduces to the *partial fraction expansion* of a rational function

$$f(z) = G(z) + \sum_{k=1}^{n} \sum_{r=1}^{m_k} \frac{c_{-r}^{(k)}}{(z-z_k)^r} \tag{A.11.2}$$

[3]This famous result was first published in the paper: H. Nyquist, "Regeneration Theory", *Bell Sys. Tech. J.*, Vol. 11, p. 126-147, Jan. 1932.

An important particular case arises when all of the poles are simple. Eq. (A.11.2) then reduces to

$$f(z) = G(z) + \sum_{k=1}^{n} \frac{c_{-1}^{(k)}}{z - z_k} \tag{A.11.3}$$

In this case, eq. (A.11.3), the coefficients of the partial fraction expansion are the residues of the function at the various poles.

The residue for a simple pole in $f(z)$ at z_k is

$$c_{-1}^{(k)} = \lim_{z \to z_k} (z - z_k) f(z) \tag{A.11.4}$$

If $f(z)$ is meromorphic (in particular rational), i.e., it can be represented as the ratio of two entire functions (in particular polynomials) $N(z)$ and $D(z)$, the evaluation of the limit in eq. (A.11.4) yields

$$c_{-1}^{(k)} = \frac{N(z_k)}{D'(z_k)} \tag{A.11.5}$$

In the case of a pole whose order is $m_k > 1$, the coefficients in (A.11.2) are given by

$$c_{-r}^{(k)} = \lim_{z \to z_k} \frac{1}{(m_k - r)!} \frac{d^{(m_k - r)}}{dz^{(m_k - r)}} [(z - z_0)^{m_k} f(z)] \tag{A.11.6}$$

Note that when $m_k = r$, the derivative operator is replaced by unity. When $r = 1$ in eq. (A.11.6) we get the expression for the residue, and it is important to note that this evaluation of the residue *at a pole is valid even if $f(z)$ is not rational.* If $m_k = 1$, eq. (A.11.6) gives the same residue expression as eq. (A.11.5). Eq. (A.11.3) with the coefficients expressed by eq. (A.11.5) is sometimes referred to as *Heaviside's partial fraction expansion.*

A.12 Analytic Continuation

The approach to analytic function theory, as sketched above, is essentially that of Cauchy. An alternative approach was developed by Weierstrass, based on power series expansions. Consider a formal power series for $f(z)$ around some point z_0 in a region where the function is analytic. The series converges at such a point. Suppose that the *same* series also converges at some other point, say z_1. Then it can be proved that the series at z_0

converges absolutely within any circle with center z_0 and radius less than $|z_0 - z_1|$. The radii of the circles of center z_0 in which the series converges have an upper bound, finite or infinite. Such an upper bound is denoted by R and is called the *radius of convergence*; the circle with center z_0 and radius R is called the circle of convergence about z_0. The series converges absolutely in the interior of the circle of convergence and uniformly over any closed set strictly contained in it. As a consequence, the series can be integrated or differentiated termwise any number of times over such a set. The last statement implies that within the circle of convergence the power series defines an analytic function.

By the technique above, the function is defined within a circle of center z_0 and radius R. Consider a point z_1 within the circle and expand the function in a new power series around this point, i.e., in ascending powers of $z - z_1$. The two series yield identical values in the overlapping regions of their circles of convergence. If the new circle of convergence about z_1 includes some portion of the complex plane not belonging to the original one, the domain of the function has actually been extended into the non-overlapping region; we say that $f(z)$ has an *analytic continuation* into the new region. Otherwise other choices of z_1 can be tested; finally, either the continuation is possible to new regions or the circumference of the original convergence circle represents a *natural boundary* for the domain of the function. The systematic application of the continuation technique finally leads to a single analytic function whose domain of definition is the overall convergence region; this domain always has one or more singularities on its boundary which belong to some of the circular peripheries but necessarily cannot lie within any convergence circle.

The method of Weierstrass can be developed to classify all types of singularities (including branch points and multivalued functions) so as to yield a comprehensive treatment of analytic functions which parallels that of Cauchy. However, the details are more cumbersome and the subject is beyond the scope of this discussion. For our purpose the importance of Weierstrass' method is in the concept of analytic continuation which is very often used to construct a single analytic function defined over a large region starting from an *analytic element*, that is, from its definition in a small open set. For example, analytic continuation allows us to extend many functions of a real variable to the complex domain by simply replacing the variable x with the variable z. Such a replacement guarantees the preservation of all formal properties, thus allowing the use of formulas such as $\exp z = \exp(x + jy) = \exp x \cdot \exp jy$ and so on.

A useful method of obtaining an analytic continuation of a complex function applies to certain cases where the function has two analytic elements which coincide on a curve.

THEOREM A.12.1

(Schwarz Theorem of Analytic Connection) *Let $f_1(z)$ and $f_2(z)$ be two functions analytic, respectively, inside regions $\mathcal{R}1$ and $\mathcal{R}2$, which are separated by the curve γ. Let the functions be continuous and equal to each other on γ. Then $f_1(z)$ and $f_2(z)$ are analytic continuations of each other, defining a single analytic function over $\mathcal{R}1$ and $\mathcal{R}2$.*

In particular, let $f_1(z)$ be an analytic function defined in the upper half-plane, continuous to the real axis and real on it. Define the function $f_2(z) = f_1^*(z)$ which takes on conjugate values to $f_1(z)$ in the lower half plane. The common contour is the real axis, and the two functions in the upper and lower half-planes satisfy the requirements of the connection theorem. Hence they define a single analytic function. This is the *Schwarz Reflection Principle.*

As a result of the reflection principle, we conclude that a function $f(z)$, analytic in a region intersected by the real axis, and real on this axis, assumes conjugate values of $f(z)$ at conjugate values of z in the region.

A.13 Calculus of Definite Integrals by the Residue Method

The theorem of residues is the basis of a powerful technique for calculating various types of definite integrals.

We first consider integrals of the type

$$f(t) = \frac{1}{2\pi} \int_{-\infty}^{+\infty} \epsilon^{j\omega t} \hat{f}(\omega) d\omega \tag{A.13.1}$$

where ω and t are real variables.

Such integrals occur in the theory of Fourier integrals as Fourier inverse transforms. In circuit and system theory ω is usually interpreted as a *angular frequency* and t as *time.* Complex integration techniques provide a versatile tool for the calculation of the function $f(t)$ from the given $\hat{f}(\omega)$. To apply contour integration methods, the independent variable ω should first be extended to a complex plane. This can be done in two ways; physicists usually associate ω with the real part of a complex variable $z = \omega + j\tau$, while it is customary for engineers to identify ω with the imaginary part of a complex variable $s = \sigma + j\omega$. While either approach obviously yields equivalent results, we shall confine ourselves to the second choice. This leads naturally to the consideration of the imaginary axis, $j\omega$, rather than the real axis, ω, as the path of integration. Redefine $\hat{f}(\omega) = F(j\omega)$ and rewrite eq. (A.13.1) as

$$f(t) = \frac{1}{2\pi} \int_{-j\infty}^{+j\infty} \epsilon^{j\omega t} F(j\omega) d\omega \qquad \text{(A.13.2)}$$

$F(j\omega)$ is assumed to have an analytic continuation $F(s)$ into the s plane and, in order to evaluate the integral of eq. (A.13.2), we first consider the following integral

$$f(t) = \frac{1}{2\pi j} \oint_{\gamma} \epsilon^{st} F(s) ds$$

where the closed curve γ is the union of the segment $[-jR, +jR]$ and of the half-circumference c, centred at the origin and of radius R, which lies entirely in the left (right) half-plane. The function $F(s)$ is assumed to have only isolated singularities inside γ. The idea of the method is to evaluate the integral over the closed contour, γ, using residues and, if the contribution over the semicircle is zero as $R \to \infty$, the result determines the infinite integral in eq. (A.13.2).

Applying the theorem of residues to the closed contour integral, we obtain

$$\frac{1}{2\pi j} \oint_{\gamma} \epsilon^{st} F(s) ds = \pm \sum_i c_{(-1)}^{(i)} \qquad \text{(A.13.3)}$$

where the sum of the residues is extended to all singular points internal to the left (right) half-plane semicircle and the sign is (+) for the left choice, (−) for the right. The integral of eq. (A.13.3) can be further broken into the sum of two integrals along the segment $[-jR, jR]$ and along c.

$$\frac{1}{2\pi j} \int_{-jR}^{+jR} \epsilon^{st} F(s) ds + \frac{1}{2\pi j} \int_{c} \epsilon^{st} F(s) ds = \pm \sum_i c_{-1}^{(i)}$$

The infinite integral of eq. (A.13.2) will therefore be equal to the sum (with the proper sign) of the residues of the integrand in the left (right) half-plane, if and only if the integral along the semicircular contour c tends to zero in the limit as $R \to \infty$.

Such a condition is clearly not satisfied for general functions $F(s)$; however, the condition is validated for a general class of $F(s)$ functions satisfying the following lemma.

LEMMA A.13.1

(Jordan's Lemma) *Let $f(s)$ be analytic for $|s| > a$, in the left (right) half-plane, a being a positive constant. Then if $F(s) \to 0$ uniformly along the semicircle c as its radius $R \to +\infty$, we have*

$$\lim_{R \to +\infty} \int_{c} \epsilon^{st} F(s) ds = 0$$

for all $t > 0$ $(t < 0)$.

The objective for each range of t is to provide exponential damping on the large semicircle c.

The infinite integral is then evaluated as follows.

THEOREM A.13.1

If Jordan's Lemma is satisfied, then for $t > 0$

$$\frac{1}{2\pi j}\int_{-j\infty}^{+j\infty} \epsilon^{st} F(s)ds = \pm \sum_i c_{-1}^{(i)} \tag{A.13.4}$$

where the sum is extended with the $(+)$ *sign to all residues in the left half-plane. For* $t < 0$, *the sum is over all right half plane residues using the* $(-)$ *sign.*

The case $t = 0$ requires special treatment. We use the following Lemma and subsequent Theorem.

LEMMA A.13.2

Let $f(s)$ *be analytic for* $|s| > a$ *in the left (right) half-plane,* a *a positive constant. Then if* $sF(s) \to 0$ *uniformly along the semicircle* c *as its radius* $R \to +\infty$, *we have*

$$\lim_{R\to+\infty} \int_c F(s)ds = 0$$

THEOREM A.13.2

If Lemma (A.13.2) *is satisfied, then*

$$\frac{1}{2\pi j}\int_{-j\infty}^{+j\infty} F(s)ds = \pm \sum_i c_{-1}^{(i)} \tag{A.13.5}$$

where the sum is extended to all residues in the left (right) half-plane and the sign is $(+)$ *or* $(-)$ *according to whether we choose (arbitrarily)* c *to lie in the left half or right half-plane. The result is the same for either choice.*

A special case arises when a first order pole is present on the imaginary axis at, say, point s_0. In such a case the integration contour is modified so as to avoid the singularity by means of a small semicircular indentation centered at the pole, and whose radius $\epsilon \to 0$. The integral along $j\omega$ now includes the contributions of these small indentations, which can be explicitly evaluated by imagining the simple $j\omega$ pole split into two, one slightly moved into the right half-plane, the other into the left half-plane, each with a residue one-half that of the original pole.

In most problems involving Fourier and Hilbert transforms, the required integration path is one which does not include the portion along the semi-circular pole detours. It is obvious that the integration path cannot include the pole itself, hence a modified definition of the infinite integral of eqs. (A.13.4) and (A.13.5) is used. This definition excludes the pole by omitting equal infinitesimal segments of path centered about the pole. The resultant integral is called a *principal value*, and indicated with "P". For instance if there is a pole at $s = 0$

$$P\int_{-j\infty}^{+j\infty} f(s)ds = \lim_{\epsilon\to 0}\left(\int_{-j\infty}^{-j\epsilon} f(s)ds + \int_{j\epsilon}^{j\infty} f(s)ds\right) \tag{A.13.6}$$

A similar expression holds for poles at $\pm j\omega_0$ (a function real on the real axis has poles at conjugate locations). In that case the integration path avoids the two equal infinitesimal segments at $\pm j\omega_0$. As an example let $f(s) = 1/s$. Then

$$P\int_{-j\infty}^{+j\infty} \frac{1}{s}ds = \lim_{\epsilon\to 0\, R\to\infty}\left(\int_{-jR}^{-j\epsilon} \frac{1}{s}ds + \int_{j\epsilon}^{jR} \frac{1}{s}ds\right) = \ln 1 = 0 \tag{A.13.7}$$

Instead of applying the limiting process of eq. (A.13.6) directly, it is often more convenient to use contour integration. Thus if the Jordan Lemma is satisfied, the total contour integral is evaluated by residues and then the contribution of the infinitesimal indentations (either right or left) around the $j\omega$ poles, computed as discussed above, is subtracted from the value of the closed contour integral. Since the contribution of the large semicircle is zero, the result is the principal value of the integral.

As an illustration, consider the generalized Inverse Fourier Transform $\mathcal{F}^{-1}[F(j\omega)]$, $F(j\omega) = 2/j\omega$ which continues into the s-plane as $2/s$. Due to the simple pole at the origin, we are concerned with the principal value

$$\frac{1}{2\pi j}P\int_{-j\infty}^{j\infty} \frac{2\epsilon^{st}}{s}ds = \frac{1}{2\pi j}\lim_{\epsilon\to 0}\left(\int_{-j\infty}^{-j\epsilon} \frac{2\epsilon^{st}}{s}ds + \frac{1}{2\pi j}\int_{j\epsilon}^{+j\infty} \frac{2\epsilon^{st}}{s}ds\right)$$

First consider $t < 0$. The integrand satisfies the Jordan Lemma and has no poles in the right (or left) half plane. Thus including a semicircular indentation to the right of the origin, the infinite integral along $j\omega$ is zero. The integration over the indentation at $s = 0$, traversed in the positive sense, is evaluated by taking twice the half residue of the pole, imagined slightly to the left, or $2\pi j$. Subtracting this from zero and dividing by $2\pi j$, the result is -1 for $t < 0$. For $t > 0$, the contour integral is closed to the left, and the indentation is taken in the left half plane. We can repeat the calculation, only now the tiny indentation is traversed in the negative sense (clockwise), so when twice the half residue is subtracted, the net result for the inverse transform is 1, $t > 0$. When $t = 0$ we can use eq. (A.13.7) and get zero for the principal value. Summarizing (also see Example 3.9.3)

$$f(t) = 1\,, t > 0;\ f(t) = -1\,, t < 0;\ f(t) = 0\,, t = 0;\ \text{ or } f(t) = \operatorname{sign} t$$

Note that the value $f(0) = 0 = 1/2[f(0^-) + f(0^+)]$ is typical of the behavior of Fourier transforms and series at a discontinuity.

Another type of integral which can be profitably treated by the techniques of contour integration is the kind representing Fourier and Laplace transforms.

$$F(j\omega) = \int_{-\infty}^{+\infty} \epsilon^{-j\omega t} f(t)dt \tag{A.13.8}$$

and

$$F(s) = \int_{-\infty}^{+\infty} \epsilon^{-st} f(t)dt$$

Since here we integrate with respect to t, we introduce a complex variable of which t is either the real or the imaginary part. If, as is usual, we make the first choice, we construct a complex plane $u = t + jt'$, which is nothing but the s plane rotated by $-\pi/2$. Thus the left half-plane is replaced by the upper half-plane, the right half-plane by the lower half-plane. But we must take into account the fact that the sign of j has changed in eq. (A.13.8) with respect to eq. (A.13.1). Thus we must integrate along the lower plane semicircle for $\omega > 0$, along the upper plane semicircle for $\omega < 0$; in the first case we move clockwise, hence we take the sum of residues with negative sign, in the second we move counterclockwise, hence for this case we take the sum of residues with positive sign.

As an example, consider the Fourier transform of $1/(1 + t^2)$. We have

$$F(j\omega) = \int_{-\infty}^{+\infty} \frac{\epsilon^{-j\omega t}}{1 + t^2} dt$$

The integrand has poles at $t = \pm j$ with residues $\epsilon^{\omega}/2j$ and $-\epsilon^{-\omega}/2j$. For $\omega > 0$ we traverse the path in the lower half-plane and obtain for the contour integral $-2\pi j(-\epsilon^{-\omega}/2j) = \pi\epsilon^{-\omega}$. For $\omega < 0$ we traverse the path in the upper half-plane and we obtain $2\pi j(\epsilon^{\omega}/2j) = \pi\epsilon^{\omega}$. Thus the result for all real ω is

$$\mathcal{F}\left(\frac{1}{1 + t^2}\right) = \epsilon^{-|\omega|}$$

The use of contour integration to solve difficult problems is far more extensive than could be presented here; our discussion has been limited to those cases which occur frequently in network theory.

Appendix B

Linear Algebra

B.1 General Concepts

A *matrix* is a set of *components* arranged in a rectangular array consisting of *rows* and *columns*. A matrix with m rows and n columns is an $(m \times n)$-*matrix*. If the numbers of rows and columns are equal, say n, the matrix is *square*. The components of a matrix may be real or complex numbers, or functions of a real or a complex variable; whatever they be, they will be called *scalars*. The scalars obey the rules of ordinary algebra, under which they still produce scalars of the same kind. The component of matrix $\boldsymbol{A}$ lying at the intersection of row i and column j is denoted as $(\boldsymbol{A})_{ij}$ or simply a_{ij}. The *transpose* of matrix $\boldsymbol{A}$ is the matrix obtained by exchanging the rows and the columns and is denoted as $\boldsymbol{A}'$. The *conjugate transpose* of matrix $\boldsymbol{A}$ is denoted as $\boldsymbol{A}^\dagger$. A matrix consisting of a single column is often called a *column vector* and is usually denoted by a boldface lower-case character, $\boldsymbol{a}$. The transpose of a column vector $\boldsymbol{a}$ is a *row vector* $\boldsymbol{a}'$. Similarly defined is the conjugate transpose of a column vector $\boldsymbol{a}^\dagger$.

The *main diagonal* of a square matrix is the set of all components whose row and column indices are equal. A square matrix is *diagonal* if all of its components are zero except those on the main diagonal. A diagonal matrix is *scalar* if all of its diagonal components are equal; if, moreover, they are equal to 1, the matrix is the *unit matrix* $\boldsymbol{I}$.

A square matrix is *symmetric* if it is equal to its transpose ($\boldsymbol{A} = \boldsymbol{A}'$), *skew-symmetric* if it is equal to the negative of its transpose ($\boldsymbol{A} = -\boldsymbol{A}'$), *hermitian* if it is equal to its conjugate-transpose ($\boldsymbol{A} = \boldsymbol{A}^\dagger$), *skew-hermitian* if it is equal to the negative of its conjugate transpose ($\boldsymbol{A} = -\boldsymbol{A}^\dagger$).

Matrices with the same numbers of rows and columns can be summed by summing the corresponding components.

$$(\boldsymbol{A} + \boldsymbol{B})_{ij} = a_{ij} + b_{ij}$$

A matrix $\boldsymbol{A}$ can be multiplied by a scalar by multiplying all of its components by the scalar.

$$(\alpha \boldsymbol{A})_{ij} = \alpha a_{ij}$$

Multiplication of two matrices $\boldsymbol{AB}$ is defined whenever the number of columns of the first factor equals the number of rows of the second; let matrix $\boldsymbol{A}$ be $(m \times r)$ and matrix $\boldsymbol{B}$ be $(r \times n)$. Then the product $\boldsymbol{AB}$ is an $(m \times n)$-matrix defined as

$$(\boldsymbol{AB})_{ij} = \sum_{s=1}^{r} a_{is} b_{sj}$$

The sum and product of matrices enjoy all the properties of the corresponding operations on scalars, except that multiplication is not commutative; in fact, the product of two matrices in reverse order is not even defined in general, because the numbers of rows of the first and of columns of the second may not be equal. But, even if products $\boldsymbol{AB}$ and $\boldsymbol{BA}$ are both defined, they generally yield different matrices.

The *determinant* of a square matrix is defined as follows

$$\det \boldsymbol{A} = \sum (-1)^p a_{1p_1} a_{2p_2} \dots a_{np_n} \tag{B.1.1}$$

Each of the row and column elements of $\boldsymbol{A}$, as indicated by the subscripts, are represented only once in the n factors that make up a term of eq. (B.1.1). It should be noted that eq. (B.1.1) sums all possible $n!$ terms chosen in this fashion from $\boldsymbol{A}$. Furthermore, if the first subscripts of the factors are in numerical order (as shown), then p is given as the number of inversions from numerical order in the second subscripts. As an example, consider the expansion of a third order determinant

$$\det \boldsymbol{A} = \begin{vmatrix} a_{11} & a_{12} & a_{13} \\ a_{21} & a_{22} & a_{23} \\ a_{31} & a_{32} & a_{33} \end{vmatrix}$$

Using eq. (B.1.1)

$$\begin{aligned} \det \boldsymbol{A} = {} & a_{11}a_{22}a_{33} + a_{12}a_{23}a_{31} + a_{13}a_{21}a_{32} \\ & -a_{13}a_{22}a_{31} - a_{11}a_{23}a_{32} - a_{12}a_{21}a_{32} \end{aligned}$$

The inversion rule gives the signs in the expansion. For instance, the third term has two inversions of the second subscript from numerical order, so it is $p = 2$ and the sign is $(+)$. On the other hand, for the fourth term, it is $p = 3$, so the sign is $(-)$.

The submatrix remaining when an equal number of rows and columns are deleted from $\boldsymbol{A}$ is a square array whose determinant is called a *minor*. In particular, if one row i and one column j are deleted and the resulting minor is prefixed by $(-1)^{i+j}$, the result is called the signed minor or *cofactor* A_{ij} of the matrix element a_{ij}. For example, in the (3×3)-matrix above the cofactor for a_{21} is $A_{21} = (-1)(a_{12}a_{33} - a_{13}a_{32})$.

The cofactors of an $(n \times n)$-matrix $\boldsymbol{A}$ satisfy

$$\sum_{s=1}^{n} a_{is}A_{js} = \det \boldsymbol{A}\delta_{ij} \tag{B.1.2}$$

where $\delta_{ij} = 1$ for $i = j$, $\delta_{ij} = 0$ for $i \neq j$. When $i = j$, eq. (B.1.2) gives the *Laplace expansion* of a determinant about row i. If i=j and the i, s subscripts are interchanged in eq. (B.1.2), we have the Laplace expansion of $\det \boldsymbol{A}$ about column i.

The *inverse* of a square matrix, $\boldsymbol{A}^{-1}$, if it exists, produces the *identity matrix* $\boldsymbol{I}$ when it multiplies $\boldsymbol{A}$. The elements of the inverse can be written

$$(\boldsymbol{A}^{-1})_{sj} = \frac{A_{js}}{\det \boldsymbol{A}} \tag{B.1.3}$$

A_{js} is the cofactor of component a_{js}. If $\det \boldsymbol{A} \neq 0$, the inverse matrix exists and $\boldsymbol{A}$ is *nonsingular*; otherwise, it is *singular* and the inverse does not exist. Referring to eqs. (B.1.1) and (B.1.2), it can be shown that if $\boldsymbol{A}$ is nonsingular, then $\boldsymbol{A}^{-1}$, eq. (B.1.3), satisfies the inversion property

$$\boldsymbol{A}^{-1}\boldsymbol{A} = \boldsymbol{A}\boldsymbol{A}^{-1} = \boldsymbol{I} \tag{B.1.4}$$

A nonsingular matrix commutes with its inverse, as in eq. (B.1.4). Eq. (B.1.3) is not applicable to nonsquare matrices. In such a case left and right inverses can be defined, but the two, when they exist, will differ.

The following properties of the product of two matrices are often useful:

$$(\boldsymbol{AB})' = \boldsymbol{B}'\boldsymbol{A}'; \quad (\boldsymbol{AB})^\dagger = \boldsymbol{B}^\dagger\boldsymbol{A}^\dagger; \quad (\boldsymbol{AB})^{-1} = \boldsymbol{B}^{-1}\boldsymbol{A}^{-1}$$

B.2 Geometrical Interpretation

The concepts discussed can be advantageously interpreted in geometrical terms. Consider the set of all n-row column vectors (*elements* of the set) whose components are scalars of some kind. Such a set has the properties that the sum of two of its elements and the product of any element by a scalar produce elements in the set. This set, consisting of *all* vectors of n components, is called a *vector space*. In addition to the above stated closure rules, the sum rule and scalar multiplication for the vector space have the same arithmetic properties as in matrix algebra, e.g., the sum rule is commutative and associative. The zero vector is $\mathbf{0} = (0, 0, \ldots, 0)'$.

Let X be a vector space. Let $\boldsymbol{x}_1, \boldsymbol{x}_2, ..., \boldsymbol{x}_r$, be any r vectors chosen from it. If the equation

$$\sum_{i=1}^{r} c_i \boldsymbol{x}_i = 0$$

is satisfied with values of the constants c_i not all zero, the vectors $\boldsymbol{x}_1, \boldsymbol{x}_2, ...\boldsymbol{x}_r$ are said to be *linearly dependent*; otherwise, they are *linearly independent.* If a space contains n linearly independent vectors, while any $n + 1$ vectors are linearly dependent, it is said to have *dimension* n (or to be n*-dimensional*). In a n-dimensional space, any set of $r < n$ linearly independent vectors defines a r-dimensional proper subspace. For example, the space of all 3-component vectors whose third component is twice the first component is of dimension $r = 2$.

Any vector in a space X of dimension n can be expressed as a linear combination of some specific and fixed set of linearly independent vectors, which are said to form a *base* for the space. The base *spans* the space. A linear vector space is of dimension n if and only if it contains a base of n vectors (*Theorem of the Base*).

As an example of a base in an n-dimensional space, consider the following set of column vectors: $\boldsymbol{e}_1 = (1 \quad 0 \quad 0 \quad ... \quad 0)'$, $\boldsymbol{e}_2 = (0 \quad 1 \quad 0 \quad ... \quad 0)'$, ..., $\boldsymbol{e}_n = (0 \quad 0 \quad 0 \quad ... \quad 1)'$. It is straightforward to verify that the number of elements in this base set is exactly n and that the elements are linearly independent; they form the so called *canonical* base.

A vector space need not to be defined in terms of column vectors; it simply bears on the existence of entities, respectively called *vectors* and *scalars*, for which the two operations of *sum* of vectors and *product* of a vector by a scalar satisfy closure and are defined through a proper set of axioms. These axioms preserve the properties of ordinary algebra appropriate for such operations.

As an example, consider the set of all sinusoidal functions of time having the same frequency. It can be described by the equation

$$x(t) = \xi_1 \sin \omega t + \xi_2 \cos \omega t \tag{B.2.1}$$

By looking at the rules by which two such sinusoids are added and one is multiplied by a (real) constant, it is immediately seen that the set is a vector space of dimension 2. It can be put in one-to-one correspondence with the space of 2-dimensional column vectors by identifying the sinusoid $x(t)$ with the vector $\boldsymbol{x} = (\xi_1 \quad \xi_2)'$. On this correspondence, which turns out to be an identification (all spaces with the same dimension are *isomorphic*), is based the replacement of sinusoid algebra with the vector and complex algebra commonly used in the study of sinusoidal steady state of electric networks.

The n-dimensional space of column vectors is called $\mathcal{R}^n$ or $\mathcal{C}^n$ depending on whether the scalars are real or complex numbers.

As we defined the vector space X of column vectors $\boldsymbol{x}$, so we can define the *dual* space X' of the row vectors $\boldsymbol{x}'$ and the *adjoint* space $X^\dagger$ of the conjugate row vectors $\boldsymbol{x}^\dagger$, the latter being employed when the scalars are complex. In the following, complex components are assumed; real vectors can be treated similarly. The space X and its adjoint $X^\dagger$ can be put into one-to-one correspondence by associating each column vector of the former with the row vector of the latter which is its conjugate transpose. Real (complex) row vectors can be multiplied by a real (complex) column vectors yielding a real (complex) *scalar*; such an operation is called the *scalar product* of the two vectors and is defined as follows.

$$\boldsymbol{x}^\dagger \boldsymbol{y} = \sum_{i=1}^{n} x_i^* y_i$$

The scalar product of a column vector by its corresponding row vector is always real and positive, and its positive square root is called the *norm* of the vector. Two vectors whose scalar product is zero are said to be *orthogonal.*

A vector space all of whose elements belong to a given vector space X, but do not necessarily constitute all of X, is said to be a *subspace* of X.

A vector space of dimension n can be decomposed into h subspaces ($1 \leq h \leq n$) which have no common elements (i.e., do not intersect), and whose union gives X. This union defines the *direct sum* ($\oplus$) of subspaces. The subspaces can always be chosen so that any vector of each subspace is orthogonal to any vector of the other subspaces.

In particular, when X is the direct sum of two orthogonal subspaces ($h = 2$), S and $S^\perp$, they are *orthogonal complements* of each other. We write $X = S \oplus S^\perp$. In this case, a base for X consists of the base vectors of S augmented by the base vectors of $S^\perp$. Thus the dimension of X is the sum of the dimensions of S and $S^\perp$.

We can denote the transformation of a vector space X into a vector space Y by the following equation

$$\boldsymbol{y} = \boldsymbol{A}\boldsymbol{x} \tag{B.2.2}$$

This is defined as a *linear transformation* (or $\boldsymbol{A}$ is a linear operator) if the following rules hold (note that the vectors need not be numeric, see eq. (B.2.1)):

$$\begin{aligned} &(1) \qquad \boldsymbol{A}(\boldsymbol{x}_1 + \boldsymbol{x}_2) = \boldsymbol{A}\boldsymbol{x}_1 + \boldsymbol{A}\boldsymbol{x}_2 \\ &(2) \qquad (\lambda \boldsymbol{A})\boldsymbol{x} = \lambda(\boldsymbol{A}\boldsymbol{x}) \end{aligned} \tag{B.2.3}$$

If both spaces X and Y have the same dimension, say n, then the operator defines a mapping of a space into itself.

The association of linear transformations of a vector space with matrix algebra is subsumed in the following result. *For given bases in Y and*

X, to every linear operator $\boldsymbol{A}$ defined as in eqs. (B.2.2) *and* (B.2.3) *there corresponds a rectangular matrix of dimensions* $(m \times n)$, *and conversely to every such matrix $\boldsymbol{A}$ there corresponds a linear operator mapping X into Y.*

A simple example illustrates this basic operator-matrix correspondence. Consider the set of all real third degree polynomials of a variable x, which we represent as $P(x) = a_0x^3 + a_1x^2 + a_2x + a_3$. It is a simple affair to verify that it is closed under polynomial addition and multiplication by a real number; i.e., these operations yield another third degree polynomial. Thus the set of all $P(x)$'s forms a four-dimensional vector space that we will call X. In such a space the first four powers of x, i.e., $x^3, x^2, x, 1$ represent the chosen base, and the coefficients a_0, a_1, a_2, a_3, the components of the vector P with respect to the base. The space X is isomorphic to the space R^4 of the column vectors $(a_0\ a_1\ a_2\ a_3)'$, endowed with the canonical base $\boldsymbol{e}_1, \boldsymbol{e}_2, \boldsymbol{e}_3, \boldsymbol{e}_4$. Thus it is seen that the space is 4-dimensional. The derivative operator with respect to x, D, is linear, i.e., satisfies eq. (B.2.3), and is defined on the whole space because any polynomial can be differentiated.

$$D(a_0x^3 + a_1x^2 + a_2x + a_3) = 3a_0x^2 + 2a_1x + a_2 = Q(x) \tag{B.2.4}$$

The equation above shows that the operator D maps the 4-dimensional space X onto a 3-dimensional space Y, which is a subspace of the former, since it contains the base $x^2, x, 1$. We note, *en passant*, that X is the direct sum of Y and of a unidimensional space whose base is x^3. If we represent the components of $P(x)$ and those of $Q(x)$ as column vectors, we immediately see that eq. (B.2.4) can be rewritten as

$$\begin{pmatrix} 3a_0 \\ 2a_1 \\ a_2 \end{pmatrix} = \begin{pmatrix} 3\,0\,0\,0 \\ 0\,2\,0\,0 \\ 0\,0\,1\,0 \end{pmatrix} \begin{pmatrix} a_0 \\ a_1 \\ a_2 \\ a_3 \end{pmatrix}$$

In the equation above, the (3×4) rectangular matrix is the representation of the derivative operator D with respect to the chosen bases of X and Y.

The subspace $N(\boldsymbol{A})$ of X such that $\boldsymbol{A}\boldsymbol{x} = 0$ for all $\boldsymbol{x} \in N(\boldsymbol{A})$ is said to be the *null-space of* $\boldsymbol{A}$ and its dimension its *nullity* ν. The subspace $R(\boldsymbol{A})$ of Y, consisting of the vectors $\boldsymbol{y}$ that can be represented as in eq. (B.2.2), is called the *range* of $\boldsymbol{A}$ and its dimension is the *rank* ρ of $\boldsymbol{A}$. Subspaces $N(\boldsymbol{A})^{\perp}$ and $R(\boldsymbol{A})$ are in one-to-one correspondence in the mapping induced by the operator $\boldsymbol{A}$. In fact, suppose there are two vectors, $\boldsymbol{x}_1$ and $\boldsymbol{x}_2$, both belonging to $N(\boldsymbol{A})^{\perp}$, whose image in $R(\boldsymbol{A})$ is a vector $\boldsymbol{y}$. Then $\boldsymbol{A}\boldsymbol{x}_1 = \boldsymbol{y}$ and $\boldsymbol{A}x_2 = \boldsymbol{y}$ implies $\boldsymbol{A}(\boldsymbol{x}_1 - \boldsymbol{x}_2) = \boldsymbol{0}$, that is, $\boldsymbol{x}_1 - \boldsymbol{x}_2$ belongs to $N(\boldsymbol{A})$, which is impossible unless such a vector is the zero vector. As a consequence, subspaces $N(\boldsymbol{A})^{\perp}$ and $R(\boldsymbol{A})$ have the same dimension and the relation holds $\nu + \rho = n$.

We note that once the range and null space of $\boldsymbol{A}$ are known, corresponding spaces for the conjugate transpose (*adjoint*) operator $\boldsymbol{A}^\dagger$ which takes the space Y into X, $\boldsymbol{x} = \boldsymbol{A}^\dagger \boldsymbol{y}$, are immediately determined. Consider the equation $\boldsymbol{y}^\dagger \boldsymbol{A}\boldsymbol{x} = \boldsymbol{0}$. If this is satisfied for all $\boldsymbol{x} \in X$, $\boldsymbol{y}$ must belong to $R(\boldsymbol{A})^\perp$. On the other hand, the equation can be rewritten in conjugate transpose form as $\boldsymbol{x}^\dagger \boldsymbol{A}^\dagger \boldsymbol{y} = \boldsymbol{0}$, which shows that since the components of $\boldsymbol{x}$ are arbitrary, it follows that $\boldsymbol{A}^\dagger \boldsymbol{y} = \boldsymbol{0}$, that is, $\boldsymbol{y}$ must belong to $N(\boldsymbol{A}^\dagger)$. Thus we have the result that $N(\boldsymbol{A}^\dagger) = R(\boldsymbol{A})^\perp$. In much the same way, it can be shown that $R(\boldsymbol{A}^\dagger) = N(\boldsymbol{A})^\perp$.

Consider the equation $\boldsymbol{A}\boldsymbol{x} = \boldsymbol{y}$ and expand the vector $\boldsymbol{x}$ in the canonical base $\boldsymbol{e}_i$ of the space X. We obtain

$$x_1 \boldsymbol{A}\boldsymbol{e}_1 + x_2 \boldsymbol{A}\boldsymbol{e}_2 + \quad \ldots \quad x_n \boldsymbol{A}\boldsymbol{e}_n = \boldsymbol{y} \tag{B.2.5}$$

Note that $\boldsymbol{A}\boldsymbol{e}_i$ is just the i-th column vector extracted from the matrix $\boldsymbol{A}$. Thus eq. (B.2.5) represents the vector $\boldsymbol{y}$ as a linear combination of the columns of matrix $\boldsymbol{A}$, and, ultimately, of a maximal linearly independent subset of them.

Alternatively, we may expand the vector $\boldsymbol{y}$ in the canonical base $\boldsymbol{d}_i$ of the space Y. We obtain

$$\boldsymbol{A}\boldsymbol{x} = y_1 \boldsymbol{d}_1 + y_2 \boldsymbol{d}_2 + \quad \ldots \quad + y_m \boldsymbol{d}_m \tag{B.2.6}$$

If we multiply both sides of eq. (B.2.6) by the elements of the canonical base of the adjoint space $Y^\dagger$, which are the transposed $\boldsymbol{d}'_i$ of the vectors $\boldsymbol{d}_i$, we obtain

$$\boldsymbol{d}'_i \boldsymbol{A}\boldsymbol{x} = y_i$$

It is immediately seen that $\boldsymbol{d}'_i \boldsymbol{A}$ is the i-th row vector extracted from the matrix $\boldsymbol{A}$. Thus the components of the vector $\boldsymbol{y}$ are the projections of the vector $\boldsymbol{x}$ on the row vectors extracted from the matrix $\boldsymbol{A}$, which include a base for the range of the adjoint operator.

B.3 Linear Simultaneous Equations

Sets of linear simultaneous algebraic equations illustrate in a practical way many of the properties of vector spaces and operator-matrix correspondence discussed in the previous section. In this context, and central to the problem of solving linear algebraic equations, are the concepts of matrix *rank* and *nullity*.

DEFINITION B.3.1 **(Matrix rank)** *The maximum number of linearly independent rows (columns) in a $m \times n$ matrix $\boldsymbol{A}$ is equal to the row (column) rank ρ of the matrix.*

DEFINITION B.3.2 **(Matrix nullity)** *The nullity of a $m \times n$ matrix $\boldsymbol{A}$ is $\nu = \min(m, n) - \rho$.*

The following theorem holds.

THEOREM B.3.1

1. *The row rank and column rank of a matrix are equal.*

2. *The rank must be less than or equal to the smaller of m and n, $\rho \leq \min(m, n)$.*

3. *The matrix rank ρ is equal to the maximum order of nonvanishing minors (Section B.1) of $\boldsymbol{A}$. Denoting all the k rowed minors as M_k, this means that at least one $M_\rho \neq 0$, and all $M_q = 0, q > \rho$.*

4. *A square $m \times m$ matrix, $\boldsymbol{A}$, is of full rank, if it is non-singular, i.e., $\rho = m$, or $\det \boldsymbol{A} \neq 0$.*

5. *Any vector $\boldsymbol{y}$ for which the simultaneous equations $\boldsymbol{A}\boldsymbol{x} = \boldsymbol{y}$ have a solution $\boldsymbol{x}$ is a linear combination of ρ linearly independent column vectors of $\boldsymbol{A}$.*

6. *Any vector $\boldsymbol{x}$ which is a solution of the homogeneous equations $\boldsymbol{A}\boldsymbol{x} = \boldsymbol{o}$ is the linear combination of ν linear independent vectors, each of which is orthogonal to ρ linearly independent row vectors of the matrix $\boldsymbol{A}$.*

The statements above have simple geometrical interpretation.

Statement 5 shows that the range of the operator $\boldsymbol{A}$ is spanned by ρ linearly independent column vectors extracted from the corresponding matrix, which thus form a base for such a range. The rank of the matrix, as defined above, is the dimension of the range of the corresponding operator.

Statement 6 shows that ν linearly independent vectors satisfying $\boldsymbol{A}\boldsymbol{x} = \boldsymbol{0}$ are orthogonal to ρ linearly independent row vectors extracted from matrix $\boldsymbol{A}$, i.e., to ρ linearly independent column vectors extracted from its adjoint $\boldsymbol{A}^\dagger$. This means that a base for $N(\boldsymbol{A})$ is orthogonal to a base for $R(\boldsymbol{A}^\dagger)$, or the two subspaces are mutually perpendicular.

The equality of column and row ranks (statement 1) follows from the fact that the dimensions of $R(\boldsymbol{A})$ and $R(\boldsymbol{A}^\dagger)$ are both equal to ρ. This in turn implies statement 2, because the number of linearly independent vectors cannot exceed their total number.

As an example of the matrix equation

$$\boldsymbol{A}\boldsymbol{x} = \boldsymbol{y} \tag{B.3.1}$$

which corresponds to a linear mapping between spaces X and Y, consider the following set of linear equations, written in matrix form.

$$\begin{pmatrix} 3 & 5 & 2 \\ -1 & 7 & 3 \\ 7 & 3 & 1 \end{pmatrix} \begin{pmatrix} x_1 \\ x_2 \\ x_3 \end{pmatrix} = \begin{pmatrix} y_1 \\ y_2 \\ y_3 \end{pmatrix} \tag{B.3.2}$$

Here the mapping is between 3-dimensional spaces X and Y.

First, we observe that in matrix $\boldsymbol{A}$ the first row multiplied by 2 minus the second row equals the third row, while any two rows are linearly independent. It is therefore clear that, whatever x_1, x_2, x_3 are, the components of vector $\boldsymbol{y}$ must satisfy the same interdependence, that is

$$2y_1 - y_2 - y_3 = 0 \tag{B.3.3}$$

This is the equation of a 2-dimensional subspace of Y, i.e., a plane, which is the range $R(\boldsymbol{A})$ of matrix $\boldsymbol{A}$, whose rank is therefore 2. Thus any vector of Y not belonging to such a plane (that is, whose components do not satisfy eq. (B.3.3)), cannot be the result of the transformation of any vector of X by the operator $\boldsymbol{A}$. This means that for such a given $\boldsymbol{y}$, the simultaneous equations (B.3.2) admit of no solution. The *inhomogeneous* equations are then *incompatible*. If, on the other hand, $\boldsymbol{y}$ belongs to the range $R(\boldsymbol{A})$, the equations admit of at least one solution $\boldsymbol{x}$. There is a straightforward rule for determining when a set of linear inhomogeneous equations is compatible, which can be illustrated by the present example. Form the *augmented matrix* by adding the column vector $\boldsymbol{y}$ to $\boldsymbol{A}$.

$$\boldsymbol{A}_{\text{aug}} = \begin{pmatrix} 3 & 5 & 2 & y_1 \\ -1 & 7 & 3 & y_2 \\ 7 & 3 & 1 & y_3 \end{pmatrix}$$

If the components of $\boldsymbol{y}$ do not satisfy eq. (B.3.3), the rank of the matrix $\boldsymbol{A}_{\text{aug}}$ exceeds that of $\boldsymbol{A}$, i.e., in this case is $3 > 2$; the equations are incompatible; if the components of $\boldsymbol{y}$ satisfy eq. (B.3.3), the rank of the matrix $\boldsymbol{A}_{\text{aug}}$ is equal to the rank of $\boldsymbol{A}$, i.e., in this case is 2; then solutions do exist. Thus we have the following result.

THEOREM B.3.2
(Consistency) *A set of linear inhomogeneous equations* (B.3.1) *admits of at least one solution if and only if the rank of the augmented matrix,* $\boldsymbol{A}_{\text{aug}}$,

equals the rank of $\boldsymbol{A}$. *(Note that for homogeneous equations,* $\boldsymbol{y} = \boldsymbol{0}$*, the rule is trivially satisfied and* $\boldsymbol{x} = \boldsymbol{0}$ *is always one solution.)*

We revisit the discussion at the end of Section B.2 concerning range and nullspace of $\boldsymbol{A}$, $\boldsymbol{A}^\dagger$, and their orthogonal complements, but now in the context of the Consistency Theorem and matrix algebra so as to establish the following theorem as a test for consistency.

THEOREM B.3.3

(Consistency Test) *The linear simultaneous equations* $\boldsymbol{Ax} = \boldsymbol{y}$ *are consistent if and only if* $\boldsymbol{y} \in N(\boldsymbol{A}^\dagger)^\perp$.

PROOF

1. Suppose the equations are consistent. Then $\boldsymbol{y} \in R(\boldsymbol{A})$ and when $\boldsymbol{y}$ augments $\boldsymbol{A}$, the rank or number of linearly independent rows, is unchanged. To examine the linear dependence of the rows, define a column vector $\boldsymbol{r} = (r_1^*, r_2^*, ...r_k^*, ..., r_m^*)^\dagger$, multiply each row of $\boldsymbol{A}$ by the associated r_k^*, and set the linear combination of rows to zero. The resultant equations are represented by $\boldsymbol{r}^\dagger\boldsymbol{A} = \boldsymbol{0}$, or equivalently $\boldsymbol{A}^\dagger\boldsymbol{r} = \boldsymbol{0}$. Therefore the solution vectors $\boldsymbol{r} \in N(\boldsymbol{A}^\dagger)$. Since the rank remains unchanged, the same r_k^* multiplicative coefficients must also define the dependence of the elements of the augmenting $\boldsymbol{y}$. Hence $\boldsymbol{r}^\dagger\boldsymbol{y} = \boldsymbol{0}$, and the vectors are orthogonal. We have
$$\boldsymbol{y} \in R(\boldsymbol{A}) = N(\boldsymbol{A}^\dagger)^\perp$$
2. For the converse we suppose vectors $\boldsymbol{y}$ form the subspace $N(\boldsymbol{A}^\dagger)^\perp$. The vectors $\boldsymbol{r}$, which form $N(\boldsymbol{A}^\dagger)$ satisfy $\boldsymbol{r}^\dagger\boldsymbol{A} = \boldsymbol{0}$. The augmented matrix is $\boldsymbol{A}_{\text{aug}} = [\boldsymbol{A}|\boldsymbol{y}]$ and since by hypothesis $\boldsymbol{r}^\dagger\boldsymbol{y} = 0$, we have $\boldsymbol{r}^\dagger\boldsymbol{A}_{\text{aug}} = \boldsymbol{0}$. In other words, both $\boldsymbol{A}$ and $\boldsymbol{A}_{\text{aug}}$ have the same linear dependence of rows. Their rank is the same, so the equations $\boldsymbol{Ax} = \boldsymbol{y}$ are consistent. ∎

We now consider the general solution of a consistent set of linear simultaneous equations. We can readily establish the following basic result.

THEOREM B.3.4

(Solution of Linear Simultaneous Equations) *The general solution,* $\boldsymbol{x}_1$*, of eq.* (B.3.1) $\boldsymbol{Ax} = \boldsymbol{y}$ *is equal to the sum of any particular solution,* $\boldsymbol{x}_0$*, and the general solution* $\boldsymbol{u}$ *of the homogeneous equations* $\boldsymbol{Ax} = \boldsymbol{0}$.

PROOF

1. Necessity: Suppose $\boldsymbol{x}_0$ is a particular solution of the inhomogeneous set, and $\boldsymbol{x}_1$ its general solution. Write $\boldsymbol{x}_1 = \boldsymbol{x}_0 + (\boldsymbol{x}_1 - \boldsymbol{x}_0)$. By linearity, eq. (B.2.3), $\boldsymbol{A}(\boldsymbol{x}_1 - \boldsymbol{x}_0) = \boldsymbol{A}\boldsymbol{x}_1 - \boldsymbol{A}\boldsymbol{x}_0 = \boldsymbol{y} - \boldsymbol{y} = \boldsymbol{0}$. Thus $\boldsymbol{u} = (\boldsymbol{x}_1 - \boldsymbol{x}_0)$ is the general solution of $\boldsymbol{A}\boldsymbol{x} = \boldsymbol{0}$. We have therefore shown that the general solution solution of eq. (B.3.1) must be expressible as $\boldsymbol{x}_1 = \boldsymbol{x}_0 + \boldsymbol{u}$.

2. Sufficiency: Let $\boldsymbol{u}$ be the general solution of $\boldsymbol{A}\boldsymbol{x} = \boldsymbol{0}$, $\boldsymbol{x}_0$ a particular solution of the equation $\boldsymbol{A}\boldsymbol{x} = \boldsymbol{y}$, and $\boldsymbol{x}_1$ the general solution of the latter. Then $\boldsymbol{A}(\boldsymbol{x}_0 + \boldsymbol{u}) = \boldsymbol{y} + \boldsymbol{0} = \boldsymbol{y} = \boldsymbol{A}\boldsymbol{x}_1$, so a particular solution of the complete equation plus the general solution of the homogeneous equation is the general a solution of $\boldsymbol{A}\boldsymbol{x} = \boldsymbol{y}$. ∎

We can now readily deduce a number of properties which are useful for the practical solution of eq. (B.3.1). Suppose that $\boldsymbol{A}$ is an $m \times n$ matrix (m equations and n unknowns) of rank ρ. Further suppose the equations are consistent, that is the prescribed $\boldsymbol{y}$ is in the orthogonal complement of $N(\boldsymbol{A}^\dagger)$. The general homogeneous solution is $\boldsymbol{u}$.

COROLLARY B.3.1

1. *As an immediate consequence of the Solution Theorem, we observe that the solution of eq.* (B.3.1) *is unique if and only if the only homogeneous solution is* $\mathbf{u} = \mathbf{0}$. *Thus the nullity* ν *of* $\mathbf{A}$ *(dimension of its nullspace) must be zero. Since* $\nu + \rho = n$, *it is evident that uniqueness requires that the rank,* ρ, *of* $\mathbf{A}$ *be equal to the number of unknowns.*

2. *Suppose* $m < n$. *Since* ρ *can be no greater than the lesser of* m *and* n, *then* $\rho < n$, *and by property 1 above, a multiplicity of solutions of eq.* (B.3.1) *exists, but there can be no unique solution.*

3. *Suppose* $m > n$. *A unique solution of eq.* (B.3.1) *exists if* $\rho = n$, *in which case* $m - n$ *equations are discarded as redundant.*

4. *Suppose* $m = n$. $\boldsymbol{A}$ *is square, and if it is of full rank,* $\rho = m = n$, *the solution of eq.* (B.3.1) *is unique and is given by* $\boldsymbol{x} = \boldsymbol{A}^{-1}\boldsymbol{y}$.

Let us come back to our example. For this case the rank of $\boldsymbol{A}$ is 2. Referring to the discussion at the start of this section, there are only two linearly independent row (or column) vectors in $\boldsymbol{A}$. Such a pair of rows must contain a nonvanishing 2×2 minor, M_2. This means we can choose any *two* equations to solve provided they are based on two linearly independent rows. Two components of $\boldsymbol{x}$ associated with M_2 are then solved in terms of arbitrary values of the third component (a *free variable).*

Thus in the homogeneous equation

$$\begin{pmatrix} 3 & 5 & 2 \\ -1 & 7 & 3 \\ 7 & 3 & 1 \end{pmatrix} \begin{pmatrix} x_1 \\ x_2 \\ x_3 \end{pmatrix} = \begin{pmatrix} 0 \\ 0 \\ 0 \end{pmatrix}$$

the first two rows of $\boldsymbol{A}$ contain a nonvanishing determinant in the upper right hand corner, i.e., the minor $\boldsymbol{A}_{31}$ is nonsingular, so we assign an arbitrary value to $x_1 = \lambda$ and solve for x_2, x_3 in terms of λ. Thus transposing terms in x_1 to the right hand side, we obtain

$$\begin{pmatrix} 5 & 2 \\ 7 & 3 \end{pmatrix} \begin{pmatrix} x_2 \\ x_3 \end{pmatrix} = \begin{pmatrix} -3x_1 \\ x_1 \end{pmatrix} = \begin{pmatrix} -3\lambda \\ \lambda \end{pmatrix}$$

Multiply both sides by the inverse of the coefficient matrix on the left hand side (it is nonsingular, with determinant equal to one), and thus obtain the solution for x_2, x_3. Including $x_1 = \lambda$, the general solution of the homogeneous equations is

$$\boldsymbol{x} = \begin{pmatrix} \lambda \\ -11\lambda \\ 26\lambda \end{pmatrix} \tag{B.3.4}$$

Referring to the example eq. (B.3.2), and according to the Consistency Test, the only prescribed vectors $\boldsymbol{y}$ that can yield a solution are those that lie in the orthogonal complement of $N(\boldsymbol{A}')$ (for real matrices $\boldsymbol{A}'$ replaces $\boldsymbol{A}^\dagger$). Thus to find an admissible $\boldsymbol{y}$ we must solve the homogeneous equations $\boldsymbol{A}'\boldsymbol{u} = \boldsymbol{0}$. The solution is carried out exactly as in eq. (B.3.4). Choose the second and third homogeneous equations of the transposed set, and nonvanishing minor M_{13} of $\boldsymbol{A}'$. Then, with $u_3 = \alpha$ arbitrary, we have

$$\begin{pmatrix} 5 & 7 \\ 2 & 3 \end{pmatrix} \begin{pmatrix} u_1 \\ u_2 \end{pmatrix} = \begin{pmatrix} -3\alpha \\ -\alpha \end{pmatrix}, \quad \boldsymbol{u} = \begin{pmatrix} -2\alpha \\ \alpha \\ \alpha \end{pmatrix}$$

Then, for orthogonality, it must be $\boldsymbol{u}'\boldsymbol{y} = 0$ or the components of $\boldsymbol{y}$ must satisfy $2y_1 - y_2 - y_3 = 0$, as earlier given by eq. (B.3.3).

To obtain the general solution of our example, the Solution Theorem calls for any one solution of the inhomogeneous set. The procedure to find this is again similar to that used for finding the null space. Since the nullity of $\boldsymbol{A}$ is one, choose one free variable, say, x_1 as before, and solve the first two independent equations. Transposing the terms in x_1 to the right-hand side and, with $\boldsymbol{y} = (y_1\; y_2\; y_3)'$ assumed to satisfy eq. (B.3.3),

$$\begin{pmatrix} 5 & 2 \\ 7 & 3 \end{pmatrix} \begin{pmatrix} x_2 \\ x_3 \end{pmatrix} = \begin{pmatrix} y_1 - 3x_1 \\ y_2 + x_1 \end{pmatrix}$$

The solution is straightforward. Thus $x_2 = 3y_1 - 2y_2 - 11x_1$, $x_3 = -7y_1 + 5y_2 + 26x_1$. For instance, if $x_1 = 1$ and the prescribed $\boldsymbol{y} = (32 \;\; 40 \;\; 24)'$, then a solution of the inhomogeneous equations is $\boldsymbol{x}_0 = (1 \;\; 5 \;\; 2)'$. To find the most general solution, $\boldsymbol{x}_1$, of the inhomogeneous equations, simply add to the general solution, eq. (B.3.4), of the homogeneous set the particular solution $\boldsymbol{x}0$.

$$\boldsymbol{x}_1 = \begin{pmatrix} 1+\lambda \\ 5-11\lambda \\ 2+26\lambda \end{pmatrix}, \quad \forall\lambda \tag{B.3.5}$$

Geometrically, the solution of the homogeneous equations, eq. (B.3.4), gives in parametric form the equation of a straight line passing through the origin. All vectors in the null space fall on this line.

Among the set of all solutions of the inhomogeneous equations corresponding to different choices of the free variable, x_1 there is one, and only one, lying in the orthogonal complement of the null space of $\boldsymbol{A}$, that is in the range of $\boldsymbol{A}'$. To find it, we observe that $N(\boldsymbol{A})^\perp$ is the set of all vectors $\boldsymbol{x}$ which are orthogonal to the straight line of eq. (B.3.4) and whose components thus satisfy the equation

$$x_1 - 11x_2 + 26x_3 = 0 \tag{B.3.6}$$

We have already determined the solution x_1, x_2, x_3 of the inhomogeneous equations leading to eq. (B.3.5), in terms of the free variable x_1. This solution is substituted into eq. (B.3.6) to obtain x_1, which then determines x_2, x_3. The result is $798x_1 - 800 = 0$, $x_1 = 1.0025063$. The unique inhomogeneous solution vector lying in $N(\boldsymbol{A})^\perp$, i.e., in $R(\boldsymbol{A}')$, is therefore

$$\boldsymbol{x} = \begin{pmatrix} 1.0025 \\ 4.9724 \\ 2.0652 \end{pmatrix} \tag{B.3.7}$$

Since the most general solution is the sum of eq. (B.3.7) and the homogeneous solution, it is clear that all solutions of $\boldsymbol{Ax} = \boldsymbol{y}$ have the same projection on $N(\boldsymbol{A})^\perp$.

B.4 Eigenvalues and Eigenvectors

Let $\boldsymbol{A}$ be a square $(n \times n)$-matrix, that we regard as a linear operator from the n-dimensional vector space X into itself.

We consider the problem of finding the nontrivial solutions of the equation

$$(\boldsymbol{A} - \lambda\boldsymbol{I})\boldsymbol{x} = \boldsymbol{0} \tag{B.4.1}$$

In geometric language, this is equivalent to require that $N(\boldsymbol{A}-\lambda\boldsymbol{I})$ does not reduce to the zero vector, or that the rank of $\boldsymbol{A}-\lambda\boldsymbol{I}$ is less than n. This implies that the maximum order of nonzero minors extracted from the matrix above must be less than n, or the determinant must be zero. Thus we are led to the equation

$$\det(\boldsymbol{A}-\lambda\boldsymbol{I})=0 \tag{B.4.2}$$

which, by developing the determinant, turns into an algebraic equation with real coefficients of degree n in λ

$$\lambda^n + a_1\lambda^{n-1} + a_2\lambda^{n-2} + ... + a_{n-1}\lambda + a_n = 0 \tag{B.4.3}$$

Such an equation has exactly n real or complex-conjugate roots $\lambda_1, \lambda_2, ..., \lambda_n$, some of which may coincide. For the moment, we suppose they are all distinct. For all values of λ which do not coincide with a root, the determinant (B.4.2) is nonzero and therefore eq. (B.4.1) has but the trivial solution $\boldsymbol{x}=0$. For each root λ_i of eq. (B.4.3), the determinant is zero, and therefore eq. (B.4.1) possesses a nontrivial solution $\boldsymbol{x}_i$. The roots λ_i's are called *eigenvalues*, the corresponding solutions $\boldsymbol{x}_i$'s *eigenvectors*. Thus with n distinct eigenvalues there are at least n eigenvectors. It can be proven that the eigenvectors are linearly independent. Thus there are exactly n, since in an n-dimensional space there cannot be more than n linearly independent vectors. Hence, for $\lambda=\lambda_i$, the rank of matrix $\boldsymbol{A}-\lambda\boldsymbol{I}$ decreases exactly by one and the corresponding eigenvector is unique within normalization. The conclusion is that $N(\boldsymbol{A}-\lambda_i\boldsymbol{I})$ for $i=1,2,...,n$ is 1, and the direct sum of all such *eigenspaces* constitutes the space X.

The discussion above shows that the n eigenvectors span the space and therefore can be assumed as a base for it. The eigenvectors $\boldsymbol{x}_i$ can be conveniently arranged in a matrix as follows

$$\boldsymbol{M}=(\boldsymbol{x}_1 \quad \boldsymbol{x}_2 \quad ... \quad \boldsymbol{x}_n)$$

In the equation above, the matrix $\boldsymbol{M}$ has n rows, because each eigenvector $\boldsymbol{x}_i$ has n components and n columns, because the eigenvectors are exactly n. Such a matrix is known as the *modal matrix*; it describes the full set of eigenvectors and these vectors span the n-dimensional space X. The inverse $\boldsymbol{M}^{-1}$ can be interpreted as a collection of row vectors $\tilde{\boldsymbol{x}}_i$, which are called *inverse eigenvectors*, with the property

$$\tilde{\boldsymbol{x}}_i\boldsymbol{x}_j=\delta_{ij}$$

Each set of eigenvectors forms a base for the space X and the two bases (eigenvectors and inverse eigenvectors) are reciprocally *biorthonormal.*

From eq. (B.4.1) is obtained

$$\boldsymbol{A}(\boldsymbol{x}_1 \quad \boldsymbol{x}_2 \quad ... \quad \boldsymbol{x}_n)=(\lambda_1\boldsymbol{x}_1 \quad \lambda_2\boldsymbol{x}_2 \quad ... \quad \lambda_n\boldsymbol{x}_n)$$

or

$$AM = M\Lambda \tag{B.4.4}$$

where Λ is the diagonal matrix of the eigenvalues,

$$\Lambda = \begin{pmatrix} \lambda_1 & 0 & 0 & \ldots & 0 \\ 0 & \lambda_2 & 0 & \ldots & 0 \\ 0 & 0 & \lambda_3 & \ldots & 0 \\ \ldots & \ldots & \ldots & & \ldots \\ 0 & 0 & 0 & \ldots & \lambda_n \end{pmatrix} \tag{B.4.5}$$

Eq. (B.4.4) can be solved with respect to matrix Λ

$$\Lambda = M^{-1}AM \tag{B.4.6}$$

which is called the *spectral representation* of matrix A. Eq. (B.4.6) connects the representations of an operator from the base, where it is described by matrix A, to the base formed by the eigenvectors, where it is described by matrix Λ.

If some eigenvalues coincide, say λ_i is a root of multiplicity n_i of eq. (B.4.3), two cases may occur:

1. $N(A - \lambda_i I) = n_i$; in this case the null-space is spanned by n_i linearly independent eigenvectors (all of whose linear combinations are still eigenvectors); since the sum of the n_i's is the dimension n of the space X, the set of the n eigenvectors still spans the space and therefore constitutes a base; all the formalism above can still be applied, with the proviso that in the matrix of eq. (B.4.5) some eigenvalues coincide and that the pertinent eigenvectors can be freely replaced by any of their linear combinations;

2. $N(A - \lambda_i I) < n_i$; in this case the direct sum of the eigenspaces does not reconstruct the whole space; the sum of their dimensions is less than the dimension of X; otherwise stated, we need to add to the set of eigenvectors spanning each null space other vectors, in order to complete a base.

A very simple example of case 2 is provided by the matrix

$$A = \begin{pmatrix} 1.5 & -0.5 \\ 0.5 & 0.5 \end{pmatrix} \tag{B.4.7}$$

It has the double eigenvalue $\lambda_1 = 1$ and the unique eigenvector $x_1 = (1 \quad 1)'$. While the space X is two-dimensional, the null-space is one-dimensional and spanned by the eigenvector above; it is evident that the direct sum of null-spaces, which here reduces to a single vector, is insufficient to reconstruct the space, so we need to complete the base, and this can be done by solving the equation

$$(\boldsymbol{A} - \lambda_1 \boldsymbol{I})^2 \boldsymbol{x}_2 = \boldsymbol{0} \tag{B.4.8}$$

Eq. (B.4.8) may evidently be satisfied by assuming

$$(\boldsymbol{A} - \lambda_1 \boldsymbol{I}) \boldsymbol{x}_2 = \boldsymbol{x}_1 \tag{B.4.9}$$

Eq. (B.4.9) admits the solution $\boldsymbol{x}_2 = (1 \; -1)'$, which, together with the eigenvector $\boldsymbol{x}_1$, provides a base for the 2-dimensional space. The two eigenvectors are thus put side by side to construct the modal matrix $\boldsymbol{M}$. Finally, we compute $\boldsymbol{M}^{-1}\boldsymbol{AM}$ and we obtain

$$\begin{pmatrix} 1 & 1 \\ 1 & -1 \end{pmatrix}^{-1} \begin{pmatrix} 1.5 & -0.5 \\ 0.5 & 0.5 \end{pmatrix} \begin{pmatrix} 1 & 1 \\ 1 & -1 \end{pmatrix} = \begin{pmatrix} 1 & 1 \\ 0 & 1 \end{pmatrix} \tag{B.4.10}$$

The procedure above can be generalized to produce k *generalized eigenvectors* associated with an eigenvalue λ_{i_k} of multiplicity k, such that $N(\boldsymbol{A} - \lambda_{i_k} \boldsymbol{I})$ is of dimension 1. One must simply solve the following chain of equations

$$\begin{aligned} &(\boldsymbol{A} - \lambda_{i_k} \boldsymbol{I}) \boldsymbol{x}_k = \boldsymbol{x}_{k-1} \\ &(\boldsymbol{A} - \lambda_{i_k} \boldsymbol{I}) \boldsymbol{x}_{k-1} = \boldsymbol{x}_{k-2} \\ &\vdots \\ &(\boldsymbol{A} - \lambda_{i_k} \boldsymbol{I}) \boldsymbol{x}_2 = \boldsymbol{x}_1 \end{aligned}$$

which yield just k linearly independent *generalized eigenvectors* that span the k-dimensional *generalized eigenspace* associated with the eigenvalue λ_{i_k}.

By constructing a modal matrix $\boldsymbol{M}$ using all generalized eigenvectors, the matrix $\boldsymbol{A}$ can be transformed according to the equation

$$\boldsymbol{M}^{-1} \boldsymbol{AM} = \boldsymbol{J}$$

where

$$\boldsymbol{J} = \begin{pmatrix} \boldsymbol{J}_1 & 0 & 0 & \dots & 0 \\ 0 & \boldsymbol{J}_2 & 0 & \dots & 0 \\ 0 & 0 & \boldsymbol{J}_3 & \dots & 0 \\ \dots & \dots & \dots & & \dots \\ 0 & 0 & 0 & \dots & \boldsymbol{J}_n \end{pmatrix}$$

and

$$\boldsymbol{J}_i = \begin{pmatrix} \lambda_i & 1 & 0 & \dots & 0 \\ 0 & \lambda_i & 1 & \dots & 0 \\ 0 & 0 & \lambda_i & \dots & 0 \\ \dots & \dots & \dots & & \dots \\ 0 & 0 & 0 & \dots & \lambda_i \end{pmatrix}$$

$\boldsymbol{J}$ is the *Jordan canonical form* of matrix $\boldsymbol{A}$; the $\boldsymbol{J}_i$'s are the *Jordan blocks*, each made of i_k rows and columns. In the example above, the Jordan form of the matrix of eq. (B.4.7) consists of a unique Jordan block as in eq. (B.4.10).

The foregoing material finds extensive application to the solution of linear ordinary differential equations in the form in which they arise when *State Space* analysis is employed. To illustrate the broad applicability of matrix algebra, we briefly consider here the constant coefficient homogeneous differential equations that describe LTI systems when there are no forcing terms. In this case the response (the *zero input* solution) is entirely due to nonzero initial conditions, e.g., initial capacitor charges and coil currents.

It is not difficult to show that any set of constant coefficient ordinary linear homogeneous differential equations can be put into the following *normal form*

$$\frac{d\boldsymbol{x}(t)}{dt} \equiv \dot{\boldsymbol{x}}(t) = \boldsymbol{A}\boldsymbol{x}(t) \tag{B.4.11}$$

The n unknown time dependent elements x_k of the column vector $\boldsymbol{x}$ are the *state variables*, and $\boldsymbol{A}$ is an $n \times n$ square matrix of constants. The solution of the homogeneous equations (B.4.11) can be put into the form

$$\boldsymbol{x}(t) = \boldsymbol{\Phi}(t)\boldsymbol{x}(0)$$

where $\boldsymbol{\Phi}(t)$ is the $n \times n$ *state transition matrix*, and $\boldsymbol{x}(0)$ gives the initial conditions ($t = 0$), or the initial state of the x_k's. The solution of the scalar equation $\dot{x} = ax$ for unit initial condition is simply $x(t) = \epsilon^{ta}x(0)$, and formally, for a vector problem, the state transition matrix can be represented in a similar fashion using an exponential matrix function.

$$\boldsymbol{\Phi}(t) = \epsilon^{t\boldsymbol{A}} \tag{B.4.12}$$

We now use the Jordan form $\boldsymbol{J}$ and the modal matrix $\boldsymbol{M}$ of $\boldsymbol{A}$ and write eq. (B.4.12) as

$$\boldsymbol{\Phi}(t) = \boldsymbol{M}\epsilon^{t\boldsymbol{J}}\boldsymbol{M}^{-1} \tag{B.4.13}$$

The explicit expression for $\boldsymbol{\Phi}$ is obtained by using[1]

$$\epsilon^{t\boldsymbol{J}} = \text{diag}\left(\epsilon^{t\boldsymbol{J}_1}, \epsilon^{t\boldsymbol{J}_2}, \ldots, \epsilon^{t\boldsymbol{J}_n}\right)$$

Each diagonal element represents a matrix block. If $\boldsymbol{J}_1$ is the Jordan block for the simple eigenvalues $\boldsymbol{J}_1 = \text{diag}\,(\lambda_1, \lambda_2, \ldots, \lambda_k)$, the exponential matrix has the form

$$\epsilon^{t\boldsymbol{J}_1} = \text{diag}\,\left(\epsilon^{t\lambda_1}, \epsilon^{t\lambda_2} \ldots, \epsilon^{t\lambda_k}\right) \tag{B.4.14}$$

[1]E. A. Coddington and N. Levinson, *Theory of Ordinary Differential Equations*, New York, McGraw-Hill, 1956, p. 75.

For a multiple eigenvalue λ_i of order $r_i > 1$, the corresponding diagonal block of eq. (B.4.14) is the $r_i \times r_i$ matrix

$$\epsilon^{t\mathbf{J}_i} = \epsilon^{t\lambda_i} \begin{pmatrix} 1 & t & t^2/2! & \cdots & t^{r_i-1}/(r_i-1)! \\ 0 & 1 & t & \cdots & t^{r_i-1}/(r_i-2)! \\ \vdots & \vdots & \vdots & \vdots & \vdots \\ 0 & 0 & 0 & .. & 1 \end{pmatrix} \tag{B.4.15}$$

As an example, consider $\dot{\boldsymbol{x}} = \boldsymbol{A}\boldsymbol{x}$ with

$$\boldsymbol{A} = \begin{pmatrix} 1 & 3 \\ -3 & -5 \end{pmatrix} \quad \& \quad \det(\boldsymbol{A} - \lambda \boldsymbol{I}) = (\lambda + 2)^2 = 0$$

Thus there is an eigenvalue of double multiplicity, $\lambda = -2$. The corresponding eigenvector (first column of $\boldsymbol{M}$) is computed as $(1 \; -1)'$, and the base is completed by using eq. (B.4.9) to find the second column of $\boldsymbol{M}$, $(1/3 \;\; 0)'$. The transition matrix may now be found using eqs. (B.4.13) and (B.4.15).

$$\boldsymbol{\Phi}(t) = \boldsymbol{M}\epsilon^{t\boldsymbol{J}}\boldsymbol{M}^{-1} = 3\epsilon^{-2t} \begin{pmatrix} 1 & 1/3 \\ -1 & 0 \end{pmatrix} \begin{pmatrix} 1 & t \\ 0 & 1 \end{pmatrix} \begin{pmatrix} 0 & -1/3 \\ 1 & 1 \end{pmatrix}$$

or

$$\boldsymbol{\Phi}(t) = \begin{pmatrix} 3t\epsilon^{-2t} + \epsilon^{-2t} & 3t\epsilon^{-2t} \\ -3t\epsilon^{-2t} & -3t\epsilon^{-2t} + \epsilon^{-2t} \end{pmatrix}$$

Index

n-port
augmented, 10, 12
general representation, 174
infinite, 178
n-port, augmented, 19
n-port, degenerate, voltage-, 17

a.c. analysis, 44
admittance
matrix, 31
nodal, 61
of a lossless two-port, 238, 239
allpass, 322
C-section, 323
D-section, 326
amplitude approximation, 296
Amstutz' elliptic filter function, 313
analytic function
reconstruction from its real part, 140
antimetric two-ports, 234
approximation, 285
attenuation function, 338
augmented
n-port, 10, 12
augmented, n-port, 19
available power, 179

bandpass filters, 291
bandstop filters, 294
Bartlett's theorem, 346
Bessel polynomials, 330
Bode's real-imaginary calculations, 148
Boucherot's Theorem, 96
branch, 47
admittance
matrix, 47
impedance
matrix, 47
branch-loop incidence matrix, 72
broadband matching, 384
Bromwich integral, 164
Brune's section, 254
Butterworth approximation
by stepped lines, 358
for broadband matching, 396
UE and stubs, 368
Butterworth filters, 299

C-section
allpass, 323
capture property, 307, 317
carrier, 283
causality, under current excitation, 19
causality, under voltage excitation, 18
chain matrix, 385
characteristic function, 235
Chebyshev approximation
by stepped lines, 358

UE and stubs, 369
Chebyshev filters, 304
Chebyshev polynomials, 304
compactness, 240
compatible systems of voltages and currents, 95
complete solution, 39
complex amplitude
from modulus, 143
complex normalization, 389
complex power
n-ports, 14
one-ports, 14
time dependent, 14
constitutive relations, 47
convolution, 122
of delayed functions, 122
cotree, 49
cut set, 48
admittance
matrix, 47
admittance matrix, 75
analysis, 47
equations, 53, 75
voltage, 74

D-section
allpass, 326
Darlington's Theorem, 244
degree
of a lossless two-port, 235
of a one-port, 211
of a rational function, 211
delay rule, 119
delta function, 105
approximation of, 110
Dirac's function, 105
directional coupler, 374
dominance of natural modes, 42
double matching
analytic procedure, 410
double matching
active devices, 433
real frequency technique, 418, 429

eigenfunction, 28, 29
analysis and Laplace transform, 161
eigenfunction analysis
and Laplace transform, 160
eigenvalue
matrix, 28, 29
matrix and transfer function, 78
elliptic filters, 312
energy integral, 11
exponential excitation
complete solution, 39
particular solution, 37

Fano's matching theory, 384
filter, 284
doubly terminated, 286
ideal, 284
stepped lines, 362
Foster synthesis, 216
Fourier integral theorem, 116
Fourier Transform, 116
L^1, 116
L^2, 117
Fourier transform
as boundary value of Laplace transform, 137
of a constant, 155
of causal functions, 134
of impulse function, 155
of the convolution of two functions, 124
of the product of two functions, 124
of the unit step, 156
operational rules, 153
frequency transformation, 289
Fujisawa Theorem, 274
fundamental cut set, 52

gain-bandwidth restrictions, 392

in integral form, 402
gain-bandwidth theory, 384
Gewertz procedure, 334
graph, 48
directed, 48
nullity, 49, 52, 55, 57
planar, 48, 67
rank, 49, 53, 57
separable, 49
simply connected, 48
group delay, 321
positivity of, 322

highpass filters, 291
Hilbert transform, 128, 418
numerical approximation by straight lines, 129
of derivatives, 128
straight line approximation, 148–150
homogeneous matrix equation solution, 33, 35
homogeneous scalar equation solution, 33
Hurwitz polynomial, 211
test, 224

immittance matrices
and scattering matrix, 183
impedance
bilinear transformation, 387
characteristic, 338
matrix, 31
of a lossless two-port, 238, 239
scaling, 289
synthesis, 248
impedance transformers
stepped line, 359
impulse function, 105
approximation, 106, 110
approximation of, 111
derivative of unit step function, 107
Fourier transform, 155
integral representation, 155
impulse response, 121
causality, 123
LTI operators, 123
LTI systems, 123
impulse sampling, 107
incidence
matrices, 47
incident wave, 181
input immittance
of a lossless terminated two-port, 239
input reflectance
of a lossless terminated two-port, 239
insertion loss theory, 384
instantaneous power
conservation, 96
interference, 283
Inverse Fourier Transform, 116
inverter
impedance, 342

junction, 47

KCL equations, 53, 57
Kuroda identity, 365
KVL equations, 55, 57

ladder synthesis, 274
transmission zeros at infinity, 279
transmission zeros at the origin, 281
Laplace transform, 159
and eigenfunction analysis, 160, 161
causal, 160
integral representations, 139
inverse, 160
of convolution, 163
of derivatives, 163
operational rules, 162

line-segment approximation, 418, 419
link, 49
loop, 48
 analysis, 47
  equations, 53, 72, 73
 fundamental, 52
 impedance
  matrix, 47, 72
loops
 fundamental, 56
lossless
 bounded real matrix, 196
 system, 16
losslessness, 386
lowpass filters, 290
LPR admittance
 partial fraction expansion, 216
LPR function
 increasing property along $j\omega$, 224
LPR immittance
 continuous fraction expansion around infinity, 221
 continuous fraction expansion around the origin, 222
LPR impedance
 partial fraction expansion, 215
LTI system
 exponential response, 27

matching
 broadband, 180
 narrow band, 180
mesh, 48
 analysis, 47
  equations, 69
 impedance
  matrix, 47, 69
 impedance matrix, 75
minimax polynomial, 307
minimum phase function, 144
minimum reactance impedance, 130
minimum resistance impedance, 130
modulation rule, 120

nodal admittance matrix, 60, 61
nodal analysis, 58
nodal analysis equations, 64, 65
node, 47
 admittance
  matrix, 47
 analysis, 47
normalized
 variables, 10
Norton's theorem, 65
nullity
 of a graph, 49, 52, 55, 57

operator, 1, 3, 4, 7
 causal, 7
 linear, 3
  time-invariant, 4
 real, 3
 time invariant, 4
outer loop, 67

Paley-Wiener Theorem, 145
paraconjugate matrix, 196
Parseval's Theorem, 125
particular solution, 37
passband, 284
passive
 circuit elements, 12
 system, 9
passive n-ports
 active power, 97
 reactive power, 97
passivity
 abstract and concrete properties, 171
phase
 from log amplitude, 143

phase equalization, 288, 321
phase function, 338
phase velocity, 339
power
  and immittance matrices, 188
  and scattering matrix, 187
power divider, 348
propagation function, 338
prototype filter, 289
pseudomeromorphic matrix, 196
pure eigenfunction response, 43, 44

rank
  of a graph, 49, 53, 57
rational function
  degree, 211
RC impedance
  continuous fraction expansion around infinity, 230
  continuous fraction expansion around the origin, 230
RC one-port
  admittance, 228
  impedance, 227
real frequency technique (RFT), 384, 416
realizability
  abstract, 189, 213
  concrete, 213
reciprocity, 98, 386
  n-ports, 100
  and symmetric matrices, 102
reflectance, 182
  LBR, 352
reflected wave, 181
residue matrix
  of a lossless two-port, 240
response by convolution, 121
Richard's Theorem, 351
Richard's transformation, 343
  modified, 344
scattering matrix, 180
  and general representations, 182
  and immittance matrices, 183
  definition, 181
  of a lossless reciprocal two-port, 234
  of a lossless two-port, 233, 237
  of a symmetric or antimetric two-port, 234
  rational, 197
  realizability, 189
sidebands, 283
sifting property, 107
signals
  square integrable, 11
single-matching, 418, 419
single-terminated filters, 332
solution
  complete with exponential excitation, 39
  homogeneous matrix equation, 33, 35
  homogeneous scalar equation, 33
  particular with exponential excitation, 37
source transportation, 66
spectral factorization, 243
square integrable signals, 11
stability
  LTI systems, 41
stepped line coupler, 376
stopband, 284
stub, 343
symmetric two-ports, 234
system, 3, 4, 7
  causal, 7
  linear, 3
    time-invariant, 4
  lossless, 16
  passive, 9

real, 3
space invariant, 338
time invariant, 4

Tellegen's Theorem, 96
Thévenin's theorem, 65
Titchmarsh theorem, 135
transducer power gain, 187, 286
realizability, 288
transfer function, 78, 79
and eigenvalue matrix, 78
poles, 79
transmission line, 337
transmission zeros, 241, 242
transmittance, 182
tree, 49
type A section, 252
type B section, 252
type C section, 262
transmission line, 381
type D section, 267

UE reactance functions, 351
unit element, 342, 345
conductance with stubs, 426
gain with stubs, 365, 425
transducer gain, 356
unit step function, 107
approximation of, 110
Fourier transform, 156

vertex, 47

wave
incident, 339
reflected, 339

Youla's matching theory, 384